MAIN GROUP ELEMENTS AND THEIR COMPOUNDS

Perspectives in Materials Science,
Chemistry and Biology

MAIN GROUP ELEMENTS AND THEIR COMPOUNDS

Perspectives in Materials Science,
Chemistry and Biology

EDITOR

V. G. Kumar Das

Springer-Verlag Berlin Heidelberg GmbH

EDITOR
Professor V.G. Kumar Das
Dean, Faculty of Science
University of Malaya
Kuala Lumpur, Malaysia

ISBN 978-3-642-52480-6   ISBN 978-3-642-52478-3 (eBook)
DOI 10.1007/978-3-642-52478-3

PREFACE

Main Group Elements have long been at the centre stage of chemistry with their far-reaching impact on products, processes and technologies that continue to underpin our chemical industries and the quality of our lives. In several fields of physics and biology, and now increasingly in the burgeoning field of Materials Science, these elements continue to show their mettle. Some like silicon in the electronics industry already enjoy an exalted position; others are strategic components of the elemental make-up of many new advanced materials designed and made with specific properties to fit well-defined needs in the marketplace. These specialized materials which embrace several new types of ceramics, alloys, polymers, composites, materials for electronics and photonics, and biomaterials are at the core of many emergent technologies, including the field of microscopic machines that could well presage a new industrial revolution.

It is well to remind ourselves that carbon, which has a vast and distinctive chemistry of its own, is also a main group element, *albeit* a very important one. The recent discovery of fullerenes as new allotropic forms of carbon has already opened up an exciting field of research in organic chemistry, and sparked creativity also among a cross-section of material scientists and physicists allured by their potential industrial applications, including as organic superconductors.

It is the purpose of this book to convey to the readers some of the fervour of frontier research in a number of areas of materials science involving main group elements, as well as to provide them useful perspectives on the synthesis and structure of some of such materials and other novel compounds, and on their biological and environmental properties.

An international team of contributors has been assembled for this volume, comprising leading researchers from academia and industry as well as younger investigators working actively in the exciting aforesaid research areas. A good mix of basic and applied research is presented, with several chapters in the book addressing both.

This book is largely the outcome of a major international conference on materials science and environmental chemistry of main group elements organised by the Asian Network for Analytical and Inorganic Chemistry under my chairmanship in Kuala Lumpur, and also includes a number of additional invited papers by leading scientists in the 'Main Group' field. It is hoped that the book will serve as a useful reference to material scientists, physicists and chemists in many diverse research and industrial organisations, besides being also of interest to administrators in governmental agencies who will find much in it to influence techno-economic policies and implement change. Readers are encouraged to contact the authors for additional information.

Acknowledgements

I would like to gratefully acknowledge the valuable contributions of the authors and record my appreciation to the publishers, Narosa Publishing House and Springer-Verlag, for their meticulous production of this volume.

It is also a pleasure to acknowledge the generous financial support received from the following sources: The Ministry of Science, Technology and Environment, Government of Malaysia, Uniphoenix Corporation Bhd., TWAS, ICI Paints (M) Sdn. Bhd. , Lee Foundation States of Malaya and Schott Glass (S) Pte.Ltd.

The text was capably typed and retyped by Miss Gnanamalar. To her, to my colleagues Drs. Ng Seik Weng and N.H. Tioh, and especially to my patient wife Ambika, I owe many thanks.

V. G. Kumar Das

Contributors

Afenya, P.M., *University of Technology, Lae, Papua New Guinea*

Agaskar, Pradyot A, *Mobil R&D Corporation, Central Research Laboratory, P.O. Box 1025, Princeton, NJ 08540 (USA)*

Ambrosini, Annarina, *Institute of Biochemistry, Medical School, University of Ancona, Via Ranieri, 60131 Ancona (Italy)*

Arakawa, Yasuaki, *Department of Hygiene & Preventive Medicine, Faculty of Health Sciences, The University of Shizuoka, 52-1 Yada, Shizuoka-shi, Shizuoka 422, Japan*

Aslanov, L.A., *Chemistry Department, M.V. Lomonosov University, Moscow, Russia*

Babonneau, Florence, *Departments of Materials Science & Engineering & Chemistry, and the Macromolecular Science & Engineering Center, University of Michigan, Ann Arbor, Michigan 48109-2136 (USA)*

Bei, Jianzhoug, *Institute of Chemistry, Academia Sinica, Beijing 100080, China*

Beletskaya, I.P., *Moscow State University, Russian Academy of Sciences, Russia*

Bertoli, Enrico, *Institute of Biochemistry, Medical School, University of Ancona, Via Ranieri, 60131 Ancona (Italy)*

Burford, R.P., *Polymer Science Department, University of South Wales, Sydney, Australia*

Chen Tianlang, *Department of Chemistry, Sichuan University, Chengdu 610064, Sichuan, P.R. China*

Chen, Wei, *Department of Materials Physics, University of Science and Technology Beijing, Beijing 100083, P.R. China*

Craig, Peter J., *Department of Chemistry, De Montfort University Leicester, The Gateway, Leicester, LE1 9BH, UK.*

Das, V.G. Kumar, *Department of Chemistry, University of Malaya, 59100 Kuala Lumpur, Malaysia*

Davydova, S.L., *Moscow State University, Russian Academy of Sciences, Russia*

Eng, George, *The District of Columbia Agricultural Experiment Station and Department of Chemistry, The University of the District of Columbia, Washington, DC 20008, USA*

Gan Fuxi, *Shanghai Institute of Optics and Fine Mechanics, Chinese Academy of Sciences, P.O. Box 800-211, Shanghai, 201800, P.R. China*

Gan Seng-Neon, *Department of Chemistry, University of Malaya, 59100 Kuala Lumpur, Malaysia*

Gielen, Marcel, *Free University of Brussels VUB, Faculty of Applied Sciences, Department of General and Organic Chemistry, Room 8G512, Pleinlaan 2, B-1050 Brussels, Belgium*

Grigor'ev, E.V., *Chemistry Department, M.V. Lomonosov University, Moscow, Russia*

Hagenmuller, Paul, *CNRS Solid State Chemistry Laboratory, University of Bordeaux I, 351 cóurs de la Liberation, 33405 Talence Cedex, France*

Haiduc, Ionel, *Facultatea de Chimie, Unviersitatea "Babes-Bolyai", Ro-3400 Cluj-Napoca, Roumania*

Hitchcock, Peter B., *School of Chemistry and Molecular Sciences, University of Sussex, Brighton, BN1 9QJ, United Kingdom*

Hosmane, Narayan S. , *Department of Chemistry, Southern Methodist University, Dallas, Texas 75275, USA*

Irgolic, Kurt J., *Institut für Analytische Chemie, Karl-Franzens Universität, Uiversitätsplatz 1, A-8010 Graz, Austria*

Jang, Eunseok, *School of Chemistry and Molecular Sciences, University of Sussex, Brighton, BN1 9QJ, United Kingdom*

Jouppi, Sarah, *Departments of Materials Science & Engineering & Chemistry, and the Macromolecular Science & Engineering Center, University of Michigan, Ann Arbor, Michigan 48109-2136 (USA)*

Jousseaume, Bernard, *Laboratoire de Chimie Organique et Organometallique, URA 35 CNRS, Universite Bordeaux 1, 351, cours de la Liberation, 33405 Talence, France*

Kalcher, Kurt, *Institut für Analytische Chemie, Karl-Franzens Universität, Uiversitätsplatz 1, A-8010 Graz, Austria*

Kannan, S.,*Department of Chemistry, Indian Institute of Technology, Madras - 600 036, India*

Kansal, Pallavi, *Departments of Materials Science & Engineering & Chemistry, and the Macromolecular Science & Engineering Center, University of Michigan, Ann Arbor, Michigan 48109-2136 (USA)*

Katbab, A.A., *Department of Polymer Engineering, Amirkabir University, Tehran, Iraq*

Kemmler, Martin, *Institut fur Anorganische Chemie der Universitat Auf der Morgenstelle 18, D-7400 Tubingen 1, Germany*

Khoo Lian Ee, *School of Science, Nanyang Technological University, 469 Bukit Timah Road, 1025 Singapore*

Kishino S., *Department of Electronics, Faculty of Engineering, Himeji Institute of Technology, Shosha, Himeji 671-22, Japan*

Kölbl, Gottfried, *Institut für Analytische Chemie, Karl-Franzens Universität, Uiversitätsplatz 1, A-8010 Graz, Austria*

Kurosasawa, Hiroki, *Research Laboratory of Resources Utilization, Tokyo Institute of Technology, Nagatuta 4259, Midori-mu, Yokohama 227, Japan*

Kuwano, Yukinori, *Functional Materials Research Center, SANYO Electric Co., Ltd. 1-18-13, Hashiridani, Hirakata, Osaka 573, Japan.*

Lappert, Michael F., *School of Chemistry and Molecular Sciences, University of Sussex, Brighton, BN1 9QJ, United Kingdom*

Laurie, Stuart H., *Department of Chemistry, De Montfort University Leicester, The Gateway, Leicester, LE1 9BH, UK.*

Lawrence, Marcus F., *Department of Biochemistry, Concordia University, 1455 de Maisonneuve Blvd. W, Montreal, Quebec, Canada H3G 1M8*

Lindner, Ekkehard, *Institut fur Anorganische Chemie der Universitat Auf der Morgenstelle 18, D-7400 Tubingen 1, Germany*

Livantsov, M.V., *Chemistry Department, M.V. Lomonosov University, Moscow, Russia*

Lo Kong Mun, *Department of Chemistry, University of Malaya, 59100 Kuala Lumpur, Malaysia*

Lorberth, J., *Chemistry Department, Philipps-University, Marburg, Germany*

Macey, D.J. , *School of Mathematical and Physical Sciences, Murdoch University, Murdoch, Western Australia 6150*

Magee, Robert J. , *Department of Chemistry, La Trobe University, Bundoora, Vic. 3083, Australia*

Mai, Ruzhang, *Department of Materials Physics, University of Science and Technology Beijing, Beijing 100083, P.R. China*

Marton, Daniele, *Dipartimento di Chimica Inorganica, Metallorganica ed Analitica, Universitá di Padova, Via Marzolo, 1, I-35131-Padova (Italy)*

May, Leopold, *Department of Chemistry, The Catholic University of America, Washington, DC 20064, USA*

Mayer, Herman A., *Institut fur Anorganische Chemie der Universitat Auf der Morgenstelle 18, D-7400 Tubingen 1, Germany*

Mehrotra, Ram C. , *Chancellery, University of Allahabad, Allahabad 21002, India.*

Mennie, Darren, *Ministry of Agriculture, Fisheries and Food, Torrey Research Station, 135 Abbey Road, PO Box 31, Torrey, Aberdeen, AB9 8DG, UK.*

Mirzadeh, H., *Department of Polymer Engineering, Amirkabir University, Tehran, Iraq*

Nakano, Shoichi, *Functional Materials Research Center, SANYO Electric Co., Ltd. 1-18-13, Hashiridani, Hirakata, Osaka 573, Japan.*

Naidu, B. Srinivasulu, *Department of Physics, S.V. University, Tirupati-517502, India*

Ng Seik Weng, *Institute of Advanced Studies, University of Malaya, 59100 Kuala Lumpur, Malaysia*

Ohnishi, Michitoshi, *Functional Materials Research Center, SANYO Electric Co., Ltd. 1-18-13, Hashiridani, Hirakata, Osaka 573, Japan.*

Ordonez, Ishmael D., *Department of Chemistry, Concordia University, 1455 de Maisonneuve Blvd. W, Montreal, Quebec, Canada H3G 1M8*

Pellerito, L., *Chemistry Department, Palermo University, Palermo, Italy*

Petrosyan, V.S., *Chemistry Department, M.V. Lomonosov University, Moscow, Russia*

Prasad, L.C., *Department of Chemistry, T.N.B. College, Bhagalpur University Bhagalpur - 812007, India*

Prischenko, A.A., *Chemistry Department, M.V. Lomonosov University, Moscow, Russia*

Quinto, Edna C., *Research Center for the Natural Sciences, University of Santo Tomas Espana, Manila, Philippines*

Radhakrishna, S., *Institute of Advanced Studies, University of Malaya, 59100 Kuala Lumpur, Malaysia*

Rao, C.N.R., *Solid State and Structural Chemistry Unit, Indian Institute of Science, Bangalore 560012, India*

Reddy, B.J., *Department of Physics, S.V. University, Tirupati-517502, India*

Reddy, P.J., *Department of Physics, S.V. University, Tirupati-517502, India*

Reed, Benjamin R., *ADA-AF, Gaithesburg, MD 20899, USA*

Scotto, Cathy, *Departments of Materials Science & Engineering & Chemistry, and the Macromolecular Science & Engineering Center, University of Michigan, Ann Arbor, Michigan 48109-2136 (USA)*

Sevilla, Fortunato III, *Research Center for the Natural Sciences, University of Santo Tomas Espana, Manila, Philippines*

Shiono, Takeshi, *Research Laboratory of Resources Utilization, Tokyo Institute of Technology, Nagatuta 4259, Midori-mu, Yokohama 227, Japan*

Siah, Lay-Foong, *Institute of Advanced Studies, University of Malaya, 59100 Kuala Lumpur, Malaysia*

Silvestru, Cristian, *Facultatea de Chimie, Universitatea Babes-Bolyai, RO-3400 Cluj-Napoca, Roumania*

Singh, Ranvir, *National Physical Laboratory, New Delhi - 110012, India*

Singh, R.N., *Department of Physics, College of Science,, Sultan Qaboos University, MUSCAT, Sultanate of Oman*

Singh, R.V, *Department of Chemistry, University of Rajasthan, Jaipur, 302004, India*

Singh, V.R., *National Physical Laboratory, New Delhi - 110012, India*

Smith, Frank E., *Department of Chemistry, Laurentian University, Sudbury, Ontario P3E 2C6, Canada*

Smith, P.J., *International Tin Research Institute, Kingston Lane, Uxbridge UB8 3PJ, Middlesex, UK*

Soga, Kazuo, *Research Laboratory of Resources Utilization, Tokyo Institute of Technology, Nagatuta 4259, Midori-mu, Yokohama 227, Japan*

Sudhana, B. Madhu , *Department of Physics, S.V. University, Tirupati-517502, India*

St.Pierre, T.G., *School of Mathematical and Physical Sciences, Murdoch University, Murdoch, Western Australia 6150*

Swamy, C.S., *Department of Chemistry, Indian Institute of Technology, Madras - 600 036, India*

Tagliavini, Giuseppe, *Dipartimento di Chimica Inorganica, Metallorganica ed Analitica, Universitá di Padova, Via Marzolo, 1, I-35131-Padova (Italy)*

Takeoka, Akio, *Functional Materials Research Center, SANYO Electric Co., Ltd. 1-18-13, Hashiridani, Hirakata, Osaka 573, Japan.*

Tanfani, Fabio, *Institute of Biochemistry, Medical School, University of Ancona, Via Ranieri, 60131 Ancona (Italy)*

Treadwell, David, *Departments of Materials Science & Engineering & Chemistry, and the Macromolecular Science & Engineering Center, University of Michigan, Ann Arbor, Michigan 48109-2136 (USA)*

Velu, S., *Department of Chemistry, Indian Institute of Technology, Madras - 600 036, India*

Wang, Sheguo, *Institute of Chemistry, Academia Sinica, Beijing 100080, China*

Wang, Zhifeng, *Institute of Chemistry, Academia Sinica, Beijing 100080, China*

Webb J., *School of Mathematical and Physical Sciences, Murdoch University, Murdoch, Western Australia 6150*

Wegner, Peter, *Institut fur Anorganische Chemie der Universitat Auf der Morgenstelle 18, D-7400 Tubingen 1, Germany*

Whalen, Deborah, *The District of Columbia Agricultural Experiment Station and Department of Chemistry, The University of the District of Columbia, Washington, DC 20008, USA*

Xiao Shexiu, *Department of Chemistry, Sichuan University, Chengdu 610064, Sichuan, P.R. China*

Yashina, N.S., *Chemistry Department, M.V. Lomonosov University, Moscow, Russia*

Yatsenko, A.V., *Chemistry Department, M.V. Lomonosov University, Moscow, Russia*

Zolese, Giovanna, *Institute of Biochemistry, Medical School, University of Ancona, Via Ranieri, 60131 Ancona (Italy)*

Contents

MAIN GROUP ELEMENTS AND THEIR COMPOUNDS

Perspectives in Materials Science,
Chemistry and Biology

Main Group Elements and Their Compounds
V.G. Kumar Das (Ed)

Silica as a Source of Chemicals and Polymers

Richard M. Laine, Florence Babonneau, David Treadwell,
Pallavi Kansal, Cathy Scotto, Sarah Jouppi

*Departments of Materials Science & Engineering & Chemistry, and the
Macromolecular Science & Engineering Center, University of Michigan,
Ann Arbor, Michigan 48109-2136 (USA)*

Carbon's intrinsic chemical behavior allows us to prepare, by a wide variety of synthetic processes, numerous simple and complex molecules with highly diverse physical and chemical properties. This enhances the value of certain carbon sources, e.g. petroleum, that are easily transformed in quantity to simple "building blocks" which in turn can be manipulated to produce more sophisticated chemical structures. As we search for new materials with new properties, efforts have focused on developing simple and complex molecules that contain elements other than carbon. These efforts have witnessed the birth of organometallic chemistry. The role of silicon containing molecules in the development of the field of organometallic chemistry has been substantial. Despite this, the actual methods whereby silicon is introduced to molecular structures is quite limited, arising primarily from the "direct process," as illustrated in equation (1) .[1]

$$\text{RCl} + \text{Si} \xrightarrow{\text{catalyst}} \text{R}_4\text{Si} + \text{R}_3\text{SiCl} + \text{R}_2\text{SiCl}_2 + \ldots\ldots\text{R}_2\text{Si}_2\text{Cl}_4.. + \text{RHSiCl}_2\ldots\ldots \quad (1)$$

We are interested in circumventing the energy and equipment intensive processes that involve the reduction of silica to silicon metal only to reoxidize the metal as shown in equation (1) and related reactions.[1] In particular, we are interested in developing Si chemistries that begin with SiO_2, but do not compete with the production and further manipulation of organosilanes such as produced in the direct process.

The primary obstacle to realizing new Si chemistries by SiO_2 depolymerization is the high silicon-oxygen bond strength of > 120 Kcal/mole. The two primary commercial methods of depolymerizing SiO_2 are by carbothermal reduction to Si metal and by dissolution processes wherein Si-O bonds are formed equivalent in energy to those broken. The latter process is actually more common than the former in that it is the method whereby SiO_2 is converted to inorganic silicates.[1]

$$\text{SiO}_2 + x\text{MOH (M = alkali metal)} \xrightarrow{-\text{H}_2\text{O}} \text{M}_2\text{O·SiO}_2 + \text{M}_2\text{O·2SiO}_2 + \ldots \quad (2)$$
$$(x = 2) \qquad (x = 4)$$

Over the past 60 years, several groups have sought to develop reactions analogous to equaton (2) by using metal alkoxides in place of metal hydroxides.[2-13] The most successful of the early routes involved the reaction of catechol with SiO_2 in a variety of forms and in a variety of solvents, including water:

$$SiO_2 + 2KOH + 3[1,2\text{-}C_6H_4(OH)_2] \xrightarrow{-4H_2O} K_2\left[\underset{O}{\overset{O}{\bigcirc}}\right]_3 Si$$

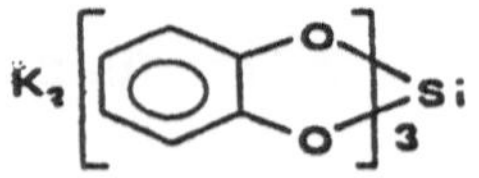

$$(3)$$

What is surprising about this compound is that it contains hexa-functional silicon rather than the much more common tetra-functional silicon. The fact that triscatecholato silicate is water soluble is a good indication of its considerable stability, which limits efforts to manipulate the molecular structure to provide access to new compounds with different properties; although some success has been achieved through reactions with strong nucleophiles such as Grignard reagents and electrophiles.[9-11]

In an effort to develop a more labile dissolution product, we sought to replace the chelating catechol ligand with the simplest analog, ethylene glycol (EG). To our surprise, silica reacts readily with EG and strong bases such as the group I metal hydroxides, e.g. KOH and group II oxides, e.g. BaO, to form simple crystalline compounds that provide access to wide variety of compounds and materials. The basic chemistry of these reactions and some of the products obtainable by manipulation of the initially formed products are described here.

Results and Discussion

If one heats a solution of silica (1 equivalent), KOH (1 equivalent) and excess ethylene glycol (EG) to the EG boiling point of $\approx 200°C$, one observes, within the first 0.5 h, the distillation of H_2O as KOH reacts with EG to form the alkoxide, reaction (4):[12-14]

$$KOH + HOCH_2CH_2OH \xrightarrow{>160°C} H_2O + KOCH_2CH_2OH \quad (4)$$

The initial rapid production of H_2O is followed by the much slower production of H_2O as silica dissolves with the formation of $Si\text{-}OCH_2CH_2OH$ bonds (reaction 5):

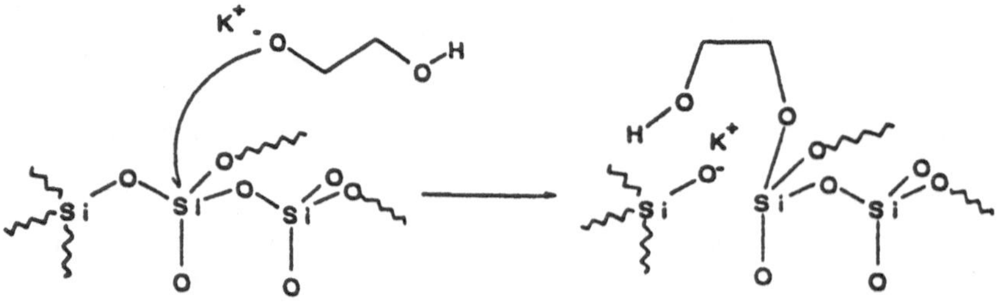

$$(5)$$

The rate of reaction was found to be first order in base concentration and in silica surface area.[14] Consequently, the rate limiting step occurs after reaction of $KOCH_2CH_2OH$ with the surface. A suitable set of events that accounts for the dissolution process requires further rearrangement of the reaction (5) product (see Scheme I) leading to the pentacoordinated anionic glycolato silicate: $KSi(OCH_2CH_2O)_2OCH_2CH_2OH$. The formation of pendant glycol groups on the surface is supported by IR evidence.[14]

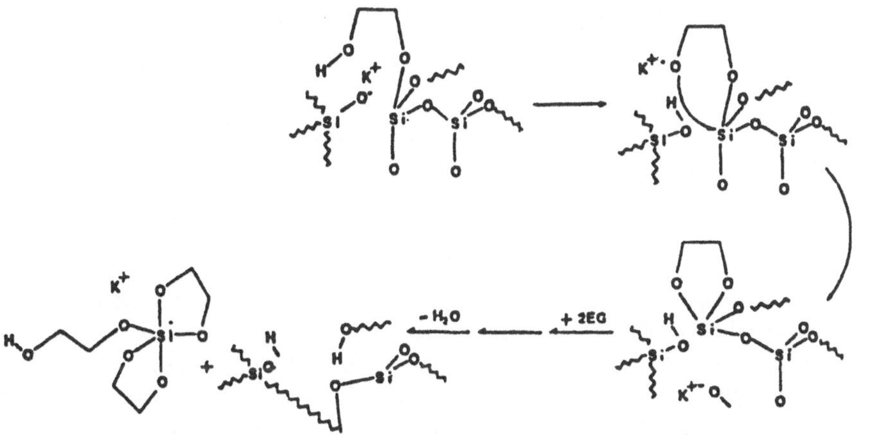

Scheme 1

One could easily envision an alternate dissolution mechanism wherein KOH first reacts with silica to form a silicate, e.g. $Si(OK)_4$, which subsequently reacts with free EG; however, the rapid removal of water via reaction (4) that must drive the formation of the alkoxide mitigates against this possibility. Further support against the intermediacy of a simple silicate comes from the zeolite synthesis literature. Bibby and Dale[15] note that in $NaOH/SiO_2/EG$ solutions, the initial removal of water followed by heating in an autoclave leads exclusively to silica sodalite (ZSM-5) rather than to complete dissolution of the silica. In an autoclave, the water produced in Scheme 1 would be trapped and would rapidly hydrolyze any pentacoordinated silicate formed. Clearly, it is extremely important to remove the water as it is formed to avoid forming zeolites. We estimate that the total water concentration during reaction is less than 0.01 M.[14] However, it is important to note that these pentacoordinated intermediates are likely intermediates in nonaqueous syntheses of zeolites.[14,15]

Given our discovery of a facile process for dissolving SiO_2 in EG with base to produce pentacoordinated glycolato silicates such as $KSi(OCH_2CH_2O)_2OCH_2CH_2OH$, and given our goal of developing new silicate chemistries, the question remains whether or not the process actually offers access to other compounds and materials. Thus, it becomes important to determine both the scope of the synthesis reaction, and the potential the initial reaction products have for further elaboration to new compounds and materials. With this in mind, we first selectively examined the types of diols and bases that could be used in the synthetic process.

3

Thus, we find that 1,2 diols (e.g. propane-1,2-diol and pinacol) and 1,3 diols such as 1,3 propanediol will react with silica to form simple complexes:

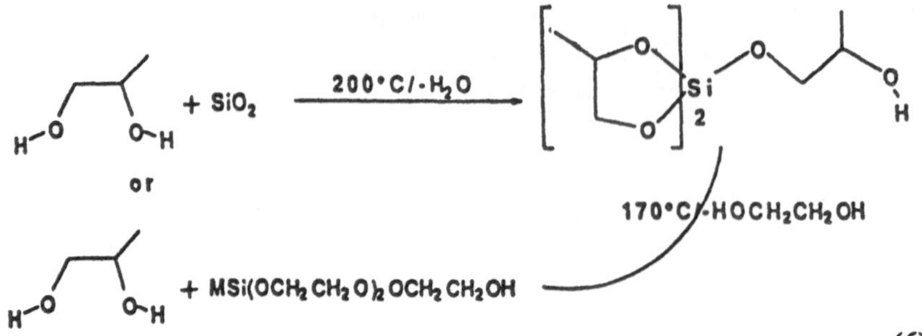

(6)

The same complexes can also be made by an exchange reaction wherein the glycol ligands on the initially formed complex are displaced, e.g. reaction (6), by heating in the presence of a large excess of 1,2 or 1,3 diol to be exchanged.

Attempts to react 1,4 and longer diols with SiO_2 were unsuccessful, likely because chelation is necessary for the depolymerization process to work. However, if: (1) the SiO_2 dissolution process is run in EG with at least 3 equivalents of a second, longer chain diol; and (2) this diol has a higher boiling point than EG; then on distillative removal of all the EG from the solution, pentacoordinated Si containing polymers remain:

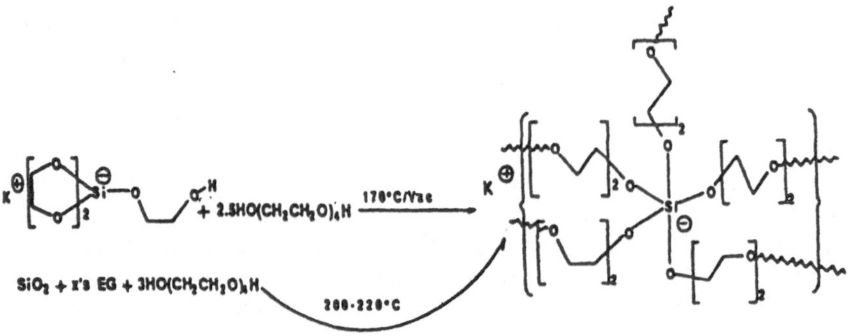

These polymers are quite interesting because they contain ethylene oxide (EO) spacer groups and pentacoordinated silicon in the backbone.[16] Preliminary studies indicate that the silicon centers remain pentacoordinated as ^{29}Si NMR shows species at -103 and -107 ppm which are in the region previously found for pentacoordinated, anionic silicon compounds.[12-14] The EO spacer groups, which are too short to act as oligomeric units by themselves, provide T_gs that are in the range of -60 to -40°C when incorporated in the polymers formed as shown above.

4

These values are typical for polyethylene oxide and are a further indication of the polymeric nature of these compounds.[16] Finally, solution and solid state ^{29}Si NMR studies for the group I metal glycolato silicates show that all of the peaks appear in the range -103 to -110 ppm, as shown in the ^{29}Si solid state NMR spectra (Figure 1).

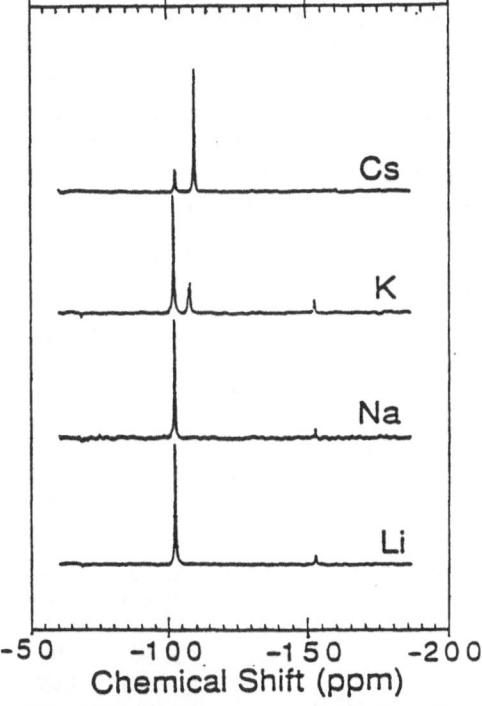

Fig. 1 : Solid state MAS ^{29}Si NMR spectra of selected glycolato silicates
(Note that the Cs peak at -109.8 ppm is shifted from the peaks of the other salts (-103 ppm), most likely because it is polymeric)

The monomeric species all exhibit peak positions at -103 ± 0.3 ppm, whereas polymeric derivatives appear to exhibit peaks at -107 to -110 ppm. In general, the peak positions are relatively independent of the counterion which suggests very loose as opposed to tight ion pairing. This in turn suggests that low T_g polymeric derivatives might exhibit ion conducting properties. The results of preliminary studies indicate that they do indeed offer ion conducting properties as shown in the Table 1.[16]

Table 1 : AC Impedance at 25°C over frequency range of 10-10^5 Hz

Polymer	Conductivities	^{29}Si NMR	T_g
Li$^+$	3 x 10^{-6} S/cm	-103 ppm	-25 to - 15
K$^+$	4 x 10^{-5} S/cm	-107 ppm	-20 to - 10
Ba^{2+}	5 x 110^{-5} S/cm		-25

5

The fact that lithium silicate, $Li_2O\cdot2SiO_2$, offers potential as a high temperature ion conductor,[17] suggested that these polymers, because they exhibit useful rheological properties which permit them to be shaped at low temperatures (*e.g.* thin films) and correct stoichiometry (2Li : 2Si), might also serve as formable precursors to high temperature ion conductors. Furthermore, because there are numerous low and high tech applications for alkali silicate glass shapes and powders, efforts were made to examine the pyrolytic decomposition of these materials to determine whether or not phase pure materials could be obtained.

As a prelude to examining the high temperature chemistry of the polymeric materials, pyrolytic decomposition studies were first conducted on the simple monomers and dimers, to better model decomposition processes. Thus, the monomers and dimers were synthesized and fully characterized by chemical analysis, solid and solution multinuclear NMR, diffuse reflectance infrared Fourier transform spectroscopy (DRIFTS), thermal gravimetric analysis (TGA), differential thermal analysis (DTA) and x-ray powder diffractometry (XRD). The details of the characterization studies are published elsewhere;[14,18] however, Table 2 shows the chemical analyses data on the monomers and the Na dimer.

Table 2 : Chemical analyses of the monomers and Na dimer [†]

Compound	MW	C (wt%)	H (wt%)	Ma (wt%)	Si (wt%)
$LiSi(OCH_2CH_2O)_2OCH_2CH_2OH$	216.6	33.30 (32.50)	6.06 (6.20)	3.21 (3.03)	13.00 (11.20)
$Na_2Si_2(OCH_2CH_2O)_5$	402.36	29.90 (29.53)	5.01 (4.93)	11.40 (11.53)	14.00 (13.91)
$KSi(OCH_2CH_2O)_2OCH_2CH_2OH$	248.32	29.00 (28.26)	5.28 (5.06)	15.70 (14.85)	14.00 (13.91)
$CsSi(OCH_2CH_2O)_2OCH_2CH_2OH$	342.16	20.72 (21.06)	3.63 (3.83)	39.38 (38.84)	8.58 (8.21)

[†]Figures given in parenthesis are experimental results while those not within parenthesis are calculated results.

The next step was to examine the low temperature behavior of monomeric, dimeric and oligomeric analogs. Surprisingly, all of the monomeric materials will, on heating to temperatures between 130 and 170°C, dimerize as shown in Scheme 2 for the potassium compound. The sodium salt is actually isolated as the sparingly soluble dimer, directly from solution. As the dimer, it has the typical -103 ppm solid state ^{29}Si NMR peak.

On heating to higher temperatures in air, all of the monomers and dimers oxidatively decompose in the region of 330-400°C to give first amorphous products and then at temperature ranging from 500-700°C, the fully crystalline, nearly phase pure $M_2O\cdot2SiO_2$ silicates in essentially quantitative yield as expected and in agreement with the calculated and found ceramic yields shown in Table 3, which provide a second proof that the as prepared compounds are pure. Figure 2 provides the XRD patterns for samples of the dimeric $Li_2Si_2(OCH_2CH_2O)_5$ heated to selected temperatures in air (2 h hold) to demonstrate the crystallization of nearly pure $Li_2O\cdot2SiO_2$.

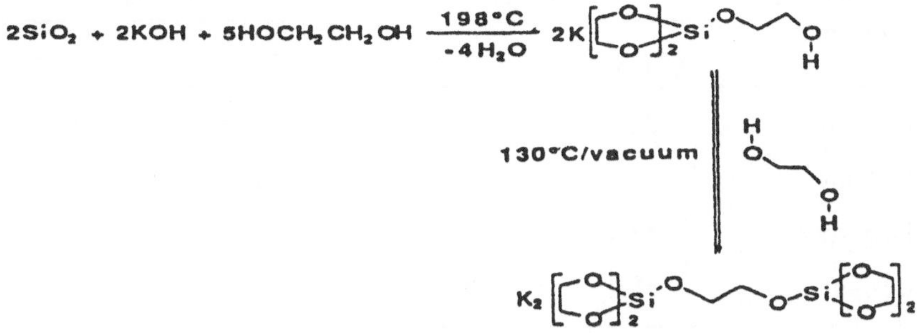

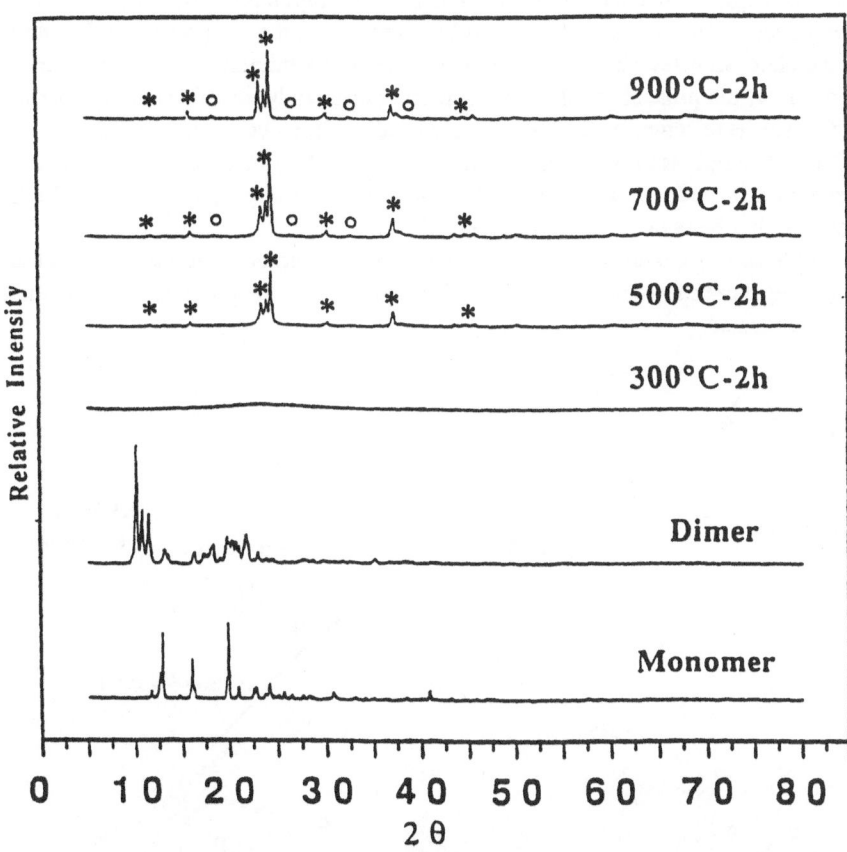

Scheme 2

Fig. 2 : XRD patterns for $Li_2Si_2(OCH_2CH_2O)_5$ heated to selected temperatures

(*) : orthorhombic α-$Li_2Si_2O_5$, JCPDS file 17-447

(°) : orthorhombic Li_2SiO_3, JCPDS File 29-829

7

Table 3 :   Yields of $M_2O \cdot 2SiO_2$ from thermal degradation of monomeric and dimeric Gp. 1 metal glycolato silicates

M glycolato silicate	Monomer		Dimer	
	Exp. (%)	Calc. (%)	Exp. (%)	Calc. (%)
Lithium	34.1	34.7	40.4	40.5
Sodium	-	39.2	45.8	45.3
Potassium	43.4	43.2	50.3	49.3
Cesium	58.2	58.6	64.3	64.6

These results provide the basis for conducting detailed studies on the higher carbon content polymers. While similar behavior might be anticipated, recent results suggest that care must be taken in extrapolating from these model compounds. For example, related pyrolysis studies on the 1,2 propanediolato and pinacolato silicates show that while the 1,2-propandiolato silicate monomer dimerizes, the pinacolato silicates do not.[19] Fortunately, this does not appear to influence the selectivity to the final ceramic product; however, work with the group II metal complexes reveals considerable incorporation of CO_2 into the ceramic product. Indeed, the amounts of CO_2 incorporated appear to affect the high temperature behavior of these materials.[20]

Still other transformations of the pentacoordinated compounds are possible including those shown in Scheme 3 , which summarizes the possible products that can be formed.

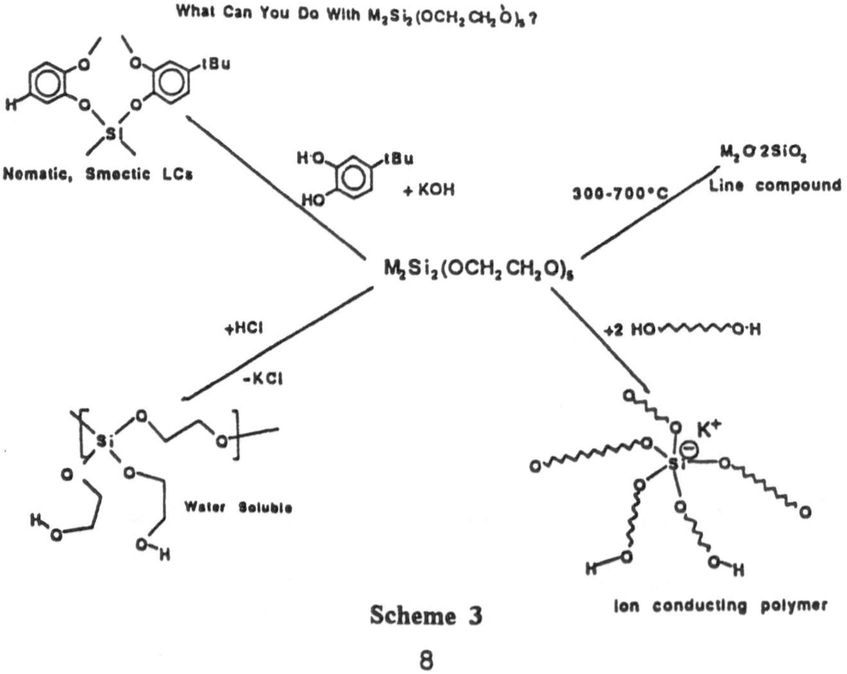

Scheme 3

8

The reactions that produce liquid crystalline materials are described by Ray et al.[21] The neutralization of the glycolates produces a neutral tetra-alkoxysilane that is a good sol-gel precursor.

As part of our efforts to expand the scope of the silica dissolution reaction, we examined replacing the group I metal hydroxides with group II metal oxides. The results of these studies were quite surprising. One objective was to make tractable precursors to $MO \cdot 2SiO_2$ silicates that might provide the basis for the synthesis of novel aluminosilicate glasses and ceramics such as $CaO \cdot 2SiO_2 \cdot Al_2O_3$ which is garnet. We envisoned using these polymer precursors to make inexpensive, garnet hard coatings. Thus, the first step was to replace the KOH in the synthesis shown in Scheme 2 with CaO or BaO.

Both of these oxides were found to dissolve in hot EG. Initial studies focused on SiO_2 reactions with dissolved BaO because it was the easiest to solubilize. As before, the BaO, EG, SiO_2 mixture was found to dissolve with distillative removal of water. Likewise, on cooling, a crystalline mass precipitated. While waiting for results from chemical analyses, efforts to pyrolyze the material in the TGA followed by XRD of the products led to the identification of a single crystalline ceramic, $BaO \cdot SiO_2$. This was quite surprising given that the initial reaction stoichiometry was chosen to give $BaO \cdot 2SiO_2$. The chemical analyses, when received, did not provide any useful information other than the Ba:Si ratio was indeed 1:1. Finally, by accident, we were able to crystallize a large enough single crystal to enable single crystal diffractometer studies to be done.

To our considerable surprise, rather than finding a pentacoordinated silicon, the diffractometry studies revealed the formation of the very first hexacoordinated, hexa-alkoxy dianionic silicate, $BaSi(OCH_2CH_2O)_3 \cdot 3.25HOCH_2CH_2OH$.[13] The presence of the EG groups of recrystallization was found to explain the chemical analyses. These results were quickly followed by the preparation of the Ca analog. Efforts to prepare the Mg analog were thwarted by the insolubility of MgO in EG. Subsequently, we found an alternate route to this material by reacting $Mg(OCH_2CH_2O)$ with $Si(OEt)_4$ in EtOH with excess EG. These materials continue to provide surprises.[20] For example, the solid state ^{29}Si NMR signals of all of the compounds appear at -140 to -155 ppm. All of the compounds crystallize with 2.5-3.25 EGs of recrystallization. Efforts to obtain reliable solution ^{29}Si NMRs provide peaks at -103 to -105 ppm, suggesting isomerization to $[Ba]^{2+}[Si^-(OCH_2CH_2O)_2OCH_2CH_2O^-]$ in solution. This may account for the very high ionic conductivity seen for the Ba polymer in Table 1 above.

Despite these apparent irregularities, we have succeeded in making a number of precursors to aluminosilicate materials, including cordierite, mullite, celsian ($BaO \cdot 2SiO_2 \cdot Al_2O_3$) as described in detail in references 22-26. Still other new materials syntheses and processes appear possible based on these novel penta- and hexa alkoxy silicate precursors formed directly from silica.

Acknowledgements

We thank the Office of Naval Research, the Air Force Office of Scientific Research, and the Army Advanced Materials and Technology Laboratory (now Army

Research Laboratories) and the Army Research Office for generous support of various aspects of the above reported work. We thank Drs. John Negrych and Jerry Snow of Minco, Inc. for helpful discussions and gifts of fused silica and virgin sand. We would also like to thank Professors R.J.P. Corriu and P. Knochel for useful discussions and suggestions.

References

1. Kirk-Othemer Encyclopedia of Chemical Technology, 3rd Ed.; Wiley-Interscience Publ., N.Y., 1979, Vol. 20, pp 750-880.
2. A. Rosenheim, B. Raibmann, G. Schendel, *Z. Anorg, Chem.*, **196** (1931) 160.
3. D.W. Barnum, *Inorg. Chem.*, **9** (1970) 1942.
4. D.W. Barnum, *Inorg. Chem.*, **9** (1970) 1424.
5. A. Weiss, G. Reiff, A. Weiss, *Z. Anorg. Allg. Chem.*, **311** (1961) 142.
6. C.L. Frye, *J. Am. Chem. Soc.*, **86** (1964) 3170.
7. F.P. Boer, J.J. Flynn, J.W. Turley, *J. Am. Chem. Soc.*, **90** (1968) 6973.
8. J.J. Flynn, F.P. Boer, *J. Am. Chem. Soc.*, **91** (1969) 5756.
9. (a) A. Boudin, G. Cerveau, C. Chuit, R.J.P. Corriu, C. Reye, *Angew. Chem. Int. Ed.*, **25** (1986) 473. (b) A. Boudin, G. Cerveau, C. Chuit, R.J.P. Corriu, C. Reye, *Organomet.*, **7** (1988) 1165.
10. (a) R.J. P Corriu, J.C. Young, *Chemistry of Organic Silicon Compounds*, eds., S. Patai, Z. Rappaport, Ch. 20, Wiley, Chichester, 1989. (b) R.J.P. Corriu, *Pure and Appl. Chem.*, **60** (1988) 99.
11. R.J.P. Corriu, *J. Organomet. Chem.*, **400** (1990) 81.
12. R.M. Laine, K.Y. Blohowiak, T.R. Robinson, M.L. Hoppe, P. Nardi, J. Kampf, J. Uhm, *Nature*, **353** (1991) 642.
13. M.L. Hoppe, R.M. Laine, J. Kampf, M.S. Gordon, L.W. Burggraf, *Angew. Chem. Int.*, **32** (1993) 287.
14. K.Y. Blohowiak, M.L. Hoppe, K.W. Chew, P. Kansal, B.L. Mueller, C.L.S. Scotto, D.R. Treadwell, T.Hinklin, F. Babonneau, J. Kampf, R.M. Laine, *JACS*, (submited).
15. D.M. Bibby, M.P. Dale, *Nature*, **317** (1985) 157.
16. (a) K.W. Chew, B. Dunn, T. Faltens, M.L. Hoppe, R.M. Laine, L. Nazar, H.K. Wu, *Am. Chem. Soc. Poly. Prprts.*, **34** (1993), 254. (b) K.W. Chew, B. Dunn, T. Faltens, M.L. Hoppe, R.M. Laine, T.R. Robinson, C.S. Scotto manuscript to be submitted.
17. S-P. Szu, M. Greenblatt, L.C. Klein, *J. Non-crystal. Sol.*, **124** (1990) 91, and references therein.
18. P. Kansal, R.M. Laine, *J. Am. Ceram. Soc.*, in press.
19. S. Jouppi, C. Scotto, and R.M. Laine, unpublished work.
20. P. Kansal, R.M. Laine, *J. Am. Ceram. Soc.*, (submitted).
21. D.J. Ray, R.M. Laine, T.R. Robinson, C. Viney, *Mol.Crystal, Liq. Crystal,* **225** (1992) 153, and references therein.
22. C. Bickmore, M.L. Hoppe, R.M. Laine, In: *Synthesis and Processing of Ceramics: Scientific Issues*, eds., W. E. Rhine, T. M. Shaw, R. J. Gottschall and Y. Chen, Mat. Re. Soc. Symp. Proc., 1991, vol. **249**, pp. 107-114.
23. Z-F. Zhang, M.L. Hoppe, J.A. Rahn, S-M. Koo, and R.M. Laine, In: *Synthesis and Processing of Ceramics: Scientific Issues*, eds., W. E. Rhine, T.M. Shaw,

R.J. Gottschall and Y. Chen, Mat. Re. Soc. Symp. Proc., 1991, vol. 249, pp. 81-86.

24. R. M. Laine, K. A. Youngdahl, and P. Nardi, Silicon and Aluminum Complexes, U.S. Patent 5,099,052 March 24, 1992.

25. R. M. Laine and K. A. Youngdahl, Silicon and Aluminum Complexes, U.S. Pat. No. 5,216,155, June 1, 1993.

26. R.M. Laine, B.L. Mueller, T. Hinklin, D. Treadwell, to be submitted to Nature.

Main Group Elements and Their Compounds
V.G. Kumar Das (Ed)
Copyright © 1996 Narosa Publishing House, New Delhi, India

Molecules to Materials: Engineering at the Nanometer Level

Pradyot A. Agaskar

*Mobil R&D Corporation, Central Research Laboratory,
P.O. Box 1025, Princeton, NJ 08540 (USA)*

Converting molecules to materials by linking them up to form linear organic or inorganic polymers has yielded, over the past sixty years, a vast number of new and useful products. Since the precursor molecules retain their structural integrity in the product, structure-property relationships can be established, and a polymer that has some predetermined desirable property can, in most cases, be 'rationally' synthesized.[1]

The other two major classes of materials, metals and ceramics are generally prepared using high-temperature processes from precursors having extended non-molecular structures. However, even though the use of molecular precursors with well-defined structures may be advantageous in many ways,[2] no structure property relationships can be established due to the high temperatures involved in the synthesis.

In recent years, however, there has been growing interest in 'hybrid inorganic/organic materials', that are expected to have properties unattainable by either of the components of the hybrid. Synthetic routes analogous to those used for the synthesis of the linear polymeric materials, Figure 1, have been devised and implemented with varying degrees of success.[3-5]

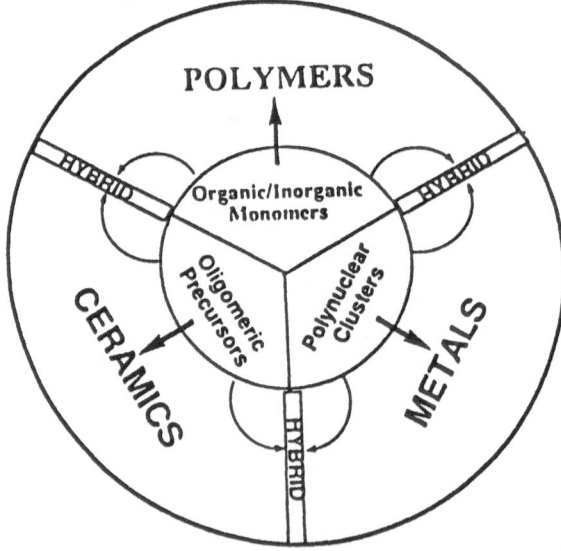

Fig. 1 : 'Rational' synthetic routes to the three major classes of materials

Organolithic Macromolecular Materials: OMMs

Hybrid materials containing an ionic silicate component and a linear siloxane component co-mingled at the molecular level were envisaged a number of years ago as a means of combining the advantageous properties of two of the three major classes of materials, polymers and ceramics.[6] The method most commonly used to synthesize these materials, appropriately termed Organolithic Macromolecular Materials (OMMs)[7], involves using sol-gel techniques to generate particles of the silicate component in the presence of the organic component.[8,9] These methods, however, do not allow for much fine control over the structure of the silicate particles, since these generally grow in an uncontrolled manner by sequential hydrolysis and condensation reactions.

An alternative route to such materials is to use preformed silicate structures, molecular or extended, and the methods of molecular chemistry to introduce covalent linkages between them and the organic components. The resultant OMMs are composed of well-defined organic and silicate moieties covalently linked to each other. The silicate components of such materials can have a variety of structures that may be classified in terms of their formal dimensionality. Thus OMMs containing the formally 2-dimensional layered silicates, such as chrysotile[10-12] have been known for a number of years and an example of an OMM containing a formally 1-dimensional tube silicate[13] was reported recently. In these materials, the inorganic component itself has an extended structure and the organic component is a small molecule. We reported a few years ago[14] the first example of an OMM containing formally 0-dimensional spherosilicates[15,16] cross-linked by a small organic moiety. Similar materials have since been reported[17] by other workers.

Microporosity of Spherosilicate Containing OMMs

We expected the OMMs, whose structures consisted of spherosilicates cross-linked by organic moieties, to be microporous materials with high-surface area because we believed that geometric restrictions would prevent a 3-D network of corner-linked polyhedra from filling space.[18] However, our results belied our expectations and the new materials far from being microporous, showed no porosity at all. We initially interpreted this result as being due to the interpenetration of two or more 3-D networks as is observed in some crystalline materials, e.g. adamantane-1,3,5,7- tetracarboxylic acid.[19]

An alternative interpretation, however, was that our OMMs were not in fact 3-D networks but had lamellar or columnar structures, as shown schematically in Figures 2a and 2b. These ordered structures might have been favoured by our choice of the cross-linking organic moiety, i.e. the 4,4'-bis(dimethylhydrosilyl)-diphenylether, which was based solely on the fact that the distance between the two reactive ends of this moiety is so great that it could not connect two vertices of the same spherosilicate moiety. However, the long lath-like shape of the diphenylether moieties might have favored the formation of precursor species such as those shown in Figures 2c and 2d. Compounds with molecules having similar structures have been synthesized and shown to be liquid crystalline.[20] Scanning electron micrographs of fracture surfaces of these OMMs showed long uniform ridges, shown in Figure 3, which suggested that there was some degree of ordering in the structure, possibly such that the inorganic moieties,

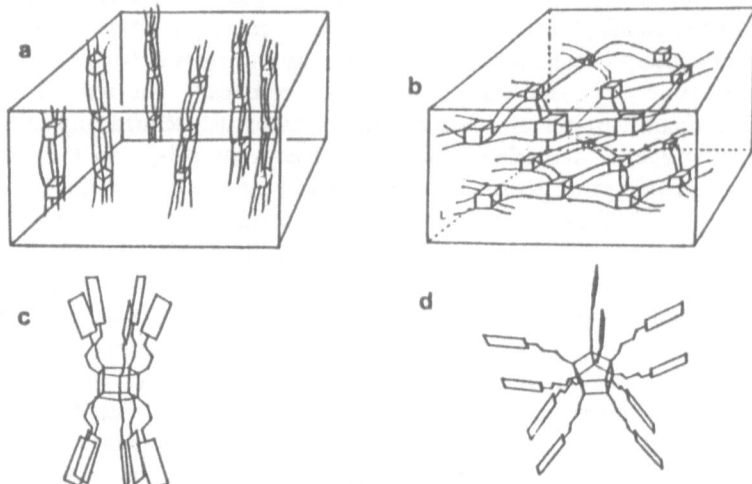

Fig. 2 : Schematic representations of possible ordered structures of the OMM derived from the cubic vinyl-functionalized spherosilicate and 4,4'-bis(dimethylhydrosilyl)diphenylether

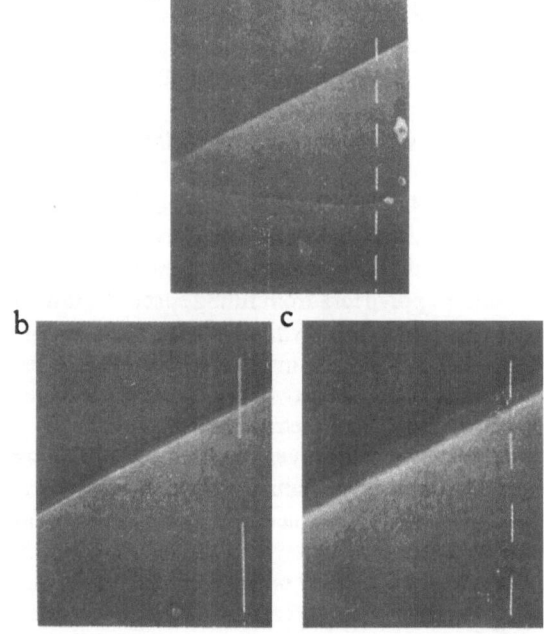

Fig. 3 : Scanning electron micrographs of a fracture surface of one of the OMMs at magnifications of (a) 1600X and (b) 6400X and (c) 25000X. The regular features observed suggest tht the secondary structure of the OMM is to some extent ordered.

the spherosilicates, and the organic moieties were aligned to form silica-rich inorganic domains. We reasoned that such an ordered structure upon pyrolysis would form a nanocomposite material, with a silica-rich phase derived from the spherosilicate moieties and a carbon rich phase derived from the organic cross-linking moieties, and further that leaching this material with hydrofluoric acid would yield a microporous material with pore sizes in the nanometer range.

Microporous Ceramic Materials

We reported earlier[21] that the OMMs containing the cubic or the pentagonal prismatic spherosilicates cross-linked by the diphenylether moieties can be converted into amorphous microporous ceramic materials by pyrolysis and subsequently leaching the pyrolysate with HF to preferentially dissolve the silica-rich component derived from the spherosilicate moieties in the starting OMMs.[22] The Type I argon adsorption isotherms observed for the two materials, indicative of microporosity, can be converted to pore size distribution plots using the Horvath-Kawazoe transform.[23] These plots, shown in Figure 4, reveal that the pore size distributions are very narrow, surprisingly similar to the distributions observed in the crystalline aluminophosphates ALPO-5[24] and VPI-5[25]. The pore volumes are related to the extent of leaching of the silica-rich component, which is shown to be greater by ^{29}Si MASNMR in the material derived from the pentagonal prismatic spherosilicate, possibly because it had been ground more finely.

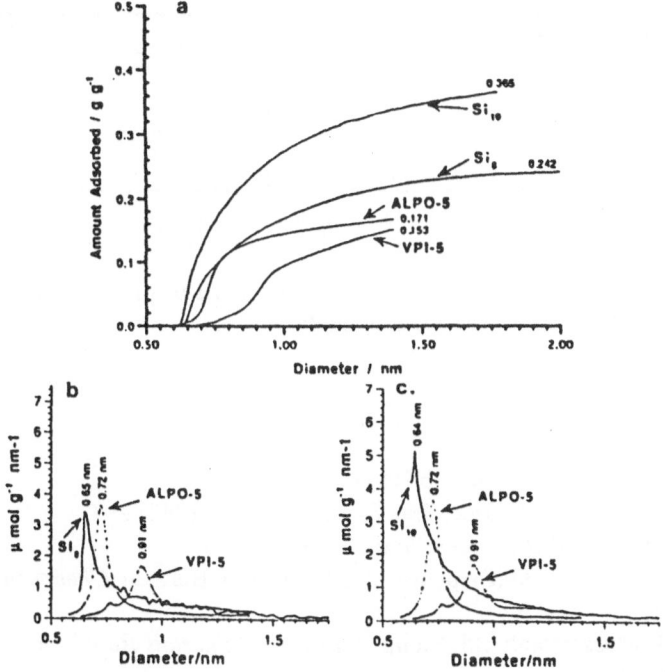

Fig. 4 : **Horvath-Kawazoe transforms of digitized data obtained from the Ar adsorption isotherms acquired using quasi-equilibrium high-resolution gas sorption techniques**

15

Conclusion

We have shown that the four-stage synthetic strategy outlined in Figure 5 yields high surface area materials with narrow pore size distributions, similar to those observed in crystalline microporous materials. The pore diameters in these materials match the dimensions of the spherosilicate moieties in the molecular precursors indicating that a structure-property relationship could be established in microporous materials prepared by this route, especially if milder conditions can be found to carry out the second step. There are a variety of potential uses of these materials, e.g. as catalyst supports[26], microporous membranes, chemical sensors etc., based both on the unusual structures/properties of these materials and their mode of preparation.

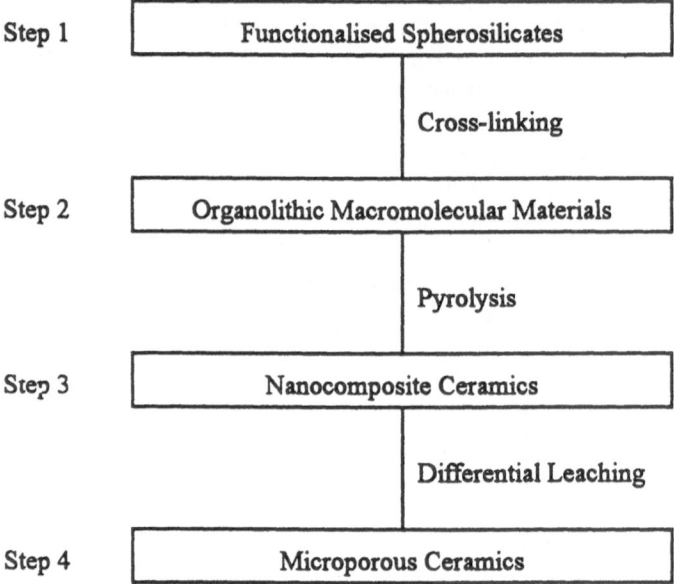

Fig. 5 : Outline of the synthetic procedure used to prepare amorphous microporous ceramic materials

References

1. H.R. Allcock, *Science*, **255** (1992) 1106.
2. K.J.Wynne, R.W. Rice, *Ann. Rev. Mater.*, **14** (1984) 297.
3. S. Komarneni, *J. Mater. Chem.*, **2** (1992) 1219.
4. C.M. Lukehart, J.P.Carpenter, S.B.Milne, K.J. Burnam, *Chemtech*, 1993, p.29.
5. P.B. Messersmith, S.I. Stupp, *Journal of Materials Research*, **7** (1992) 2599.
6. L. Holliday, *Chem. & Ind.*, **1972** (1972) 921.
7. A.D.Wilson, S.Crisp, *Organolithic Macromolecular Materials*, Appl. Sci. Publ., London, 1977.

8. G.L. Wilkes, *Polymer Preprints*, **26** (1985) 300.

9. B.M. Novak, *Adv. Mater.*, **5** (1993) 422.

10. C.Bleiman, J.P. Mercier, *Inorg. Chem.*, **14** (1975) 2853.

11. E. Ruiz-Hitzky, J.M. Rojo, *Nature*, **287** (1980) 28.

12. T. Yanagisawa, M. Harayama, K. Kuroda, C. Kato, *Solid State Ionics* , 42 (1990) 15.

13. B.A. Harrington, M.E. Kenney, *Coll. & Sur.*, **63** (1992) 139.

14. P.A. Agaskar, *J. Am. Chem. Soc.*, **111** (1989) 6858.

15. P.A. Agaskar, *Inorg. Chem.*, **29** (1990) 1603.

16. P.A. Agaskar, *Synth. Reactivity in Inorg. & Metal-Org. Chem.*, **20** (1990) 483.

17. D. Schultze, P. Kölsch, M. Noack, P. Toussaint, I. Pitsch, D. Hoebbel, *Zeitschrift für anorganische und allgemeine Chemie*, (1992)

18. V.W. Day, W.G. Klemperer, V.V. Mainz, D.M. Millar, *J. Am. Chem. Soc.*, **107** (1985) 8262.

19. O. Ermer, *J. Am. Chem. Soc.*, **110** (1988) 3747.

20. F.-H. Kreuzer, R. Maurer, P.Spes, *Makromol. Chem., Makromol. Symp.*, **50** (1991) 215.

21. P.A. Agaskar, *Coll. & Surf.*, **63** (1992) 131.

22. P.A. Agaskar, *J. Chem. Soc., Chem. Commun.*, **1992** (1992) 1024.

23. G. Horváth, K. Kawazoe, *J.Chem. Eng. Jpn.*, **16** (1983) 470.

24. S.T. Wilson, B.M. Lok, C.A Messina, T.R. Cannon, E.M. Flanigen, *J. Am. Chem. Soc.*, **104** (1982) 1146.

25. M.E. Davis, C. Montes, P.E. Hathaway, J.P. Arhancet, D.L. Hasha, J.M. Garces, *J. Am. Chem. Soc.*, **111** (1989) 3919.

26. G.J. Hutchings, C.S McKee, *Appl. Catal.*, **93** (1992) N3.

Main Group Elements and Their Compounds
V.G. Kumar Das (Ed)
Copyright © 1996 Narosa Publishing House, New Delhi, India

New Materials and Nanoscale Structures derived from Biominerals

J. Webb, T.G. St.Pierre and D.J. Macey
School of Mathematical and Physical Sciences
Murdoch University, Murdoch, Western Australia 6150

Biominerals, the inorganic materials of biology, are found throughout the natural world, including microorganisms, plants and animals. Many familiar objects are instances of biominerals. Perhaps the most dramatic are the coral reefs and sea shells of the marine environment (calcium carbonate) and human bone and teeth (calcium hydroxyapatite), but there are many other examples. In the past 10 years, an increasing number of biominerals has been reported (Table 1). Interest in the biological and chemical processes that lead to biomineralization, however, has only developed comparatively recently. Early observations were made by paleontologists who were interested in the preservation, through geological time, of the hard parts of organisms such as shells and skeletons. The year 1989 saw the almost simultaneous publication of three monographs covering current knowledge of the biological, biochemical, chemical and taxonomic aspects of biomineralization. [1-3]

Biominerals exhibit a wide variety of functions as shown in Table 1. Understanding how biominerals perform these functions is a major new frontier in interdisciplinary scientific research. Much of the current interest derives from the inspiration to be found in biominerals for the design of novel materials. As Derek Birchall of ICI (UK)'s Advanced Materials Group has put it:[4]

> "Biology does not waste energy manipulating materials and structures that have no function and it eliminates those that do not function adequately and economically. The structures that we observe 'work' and their form and microstructure have been developed and refined over millions of years. The task of the modern materials technologist is to develop, within a very short timescale, materials the microstructure of which is such that they perform a function efficiently. It is as well, then, to look for fresh insights to biology at the wisdom encapsulated in the materials it uses."

Biomineral phases can be crystalline, amorphous or partly crystalline/partly amorphous. They contain individual biomineral deposits that are generally less than 1 μm and may be as small as 10 nm. Such nanospace materials are of interest to emerging technologies. Furthermore, in many biological systems, the biominerals are present as components of composite materials, where the inorganic components are organized together with organic components such as proteins or polysaccharides to form the final biological structure.

18

Table 1 : The types and functions of the main inorganic solids found in biological systems

Mineral	Formula	Organism/Function
Calcium carbonate:		
Calcite	$CaCO_3^*$	Algae/exoskeletons Trilobites/eye lens
Aragonite	$CaCO_3$	Fish/gravity device Molluscs/exoskeleton
Vaterite	$CaCO_3$	Ascidians/spicules
Amorphous	$CaCO_3.nH_2O$	Plants/Ca store
Ca phosphate:		
Hydroxyapatite	$Ca_{10}(PO_4)_6(OH)_2$	Vertebrates/endoskeletons teeth Ca store
Octa-calcium phosphate	$Ca_8H_2(PO_4)_6$	Vertebrates/precursor phase in bone?
Amorphous	?	Mussels/Ca store Vertebrates/precursor phases in bone?
Calcium oxalate:		
Whewellite	$CaC_2O_4H_2O$	Plants/Ca Store
Weddellite	$CaC_2O_4.2H_2O$	Plants Ca store
Group IIA metal suphates:		
Gypsum	$CaSO_4$	Jellyfish larvae gravity device
Barite	$BaSO_4$	Algae/gravity device
Celestite	$SrSO_4$	Acantharia/cellular support
Silicon dioxide:		
Silica	$SiO_2.nH_2O$	Algae/exoskeletons
Iron oxides:		
Magnetite	Fe_3O_4	Bacteria/magnetotaxis Chitons/teeth
Maghemite	$-Fe_2O_3$	Chitons/teeth
Geothite	$-FeOOH$	Limpets/teeth
Lepidocrocite	$-FeOOH$	Chitons (Mollusca) teeth
Ferrihydrite	$5 Fe_2O_3.9H_2O$	Animals and plants/ Fe storage/transport proteins
Iron sulfides:		
Greigite	Fe_3S_4	Bacteria/magnetotaxis
Pyrite	FeS_2	

* A range of magnesium-substituted calcites is also formed.

In this paper, two instances will serve to illustrate the link between the structure of biominerals and the design of novel materials. A third example relates to our own studies of the biological mineralization of iron in some marine molluscs (limpets and chitons), specifically, in the teeth of the radula, a tongue-like organ used for feeding. Finally, the protein ferritin will be used to illustrate how a biological macromolecule allows selective synthesis of nanoscale particles that are of interest to new nanotechnologies.

Flexible Concrete

The microarchitecture of the composite material that is the molluscan shell has been used to inspire the preparation of what at first glance appears to be a contradiction in terms: flexible concrete, or, more accurately, macrodefect-free concrete.[4] Molluscan shells are composed of microcrystals of $CaCO_3$ organized with a protein-rich organic matrix (Fig. 1). Such shells have strengths in tension that are an order of magnitude greater than that of concrete. The innovative inclusion of a water-soluble polymer in the concrete formulation was derived directly from the analogy of the structural architecture within the shells. This proved to be critical in obtaining a plastic rheology for the concrete. This novel concrete material has been used to fabricate building materials and automobile components, including car springs, although not yet on an industrial mass production scale. Difficulties persist, in part due to the hydrophilic character of the included polymer, which thus absorbs atmospheric water.

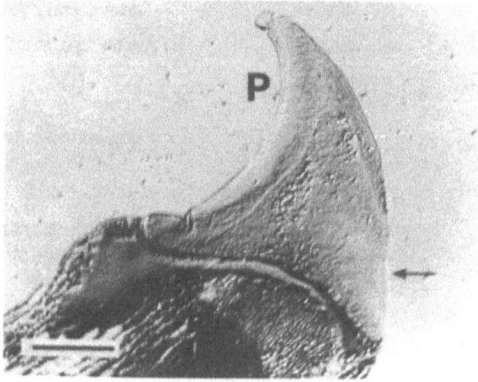

Fig. 1 : **Light micrograph of a longitudinal section through a mature tooth of** *Plaxiphora albida*. **The distinct magnetite cap can be seen as well as the large fibres that run through the teeth. P indicates the posterior (cutting) surface while the arrow indicates a 'window' region on the anterior surface. The section was stained with Toluidine blue. Scale bar is 50 μm.**

Biocoral

The natural coral skeleton is a porous structure with many channels and spaces, whose dimensions in certain coral species match those of human bone. This similarity to, and compatibility with, bone has been exploited in that coral can be used as an implant in bone tissue. In France, such a material, known as Biocoral® has been prepared by selective treatment of natural coral and used for some years as a material for bone grafts and in plastic surgery.[5] The Australian distributors (3M Company) indicate that Biocoral has been approved for use in Australia and already surgical procedures have been carried out in Sydney and Melbourne. The mechanical properties of Biocoral can be controlled by changing its porosity, giving materials that match or exceed bone in compressive strength and elasticity. It is the chemistry of the coral preparation that is particularly appropriate for this surgical use. As noted above, bone is made of a phosphate salt of calcium, while coral is a carbonate salt. The particular form of $CaCO_3$ in coral is aragonite, a metastable form of $CaCO_3$. This metastable nature means that, when implanted, the $CaCO_3$ gradually dissolves (i.e. is resorbed) and is replaced by newly formed bone, synthesized by cells that migrate into the porous structure from adjacent bone. The other common form of $CaCO_3$, calcite, does not dissolve under these conditions. The implant material contains some amino acids but no protein and thus differs from natural bone. The absence of foreign protein however is particularly attractive in an implant since this avoids any adverse immunological reaction and hence rejection. In this instance, the structure and composition of the natural biomineral, coral, can be used to prepare a highly useful biomaterial for surgery.

Chitons and Limpets

The biological structure that has engaged our particular interest is the radula of two kinds of marine molluscs, namely, chitons and limpets, that are widely distributed in intertidal environments.[6,7] Chitons and limpets have highly mineralized teeth on the radula (a long tongue-like organ) to scrape from rocky substrates the crustose coralline algae on which they feed. The hardened mineralized teeth have an overall tooth morphology that is appropriate for excavating rocky surfaces. This biomineralized structure, fitted to its biological role through evolutionary processes, can be used to inspire new technologies. Thus, at the University of Groningen in The Netherlands, the chiton radula is being used to develop new designs for dredging machinery.[8]

The macroscopic function (feeding) can be related to the microarchitecture of the tooth. Both chitons and limpets use iron to harden their teeth, depositing within them a variety of iron oxides. These, together with other inorganic solids, reinforce and harden the organic matrix present in the tooth. In the limpet, silicon is also deposited, as amorphous silica SiO_2, while in the chiton, calcium is deposited as crystalline forms of hydroxyapatite.

The organic matrix of teeth and mineralizing elements for subsequent hardening are secreted by epithelial cells as the teeth migrate slowly along the radula membrane. This migration results from the continual wearing away of the teeth during feeding. The radula is thus a 'conveyor belt' of teeth, with the early teeth being soft and unmineralized and the fully mature teeth being hardened and functional composite structures.

The availability of teeth at varying stages of mineralization allows us to study the process of mineralization and the mechanisms involved. The products, i.e. the biominerals together with associated organic material, are also of great interest, since their form, morphology and architectural organization provide the composite structure that is so suited to its overall biological function. A distinct mineralized layer, composed predominantly of magnetite (plus, in some species of chiton, as many as four other iron oxides) covers the surface of the chiton tooth that strikes the rocky substrate during feeding. The molecular mechanisms that lead to their formation in the species *Acanthopleura hirtosa* have been clarified only recently.[9,10] Behind the magnetite layer are densely packed fibres that are rope-like strands of the partially deacetylated α-polymorph of the polysaccharide, chitin.[11] Most of the protein of the tooth is present in the calcified region, suggesting that proteins play a key role in initiating and controlling mineralization.[12] For iron mineralization, this matrix-mediated control is less evident although in other species, extensive arrays of fibres do occur in the magnetite region. This is illustrated in Figures 1 and 2 for the species *Plaxiphora albida*. In Fig. 3, all biominerals have been removed by chemical treatment, revealing in this transmission electron micrograph the complex arrays of fibres: stippled near the surface, changing orientation just below the surface and appearing as long oriented fibres that develop into a basketweave array further into the tooth. Related structures have been observed in the organic matrix of another chiton species, *Acanthopleura hirtosa*.[13]

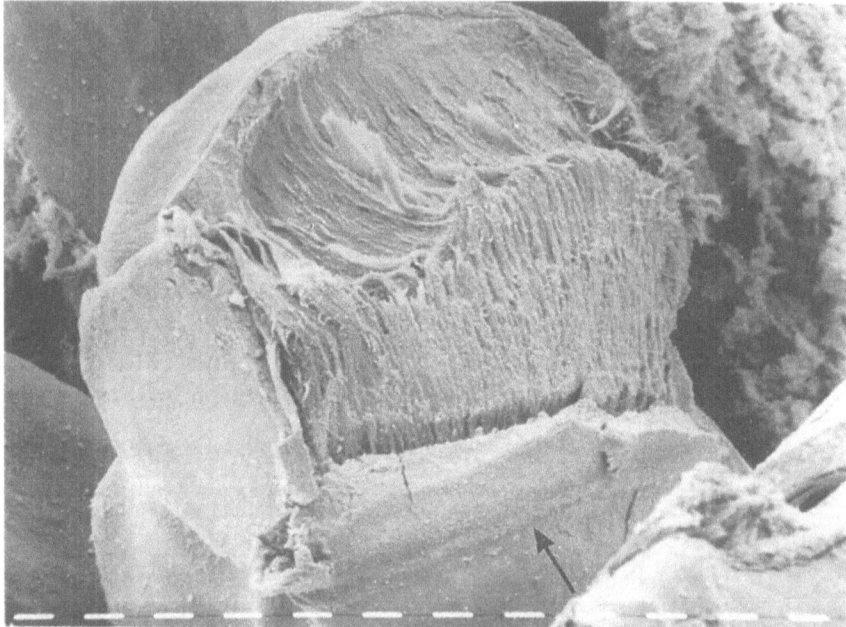

Fig. 2 : Scanning electron micrograph of a fractured tip of a mature tooth of *Plaxiphora albida* showing the prominent fibres. The arrow indicates the junction of tooth base and cusp. Scale bar is 10 μm.

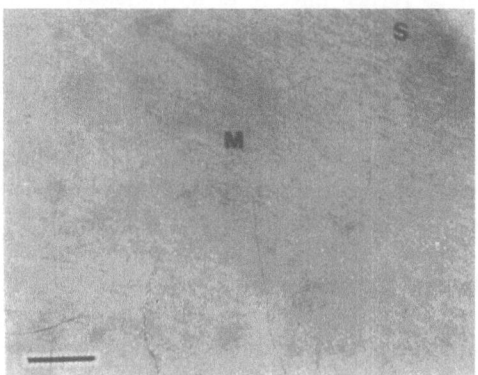

Fig. 3 : Transmission electron micrograph of an area through the posterior region of a mature tooth (demineralized) of *Plaxiphora albida* showing the arrangement of fibres in the surface (S) and magnetite (M) regions. Scale bar is 1 μm.

As reported by Lowenstam in a series of papers that initiated such studies (see Lowenstam & Weiner[2] for an overview and detailed references), iron phases other than magnetite can constitute major biomineral components of chiton teeth. This can be illustrated by the X-ray analysis spectrum (Fig. 4) of a region of a tooth from the species *Cryptoplax striata*. The high levels of Fe and P (but not Ca) are strongly suggestive of the presence of iron(III) phosphate in this species, as in others. Note also the presence of silicon (presumably as silica) as a minor component of this region of the tooth analyzed.

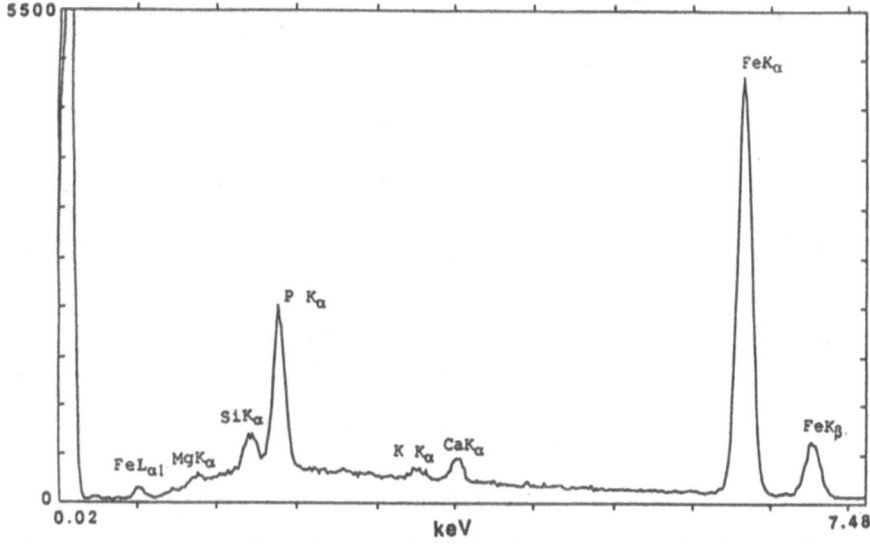

Fig. 4 : X-ray spectrum from a region of the mineralized tooth of *Cryptoplax striata*.

The microarchitecture of the chiton tooth as revealed by detailed studies of several species can be related directly to its biological function as a feeding implement. Magnetite, with a hardness of 6-6.5 on the Mohs scale (i.e. harder than apatite but not as hard as quartz) is densely packed as a surface layer of the tooth cusp that impacts on the substrate during feeding. Behind this layer are chitin fibres, reinforced with hydroxyapatite or other biomineral deposits, that act as 'shock absorbers' supporting the tooth during the physical stress of feeding. In summary, a biomineral-rich region serves as the cutting edge that is complemented by a matrix-rich mineralized region that gives it support.

Nanoscale particles and ferritin proteins

The final biomineralization system considered here involves the synthesis of nanoscale particles within the supramolecular protein cage of ferritin, a protein that consists of 24 subunits arranged as a roughly spherical shell. The shell encloses a cavity of approximately 8 nm diameter coated by the polypeptide shell, approximately 2 nm in thickness. A variety of ferritins occurs in Nature, exhibiting a range of structural order and size of their mineral cores. We have investigated these nanoscale particles using Mössbauer spectroscopy, electron diffraction and extended X-ray absorption fine structure (EXAFS) measurements. These data have been reviewed recently.[14] Illustrative data are shown in Figures 5 and 6.

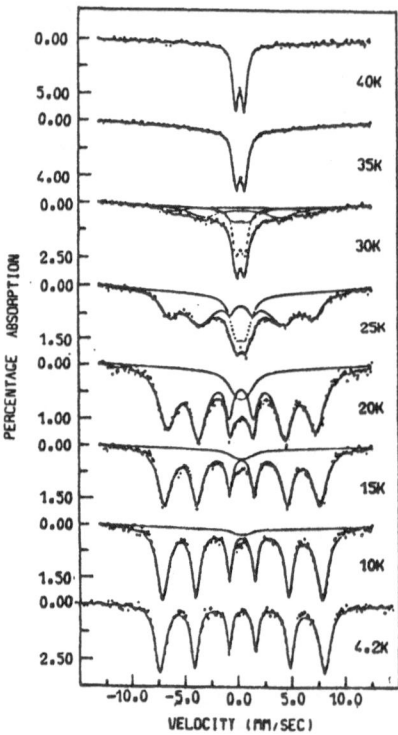

Fig. 5 : **Mössbauer spectra over the range 40-4.2 K for the ferritin protein isolated from the hemolymph (blood) of the chiton** *Acanthopleura hirtosa.*

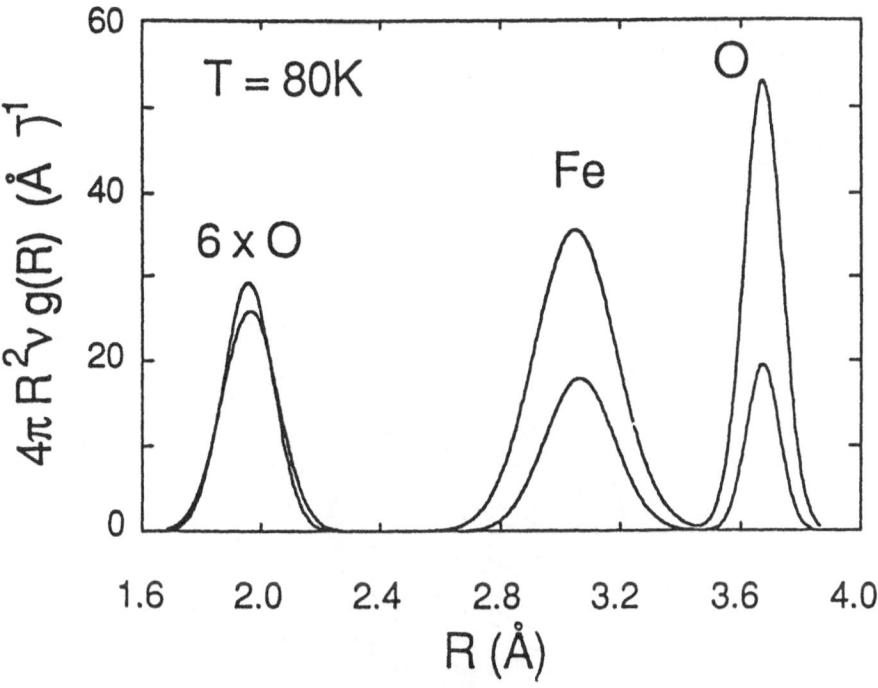

Fig. 6 : Radial distribution functions determined from EXAFS data recorded at 80 K for two ferritins of different crystallinities.

The Mössbauer spectrum of nanoscale iron(III) oxyhydroxide (ferrihydrite) particles exhibits temperature-dependence, with a low temperature sextet converting to a doublet at higher temperatures. This behaviour, characteristic of super-paramagnetism, is shown in Fig. 5 for ferritin isolated from chiton hemolymph (blood) and is discussed in detail elsewhere.[15] The ferritin core particles show only limited diffraction behaviour, partly due to limited ordering and partly due to their nanoscale dimensions. We have used EXAFS measurements to probe the structure of these particles. Data for two ferritins are shown in Fig. 6, indicating how the particles can have essentially identical local environments around the absorbing atoms (Fe) but different next-nearest neighbouring shells according to the relative amount of ordering within the particle. In addition to the variety of native cores available, the ferritin cavity can act as a reaction cage to engineer, *in vitro*, novel inorganic nanospace structures. Ferritin has been used to synthesize discrete nanoscale particles containing iron sulfides, manganese oxides, uranium oxides and ferrimagnetic iron oxides.[16] The complex composite structure of the iron oxide-iron sulfide particle formed within the ferritin nanoscale cavity is illustrated schematically in Fig. 7. Of particular interest is the controlled synthesis within this cage of nanoscale structures involving main group elements. Such materials would be of great relevance to the development of new optoelectronic materials.

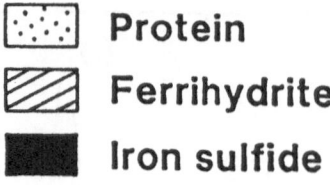 Protein
Ferrihydrite
Iron sulfide

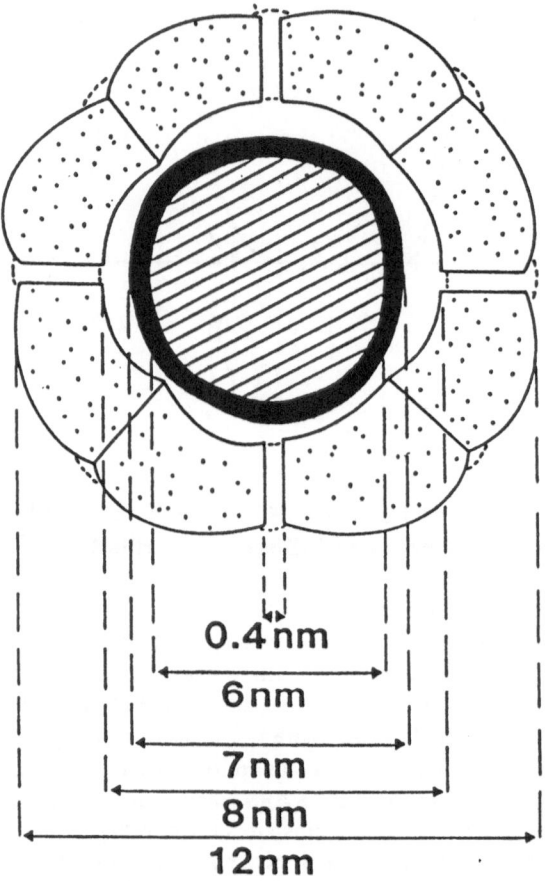

0.4 nm
6 nm
7 nm
8 nm
12 nm

Fig. 7 : **Schematic cross-section of the ferritin protein after rection with H₂S.**

Acknowledgments

We acknowledge with gratitude financial support over several years by the Australian Research Council, Murdoch University's Special Research Grant and the Australian National Beamline Facility. The support and contributions by many colleagues and students over the years are gratefully acknowledged.

References

1. S. Mann, J. Webb and R.J.P. Williams, *Biomineralization: Chemical and Biochemical Perspectives*. VCH Verlagsgesellschaft, Weinheim, 1989, p. 541.
2. H.A. Lowenstam and S. Weiner, *On Biomineralization*, Oxford University Press, 1989.
3. K. Simkiss and K.M. Wilbur, *Biomineralization*. Academic Press, 1989.
4. J.D. Birchall, The importance of the study of biominerals to materials technology. In: *Biomineralization: Chemical and Biochemical Perspectives*, eds., S. Mann, J. Webb and R.J.P. Williams, VCH Verlagsgesellschaft, 1989, p. 491.
5. Inoteb, From coral ... to biocoral ® *Inoteb* 56920 Noyal-Pontivy, France, 1989, and references therein.
6. J. Webb, D.J. Macey and S. Mann, Mineralization of iron in molluscan teeth. In: *Biomineralization: Chemical and Biochemical Perspectives*, eds., S. Mann, J. Webb and R.J.P. Williams, VCH Verlagsgesellschaft, 1989, pp. 348-388.
7. J. Webb, T.G. St. Pierre and D.J. Macey, Iron biomineralization in invertebrates. In: *Iron Biominerals*, eds., R.B. Frankel and R.B. Blackemore, Plenum Press, New York, 1990, pp. 193-220.
8. Van der Wal, P., Structure and formation of the magnetite-bearing cap of polyplacophoran tricuspid radular teeth. In: *Iron Biominerals*, eds., R.B. Frankel and R.P. Blackemore, Plenum Press, New York, 1990, pp. 221-229.
9. K.S. Kim, D.J. Macey, J. Webb and S. Mann, *Acanthopleura hirtosa. Proc. Roy. Soc. Lond. B.*, **237** (1989) 335.
10. J. Webb, L.A. Evans, K.S. Kim, T.G. St. Pierre and D.J. Macey, Controlled deposition and transformation of iron biominerals in chiton radula teeth. In: *Biomineralization '90*, eds., S. Suga and H. Nakahara, Springer-Verlag Tokyo, 1991, pp 283-290.
11. L.A. Evans, D.J. Macey and J. Webb, *Phil. Trans. Roy. Soc. Lond. B.*, **329** (1990) 87.
12. L.A. Evans, L.A. Macey and J. Webb, *Acanthopleura hirtosa. Mar. Biol.*, **109** (1991) 281.
13. L.A. Evans, D.J. Macey and J. Webb, *Acanthopleura hirtosa. Acta. Zoologica (Stokh.)* **75** (1994) 75.
14. T.G. Pierre, P. Sipos, P. Chan, W. Chua-anusorn, K.R. Bauchspiess and J. Webb, Synthesis of nanoscale iron oxide structures using protein cages and polysaccharide networks. In *Nanophase Materials*, eds., G.C. Hadjipanayis and R.W. Siegel, Kluwer Academic Publishers, Dordrecht, 1994, pp. 49-56.
15. T.G. St.Pierre, J. Webb and S. Mann, Ferritin and hemosiderin: structural and magnetic studies of the iron core. In: *Biomineralization: Chemical and Biochemical Perspectives*, eds., S. Mann, J. Webb, and R.J.P. Williams, VCH, Weinheim, 1989, pp. 295-344.
16. T.G. St.Pierre, W. Chua-anusorn, P. Sipos, I. Kron and J. Webb, *Inorg. Chem.* **32** (1993) 4480, and references therein.

Main Group Elements and Their Compounds
V.G. Kumar Das (Ed)
Copyright © 1996 Narosa Publishing House, New Delhi, India

Advanced Photovoltaic Technologies and the GENESIS Project

Shoichi Nakano, Michitoshi Ohnishi, Akio Takeoka
and Yukinori Kuwano

Functional Materials Research Center, SANYO Electric Co., Ltd.
1-18-13, Hashiridani, Hirakata, Osaka 573, Japan.

Society throughout the world today has become increasingly aware of the several forms of environmental pollution that, if left unarrested, would have a disastrous impact on life on earth. Problems such as ozone depletion, greenhouse warming, acid rain and chemical smog have especially received much publicity and remain in the forefront of public concern. Of these, the greenhouse effect and acid rain are caused primarily by massive consumption of fossil fuels such as coal and oil. Both these problems could be resolved, in principle, through serious efforts at developing 'clean' energies. Solar cells, which convert sunlight directly into electricity through the photovoltaic effect of semiconductors, are in this respect a key technology for the conquest of global environmental problems. This paper gives a brief review of the developmental history of solar cells, their current status and application technologies, and also expands on the futuristic concept of a global-scale energy supply system based on solar cells.

Fossil fuels

The earth that nurtures us is surrounded by an atmosphere that maintains the temperature and creates clouds and rain. If the diameter of the earth, around 12,700 km, is reduced on a hypothetical scale to 1 m, then the atmosphere that covers its surface (12 km in altitude) would extend to only 1 mm in height. It is into this ultra-thin atmosphere that man has dared to release the gases from burning fossil fuels. This must clearly be seen as a suicidal act.

The fossil fuels were created in ancient times by carbon dioxide assimilation by plants using the energy of the sun in a process lasting around 200 million years. At the present rate of use, the fossil fuels will be exhausted by man within a span of 100 to 150 years.[1] This unbridled use of fossil fuels is responsible for the large amounts of carbon dioxide and sulfur oxides present in our atmosphere, and it is only natural that man should be experiencing the consequences of his action on the environment (see Fig. 1).

Notwithstanding the scenario of dwindling fossil fuel reserves, the world demand for energy continues its insatiable steep upward trend (Fig. 2), as the global poulation and the energy consumption per capita increase. Thus in the decades beyond 2020, man's energy requirements would be so scantily met by fossil fuels that there will be an energy shortfall ("energy gap"). It is projected that the world poulation will reach a staggering 6.1 billion by the year 2000 (Fig. 3), that is , about twice that of the population in 1960. As we approach this new millenium, the forces of modernization will also increasingly extend their reach to wider and wider levels of the population. Therefore it can be expected that worldwide energy demand will cumulatively increase by multiples of population increase as well as with the expansion of modernization.

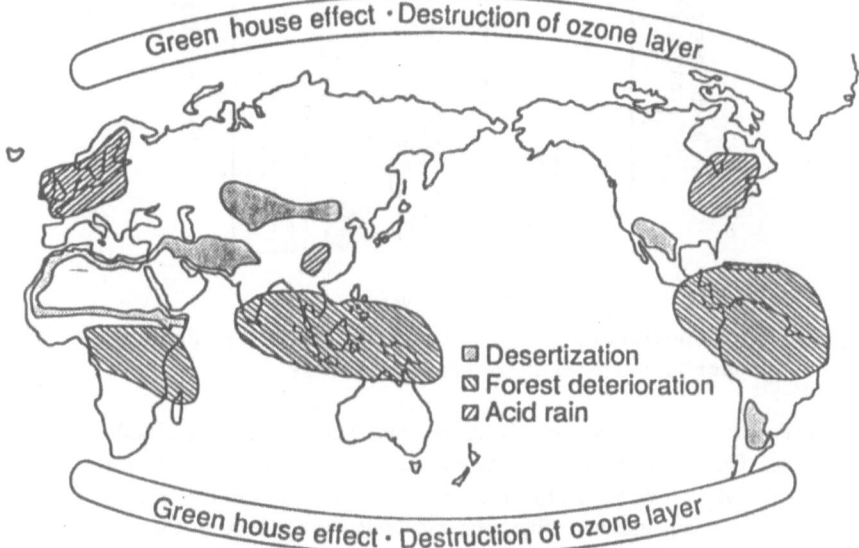

Fig. 1 : Degradation of the global environment

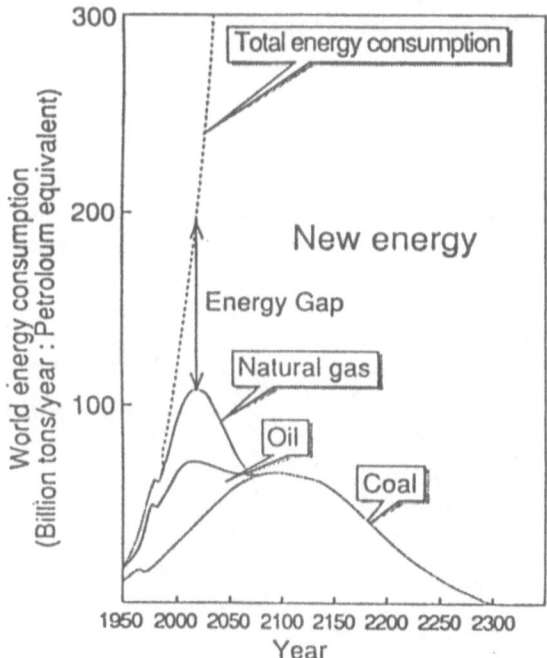

Fig. 2 : Progress and forecasts for consumption of various energies[2]

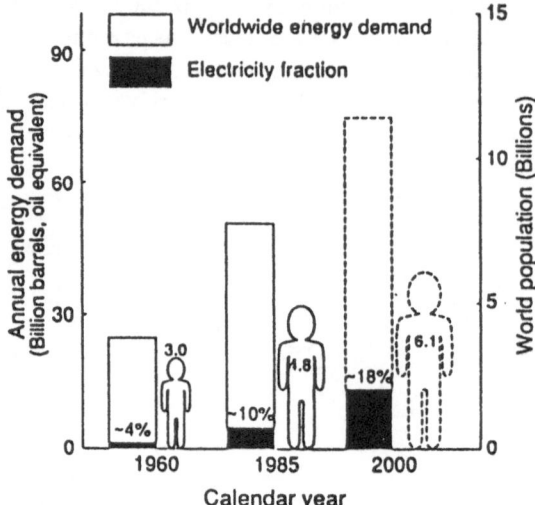

Fig 3 : World population growth, worldwide total energy demand, and the
proportionate demand for electrical energy[3]

But the demand for electricity will be more critical and far outpace all other energy
demands. This means that an alternate source of energy to meet future electricity
needs is urgently required. In this respect, solar energy is the ideal form of energy
because it is clean, inexhaustible, and available everywhere in the world. The
amount of solar energy that showers the Earth per day is about 170 billion MW.
This quantity is so massive that one hour's worth of solar energy could supply the
energy needs of the entire world for one year.

Features of solar cells

Among the methods available for utilizing the sun's energy, the most prominent is
that based on solar cells, which use the photovoltaic effect of semiconductors to
convert solar light energy to electrical energy. The process is depicted in Fig. 4.

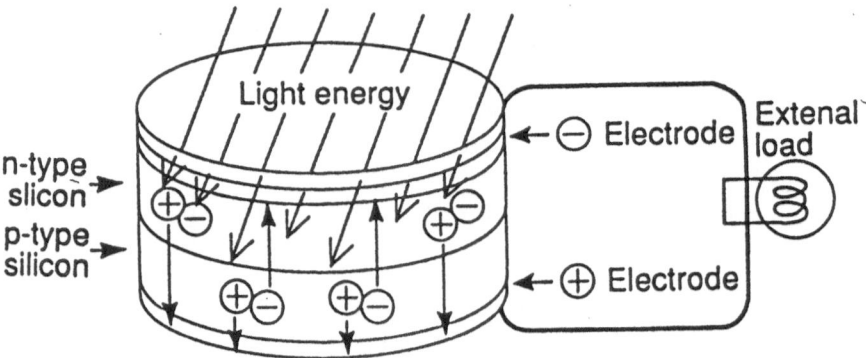

Fig. 4 : Principles of photovoltaics

30

When light enters a semiconductor which has a p-n junction, a hole (electron hole) with a positive charge and an electron with a negative charge are produced. These are separated along the p-n junction, and the positive and negative charges collect at both electrodes. When these two electrodes are connected, electric current flows and work is performed. Since sunlight is used as the energy source, the result is a power generating element which does not require fossil fuels, does not produce exhaust gases, and does not have moving parts.

In addition to providing the many features of solar energy, solar cells also offer the following: they provide electrical energy directly; the conversion efficiency is the same irrespective of the scale of power generation, that is, whether it is 1 W or 1 MW; power is generated even with diffused light such as on cloudy days; the service life is basically semi-permanent because there are no movable parts. And finally, silicon - the main material comprising the solar cells - is the second most abundant element on Earth, and hence should pose no problem from the standpoint of resource availability.

Progress in solar cells

History of solar cells

The history of solar cells began with Pearson and others in the United States in 1954. In 1958, solar cells were sent up into orbit on the U. S. Vanguard 1 satellite and used to provide power for telecommunications. Since then, they have been used in radio relay stations, lighthouses and other locations, but their high price prohibits them from achieving widespread use. Following the so-called "oil shocks" of 1973, however, the superior characteristics of solar cells caused attention to be focused on them as an alternative source of energy. Technical development was pursued in the United States under the auspices of the Department of Energy and in Japan under the New Sunshine Program (Formally, the Sunshine Project) of the Ministry of International Trade and Industry (MITI). The history of solar cells is summarized in Table 1.

Table 1 : The history of solar cells

1954	Single crystalline silicon solar cell (Pearson)
1958	Satellite equipped with solar cells (Vanguard 1)
1973	Oil crisis
1974	Sunshine projects started in Japan
1976	Amorphous silicon (a-Si) solar cell (Japan)
1980	Mass production of a-Si solar cells (Japan) Consumer electronics powered by a-Si solar cells (Japan)
1984	7MW solar power generation plant (U.S.)
1988	Global environmental issue
1992	Practical reverse-flow solar power generation system (Japan)
1993	New Sunshine Program (Japan)

(A) Types of solar cells

There are many different types of solar cells, depending on the material used (silicon, compound semiconductors, organic semiconductors, etc.) and the shape of the crystal material (single crystal, poly-crystal, amorphous etc., or a combination). Silicon is the main material for solar cells. Some representative methods for manufacturing Si solar cells are shown in Fig. 5.

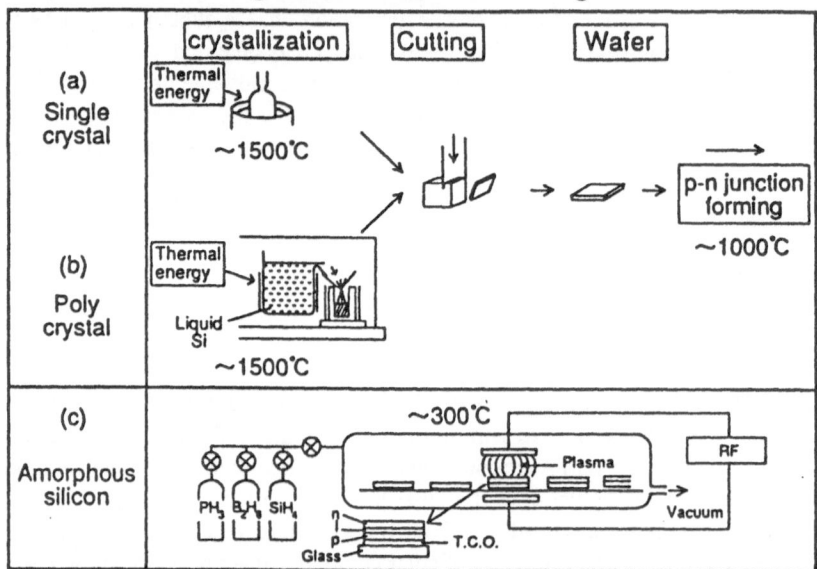

Fig. 5 : Comparison of fabrication methods for various solar cells

Single crystalline silicon (c-Si) solar cells were the first to be developed. They enable a high energy conversion (defined as the percentage of incident energy that can be converted to electrical energy) of 20% or greater for a small area. There are two major problems with this type of solar cell, namely, a complex manufacturing process and high costs.

In order to solve these problems, a method was developed in which molten silicon is hardened in a mold and then sliced into wafers to form poly-crystalline silicon (poly-Si) solar cells. These cells have a conversion efficiency of around 16%, lower than single crystal cells, but costs are also lower.

The production method for amorphous Si (a-Si) solar cells differs greatly from that of the two solar cells described above. Because a gas such as SiH_4 is decomposed using glow discharge and then deposited on a substrate such as glass, (i) the production processes for a-Si solar cells are simple, (ii) the energy required for production is low with processes demanding less than 300°C, (iii) the quantity of materials used is low with thicknesses less than 1 μm, whereas with crystal-based silicon the thickness is about 300 μm, (iv) the use of gas reaction facilitates larger sizes, (v) a high output voltage can be drawn from a single substrate using the integrated-type structure unique to a-Si solar cells. These and other characteristics make a-Si solar cells ideal low-cost solar cells.

(B) Energy payback time (EPT)

One of the most important concepts when evaluating solar cells as a new energy source is energy payback time (EPT). EPT is the number of years that are required for solar cell modules to generate the same amount of electric power as was consumed in their fabrication and this depends on the conversion efficiency and production volume of solar cells. EPT calculations for a-Si and poly-Si solar cells are shown in Fig. 6.

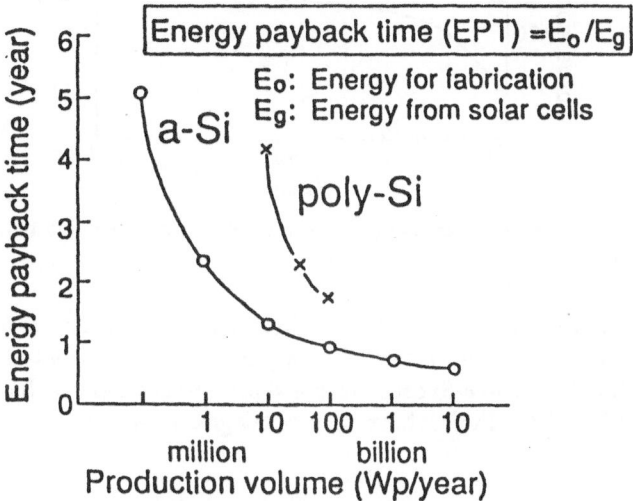

Fig. 6 : Estimation of the EPT for a-Si and poly-Si solar cells[4]

EPT decreases as the production volume increases. For a production volume of 10 million W/year of solar cells of 8% conversion efficiency, the EPT is estimated to be 1.2 years for a-Si solar cells, and 4.2 years for poly-Si solar cells. The EPT of a-Si solar cells is much shorter than that of poly-Si solar cells. This is because of the lower fabrication temperature, simpler manufacturing process, and other a-Si solar cell features. In any case, the EPTs of these solar cells are very much shorter than their lifetime, which is estimated at longer than 20 years. Solar cells are thus quite suitable as a new energy source.

(C) Improvements in conversion efficiency

Figure 7 shows the progress and forecasts for the conversion efficiency of silicon-based solar cells. In the past 10 years we have seen the conversion efficiency for small-area cells improve from 18 to 23% for c-Si based cells, from 12 to 18% for poly-Si based cells, and from 5 to 13% for a-Si based cells. Much improvement continues in the conversion efficiency of modules for practical use as follows :

c-Si   : 7 - 8 % → 12 -14 %
poly-Si : 6 - 7 % → 11 -13 %
a-Si   : 2 - 3 % → 6 -9 %

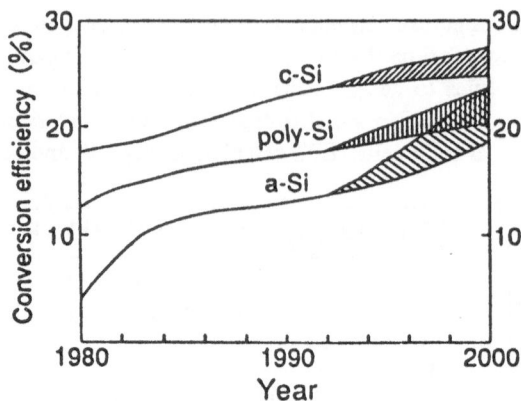

Fig. 7 : Progress and forecasts in conversion efficiency for silicon-based solar cells

We expect that research and development in this area will accelerate to the point where we should see the conversion efficiency improve to 27% for c-Si cells, and 24% for both poly-Si and a-Si cells. National projects such as the New Sunshine Program and U.S. DOE Project have contributed greatly to these achievements.

Applications and systems using solar cells

The production volume of solar cells has increased rapidly in recent years (Fig. 8). The total production volume in the world was about 58 MW in the year 1992.

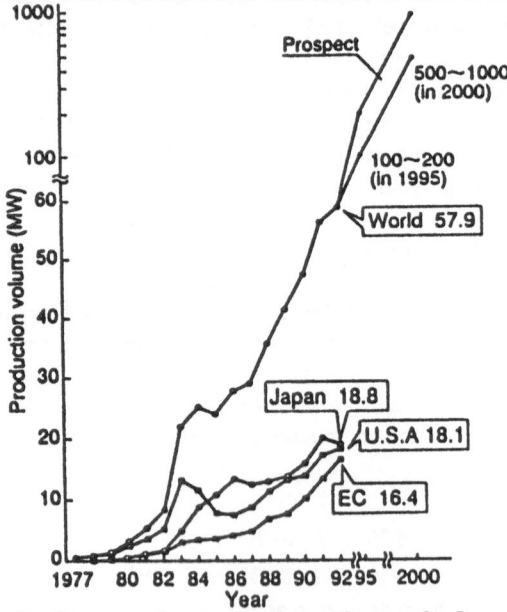

Fig. 8 : Progress in production volume of solar cells[5]

Figure 9 shows trends in the actual and projected cost of solar cells. The sharp fall in cost is a consequence of the increased production volume and the development of manufacturing technologies.

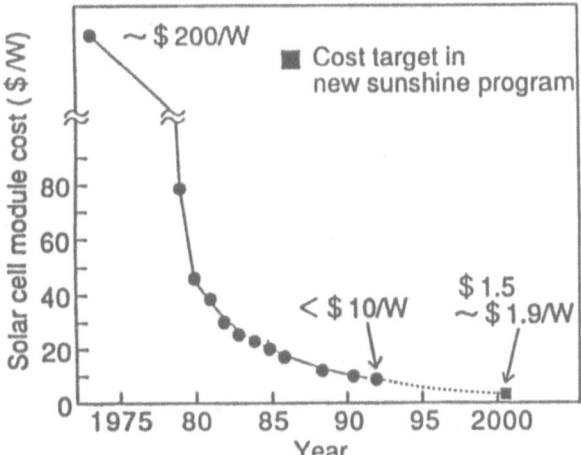

Fig. 9 : Actual and projected cost for solar cells[6]

The application of solar cells in electronic goods, i.e., consumer electronics, has progressed rapidly since 1980. This has been achieved as a consequence of vastly lower power consumption in the electronic goods themselves through advances in ICs and LSIs, and as a result of integrated-type amorphous silicon solar cells coming into practical use. Applications in calculators, radios, watches, and chargers have been growing. Most of those presently found in card-type calculators are amorphous silicon solar cells (Fig. 10).

Fig. 10 : Applications to consumer products

Photovoltaic power generation systems of several tens of watts to several kilowatts have already achieved practical use. In the past, solar cells have served as power supplies for various kinds of electrical equipment located in places where people could not get power (electricity) easily, such as radio relay stations on mountain tops and remote lighthouses. Recently, stand-alone systems such as solar streetlights, solar lantern, a pumping system, and solar refrigerators for storing vaccines (Fig. 11 (a), (b), (c), (d)) have come into practical use. Applications in vehicles, solar boats, solar cars and solar planes have also been developed.

(a) Solar streetlight   (b) Solar lantern

(c) Pumping system   (d) Solar refrigerator

Fig. 11 : Stand-alone systems

Furthermore, rural area systems have been developed, for example, house lighting systems, etc. for villages (Fig. 12), electrodialysis desalination systems (Fig. 13), and so on.

Fig. 12 : Rural area system (Senegal in Africa)

Fig. 13 : Desalination system

To further expand the solar cell market, the "see-through" solar cell has recently been developed. This is a translucent solar cell with many uniformly-spaced microscopic holes on an integrated-type a-Si solar cell submodule. Incident light can pass naturally through this submodule, making it suitable for use in home windows and sunroofs, as shown in Fig. 14.

Systems for use in ordinary homes are critical to expand the market for solar power generation. A positive step in this direction has recently been taken by Japan through interconnection of solar power generating systems. Figure 15 shows the first practical application in Japan of a reverse flow 1.8kW solar power generating system in an actual home. Reverse flow means that surplus power is fed back to the power system. This solar power generating system has no storage batteries.

Figure 16 depicts some actual data obtained in operating such a reverse flow photovoltaic power generation system. As can be seen, the system amply caters for the peak noon-time demand for power; approximately 8.1 KWh was actually generated on the particular date indicated. Of this, around 5.6 kWh, or 69% of the total power generated, was sold back to the Company.

Fig. 14 : Car sunroof using a see-through solar cell (Photo provided by Mazda)

Fig. 15 : The first practical use of a reverse-flow 1.8 kW solar power generating system in a Japanese home (Kuwano's house)

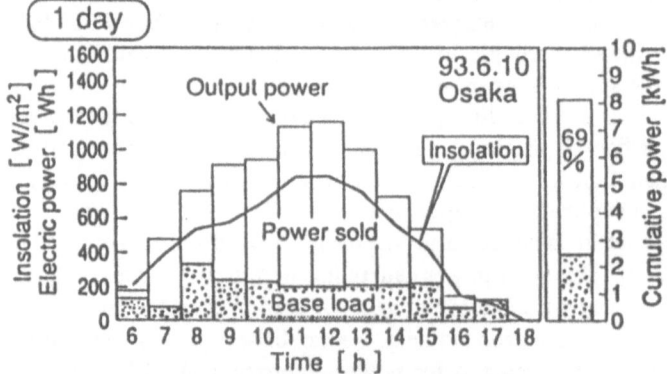

Fig. 16 : Operational data of reverse flow solar power generating system (courtesy, The Kansai Electric Power Co. Inc.)

38

Environmental problems have prompted the world's governments to expand their plans to promote solar cell applications. In Japan, subsidies have been enacted whereby the government pays for 2/3 of the installation costs when regional public organizations install photovoltaic power generation systems. Using this subsidy in 1992, a 25 kW system supported by NEDO as a part of the Sunshine Project in Hyogo Prefecture (Fig. 17) was installed along with 11 other systems in additional sites. More are expected in the future. In addition to these, a 200 kW system was installed by The Kansai Electric Power Co. Inc. in Hyogo Prefecture (Fig. 18) as a dispersible photovoltaic power generation system. It is also supported by NEDO.

Fig. 17 : Air conditioning system for green house (Grid-connected system with reverse-flow: 25 kW)

Fig. 18 : 200 kW power generation system (Rokko-island, Kobe, Japan; NEDO Project)

In the United States, the world's largest PV plant, having a 7 MW output power, started operations in Carrisa Plain, California in 1984 (Fig. 19). A PV power station is also being evaluated as a practical power source under the so-called PVUSA Plan.

In this context, a variety of technological studies required for the industrialization of centralized solar power generating systems are now in progress.

Fig. 19 : 7 MW Photovoltaic plant in Carrisa Plain, California; by ARCO-Solar (SIEMENS-Solar) Inc. Pacific Gas & Electric[7]

In order to bring solar power-generation systems for electric power into practical use, it is, of course, important to keep costs low. For this to be the case, it is necessary to cause a "chain reaction" (positive feedback) in the sense shown in Fig. 20, featuring the following key factors:
a) Performance increases of solar cells, the development of peripheral technologies, the development of low-cost mass-production technologies
b) Creation of new demand
c) Increase of production volumes

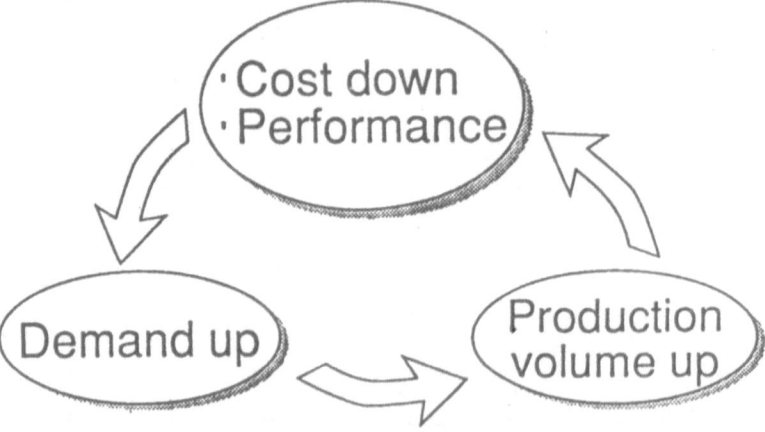

Fig. 20 : Positive feedback for the low-cost of solar power generation

With regard to technology development, efforts to increase efficiency and reduce costs have already begun in Japan, particularly within the New Sunshine Program. The cost of power at the beginning of the 21st century has been forecast as being the same as today's electricity rate at nearly $ 0.2 / kWh. Innovative technology development is necessary to realise large-scale power generation by the year 2010 with further reduced costs. The importance of ultra-high efficiency for solar cells is also to be stressed.

Future Prospects

The main disadvantages of solar cells are their inability to generate power at night, and the fact that their output fluctuates dramatically with the sunlight conditions. These problems are the source of some concern to those who feel that solar energy is unstable as a prime energy source.

As outlined in Fig. 21, a Global Energy Network Equipped with Solar cells and International Superconductor grids (GENESIS), is our proposal for resolving these problems. A satellite view of the Earth from outer space shows that rainy and cloudy areas cover less than 30% of the total land mass, and that it is always daylight on the opposite side of the globe to those areas under the shade of night. A worldwide photovoltaic power generating system connected by superconducting cables with no transmission losses would enable daylight areas to provide clean solar energy around the clock to those areas where it is night time, rainy or cloudy. This would ensure that no area on the Earth is without power.

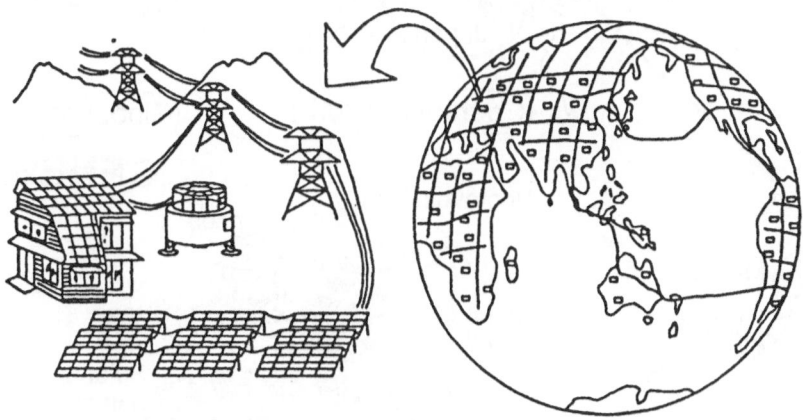

Fig. 21 : Conceptual view of GENESIS (Global Energy Network equipped with Solar cells and International Superconductor grids)[s]

It is forecast that in the year 2000, energy demands will be the equivalent of 14 billion kilo litres of crude oil per year. To meet this requirement, 800 km square of solar cells would be needed, assuming a conversion efficiency of 10%. Table 2 presents our estimates on world energy consumption and of the required solar cell system area. Conceptually, the plan is quite feasible as the area involved constitues no more than 4% of the world's desert area.

Table 2 : World's consumption and requried solar cell system area

Year	Primary energy consumption (billion kl/yr oil equivalent)	Conversion efficiency of solar cell system (%)	Required solar cell system area (100% solar cell) (km square)
1990	10	10	710
2000	14	10	807 4% of desert area
2050	62	15	1367
2100	270	15	2880

With worsening of the energy crisis and growing environmental concerns, there may be a need in the 21st century to apply the plan. The following steps are relevant to the successful development of the plan (Fig. 22).

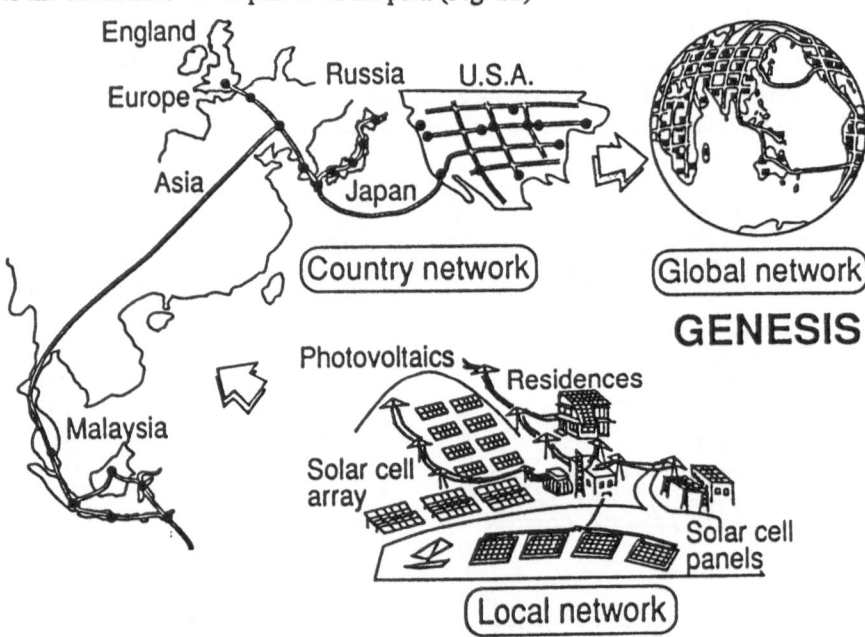

Fig. 22 : Steps in the GENESIS Project

Step 1 : When a large number of people install photovoltaic power generation systems in their homes and factories, and these then are connected to the electric power grid, a scenerio not unrealistic for Japan, the whole country will be made into a network through photovoltaic power generation power lines. If the same pattern is repeated in several countries, photovoltaic power generation networks will be created in each of these countries.

Step 2 : Each country's transmission lines are subsequently connected. For example, South Korea and Japan (Kyushu) are separated by only 200 km. If the countries' transmission lines are connected, a country network will be created. Such country or interstate networks exist in Europe and America.

Step 3 : If the country network is widely expanded, a global network can be created. In the interim period, until superconductive cables become available, it may be possible to use high-voltage direct-current transmission technology.

Conclusions

Solar cells, which convert solar light directly into electrical energy, are the most prominent candidates for a new, clean energy source. Research on solar cells is progressing at an astonishingly fast pace in the world. In order for us to resolve the energy problems facing us today and live comfortable lives in the 21st century, we must build a global solar power generating system with solar cells. Whether we choose to concentrate our efforts in this direction right now or not will determine the future of humankind. We believe that the GENESIS Project should be realized through international cooperation transcending economic and political barriers.

Acknowledgement

This work is supported in part by the New Energy and Industrial Technology Development Organization (NEDO) as a part of the New Sunshine Program under the Ministry of International Trade and Industry.

References and Notes

1. IEA estimates (1986).
2. Estimates quoted in "World Population Prospects 1990", "Energy Statistics Yearbook", etc.
3. Y.Hamakawa, *Solar Energy Materials*, **23** (1991) 139.
4. Annual Report of an Investigation by the Committee on Solar Photovoltaic Utilities Systems sponsored by the SUNSHINE PROJECT, March (1980) 88.
5. *PV News*, Feb. (1993).
6. *Sunshine Journal*, **13** (1992) 2.
7. D. D. Sumner and L. E. Schlueter, *Proc. 20th IEEE Photovol. Specialists Conf.*, (1988) 1289.
8. Y. Kuwano, Proc. *4th Int. Photovol. Science and Engineering Conf.*, Sydney (1989) 557.

Main Group Elements and Their Compounds
V.G. Kumar Das (Ed)
Copyright © 1996 Narosa Publishing House, New Delhi, India

A New Class of Frontier Materials :
The Oxyfluorides

Paul Hagenmuller

CNRS Solid State Chemistry Laboratory, University of Bordeaux I
351 cours de la Libération, 33405 Talence Cedex, France

The O^{2-} and F^- anions have radii 1.40 and 1.32 Å, respectively on the Shannon and Prewitt classification.[1] This close similarly in size suggests the possibility that the anions may mutually participate in the formation of substitution solid solutions, as has indeed been observed. On the other hand, the large electronegativity difference between oxygen and fluorine atoms (3.5 and 4.0 on the Pauling scale) can lead to ordered structures for simple composition ratios whose physical properties differ markedly from those of the related pure oxides and fluorides. Whereas, the oxides are rather covalent, the fluorides tend to be more ionic. Thus, many of the specific structural and physical properties of the intermediate oxyfluorides may be explicable in terms of bond competition involving the O and the more electronegative F atoms.[2]

Preparation of the oxyfluorides

Several methods are available for obtaining the oxyfluorides. Of these, the most common is the direct synthesis which is performed either in a Ni or monel vessel or in an Au or Pt tube sealed under argon, if the fluoride is volatile. The action of oxygen under pressure on fluorides, partial thermolysis of hydrates, and fluorine-chlorine substitution (e.g. in InOCl) are among other methods that can be used.[2]

Single crystals of oxyfluorides may be grown, for example, either by contacting a fluoride powder on an oxide crystal or by chemical transport (e.g. growth of $V_2O_{5-x}F_x$ from a V_2O_5 powder in a hydrofluoric solution).

The covalency difference between the oxygen and fluorine bonds explains why some systems involve many different phases. A good illustration is given by the Na(or K)-Ta(V)-O-F system where the coordination number of tantalum (V) may vary from 6 in oxygen-rich compounds to 8 (square based antiprism) in fluorides (whose bonds are more ionic), with other possibilities in the intermediate category.[3] In the synthetically prepared Tl_2OF_2, the same element (thallium) shows two different oxidation states: 1+ and 3+.[4]

Structural features

The structures depend on the F/O ratio and the size of the cation. In most phases the anions are disordered but anisotropy can favour O-F ordering.[5]

MOF formulation for some 3d-cations (M = Ti, V, Fe)[6,7] corresponds to the rutile structure, whereas for trivalent rare earths a fluorite-type lattice with rhombohedral or cubic symmetry is indicated according to anion ordering. CrOF could not be prepared but has been obtained with an excess of oxygen under pressure as $CrO_{2-x}F_x$. ScOF on account of O-F ordering and the relative small size of Sc^{3+} has the fluorite-type derived baddeleyite structure as revealed by NMR second moment determinations.[8]

Fluorine excess in rare earth oxyfluorides, LnO_xF_{3-2x}, leads to ordered tetragonal symmetry phases. In the PbFCl lattice intercalated "extra" fluoride anions form parallel planes which can be equidistant for simple chemical compositions (Bevan defects).[9] Under high pressure, large cation LnOF phases may lead to $PbCl_2$-type phases with coordination number 9.

Localization of fluorine in complex oxyfluorides is sometimes difficult due to the close X-ray and neutron scattering factors of the O^{2-} and F^- ions. An interesting example is given by $NaMoO_3F$ characterized by double chains of edge sharing octahedra.[10] The position of fluorine from among 4 different anionic positions has been determined by ^{19}F NMR (second moment evaluations), Raman spectroscopy (by comparison with MoO_3), electrostatic energy and site potential calculations. It can be shown that the detected F-position corresponds to the longest Mo-anion distance. The O2-O2 edges allow the highest possible screening between neighbouring Mo(VI) atoms; terminal O atoms give rise to π-reinforcement of the corresponding Mo-O1 and Mo-O4 bonds (Fig. 1).

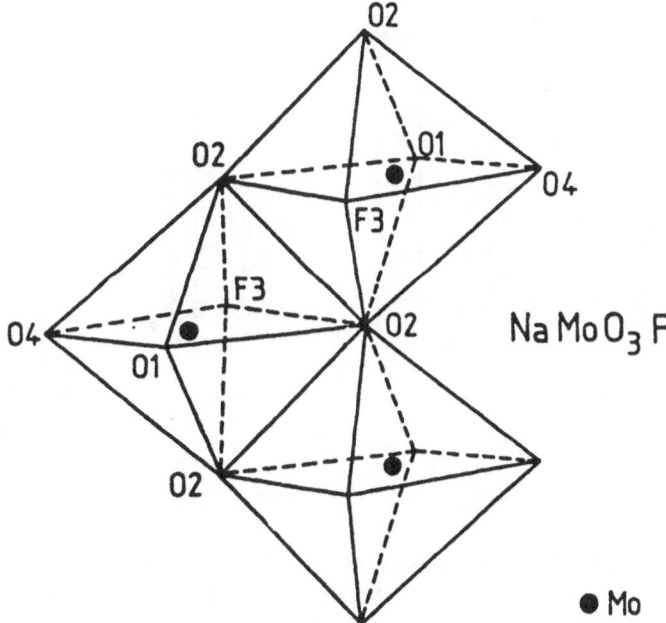

Fig. 1 : MoO_3F framework of $NaMoO_3F$.

45

Sometimes physical verifications are necessary to understand the O-F substitution mechanism. Consider, for example, the solid solution between isostructural α'-NaV$_2$O$_5$ and NaV$_2$O$_4$F in the V$_2$O$_5$-NaV$_2$O$_5$-NaV$_2$O$_4$F system (Fig. 2).[11] The former phase has two A and B sites containing respectively V^{4+} and V^{3+}, the latter has only two A sites. Two possibilities may arise when F replaces O, viz:

i)   there is a progressive change of B into A

ii) a statistical distribution of V^{4+} is present at both sites with formation of D rows (Fig. 3).

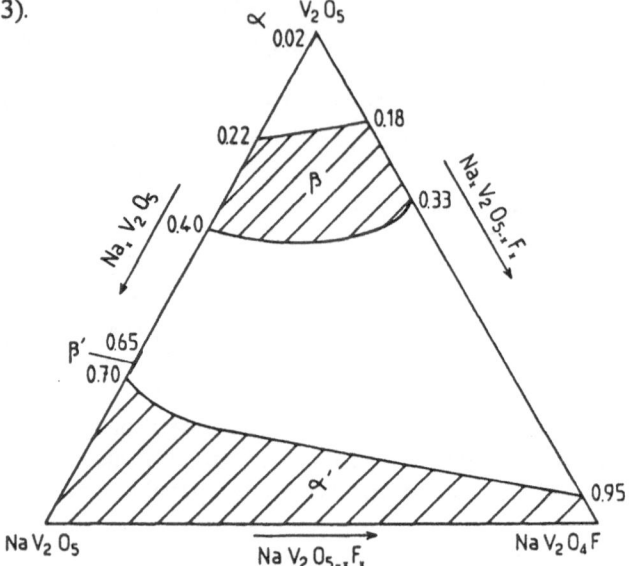

Fig. 2 : The V$_2$O$_5$-NaV$_2$O$_5$-NaV$_2$O$_4$F system at 823 K

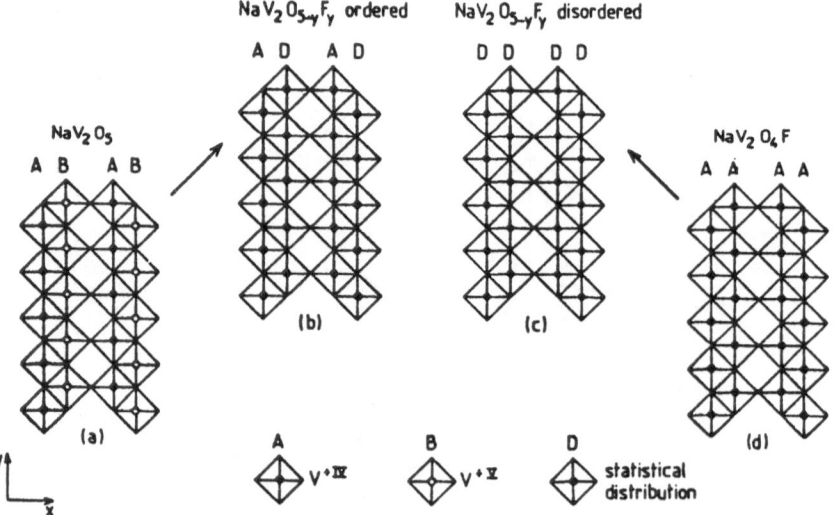

Fig. 3 : Distribution assumptions of V^{4+} in the octahedra files during progressive substitution either of O by F in α'-NaV$_2$O$_5$ or of F by O in NaV$_2$O$_4$F. The second mechanism occurs readily if small amounts of fluorine are present in NaV$_2$O$_5$.

46

In the absence of single crystals a direct study of the system is not possible, but the dramatic change of the physical properties of $NaV_2O_{5-x}F_x$ (e.g. Neel temperature, activation energy of the electronic conductivity) for small values of x points clearly to the second possibility. The detected substitution mechanism is confirmed by the p-type conductivity in the whole range of solid solutions.

In many oxides substitution of fluorine for oxygen does not modify the structure but influences the physical properties as a consequence of the strong electronegativity of fluorine. Three effects may be mentioned: electron trapping with formation of F^- anions, weakening of the bond covalency, compensation with a cation of lower oxidation state.

Substitution of oxygen for fluorine in fluorides with formation of a vacancy ($2F^- = O^{2-} + \square$) occurs often as a consequence of hydrolysis by air moisture but the substitution range is generally quite limited.[2]

Magnetic properties

The influence of O-F substitution has been extensively studied on ferrimagnetic oxides where a coupled substitution mechanism, mainly of the type $Fe^{3+} + O^{2-} = M^{2+} + F^-$, is operative, especially in spinels.[12]

In fact, two main effects occur: the first, which is the prevailing, results from the modification of the composition of one of the two competing cationic sublattices (A and B), and the second from the weakening of A-B interactions as a consequence of lower covalency of the F-bonds. The substitutions lead to variations of the O K saturation moments and to lower Curie temperatures.

A typical example is given by the inverse $Fe^{3+}[Ni^{2+}Fe^{3+}]O_4$ spinel. Substitution of Fe^{3+} by diamagnetic Zn^{2+} in the A sites (and subsequent replacement of O^{2-} by F^-) increases σ_{O_s} as long as there are not too many Zn^{2+} cations present in the A sublattice (and F^- anions in the material).

Conversely, substitution of Fe^{3+} by Ni^{2+} in the B sublattice of $Fe^{3+}[Fe^{2+}Fe^{3+}]O_4$ decreases σ_{O_s} due to weakening of the predominant B-B interactions in $Fe^{3+}[Fe^{2+}Fe^{3+}_{1-x}Ni^{2+}_x]O_{4-x}F_x$. σ_{O_s} is likewise lowered in $Fe^{3+}[Fe^{3+}_{1-x}Fe^{2+}_{1+x}]O_{4-x}F_x$ (Fig. 4). T_c decreases linearly.[13]

Whereas A-site substitution does not modify significantly the electric conductivity, it is weakened (and its activation energy enhanced) by B-site substitution as the conduction mechanism involves $F-t_{2g}$-orbitals in the common edge octahedra B-sublattice.[12-15]

Similar magnetic moment variations are observed in the garnet-type fluorine substituted ferrites such as $Gd_3Fe_5O_{12}$: substitution of Ni^{2+} for Fe^{3+} decreases σ_{O_s} while that of Zn^{2+} enhances it. The compensation points vary in opposite directions.[16,17]

The low temperature Verwey point of Fe_3O_4 decreases upon O-F substitution as a result of the perturbation of the $Fe^{3+}-Fe^{2+}$ 1/1 ordering in the B sublattice induced by the presence of excess Fe^{2+} (Fig. 5).

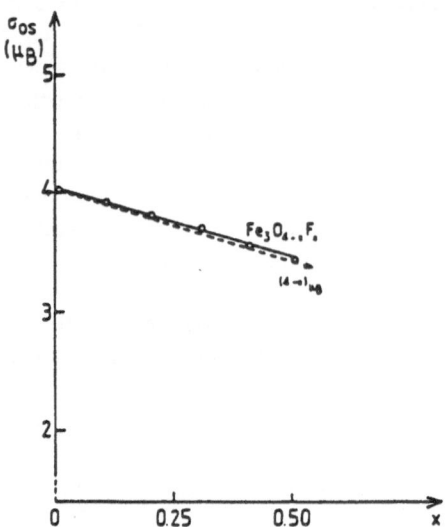

Fig. 4 : Variation of the O K saturation magnetic moment *vs.* composition of the $Fe_3O_{4-x}F_x$ spinel (decreasing σ_{Os} results mainly from a Fe^{3+}-F^{2+} substitution in the B-sites). The solid line gives the experimental results, the dotted line the calculated variation resulting from the cationic substitution

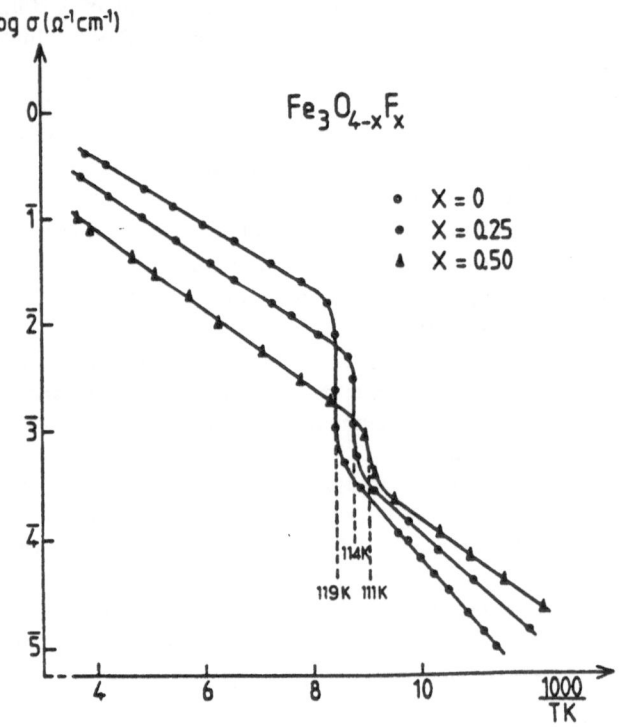

Fig. 5 : Decrease of the order-disorder Verwey temperature by oxygen-fluorine substitution in the $Fe^{3+}[Fe^{3+}_{1-x}F^{2+}_{1+x}]O_{4-x}F_x$ spinel

Metal-non metal transitions

Oxygen-fluorine substitution modifies often the phase diagram of the concerned materials, e.g. in $Na_xWO_{3-y}F_y$ where all the phases present are derived from the cubic ReO_3-type structure (in contrast to the Na_xWO_3 system where new phases appear in the semi-conductor-metal transition region and overshadow easy understanding of the transition mechanism).[18]

An interesting example is given by $VO_{2-x}F_x$. The VO_2 monoclinic-tetragonal rutile-type transition at 340 K characterized by breaking of a V^{4+}-V^{4+} pair is reduced by F^- substitution, a consequence of both a smaller gap in the band diagram for the low temperature phase (the V-F bonds being less covalent than the V-O bonds) and the introduction of extra d-electrons into the conduction band (corresponding to the appearance of V^{3+}) (Fig. 6).[19]

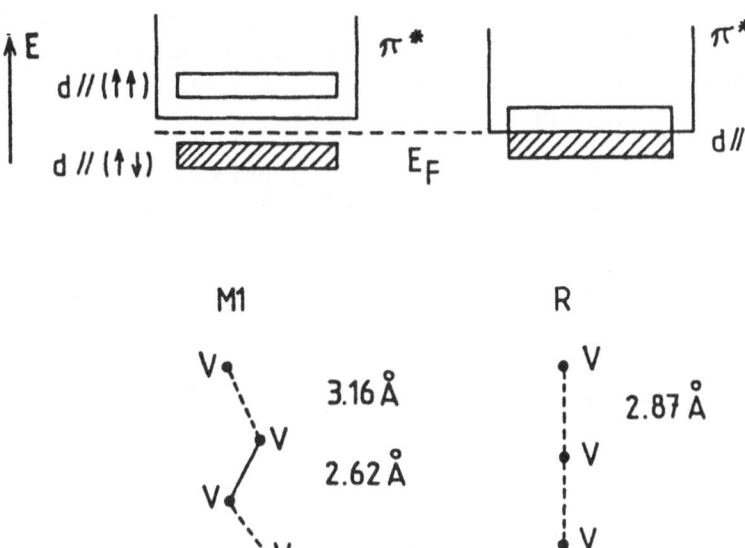

Fig. 6 : **Band structure and vanadium positions for both low and high temperture $VO_{2-x}F_x$ phases**

Phase changes in bronze-type materials result more from a modification of the phase composition than from an increase of the number of more or less delocalized d-electrons present: in the $K_xNbO_{2+x}F_{1-x}$ series the phase succession is exactly the same as in K_xWO_3 despite the absence of any mobile electrons. Niobium has only oxidation state 5+ and the appearing phases are now all white.[20] The structural changes observed result from the filling of the potassium tunnels at rising values of the insertion rate x.

Fluoride ion conductivity

One may expect that the presence of small amounts of oxygen will reduce the F^-

ion conductivity in fluorides due to stronger M-O bond formation. This is the case for tysonite-type BiO_xF_{3-x} for small x-values.[21]

On the contrary with highly polarizable cations, for example those containing lone pairs, the conductivity (σ) may be enhanced. Fig. 7 shows the variation of log σ for the $Pb_{1-x}Bi_xO_xF_{2-x}$ solid solution in the PbF_2-BiOF system. With traces of oxygen, the conductivity is slightly blocked by Bi-O pairs but further substitution leads to conductivity increase as long as the structure is of the cubic fluorite-type (x < 0.66). When a rhombohedral distortion analogous to that of BiOF appears ($0.66 \leq x \leq 0.80$), the conductivity decrease observed corresponds to the formation of localized Bi_2O_2 layers in a PbFCl-type BiOF related matrix, which hinder the F^- ion mobility (Fig. 7). The profile of the activation energy fits with the F^- ion conductivity variations. At 423 K the conductivity maximum is of the same order of magnitude as that of $Pb_{0.75}Bi_{0.25}F_{2.25}$, but with a higher thermal stability.[22]

Most fluorides and fluorine-rich oxyfluorides are very good electronic insulators due to the large bandgap between valence and conduction bands (it reaches 12 eV in MgF_2, for instance). This is a consequence of the strong electronegativity of fluorine and the efficient electronic trapping by F^- anions.[2]

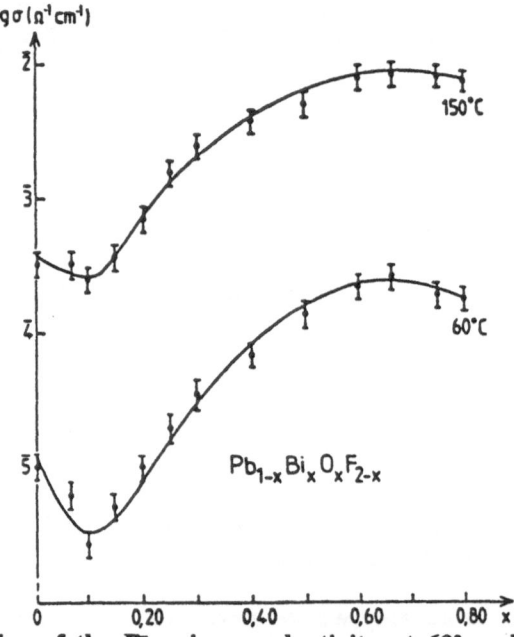

Fig. 7 : Variation of the F^- anion conductivity at 60° and 150 °C with composition for the $Pb_{1-x}Bi_xO_xF_{2-x}$ solid solution

Dielectric properties and non-linear behaviour of the polarization

Traces of LiF or NaF may increase the compactness of ferroelectric ceramic oxides such as $BaTiO_3$ or $Pb(Mg_{1/3}Nb_{2/3})O_3$ after sintering, and hence their permittivity even at relatively low temperatures on account of easy difussion of the F^- anions.

Oxygen-fluorine substitution decreases strongly spontaneous polarization P_s and Curie point T_c as a result of the weaker covalency of the new bonds which allows easier cancellation of the ferroelectric distortions. A significant example is the TTB-type $Sr_2KNb_5O_{15}$. This has a Curie temperature that decreases only from 437 to 403 K in $Sr_3Nb_4TiO_{15}$, but down to 80 K in $Sr_2KNb_4TiO_{14}$.[23,24]

Two oxyfluoride families are characterized by ferroelectric-ferroelastic coupling (with domain rotation induced by an applied voltage). These are:

a) the $(NH_4)_3FeF_6$ cryolite-type phases of formulation $A_2BMO_3F_3$, where A is a large alkali atom (coordination number 12) and B a smaller one (coordination number 6); M represents Mo or W.[25]

b) the $Na_3W_3O_9F_3$ chiolite-type oxyfluorides.[26]

The related pure oxides and fluorides have a linear behaviour which singularizes the behaviour of the intermediate mixed compounds. O-F substitution in the stoichiometric oxyfluorides leads to lowering of T_c.[27,28]

Small amounts of fluoride as a rule increase the compactness of the ferroelectric ceramics. At Curie point one observes an enhancement and a widening of the permittivity maximum probably due to formation of composition domains. If T_c is close to room temperature it may lead to applications (high capacities independent of frequency and temperature) (Fig. 8). This phenomenon is not reflected by the intrinsic behaviour of the single crystals (Fig. 8).[29]

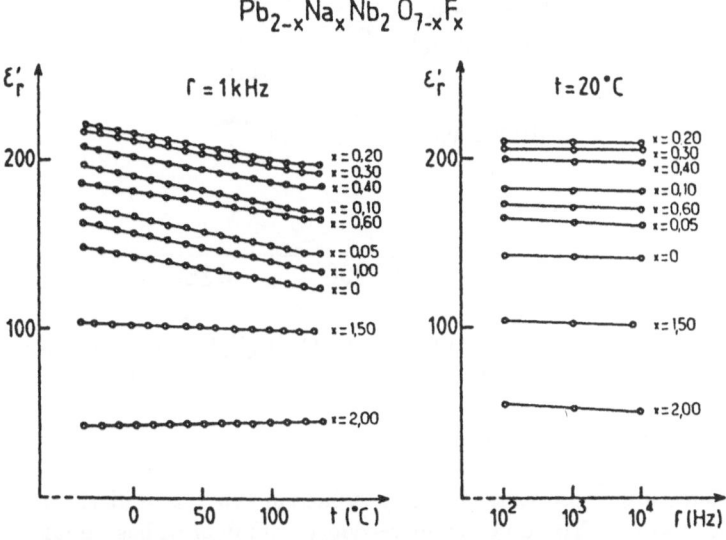

Fig. 8 : **Variation of the permittivity of the $Pb_{2-x}Na_xNb_2O_{7-x}F_x$ phases with temperature and frequency**

Some optical properties

Oxyfluorides containing tetrahedral groups such as $P(O,F)_4$ allow the formation of glasses less transparent in the IR than fluorides, but possessing a practical advantage in being easier to draw as optical fibres.

Fluorine-oxygen substitution shifts the absorption peaks of the oxides toward higher frequencies due to weakening of the nephelauxetic effect.

An interesting investigation has been reported[30] on the $Ln(OH)_{3-x}F_x$ rare earth hydroxyfluorides, prepared from LnF_3 by sealed tube hydrofluoric synthesis. Their formation is especially favoured with decreasing size of the rare earths due to easier substitution of F^- ions by OH^- groups, i.e. hydrolysis.

For increasing x-values, the absorption spectrum of $Nd(OH)_{3-x}F_x$ shows that the unique Stark line characterizing the $^2P_{1/2}$ level in D_{3h} symmetry gives rise to only one $^4I_{9/2} \rightarrow {}^2P_{1/2}$ transition progressively shifted toward smaller wave lengths. Widening of this component implies increasing anion disorder in a unique site. Its position characterizes the average covalency of the bonds (Fig. 9).

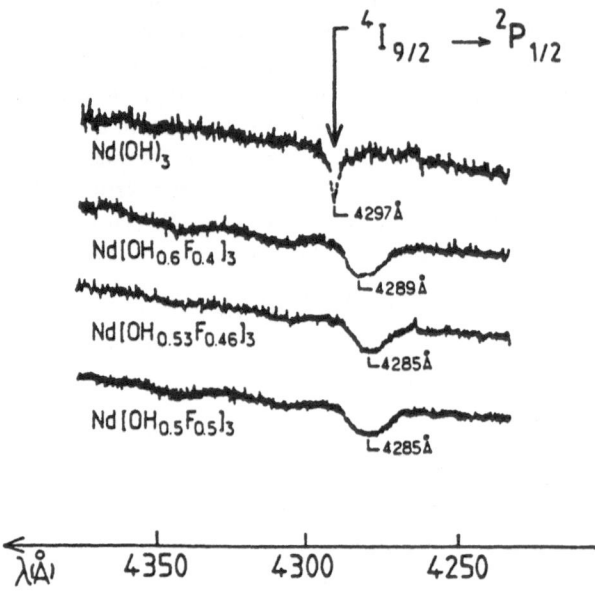

Fig. 9 : Absorption spectrum of some $Nd(OH)_{3-x}F_x$ hydroxyfluorides

Superconducting properties

The claim for the observation by some authors in 1987[31] of critical temperatures T_c above 150 K for $YBa_2Cu_3O_{7-\delta}$ samples doped with fluorine has stimulated many investigations. Due to the diversity of the sample treatment considered, in particular the fluorination process used, the properties measured vary from laboratory to laboratory. It appears, however, that gaseous fluorination at 373 K cleans the grain surface with formation of a copper(III)[1-6] containing amorphous layer but favours diffusion of oxygen in the grain bulk, giving rise eventually after annealing to a slightly higher T_c and a larger critical current density.[32] However, the T_c in no case exceeds the value obtained for $YBa_2Cu_3O_7$ itself. Thus the observed incremental changes in T_c may result from an oxygen diffusion from the surface into the grain bulk during the fluorination process.[33] The formation of a passivating surface layer results in the enhancement of the Meissner effect.

When the antiferromagnetic insulator La_2CuO_4 with K_2NiF_4-type structure is treated with pure fluorine at 473 K, one obtains a $La_2CuO_4F_\delta$-phase with an orthorhombic distortion and superconducting behaviour below 40 K. This property has been ascertained by the negative value of the magnetic susceptibility. The superconducting volume fraction of the sample has been determined at 6 K from the relation $H = -4 \pi M$, where M is calculated using the theoretical density value of La_2CuO_4. This fraction decreases from 54% to 16% to 8% with rising magnetic field from 1 to 20 to 100 Oe. The critical field at 6 K is about 700 Oe (Fig. 10).[34]

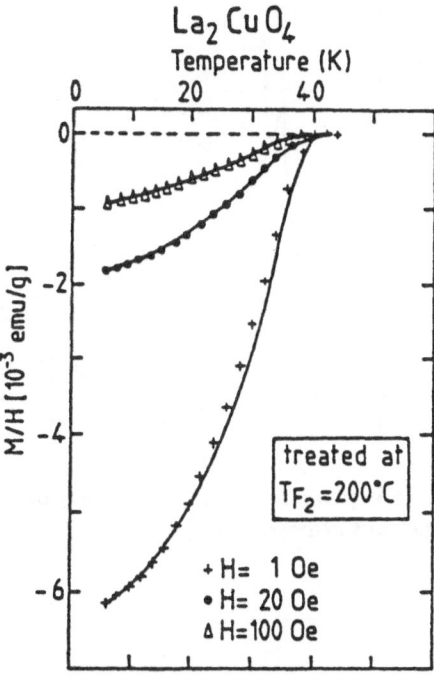

Fig. 10 : **Variation with temperature of the magnetic susceptibility of a La_2CuO_4 sample treated with fluorine gas at 200 °C in different magnetic fields**

Fig. 11 compares the electrical resistivity variation with temperature from several samples. Stoichiometric La_2CuO_4 exhibits a typical semi-conducting behaviour. Higher conductivity and traces of superconductivity below 40 K appear after treatment in an oxygen flux at 773 K. After 20 h exposure to fluorine at 473 K a bulk superconductor is obtained with a metallic behaviour above 40 K. The positive value of the thermoelectric power ($\alpha \approx 55$ $\mu VK-1$ at 300 K) is consistent with the formation of copper(III). Extra anions occupy a tetrahedral site in the NaCl-type layers of the lattice similar to that observed in $La_2CuO_{4+\delta}$, with relatively short anion-anion distances. An analogous F_2-treatment at above 523 K leads to the destruction of La_2CuO_4 and the formation of solid fluorides.

When heated in vacuum above 523 K $La_2CuO_4F_\delta$ loses oxygen and gives rise to a new phase with an anion/Cu ratio close to 4 like that of La_2CuO_4 and without any superconducting behaviour. However, the cell parameter difference confirms that fluorine is still incorporated in the lattice.[35] This example illustrates clearly that fluorine and oxygen may have behaviours which are both similar and different when they react with La_2CuO_4.

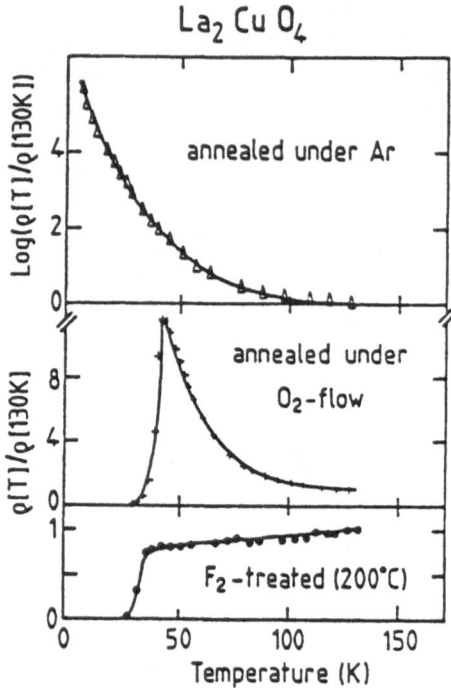

Fig. 11 : **Temperature dependence of the resistivity of La_2CuO_4 heated in different gaseous atmospheres**

References

1. R.D. Shannon and C.T. Prewitt, *Acta Crystallogr.*, **B25** (1969) 925.
2. P. Hagenmuller, "Inorganic solid fluorides: Chemistry and Physics", Academic Press, 1985.
3. J.P. Chaminade, Ph.D. Thesis, University of Bordeaux I, 1974.
4. G. Demazeau, J. Grannec, A. Marbeuf, J. Portier and P. Hagenmuller; *C.R. Acad. Sci.*, **269** (1969) 987.
5. B.L. Chamberland, 'The crystal chemistry of transition metal oxyfluorides', in ref. (2), p. 205.
6. P. Hagemuller, J. Portier, J. Cadiou and R. de Pape, *C.R. Acad Sci.*, **260** (1965) 4768.
7. J. Chappert and J. Portier, *Sol. St. Comm.*, 4 (1965) 185; 4 (1966) 395
8. M. Vlasse, M. Saux, P. Echegut and G. Villeneuve, *Mat. Res. Bull.*, **14** (1979) 807.
9. D.J.M. Bevan and A.W. Mann, *Proc. 7th Rare Earth Res. Conf.*, **1** (1968) 149.
10. J.P. Chaminade, J.M. Moutou, G. Villeneuve, M. Couzi, M. Pouchard and P. Hagenmuller, *J. Sol. St. Chem.*, **65** (1986) 27.

11. A. Carpy, A. Casalot, M. Pouchard, J. Galy and P. Hagenmuller, *J. Sol. St. Chem.* **5** (1972) 229.

12. J. Claverie, J. Portier and P. Hagenmuller, *J. Phys. C1*, **38** (1977) 169.

13. J. Portier, J. Claverie, R. Olazcuaga, H. Dexpert and P. Hagenmuller, *C.R. Acad. Sci.*, **270** (1970) 2142.

14. J. Claverie, J. Portier and P. Hagenmuller, *Z. Anorg. Allg. Chem.*, **393** (1972) 314.

15. A. Casalot, J. Claverie and P. Hagenmuller, *J. Phys. Chem. Sol.*, **34** (1973) 347.

16. J. Portier, A. Morell, R. Pauthenet, R. Olazcuaga and P. Hagenmuller, *C.R. Acad. Sc.*, **270** (1970) 821.

17. A. Morell, B. Tanguy, F. Menil and J. Portier, *J. Sol. St. Chem.*, **8** (1973) 293.

18. J.P. Doumerc, "The metal-non metal transition in disordered systems", *Proc. 9th Scottish Univ. Physics Summer School*, St. Andrew, 1978.

19. M. Bayard, M. Pouchard, P.Hagenmuller and A. Wold, *J. Sol. St. Chem.* **12** (1975) 41.

20. R. de Pape, G. Gauthier and P. Hagenmuller, *C.R. Acad Sci.*, **266** (1968) 803.

21. J. Schoonman, G.J. Dirkson and R.W. Bonne, *Sol. St. Chem.* **9** (1976) 783.

22. S. Matar, J.M. Reau and P. Hagenmuller, *J. Fluor. Chem.*, **20** (1982) 329.

23. J. Ravez, D. Tourneur and P. Hagenmuller, *Mat. Res. Bull.*, **7** (1972) 473.

24. J. Ravez, *Revue Chimie Minerale*, **23** (1986) 460.

25. G. Peraudeau, J. Ravez, P.Hagenmuller and H. Arend, *Sol. St. Chem.*, **27** (1978) 591.

26. J.P. Doumerc, M. Elaatmani, J. Ravez, M. Pouchard and P. Hagenmuller, *Sol. St. Chem.*, **32** (1979) 111.

27. M. Lorient, A. Tressaud and J. Ravez, *Rev. Chimie Minerale*, **19** (1982) 128.

28. M. Elaatmani, J. Ravez, J.P. Doumerc and P. Hagenmuller, *Mat. Res. Bull.*, **16** (1981) 105.

29. G. Campet, J. Claverie, M. Perigord, J. Ravez, J. Portier and P. Hagenmuller, *Mat. Res. Bull.*, **9** (1979) 1589.

30. A. Marbeuf, G. Demazeau, S. Turrell and P. Hagenmuller, *Sol. St. Chem.* **3** (1971) 637.

31. S.R. Ovshinsky, R.T. Young, D.D. Allred, G. De Maggio and G.V. Van der Leeden, *Phys. Rev. Lett.*, **58** (1987) 2579.

32. C. Margro, J.M. Heintz, A. Tressaud, J.P. Bonnet and J. Etourneau, *J. Phys. III France*, **1** (1991) 1751.

33. A. Tressaud, B. Chevalier, B. Lepine, K. Amine, L. Lozano, E. Marquestaut and J. Etourneau, *Eur. J. Sol. St. Chem.*, **27** (1990) 309.

34. B. Chevalier, A. Tressaud, B. Lepine, K. Amine, L. Lozano, E. Hickey and J. Etourneau, *Physica C*, **177** (1990) 97.

35. B. Chevalier, A. Tressaud, B. Lepine, C. Robin and J. Etourneau, *J. Less Common Metals*, **164 & 165** (1990) 832.

Main Group Elements and Their Compounds
V.G. Kumar Das (Ed)

Main Group Metal Alkoxides as Precursors for Novel Ceramic Materials

Ram C. Mehrotra

Chancellery, University of Allahabad, Allahabad 21002, India.

Studies on the alkoxy derivatives of elements up to the mid-1940s were mostly confined to those of silicon, boron aluminium, magnesium, beryllium and the alkali metals, but the field has considerably expanded in the last 3-4 decades to include a wide range of transition and inner transition metals.[1] The exceptional suitability (in terms of their volatility and solubility in organic solvents) of the alkoxyl derivatives of elements for the preparation of oxide-ceramic materials by the MOCVD and Sol-Gel methods[2] has led to a rapid, synergistic development of these processes.

Alkoxides of main group elements

Continued and active interest in the alkoxides of main group metals owes much to the observation of superconducting properties among several alkaline earth, thallium and bismuth alkoxides. Interest is especially centred on volatile and soluble alkoxides which may be secured using chelating alcohols like alkoxyalkanols,[3] which tend to reduce the degree of undesired association in the resulting alkoxy derivatives.

The solubilisation of many a simple polymeric bivalent metal alkoxide has also been achieved by combining these with chelating ligands[4,5] such as $\{Al(OPr^i)_4\}^-$, $\{Sb(OPr^i)_4\}^-$, $\{Nb(OPr^i)_6\}^-$, $\{Sn_2(OPr^i)_9\}^-$ and $\{Zr_2(OPr^i)_9\}^-$:

$$M + 2Pr^iOH \xrightarrow{\text{very slow}} [M(OPr^i)_2]_n + H_2$$

$$\text{polymeric, insoluble, non-volatile}$$

$$M + 2Pr^iOH + 2Al(OPr^i)_3 \xrightarrow{\text{fast}} M[Al(OPr^i)_4]_2 + H_2$$

$$M + 2Pr^iOH + 2Sb(OPr^i)_5 \xrightarrow{\text{fast}} M[Sb(OPr^i)_6]_2 + H_2$$

$$M + 4Pr^iOH + 2Sn(OPr^i)_4 \xrightarrow{\text{fast}} M[Sn^2(OPr^i)_9]_2 + H_2$$

$$\text{monomeric, volatile and soluble in organic solvents}$$

Alkoxides containing two metals were initially described as alkoxo-salts,[6] but these were later classified as 'double alkoxides'[7] in view of their covalent bonding characteristics.

Another general synthesis of the alkoxides of less electropositive metals (zinc, cadmium, gallium, germanium, tin(II), tin(IV), arsenic, antimony and bismuth) involves the reaction of their anhydrous chlorides[1] (fluorides[8] in the case of lead) with alcohols in the presence of a proton acceptor base (ammonia, lithium, sodium and potassium):

$$MCl_2 + 2\,LiOPr^i \xrightarrow[\text{C_6H_6}]{\text{Pr^iOH}} M(OPr^i)_2 \downarrow + 2LiCl$$

$$M'Cl_3 + 3\,KOPr^i \longrightarrow M'(OPr^i)_4 + 3KCl \downarrow$$

$$M''Cl_4 + 4\,KOPr^i \longrightarrow M''(OPr^i)_4 + 4KCl \downarrow$$

(M = Be, Mg, Zn, Cd; M' = Ga, In; M'' = Ge, Sn)

$$M'Cl_3 + 3\,Pr^iOH \xrightarrow{\text{NH_3}} M'(OPr^i)_3 + 3NH_4Cl \downarrow$$

(M' = Ga, In)

$$GeCl_4 + 4\,EtOH \xrightarrow{\text{NH_3}} Ge(OEt)_4 + 4NH_4Cl \downarrow$$

Homoleptic tetraisopropoxoaluminates and other isopropoxometallates can be prepared by similar reactions with the corresponding sodium/potassium alkoxometallates:

$$ZnCl_2 + 2\,KAl(OPr^i)_4 \rightarrow Zn[Al(OPr^i)_4]_2 + 2KCl \downarrow \qquad [9,10]$$

$$SnCl_4 + 4KAl(OP^i)_4 \rightarrow Sn[Al(OPr^i)_4]_4 + 4KCl \downarrow \qquad [10]$$

Novel heterolepic chloride alkoxometallate derivatives have been prepared recently.

$$SnCl_4 + n\,Al(OP^i)_4 \rightarrow SnCl_{4-n}[Al(OPr^i)_4]_n + nKCl \downarrow \qquad [11,12]$$

$$2\,CdCl_2 + 2\,KZr_2(OPr^i)_9 \rightarrow [Zr_2(OPr^i)_9]Cd(\mu-Cl)_2[CdZr_2(OPr^i)_9] + 2KCl \quad [9]$$

The crystal structure[13] of the cadmium derivative shows chloride bridging interactions.

Heterometallic alkoxides of main group metals

Interestingly, these intermediate chloride alkoxometallate products can be used as synthons for a wide variety of tri- and tetraheterometallic alkoxides:

$$[Al(OPr^i)_4]_2 \, SnCl_2 \; + \; n \, KNb(OPr^i)_6$$

$$\longrightarrow \quad [Al(OPr^i)_4]_2 SnCl_{2-n}[Nb(OPr^i)_6]_n + n \, KCl \downarrow \qquad\qquad [11,12]$$

$$(n = 1 \text{ or } 2)$$

$$[Al(OPr^i)_4] \, SnCl + \; K[Ga(OPr^i)_4] \quad \xrightarrow[\;C_6H_6\;]{\;Pr^iOH\;}$$

$$[Al(OPr^i)_4] \, Sn \, [Ga(OPr^i)_4] + KCl \downarrow \qquad\qquad\qquad [10]$$

$$(M = Al \text{ or } Ga)$$

The products of the above reactions[10-12] possess three metal atoms in the molecular formulation. Most of these are monomeric compounds in organic solvents and they can be crystallized, or even volatilized unchanged under reduced pressure. These novel heterometal derivatives appear to be stable to heat, and their chemistry adds a new dimension to our knowledge of heterometal systems; they appear to derive their stability from alkoxy bridges between dissimilar metal atoms only. Metal-metal bonds or an auxiliary ligand such as carbon monoxide, which are sometimes a requirement for the stability of such heterometal coordination systems are absent.

In contrast to the monomeric nature of most of the homoleptic heterometallic alkoxometallates, the corresponding chloride alkoxometallates tend to dimerize, typically through chloride bridges. The crystal structures of two such derivatives, $[Zr_2(OPr^i)_9]Cd(\mu\text{-}Cl)_2Cd[Zr_2(OPr^i)_9]$[13] and $\{Pr[Al(OPr^i)_4]_2(Pr^iOH)(\mu\text{-}Cl)\}_2$[14] have been elucidated.

The first alkoxide systems containing three different metals was synthesized in 1985.

$$(Pr^iO)_2Al(\mu\text{-}OPr^i)_2Be(OPr^i) \quad + \; Zr(OPr^i)_4$$
$$\rightarrow \quad (Pr^iO)_2Al(\mu\text{-}OPr^i)_2Be(\mu\text{-}OPr^i)_2Zr(OPr^i)_3$$

$$(Pr^iO)_2Al(\mu\text{-}OPr^i)_2BeCl \; + \; KNb(OPr^i)_6$$
$$\rightarrow \quad (Pr^iO)_2Al(\mu\text{-}OPr^i)_2Be(\mu\text{-}OPr^i)_2Nb(OPr^i)_4$$

As the stability of the above trimetallic systems was initially ascribed to the small size of central beryllium atom, these derivatives were termed as 'polymetallic alkoxides'.[16] The name corresponds to the label 'double alkoxides'[7] for the two-metal alkoxides. However, as this definition could be confused with the that used for the polymeric alkoxides of single metals, systems containing more than two metals in the same molecule were renamed as 'heterometallic alkoxides',[4] with the prefixes bi-, tri- and tetra- indicating the number of metal atoms.

The heterometallic alkoxides of transition metals, e.g., Mn(II)[17], Fe(II) [18], Fe(III)[18], Co(II)[19], Ni(II)[21] and Y(III)[22], the inner transition metals, lanthanons[22,24], Zn, Cd[25], Sn(II)[10] and Sn(IV)[11,12] have been synthesized, generally via reactions involving the intermediate chloride alkoxometallates. Research on the heterometal alkoxides of other main group metals, Ba, In and Pb, is being pursued:

$$MCl_n + xKAl(OP^i)_4 \xrightarrow{-xKCl} MCl_{n-x}[Al(OPr^i)_4]_x$$

$$MCl_{n-x} + y\,KSn_2(OPr^i)_9 \xrightarrow{-yKCl} MCl_{n-x-y}[Al(OPr^i)4]_n\,[Sn_2(OPr^i)_9]_y$$

$$MCl_{n-x-y} + zKSb(OPr^i)_4 \xrightarrow{-zKCl}$$

$$Mcl_{n-x-y-z}[Al(OPr^i)_4]_x\,[Sn_2(OPr^i)_9]_y[Sb(OPr^i)_4]_z + (n-x-y)KOPr^i$$

$$\downarrow -(n-x-y-z)\,KCl$$

$$M(OPr^i)_{n-x-y-z}\,[Al(OPr^i)_4]_x[Sn_2(OPr^i)_9]_y[Sb(OPr^i)_4]_z\,.$$

The synthesis of heterometallic alkoxides has recently been reviewed[26,27] elsewhere.

Main group metal alkoxides as precursors in the sol-gel process

In his pioneering work on the synthesis of $Si(OEt)_4$ in 1846, Ebelman observed that $Si(OEt)_4$ upon exposure to atmospheric moisture, was converted into silica gel.

The preparation of glasses and ceramic materials by the hydrolysis of metal alkoxides in alcoholic solutions initially gives a sol, which is converted on ageing into a gel, the sintering of which at temperatures comparatively lower than those required in the conventional fusing of oxides of metals affords the desired oxide ceramic materials. Beside the much lower sintering temperatures, another advantage of the Sol-Gel process is that the final material can be obtained directly by controlling the reaction conditions to form powders, thin wires or coatings on a substrate.[2,4,28]

In view of the stability of the basic framework of bimetallic alkoxides during alcoholysis reactions, these heterometallic alkoxides should be even better ceramic precursors than a mixture of component alkoxides.[29] The ultrahomogeneity of the final ceramic product obtained by the use of a number of different metal alkoxides has been ascribed to the formation of new chemical bonds.

The first reported synthesis of a spinel-type ceramic, $Mg\,Al_2O_4$, involved the hydrolysis of the bimetallic alkoxide $\{Mg[Al(OPr^i)_4]_2\}$[31] in the presence of triethanolamine. A NMR study showed that the basic framework of the initial $\{Mg[Al(OPr^i)_4]_2\}$ remained unaltered. Similar results were obtained in the hydrolysis of $\{Sn[Al(OPr^i)_4]_2\}$;[16] the possibilities of the use of heterometal alkoxides have been discussed.[32]

An easily-synthesized bimetallic alkoxide, $NaSn_2(OPr^i)_9$, can be used in place of $Sn(OPr^i)_4$, the preparation of which even in low yields is complicated as it involves many steps. The possible use of tin(IV) alkoxides as precursors for tin oxide has been presented in a recent review[33]. These alkoxides are, however, extremely sensitive to hydrolysis even by atmospheric moisture. The preparation and properties of a series of oxide-alkoxides with the general formula. M{O-

Al(OPr^i)$_2$}$_2$ (M = Mg, Ca, Ba, Pb) have been reinvestigated[34] as these are less hydrolysable.

$$M(OOCCH_3)_2 + 2Al(OPr^i)_3 \rightarrow M\{O\text{-}Al(OPr^i)_2\}_2 + 2CH_3COOPr^i$$

The results on the methanolysis and hydrolysis of the Ca and Mg derivatives have been published[35]. The reactions with other protic reagents such as acetic acid, β-diketones and glycols, involve only 3 of the 4 isopropoxy groups of CaO$_2$Al$_2$(OPr^i)$_4$ implicating an interesting tetrameric structure[36]. Other heterometal oxide-alkoxides/oxides have been isolated, included among which are the following:

Pb$_6$Nb$_4$(μ-O)$_4$(μ_2-OEt)$_{12}$(OEt)$_8$ [37];

Ba$_4$Ti$_{13}$(μ_3-O)$_{12}$(μ_5-O)$_6$(μ-OCH$_2$CH$_2$OCH$_3$)$_{24}$ [38];

Ni$_5$Sb$_{31}$O$_2$(OEt)$_{15}$ (HOEt)$_4$[39] ; (Mg$_{1/3}$Nb$_{2/3}$)O$_2$[40]; and

2MgO.2Al$_2$O$_3$. 5SiO$_2$ (cordierite) [40].

Other reports on the formation of oxide-alkoxide systems in the absence of moisture have led to a study[41] on non-hydrolytic sol-gel routes; hydrocarbon-soluble Zn(OCEt$_3$)$_2$ is converted to ZnO by using acetone as condensing agent.

As with the preparation of spinel from Mg[Al(OPr^i)$_4$]$_2$[31] and Sn[Al(OPr^i)$_4$]$_2$[16], heterometal alkoxides give more homogeneous materials than mixtures of randomly complexing constituent metal alkoxides.[16,19,32] Thus, ferroelectric and piezoelectric LiNbO$_3$ fibres[42] are prepared from [LiNb(OEt)$_6$], LiNb$_x$Ta$_{1-x}$O$_3$ (0 < x < 1) films[43] from mixtures of [LiNb(OEt)$_6$] and [Li Ta(OEt)$_6$] (a relaxor ferroelectric material[44]) and Pb(Mg$_{1/3}$Nb$_{2/3}$)O$_3$[45] from the co-hydrolysis of {MgNb(OEt)$_6$]$_2$}[46] and PbIOBu^t)$_2$[8].

A study[47] of the preparation of KNO$_3$ discs from [KNb(OEt)$_6$]^1 has shown that a higher homogeneity is obtained by the use of the preformed bimetallic alkoxide rather than from a mixture of component KOEt and Nb(OEt)$_5$ precursors.

Efforts are being made at synthesising superconducting materials[48] such as YBa$_2$Cu$_3$O$_{7-x}$[49] and Bi-Sr-Ca-Cu-O systems[50] using soluble alkoxides.

Following the application of "Mg[Al(OPr^i)$_4$]$_2$" type materials in CVD and hydrothermal materials synthesis, the structure of a crystalline, Mg$_2$Al$_3$(OPr^i)$_{13}$ has been determined.[51] It involves triply bridging [Al(OPr^i)$_4$]$^-$ groups; this product is, however,unstable to heat and it decomposes on attempted distillation to give Mg[Al(OPr^i)$_4$]$_2$ and other uncharacterized non-volatile materials. The crystal structure of Mg[Al(OPr^i)$_4$]$_2$.2Pr^{i}OH has also been elucidated[52] recently.

The sol-gel synthesis of ternary metal oxides, M Al$_2$O$_4$ (M=Mg, Ni, Co, Cu, Fe, Zn, Mn, Cd, Hg, Sr and Ba) by a route involving the hydrolysis of Al(OPr^i)$_3$ to a sol form and treating this with the salts of the bivalent metals in aqueous medium has been reported.[53]

In the synthesis of heterometal alkoxides with 2, 3, 4 and 5 different metals in the molecule, the basic framework appears to remain intact during the initial hydrolytic reaction.[16,31,32,35,54] Attempts to design a 'single molecular precursor'[55,56] that corresponds to the composition of desired material should be successful at least for some systems.

Acknowledgements

The author would like to thank his large number of senior coworkers, specially Prof. A. Singh, and numerous other research colleagues. Thanks are due to the D.S.T. (Government of India) for financial support.

References

1. D.C. Bradley, R.C. Mehrotra and D.P. Gaur, *Metal Alkoxides*, Academic Press, London, 1978.
2. R.C. Mehrotra, *Structure and Bonding*, 77 (1992) 1.
3. R.C. Mehrotra, In: *Sol-Gel, Science and Technology*, eds, M.A. Aegerter *et al.*, World Scientific, Singapore, 1989.
4. R.C. Mehrotra, *Natl. Acad. Sci. Lett.*, 16 (1993) 77.
5. R.C. Mehrotra, In: *Chemistry and Technology of Silicon and Tin*, eds., V.G. Kumar Das, S.W. Ng and M. Gielen, Oxford University Press, 1992, pp. 93-109.
6. H. Meerwein and T. Bersin, *Ann.*, 476 (1929) 113.
7. R.C. Mehrotra and A. Mehrotra, *Inorg. Chim. Acta Rev.*, 5 (1971) 127.
8. A.K. Jain, A.K. Rai and R.C. Mehrotra, *Polyhedron*, 10 (1991) 1103.
9. S. Sogani, A. Singh and R.C. Mehrotra, *Polyhedron*, 11 (1992) 341; *Ind. J. Chem.*, 32 (1993) 585.
10. S. Mathur, A. Singh and R.C. Mehrotra, *Polyhedron*, 12 (1993) 1073.
11. S. Mathur, A. Singh and R.C. Mehrotra, *Polyhedron*, 11 (1992) 341; *Ind. J. Chem.*, 32 (1993) 585.
12. S. Mathur, A. Singh and R.C. Mehrotra, *New J. Chem* (in press).
13. S. Sogani, R. Bohra, M. Nottemeyer, A. Singh and R.C. Mehrotra, *J. Chem. Soc., Chem. Commun.*, (1991) 738.
14. U.M. Tripathi, A. Singh, R.C. Mehrotra, S.C. Goel, M.Y. Chiang and W.E. Buhro. *J. Chem. Soc., Chem. Commun.*, (1992) 152.
15. R.C. Mehrotra and M. Aggarwal, *Polyhedron*, 4 (1985) 845.
16. R.C. Mehrotra, *Mater. Res. Soc. Symp. Proc.*, 121 (1988) 81.
17. R.K. Dubey, A. Singh and R.C. Mehrotra, *Ind. J. Chem.*, 31A (1992) 156.
18. R. Gupta, A. Singh and R.C. Mehrotra, *Ind. J. Chem.*, 30A (1991) 592; 32A (1993) 310; *New J. Chem.*, 15 (1991) 665.
19. G. Garg, A. Singh and R.C. Mehrotra, *Ind. J. Chem.*, 30A (1991) 688,866; *Synth. React. Inorg. Met.-Org, Chem.*, 21 (1991) 1047.
20. G. Garg, R.K. Dubey, A. Singh and R.C. Mehrotra, *Polyhedron*, 10 (1991)1733
21. R.C. Chhipa, A. Singh and R.C. Mehrotra, *Synth. Inorg. Met.-Org. Chem.* 20 (1990) 989; *Ind. J. Chem.*, 30A (1991) 1024.
22. U.M. Tripathi, A. Singh and R.C. Mehrotra, *Polyhedron*, 12 (1993) 1947.
23. U.M. Tripathi, A. Singh and R.C. Mehrotra, *Polyhedron*, 10 (1993) 949.
24. R.C. Mehrotra, A. Singh and U.M. Tripathi, *Chem. Rev.*, 91 (1991) 1287.
25. S. Sogani, A. Singh and R.C. Mehrotra, *Main Group Met. Chem.*, 15 (1992) 97; *Ind. J. Chem.*, 32A (1993) 345.
26. D.C. Bradley, *Chem. Rev.*, 89 (1989) 1317.

27. K.G. Caulton and L.G. Pfalzgraf, *Chem. Rev.*, **90** (1990) 969.
28. S. Sakka, *Trans. Ind. Ceram. Soc.*, **46** (1987) 1.
29. R.C. Mehrotra, Discussions with H. Bowen's group at M.I.T., U.S.A., 1985.
30. H. Dislich, *Angew. Chem. (Intl. Ed.)*, **10** (1971) 363.
31. K. Jones, T.J. Davies, H.G. Emblem and T. Parkes, *Mater Res. Soc. Symp. Proc.*, **73** (1986) 111.
32. R.C. Mehrotra, *Chemtracts*, **2** (1990) 389.
33. M.J. Hampden Smith and T.A. Wark, *Chem. Rev.*, **112** (1992) 116.
34. J. Rai, Ph.D. Thesis, University of Rajasthan, 1992.
35. J. Rai and R.C. Mehrotra, *Main Group Met. Chem.*, **15** (1992) 19; *J. Non-Cryst. Solids*, **152** (1993) 118.
36. J. Rai and R.C. Mehrotra, *J. Non-Cryst. Solids*, **134** (1992) 23.
37. R. Papiernik and L.G. Hubert-Pfalzgraf, J. Doran and Y. Jeannin, *J. Chem. Soc., Chem. Commun.*, (1990) 696.
38. J. Campion, D.A. Payne, H.K. Chae, J.K. Mausin and S.R. Wilson, *Inorg. Chem.*, **30** (1991) 3245.
39. U. Bern, R. Norrestam, M. Nygren and G. Westen, *Inorg. Chem.*, **31** (1992) 2050.
40. T. Fukui, *J. Non-Cryst. Solids*, **134** (1991) 293; **139** (1992) 205.
41. S.C. Goel, M.Y. Chiang, P.C. Gibbons and W.E. Buhro, *Mater. Res. Soc. Symp. Proc.*, **271** (1992) 3; R.C. Mehrotra and A. Singh, *Chemtracts*, **4** (1992) 350.
42. S..Hirano, T. Hayashi, K. Nosaki and K. Kato, *J. Am. Ceram. Soc.*, **72** (1989) 707.
43. S. Hirano and K. Kato, *Bull. Chem. Soc. Jpn.*, **62** (1989) 429.
44. F. Chaput, J. Boilot, M. Lejeune, R. Papiernik and L.G. Hubert-Pfalzgraf, *J. Am. Ceram. Soc.*, **77** (1989) 1335.
45. R.C. Mehrotra, *J. Non-Cryst. Solids*, **145** (1992) 1.
46. S. Govil, P.N. Kapoor and R.C. Mehrotra, *J. Inorg. Nucl. Chem.*, **38** (1976) 172.
47. A. Nazeri-Eshigi, A.X. Kuang and J.D. Mackenzie, *J. Mater Sci.*, **25** (1990) 3333.
48. S. Sakka, *J. Non-Cryst. Solids*, **121** (1990) 417.
49. S. Sakka, *J. Non-Cryst. Solids*, **121** (1990) 436; *J. Mater. Chem.*, **1** (1991) 1031; *J. Appl. Phys.*, **71** (1992) 2795.
50. S. Katayama and M. Sekine, *J. Mater. Chem.*, **6** (1991) 151.
51. J.A. Messe-Marktscheffel, R. Fukuchi, M. Kido, G. Tachibana, C.M. Jensen and J.W. Gilje, *Chem. Mater.*, **5** (1993) 755.
52. J. Sassmannschausen, R. Riedel, K.B. Pflanz and H.Z. Chmiel, *Naturforsch.*, **48b** (1973) 7.
53. L.Y. Kurihara and S.L. Siub, *Chem. Mater*, **5** (1993) 609.
54. R. Kuhlman, B.A. Vaaristra, W.E. Streib, J.C. Huffman and K.G. Caulton, *Inorg. Chem.*, **32** (1993) 1272; R.C. Mehrotra and A. Singh, *Chemtracts*, **5** (1993) 56.
55. R.C. Mehrotra, *Ind. J. Chem.*, **31A** (1992) 492.
56. R.C. Mehrotra (Lecture at VII Intl. Workshhop on Ceramics from Gels,Paris, July 1993), *J. Sol-Gel Sci. Tech.* (in press).

Main Group Elements and Their Compounds
V.G. Kumar Das (Ed)
Copyright © 1996 Narosa Publishing House, New Delhi, India

Rare Earth Alloy and Oxide Thin Films for Optical Storage

Fuxi Gan

Shanghai Institute of Optics and Fine Mechanics, Chinese Academy of Sciences, P.O. Box 800-211, Shanghai, 201800, P.R. China

The emergence of optical disk recording technology marked a major milestone in the field of digital information storage and retrieval. Optical disk storage has three advantages over other kinds of mass storage: a very large storage capability, ability to access multiple disks per drive, and rapid access to information. Advanced thin films are of interest for use in optical data storage. For erasable optical storage media the principal contenders are magneto-optical (M-O) and phase change (P-C) recording thin films. M-O optical disk recording has been already commercialized, and among its advantages are low writing and erasing laser power (<10 mW), high time response (< 50 ns), high write-erase cycles (>10^6) and high thermal and light stability. For higher density recording to decrease the magnetic domain size and to increase the chemical stability, further improvements of M-O storage media are required.

We have previously reported[2] on the laser-induced magnetic micro-domain formation in rare-earth-transition metal (RE-TM) alloy films and their recording properties.[1] In this paper, we present a brief review of our work on magneto-optic disk storage media (amorphous RE-TM) for practical applications, spanning such areas as the optimum design of multilayer structures, the influence of film surface oxidation on magneto-optical properties, material thermal stability, crystallization kinetics and the preparation of magneto-optical disks with RE-TM films. This paper also includes a coverage of new Re-TM alloy films for higher density recording at short laser wavelength, in particular rare-earth substituted garnet films. These latter are among the most promising new magneto-optical recording media because of their high corrosion resistance and strong Faraday rotations at short wavelengths. The new garnet films, such as Al or Ga substituted Bi:DyIG, have been prepared by sputtering and pyrolysis methods. Their optical absorption, Faraday rotation and magnetic coercivity, as well as recording properties were studied in some detail. The mechanism of the perpendicular magnetic anisotropy of garnet films on glass substrates has also been investigated. Highly applicable Bi:DyIG films with high Faraday rotation angle and high figure of merit for optical recording have been obtained by suitable substitution of Al or Ga for Fe.

Principles of magneto-optical storage

In magneto-optical (M-O) recording the information is stored in the form of thermally induced magnetic domains, and read-out is accomplished by sensing the polarization change (optical Kerr effect) in the optical beam. Fig. 1 shows the schematic diagram of the M-O storage process.

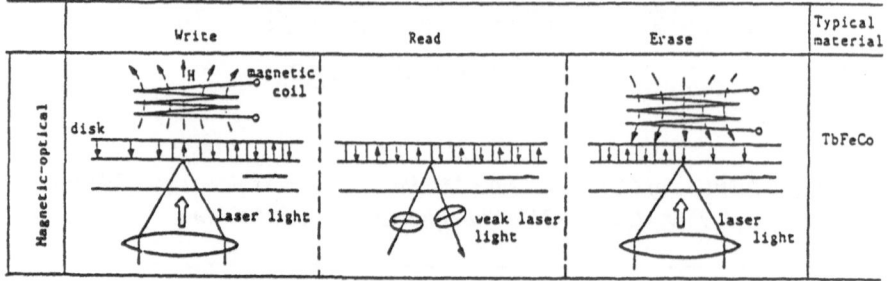

Fig. 1 : Schematic diagram of magneto-optical recording

For high density M-O recording, it is required that the film has an axis of easy magnetization perpendicular to the film plane. The following condition should be satisfied:[2]

$$K_u > 4\pi M_s \qquad (1)$$

Here K_u is magnetic anisotropy for perpendicular magnetization and M is saturated magnetization intensity. To stabilize the written bit (magnetic domain), it requires:

$$4\pi M_s/[1/(2d) - 1/(l + (3d/2h))] < H_c \qquad (2)$$

where H_c = coercivity, d = diameter of written domain, h = thickness of the film l = characteristic length of medium [$l \equiv \sigma_w/4\pi M_s$; σ_w = energy density of domain wall]. When the written domain diameter is much larger than the film thickness, the minimum domain size is given by the following equation

$$d = \sigma_w/2M_s H_c \qquad (3)$$

Hence, to stabilize the perpendicular magnetic domain and to minimize its size, the values of H_s and K_u should be high.

For optical recording, high signal-to-noise ratio (S/N) is needed. The S/N ratio can be expressed simply as

$$S/N = ARP_o \sin^2\theta_k \qquad (4)$$

where R is the reflectivity of the multilayer film, P_o is the laser power, θ_k is the Kerr rotation angle and A is the constant of the sensing system. We propose an evaluation factor F, such that for Kerr angles less than 5°,

$$F = (P_o R)^{1/2}\theta_k \qquad (5)$$

While from equation 4 it would seem that a high S/N ratio could be attained by increasing the laser power, this strategy has limitations in that a higher laser power could lead to a rise in temperature at the beam spot, and therefore to a decrease in the Kerr angle. High values of θ_k (and R) are equally necessary for good M-O recording performance.

Rare earth alloy thin films

Rare earth alloy thin films are the most widely used M-O media for erasable optical disks. The RE-TM thin films are amorphous, and being without grain boundary allow low noise levels to be achieved.

Optical and magneto-optical properties of single layer and multilayer RE-TM films

An optimal Kerr rotation angle (θ_k) is the most important requirement for M-O media. Fig. 2 illustrates the wavelength dependence of θ_k for amorphous ternary RE-TM films. It is seen from the Kerr rotation spectra that θ_k increases slightly with longer wavelength laser light. This trend suggests that a near IR semiconductor laser would be most desirable for reading signals.

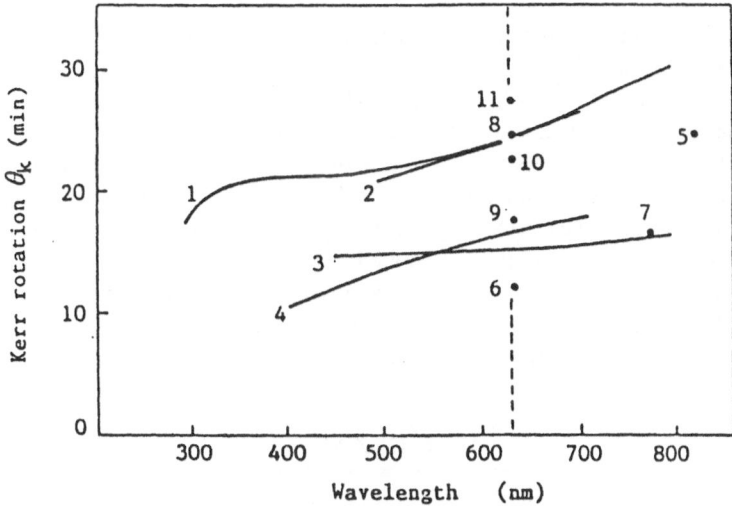

Fig. 2 : **Wavelength dependence of Kerr rotation angle of ternary amorphous RE-TM alloys**
1. $Gd_{26}(Fe_{81}Co_{19})_{74}$; 2. $Tb_{21}(Fe_{50}Co_{50})_{79}$; 3. $(Gd_{50}Tb_{50})_{28.6}Co_{71.4}$; 4. GdTbFe; 5. $Gd_7Tb_{13}Fe_{80}$; 6. TbDyFe; 7. GdTbDyFe; 8. GdFeBi; 9. $Gd_{26}Fe_{74}$; 10. $(Gd_{26}Fe_{74})_{84}Sn_{16}$; and 11. $Gd_{28}(Fe_{90}CO_{10})_{68}Bi_4$

For increasing the Kerr angle θ the multilayer interference enhancement effect can be used. By interference enhancement of dielectric layers the Kerr angle increases up to $1°$. Therefore, for securing a high recording performance it is important to design a multilayer structure such that optimization of S/N, P_o, R and θ_k could be had. A computational method that could be used to assess the inter-relationships among these values has been previously developed by us.[3] By way of example, consider the four-layered structure films listed in Table 1. The TbFeCo films are now commercially used as media for M-O recording. The measured optical and magneto-optical values of the quadrilayer SiO/TbFeCo/SiO/Al film are as follows:

Q_K (deg) 1.2,   R(%) 24%, $F^* = R^{1/2}Q_K$ (min) 35.3

The carrier to noise (C/N) ratio measured is about 45 dB at carrier frequency 2 MHz.

Table 1 : Thickness and refractive index of quadrilayer films

Material	Thickness (nm) l_i	Refractive index n_k
Antireflective film (SiO)	90	1.5
Magneto-optical (TbFeCo)	20	$\sqrt{\varepsilon_0} = 3\text{-}3.61$ $g = 0.44 + 0.012i$ (20 °C)
Intermediate film (SiO)	150	1.5
Reflective film (Al)	300	1.2-6.9i
Substrate (PMMA)	∞	1.46

For increasing record density the magnetic domain size should be minimized. As shown in Fig. 3, NdDyFeCo amorphous films possess smaller domain sizes than DyFeCo films under the same recording conditions. These results can be explained by the temperature dependence of coercivity of the thin films. The addition of Nd to DyFeCo films has the effect of increasing the coercivity (Hc) of thin films at high temperatures with concomitant influence on the recording domain sizes (Fig. 4).

In order to increase the recording density from present day levels, use of shorter wavelength (687 or 532 nm) light is required. This is expected to dominate the next generation M-O recordings. Typically, now, as shown in Fig. 2, the Kerr rotation angles of RE-TM films decrease with decreasing wavelength. The development of new media that show large Kerr rotations at short wavelengths is being eagerly awaited.

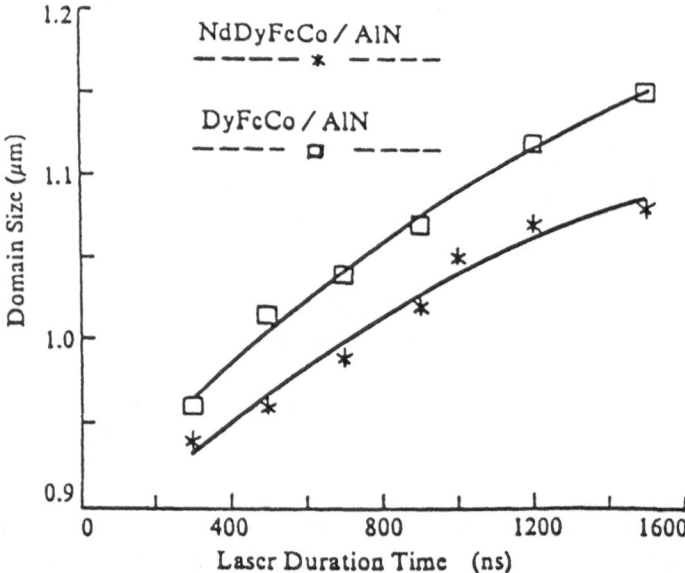

Fig. 3 : **Plots of domain size *vs* laser duration for amorphous NdDyFeCo and DyFeCo films [laser power P_w 8.5 mW, bias field H_b 300 Oe]**

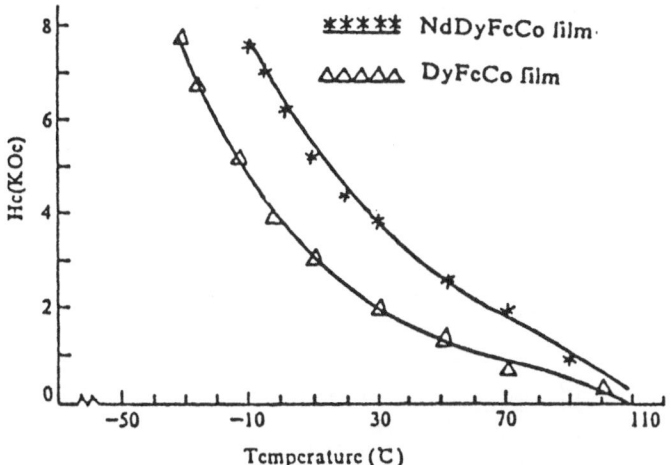

Fig. 4 : Dependence of coercivity of amorphous NdDyFeCo and DyFeCo films on temperature

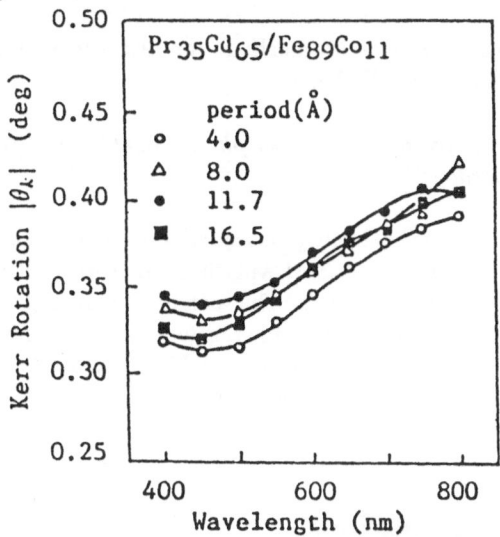

Fig. 5 : Influence of bilayer period on Kerr rotation spectra of PrGd/FeCo MLFs

M-O effects of Nd-Co and Pr-Co amorphous films are larger than those of heavy RE-Co films at shorter wavelengths around 500 nm. However, as the magnetization of (Nd,Pr)-TM films is larger than those of Tb-FeCo with similar Kerr rotation, it is difficult to prepare films with perpendicular magnetization. It has been found that artificially superstructured multilayer films (MLFs) have an axis of easy magnetization perpendicular to the film plane, and display strong interfacial magnetic anisotropy. Examples include PrGd/FeCo and Nd/FeCo superstructure multilayer films.[4] The influence of bilayer period on Kerr rotation spectra of PrGd/FeCo MLFs is shown in Fig. 5. The optimized bilayer period is about 1 nm. Comparison spectra of figure of merit F for bilayer periodic films NdGd/FeCo and

PrGd/FeCo and the single layer film of TbFeCo are given in Fig. 6. The enhancement of F value for the bilayer periodic films is obvious from the spectra.

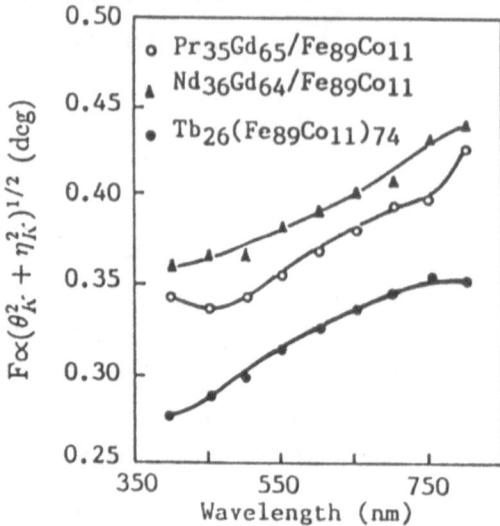

Fig. 6 :   **Spectra of figure of merit F for the three films NdGd/FeCo, PrGd/FeCo and TbFeCo**

Analysis of the microstructure of amorphous RE-TM alloy films by scanning tunnelling microscopy

The property of perpendicular magnetic anisotropy exhibited by amorphous RE-TM alloys is known to be associated with their micro-structure, which in turn is controlled by deposition parameters. The scanning tunnelling microscope (STM) can directly image the structure at atomic resolution of thin films without the need for any special treatment. We have used STM to study the microstructure of amorphous TbFeCo films that may shed light on the origins of their perpendicular magnetic anisotropy.[5,6]

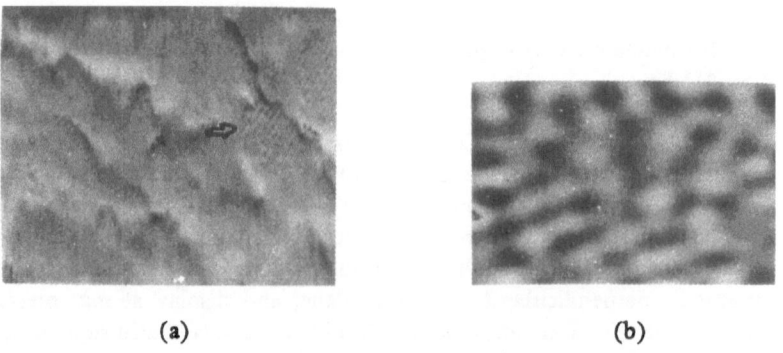

(a)   (b)

Fig. 7 :   **STM images of amorphous TbFeCo films.   (a) Scan area 10.95 x 10.95 nm; (b) Scan area 1.88 x 1.88 nm.**

Fig. 7(a) shows the microstructure of TbFeCo analysed by STM. The scan area is 10.95 x 10.95 nm. The morphology of the amorphous thin film (prepared by RF-magnetron-sputtering) comprises a cluster structure; the dimension of the clusters is in the 10-30 nm range . It is also seen that the clusters are composed of nano-scale fine crystallites of size 2.2-2.8 nm. The atomic resolution lattice image is shown in Fig. 7(b). The atomic lattice spacing of the nanocrystallites is about 0.2 nm. Based on the experimental evidence, the sputtered amorphous TbFeCo thin films are not strictly amorphous; there exist nanocrystallites in these films. These nanocrystallites are oriented randomly, and their presence may contribute to the origin of perpendicular magnetic anisotropy.

Thermal and chemical stability of RE-TM alloy films

The long life time (>10 years) for data storage in optical recording films constitutes the most important advantage that optical storage provides over magnetic storage. Therefore, material fatigue and stability considerations are of crucial relevance for optical storage media.

There are two principal factors which cause the deterioration of RE-TM M-O films. Firstly, local atom rearrangement and reaction caused by thermal stability. This process is accelerated upon increasing the temperature. Secondly, local defects and oxidation; the latter is related to humidity, which plays an important role in the chemical stability.

Fig. 8 shows the attendant variations in K_u, H_c and θ_k values with changes in annealing temperatures and times of amorphous TbFeCo films. The results do not support a high thermal stability TbFeCo as had been hoped, but it is noteworthy that K_u, H_c and θ_k values did not change at the annealing temperature of 170 °C over 280 min (in a protecting atmosphere). The duration of thermo-magnetic recording is very short (~100 ns). There is therefore little likelihood that a million times recording would alter the performance of the M-O media. XPS measurements revealed no alterations in the peak positions of Tb 4f, Fe 2p, Co 2p. This shows that the oxidation state and chemical structure of the TbFeCo films remain intact. Thus, the deterioration of magnetic properties conceivably arise from stress relaxation and atomic orientation changes. The activation energy of amorphous TbFeCo films is 0.86 eV.[7]

AES and XPS measurements show that in TbFeCo films the element Tb suffers ready oxidation which destroys the uniaxial anisotropy established by it. With increasing oxidation, the saturation magnetization M_s of the medium gradually changes direction from a perpendicular plane to an in-plane. The diffusion of oxygen is promoted by the lower density boundary regions of the columnar microstructure. The relationship between the magnetic and magneto-optical properties and the oxidation status of amorphous GdTbFe films is indicated in Table 2. The in plane residual saturation magnetization M_γ'' increases and the perpendicular $M_\gamma^\perp$ decreases with increasing oxidation temperature, while θ_k drops markedly with increasing temperature.[8]

Table 2 : The relationship between magnetic and magneto-optical properties and the oxidation status of amorphous GdTbFe films

Oxidation temp. (°C)	120	140	160
Oxidation time (h)	8	8	4
FeO_x (50 nm) (a.u.)	1624	2380	2500
Pure Fe (50 nm) (a.u.)	23136	20060	17100
$\theta_k(t)/\theta_k(0)$	0.75	0.15	0.00
$M_\gamma^\perp(t)/M_\gamma^\perp(0)$	1.5	1.0	1.0
$M_\gamma''(t)/M_\gamma''(0)$	5.0	8.5	7.8

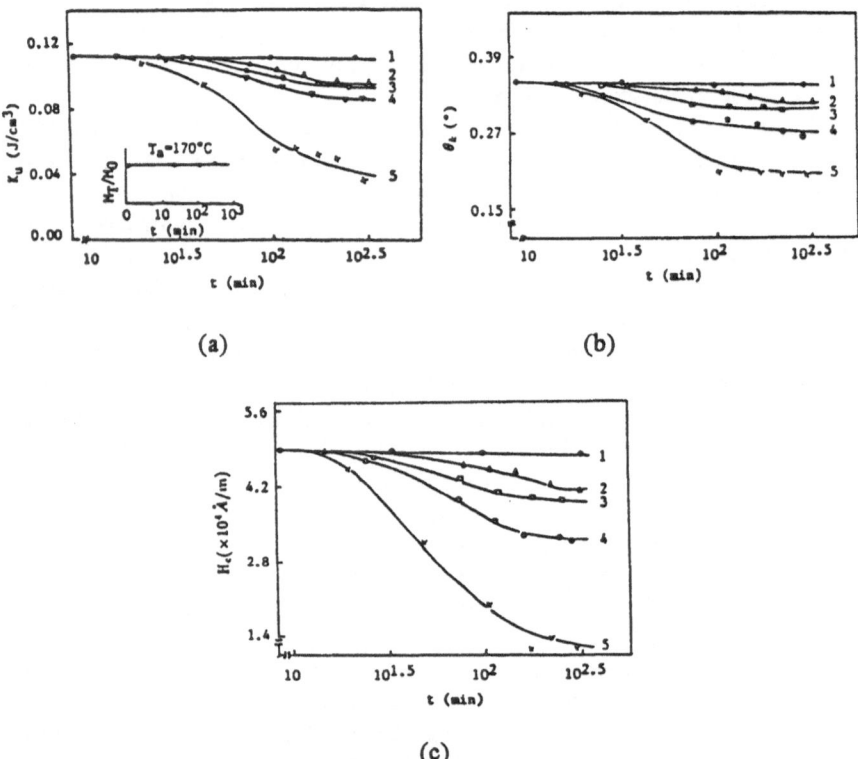

(a)

(b)

(c)

Fig. 8 : Variations of K_u, H_c and θ_k values of amorphous TbFeCo films. (a) K_u (b) θ_k (c) H_c; (1) 100 °C, (2) 120 °C, (3) 150 °C, (4) 170 °C, (5) 200 °C

Rare earth containing garnet thin films

It is well known that the Yttrium-Iron-Garnet (YIF) film is a very stable magneto-optical material, but its Kerr rotation angle (θ_k) and coercive force (H_c) are too low.

70

In recent years, bismuth-substituted garnet films have been shown to be a highly promising class of magneto-optical materials with their good corrosion resistance and large Faraday rotations at short wavelengths.[9-12] As a result of high saturation magnetization (M_s) and low H_c, it is difficult to make the film with an axis of easy magnetization perpendicular to the film surface. For improving the magnetic and magneto-optical performance of garnet films, we have systematically investigated[13] the magnetic and magneto-optical properties of Al or Ga and Cu substituted garnet films.

Methods of preparation

The magnetron sputtering and sol-gel pyrolysis have been used for preparation of different garnet thin films. By way of illustration, Table 3 shows the typical sputtering conditions used. The schematic diagram for preparing garnet films by the sol-gel pyrolysis is shown in Fig. 9.

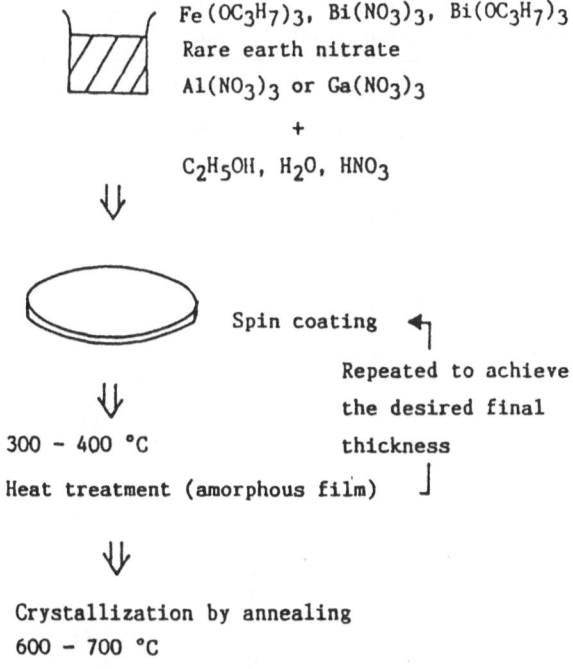

$Fe(OC_3H_7)_3$, $Bi(NO_3)_3$, $Bi(OC_3H_7)_3$
Rare earth nitrate
$Al(NO_3)_3$ or $Ga(NO_3)_3$

+

C_2H_5OH, H_2O, HNO_3

Spin coating

Repeated to achieve
the desired final
thickness

300 – 400 °C

Heat treatment (amorphous film)

Crystallization by annealing
600 – 700 °C

Fig. 9 : Schematic diagram of film preparation process by sol-gel pyrolysis

The as-deposited films prepared by both methods are amorphous. Transformation to the crystalline form is accomplished by heat treatment. Fig. 10 shows the X-ray diffraction curves of $Cu:Bi_{1.2}Dy_{1.8}Fe_4AlO_{12}$ film before and after heat treatment.

Rapid recurrent annealing (heat treatment) can effectively improve the morphology and structure of garnet films for M-O recording. Table 4 lists the parameters of rapid recurrent annealing and Fig. 11 shows the incremental changes in θ_k and H_c with the recurrent number and resultant improvements in the M-O hysteresis loops.

Table 3 : Typical sputtering conditions for Bi:DyIG thin films

Target	$Bi_{1.5}Dy_{1.5}Fe_{4.2}Ga_{0.8}O_{12}$
	$Bi_{1.5}Ho_{1.5}Fe_{4.2}Ga_{0.8}O_{12}$
	$Bi_{1.5}Ho_{1.5}Fe_{4.2}Ga_{0.8}O_{12} + 1.8\%$ Cu
Sputtering gas (Ar) pressure	30 m torr
Substrate temperature	400-450 °C
Deposition rate	2 nm/min

Table 4 : Parameters of rapid recurrent annealing

No.	Recurrent number	Annealing temperature (°C)	Heat ramp-up rate (°C/s)	Duration in one period (min)
A1	1	630-670	28-50	6
A2	2	630-670	28-50	3
A3	3	630-670	28-50	2
A4	4	630-670	28-50	1.5
A5	5	630-670	28-50	1.2
A6	6	630-670	28-50	1

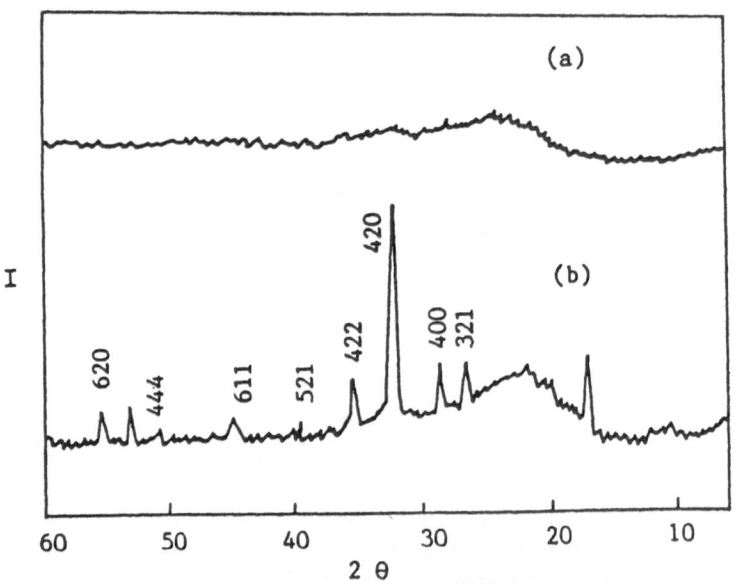

Fig. 10 : X-ray diffraction patterns of $Cu:Bi_{1.2}Dy_{1.8}Fe_4AlO_{12}$ film
(a) as-deposited; (b) after crystallization

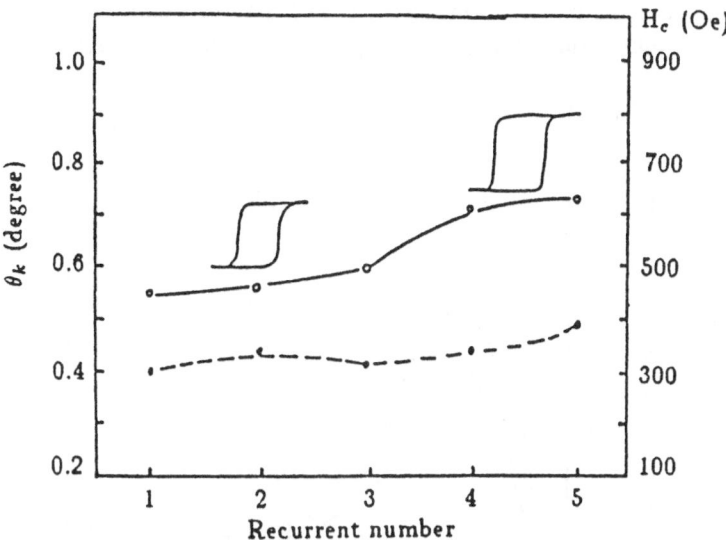

Fig. 11 : **Relationship between Faraday rotation angle and recurrent number of garnet films**

Al or Ga substituted Bi:DyIG thin films

Al or Ga-substituted Bi:DyIG garnet films have been prepared on glass substrates by the sputtering and pyrolysis methods. With the replacement of Fe by Al or Ga, there is a reduction in the values of saturation magnetization (M_s), demagnetizing field and Curie temperature (T_c), with beneficial effects for M-O recording.[13] The optimum values of Al and Ga in $Bi_{1.2}Dy_{1.8}Fe_{5-x}Al(Ga)_xO_{12}$ films were 1 and 0.7, the best annealing temperature was about 700° and 675 °C, respectively for Al and Ga substitution, and the annealing time was 30 minutes. Both Al and Ga-substituted garnet films have good Faraday rotation hysteresis loops and high coercivity, and the square is unity. Figs. 12 and 13 show the spectral

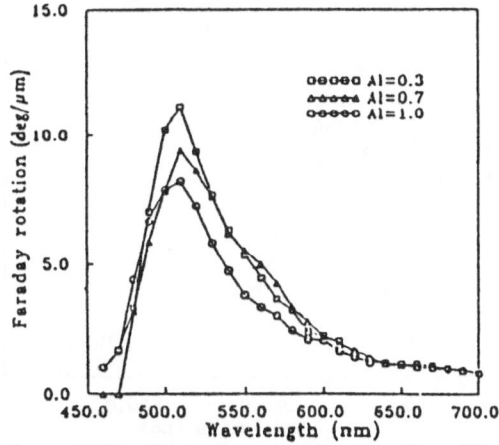

Fig. 12 : **Variation of the Faraday rotation angle with wavelength of $Bi_{1.2}Dy_{1.8}Fe_{5-x}Al_xO_{12}$ thin films**

dependence of the Faraday rotation angle θ_F of Al and Ga-substituted Bi:DyIG films. The peak wavelength at 530 nm matched the Nd:YAG double frequency laser wavelength. The θ_F value reached about 8°/μm for the optimum Al and Ga content.

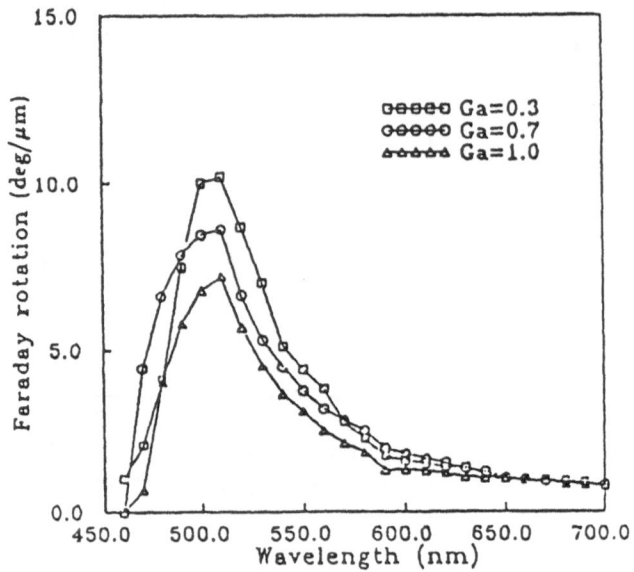

Fig. 13 : Variation of the Faraday rotation angle with wavelength of $Bi_{1.2}Dy_{1.8}Fe_{5-x}Ga_xO_{12}$ thin films

Cu-doping effect in Al or Ga substituted Bi:DyIG films

Fig. 14 shows the doping effect of Cu on Faraday rotation spectrum of Al-substituted DyIG films. These curves are similar to those of Al and Ga-substituted DyIG films. The Q_F value increased and the peak wavelength shifted somewhat to shorter wavelength.

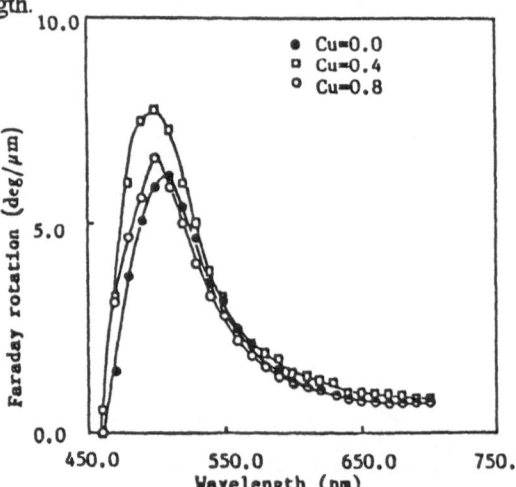

Fig. 14 : Variation of Faraday rotation angle with wavelength of $Bi_{1.4}Dy_{1.6}Fe_{4-x}AlCu_xO_{12}$ (x = 0.0, 0.4, 0.8) films

74

The strong doping effect of Cu on coercivity (H_c) of Al and Ga-substituted DyIG films is evidenced in Fig. 15. The H_c increased in amplitude upon adding more Cu, reaching 13 KOe and 7 KOe, respectively, in Cu-doped Bi,Ga:DyIAG and Bi,Al:DyIG films.

Shimokawa et al.[14] have reported on Cu-added Bi,Ga:DyIG films prepared by the sputtering method, using metal chips of Cu as additive. The maximum H_c of 5.4 KOe was obtained at the area ratio of 3.4% in which the increased H_c was attributed to the pinning effect. We consider that the increase in H_c cannot be explained simply by the pinning effect; it may be associated also with Cu entering the lattice sites. Indeed, based on a published report,[15] the Cu ion can enter the 24C site in garnet and contributes to the magnetostriction coefficient. Considering the radius of Cu^{2+} ion, it may enter 16a and 24d sites.

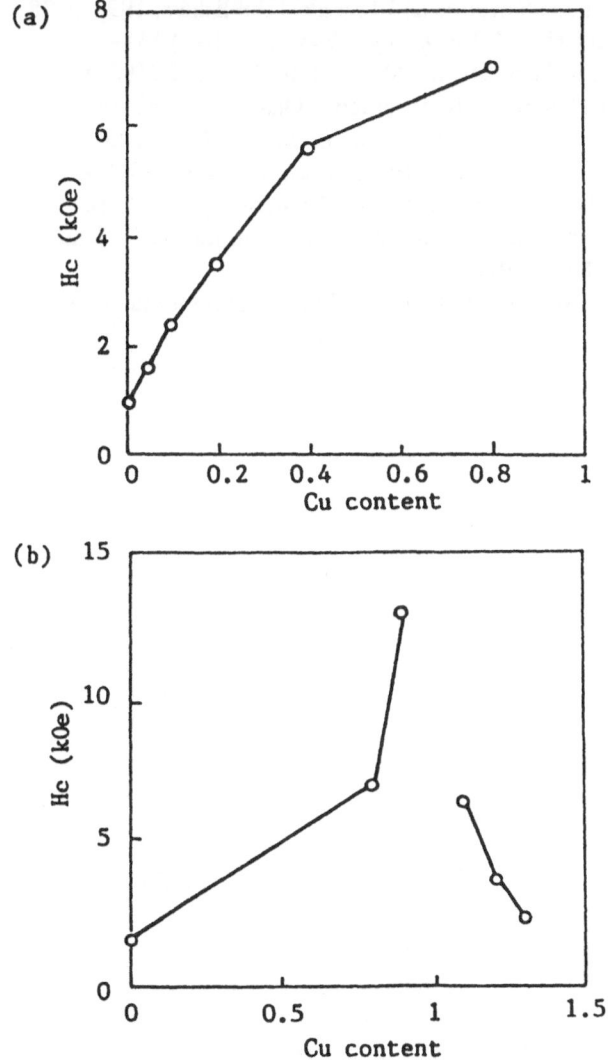

Fig. 15 : Dependence of coercivity of (a) $BiDy_2Fe_{4.3-x}Cu_xGa_{0.7}O_{12}$ and (b) $Bi_{1.4}Dy_{1.6}Fe_{4+x}Cu_xAlO_{12}$ films on Cu content

References

1. Fuxi Gan, *Proceedings SPIE*, **1519** (1991) 530.
2. A.E. Bell *et al.*, *Appl. Phys. Lett.*, **31** (1979) 275.
3. Qi Jiang and Fuxi Gan, *Acta Optica Sinica*, **9** (1989) 25.
4. X.Y. Yu, H. Watable, S. Iwata, S. Tsunashima and Ychiyama, SPIE Proceedings on Third International Symposium on Optical Storage, ed., Fuxi Gan, Yunnan, China, 2053 (1993) 2.
5. Zhanghua Wu and Fuxi Gan, *Proceedings of Magneto-Optical Recording International Symposium 1992*, Tucson, Arizona, 1993, p. 68.
6. Zhanghua Wu and Fuxi Gan, *Chinese Bulletin*, **36** (1993) 1082.
7. Y.J. Wang *et al.*, In: *Digital optical disks and optical storage media*, ed., Fuxi Gan, Shanghai Scientific and Technical Publishers, 1992, p. 28.
8. D.F. Shen and X.Y. Yu, *J. Appl. Phys.*, **67** (1990) 5319.
9. M. Gomi, K. Satoh and M. Abe, *J. Appl. Phys.*, **63** (1988) 3642.
10. K. Shono, H. Kano, N. Koshino and S. Ogawa, *J. Appl. Phys.*, **63** (1988) 3639.
11. A. Itoh, M.H. Kryder, *Appl. Phys. Lett.*, **53** (1988) 1125.
12. D.F. Shen and T.D. Du, *Acta Physica Sinica*, **40** (1991) 653.
13. Y. Zhou, D.F. Shen and F.X. Gan, *Chinese Phys. Lett.*, 10 (1993) 186.
14. K. Shimokawa, S. Takebayashi, N. Kawamura and T. Tamaki, *J. Mag. Soc. Jpn.*, **15** (1991) 209.
15. P. Hansen, W. Tolksdorf and J. Schuldt, *J. Appl. Phys.*, **43** (1972) 4253.

Main Group Elements and Their Compounds
V.G. Kumar Das (Ed)
Copyright © 1996 Narosa Publishing House, New Delhi, India

The Silicon Crystal in Semiconductor Technology

S. Kishino

Department of Electronics, Faculty of Engineering
Himeji Institute of Technology, Shosha, Himeji 671-22, Japan

Silicon crystal is a central and indispensable material in the semiconductor industry, which is currently dominated by VLSI (very large scale integration) devices. Up till now, the global growth rate of the LSI industry has been estimated at 20% per year, and the forecast is that this will continue at a rate higher than 15% or so in the years ahead. The optimism for this stems from the fact that there is a great demand for metal oxide semiconductor (MOS) LSI's in which an MOS transistor is used as the main active device. MOS devices include MOS memory LSI, MOS logic LSI and MOS micro-processor. These are widely used in many kinds of electrical appliances.

Furthermore, it is also very economical to use a LSI device with high integration[1] because the cost per bit has rapidly decreased over the last decade as shown in Figure 1. It is expected that the cost per bit of DRAM (dynamic random access memory) would soon match that of the magnetic disc which is presently the most economical memory device.

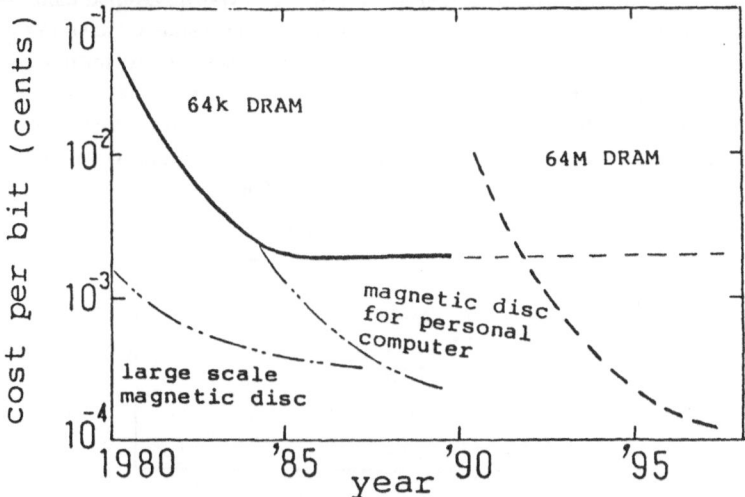

Fig. 1 : Exponential reduction of cost per bit for DRAM chips

The high integration of LSI is attainable by the miniaturization of devices as shown in Fig. 2, in which the degree of the miniaturization is indicated using a design rule. Recently, the design rule has entered into the submicron range as seen in the figure.

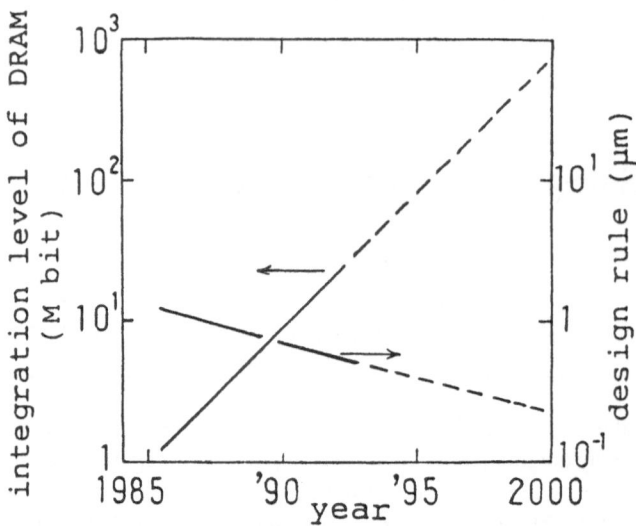

Fig. 2 : Relationship between LSI integration levels and miniaturization as expressed by the design rule

History of the semiconductor technology

Here, we discuss three representative semiconductor materials, namely, silicon (Si), germanium (Ge), and gallium arsenide (GaAs). From a historical standpoint, the first transistor (a representative semiconductor device) was developed from a Ge crystal (see Fig. 3). Even the original integrated circuit (IC) was developed using the Ge crystal.[2] However, for reasons that will be explained, Ge crystals were soon replaced by Si crystals which came to be used in a wide range of semiconductor devices such as transistor, diode and IC's (LSI's). The integration level of IC has progressed through miniaturization of the transistor in a stepwise fashion; that is, from SSI (small scale integration), through MSI (medium scale integration), and LSI (large scale integration), to VLSI (very large scale integration).[1]

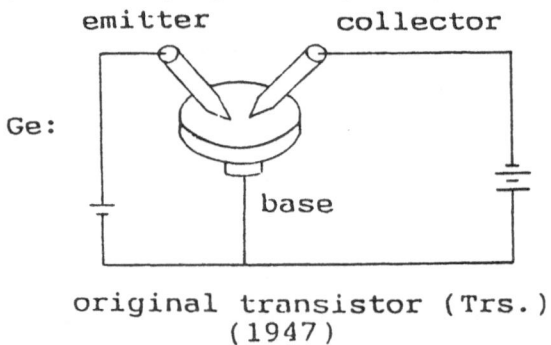

Fig. 3 : Schematic view of original transistor developed using a Ge crystal

In tandem with this, the diameter of the Si wafer, and hence that of the parent Si crystal ingot, has progressively increased as shown in Fig. 4. This is because the

chip size of the VLSI device has increased year by year.[3]   Without the increase in
wafer size, it would not be possible to increase the number of chips on the wafer.

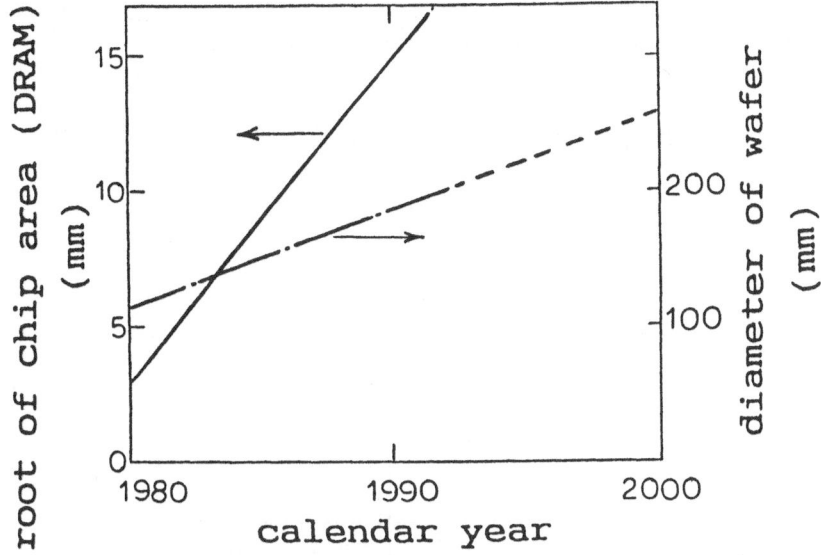

**Fig. 4 : Plot showing the increases in the wafer diameter of the Si crystal and
root of LSI chip area with calender year**

In contrast to Si, the GaAs crystal is used for optoelectronic devices such as laser
diodes (LD) and a light emitting diodes (LED).   This is because GaAs is a
semiconductor of the direct transition type with high radiation efficiency. However,
the use of GaAs crystal for VLSI devices poses problems which are yet to be fully
resolved, although this material has splendid   potential capability such as high
electron mobility and a large energy bandgap.

Why is the silicon crystal still prevailing in the semiconductor industry?

VLSI is composed of a vast number of semiconductor devices by which complicated
electric circuits are constructed. The fault of any device in a VLSI chip causes the
failure of the overall circuit of the VLSI chip and results in the failure of VLSI
production. Successful VLSI production requires the semiconductor material to
possess properties that meets the conditions of reliability, superior electric,
mechanical and chemical properties, etc. At the same time, a large diameter wafer
must be prepared as a substrate to accomodate many VLSI chips.
      Table 1 shows the list of material properties necessary for VLSI production. Of
these properties, the reliability of semiconductor device depends on the energy
bandgap of the semiconductor, a larger bandgap being more desirable. This is
because electrons and holes are easily formed by raising the temperature or by
exposing radiation (Fig. 5a). The electrons and holes produced by this mechanism
cause a leakage of current, resulting in reverse characteristics (shown in Fig. 5b with
a broken line). Such degradation is detrimental to miniaturized VLSI devices with
small signal charges.

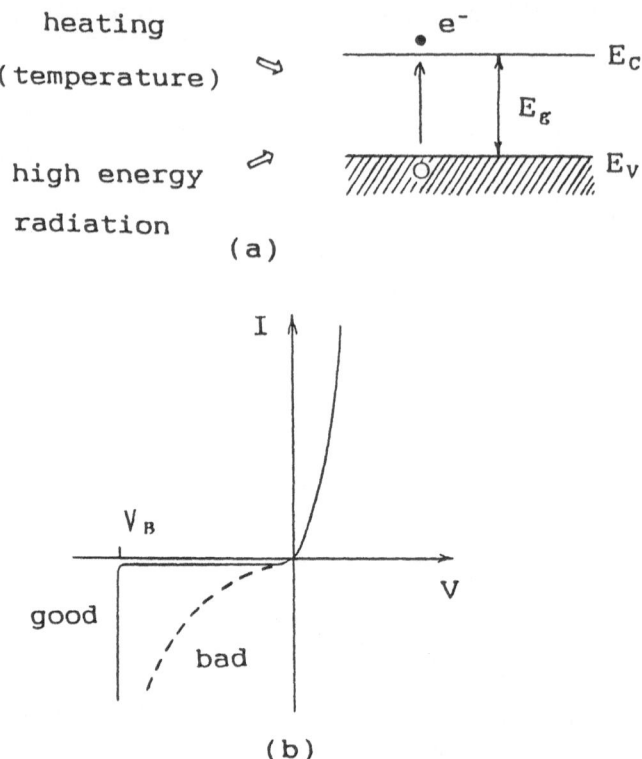

(a)

(b)

Fig. 5 : **(a) Increase of leakage current and (b) degradation of reverse characteristics of p-n junction**

The switching time (speed) of the device is also very important. In the case of a MOS transistor, the switching time τ is expressed as

$$\tau = L^2/\mu_n V_D \qquad (1)$$

where L, μ_n and V_D are respectively the channel length of a MOS transistor, electron mobility and drain voltage. A small channel length L and high electron mobility μ_n are desirable values. Silicon crystals have the advantage of small channel lengths through the micro-fabrication of devices. Mechanical strength and chemical stability are also required[2] for VLSI because semiconductor wafers are repeatedly exposed to high temperatures during the oxidation and impurity-diffusion processes, and are frequently subjected to chemical processes such as etching and cleaning.

Of the material properties listed in Table 1, high-quality oxide films are essential, especially for a MOS transistor. Oxide films are widely used for the fabrication of semiconductor devices as amplified in Table 2. It should be emphasized that the development of Si planar technology began with the use of SiO_2 films as diffusion masks against dopant impurity. In device fabrication, however, all electric charges must also be removed from the interior of the oxide film, as the existence of any electric charge is tantamount to adding extra voltage to the gate electrode below which the semiconductor substrate exists.

Table 1 : Properties of semiconductor materials necessary for VLSI production[@]

Property \ Material	Si	GaAs	Ge
Reliability (bandgap, eV)	++ (1.12)	+++ (1.42)	+ (0.68)
Switching speed*	+++	+++	+
Mechanical strength	+++	+	+
Chemical stability	+++	+	++
Oxide and interface	+++	+	+
High purity**	+++	+	++
Cost	+++	+	+

[@]score ratings indicated by +++ , ++ and +
*mobility and channel length
**low density of deep level states

Table 2 : Use of oxide films

Use	Properties
Diffusion mask	Shield against impurity diffusion
Gate oxide	No electric charges, Uniform thickness
Field oxide	No electric charges
Insulation film	Insulating
Passivation mask	Insulating, Water-proof, No electric charges

With regard to the oxide film and the interface property between the oxide film and the semiconductor, Si crystals possess more advantageous features than other semiconductor materials. Ultra-pure semiconductor materials are required in VLSI. Even a very low density of heavy-metal impurity degrades the characteristics of VLSI devices. Production cost should be desirably as low as possible in VLSI technology, but there should, however, be no compromise on the properties of the semiconductor material which should be as high as possible (Fig. 6).

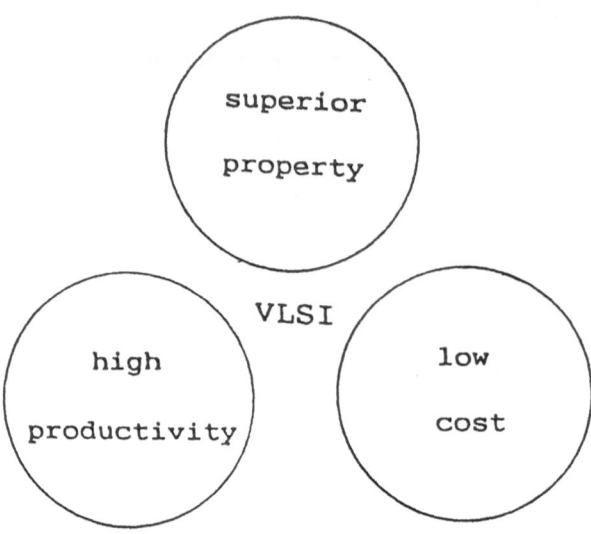

Fig. 6 : The three important factors in VLSI technology

Large diameter wafers

At present, large-sized Si wafers are used for VLSI production (Fig. 4). These large-sized Si wafers should show uniform wafer flatness and small wafer warpage, both features being important for the fabrication of miniaturized VLSI devices. Many VLSI chips (one of which is indicated with cross hatches) are designed within one sheet of Si wafer (Fig. 7a); a large number of miniaturized VLSI devices are fabricated within one VLSI chip at the same time. The number of such devices exceeds one million at present. The size of the device (minimum feature size) is about 0.5 μm, the size being defined by the gate length of a MOS transistor. Miniature sized MOS transistors, without which we could not expect such high production of VLSI, are fabricated simultaneously by using an optical lithographic method. Highly accurate lithography requires high optical resolution (δ). A long focal length Δ is also desirable for the manufacturing process. The resolution (δ) and focal length Δ are given by:

$$\delta = \gamma/1.28NA \tag{2}$$

and $\qquad \Delta = \gamma/2(NA)^2 \tag{3}$

where γ and NA are respectively the wavelength of light and numerical aperture.

In order to obtain a high resolution (i.e., as small a value of δ as possible), the wavelength should be small and the NA value large, but this clearly would be at the expense of a reduced focal length. However, resolution is a more important consideration. As a measure against short focal length, uniform flatness is required for Si wafers. Large size and perfect flatness are thus two important criteria that must be simultaneously fulfilled in the manufacture of Si wafers.

The importance of wafer warpage is understood in terms of the details of the VLSI processes. The Si wafers are frequently subjected to high-temperature heat treatment in the oxidation and impurity diffusion process. Wafer warpage is caused by dislocation induced by thermal stress during heat treatment, and is enhanced by the frequency of the thermal cycle. Wafer warpage is thus a function of heat cycle, as shown in Fig. 8. Wafer warpage is serious only when heat treatments are frequently carried out, and occurs easily for the VLSI processes having many heat treatments.

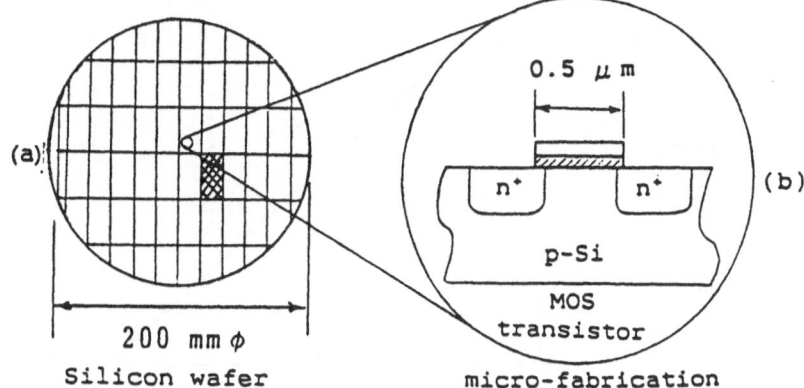

Fig. 7 : **VLSI chips designed within one Si wafer (a), and enlarged VLSI (MOS) device within one chip (b).**

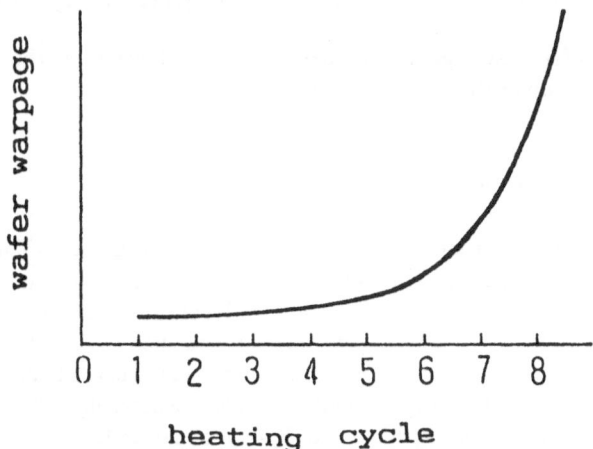

Fig. 8 : **The relationship of wafer warpage to heating cycles**

Two methods are used to grow Si crystals: the Floating Zone (FZ) and Czochralski Growth (CZ) methods. In the FZ method, high-purity Si crystals are produced, whereas for the CZ method, the Si crystal grown normally contains a high proportion of impurity elements.[5] The CZ grown Si crystals contain oxygen that comes from the quartz crucible during growth. The oxygen in the CZ crystals causes oxygen precipitates and related microdefects during high-temperature heat treatments, and is detrimental to VLSI devices.

However, oxygen impurity is useful in that the oxygen atom acts as a pinning center of dislocations and the impurity decreases wafer warpage during the VLSI

83

process. This is shown in Fig. 9 where the wafer warpage is indicated as a function of oxygen content in the wafer. If the Si wafer is of a sufficiently high oxygen content, the Si wafer does not suffer serious warpage even when subjected to high temperature heat treatment.

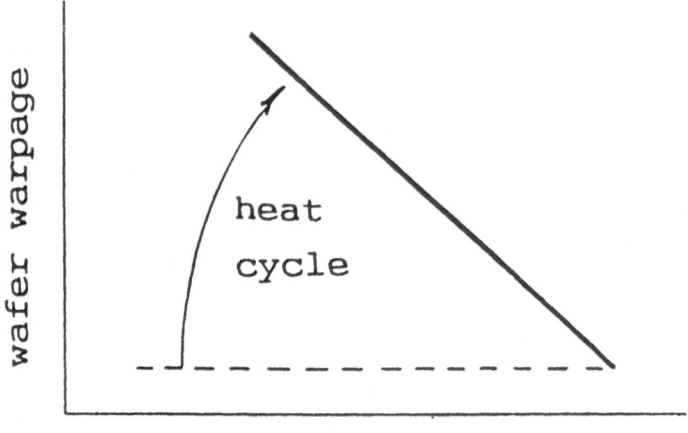

suppresion of wafer warpage by oxygen

Fig. 9 : **Wafer warpage versus oxygen concentration in the crystal with heating cycle as parameter**

Heavy metal impurities

Heavy metal impurities are injurous to semiconductor devices. They cause the carrier lifetimes to be shortened and increase the incidence of current leakage. These effects arise because of the occurrance of deep levels in the energy bands (Fig. 10) resulting from the inability of heavy metal impurities such as Au, Pt, Fe and Cu that may be present to form stable bonds with Si atoms in the crystal lattice.

The existence of either deep or shallow levels in a semiconductor is determined by the magnitude of the energy band gap, E_C-E_V. The deep levels act as generation-recombination centres for carriers (electrons or holes) thereby reducing their lifetimes. The presence of deep levels due to heavy metal impurities at p-n junctions are especially conductive to current leakage and result in the failure of such junctions in VLSI devices. (see Fig. 5(b)).

Concerning the harmful effects of heavy metal impurities on VLSI devices, two concommittant problems are attracting much attention. One is the decrease of signal to noise (S/N) ratio. The other is decrease of the production yield of VLSI devices. With increasing miniaturization of VLSI devices, the signal charge per device has decreased greatly , and a small number of noise charges can easily degrade the S/N ratio. This is a serious problem for the fabrication of miniaturized VLSI devices.

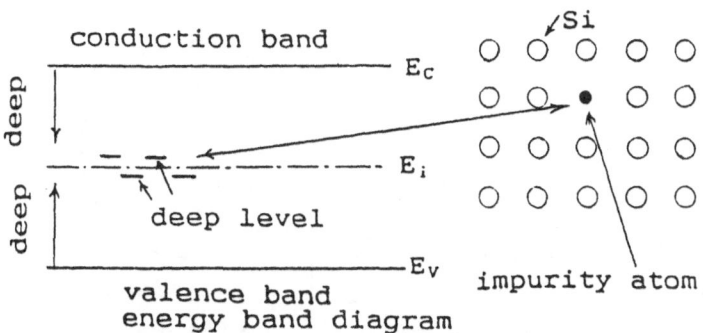

Fig. 10 : **Deep levels in the energy bandgap caused by heavy metal impurity atoms in the crystal lattice**

A high production yield is essential for VLSI devices. A poor yield may deny the production of even a single VLSI chip. Table 3 shows the relationship between the density of heavy metal impurities and the device yield.[6] The yield largely depends on the integration level of the LSI; however, for a discrete device or an LSI with a low integration level, a high production yield could be easily achieved even when the impurity density is high. On the other hand, for a VLSI with an integration level higher than 10^4, a high production yield is not attainable even where there is only a low density of heavy metal impurities.

Table 3 : **Effect of heavy metal impurities on LSI yield[†]**

D_{imp} (cm^{-3}) (deep level)	Y_{disc}	Y_{LSI}		
		10^2	10^4	10^{6*}
10^{15}	0	0	0	0
10^{13}	0 99	0.37	0	0
10^{12}	0.999	0.99	0.37	0
10^{11}	0.999999	0.9999	0.99	0.37

Footnote:
[†]D_{imp} = density of heavy metal impurity;
Y_{disc} = discrete device yield;
Y_{LSI} = LSI yield
*Integration level

Besides degrading VLSI devices through carrier lifetime shortening and increasing current leakage, heavy metal impurities also enhance the number of interface states existing at the Si/SiO$_2$ interface.[7] If the density of interface states is very high, an MOS device would not work accurately because the interface states capture the carriers induced by the field effect (as shown in Fig. 11 using a cartoon).

This is the reason why an MOS transistor using Group III-V semiconductors such as GaAs cannot be fabricated.

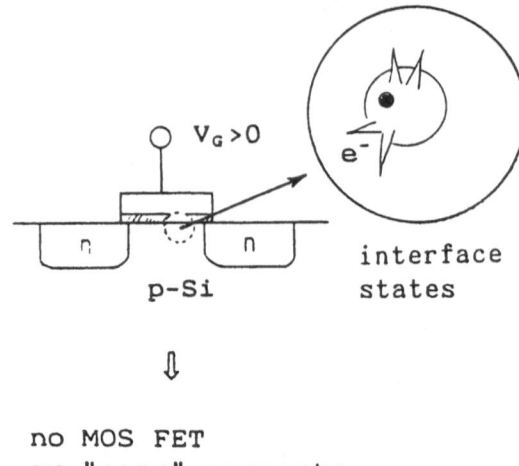

no MOS FET
or "poor" property

Fig. 11 : Harmful effects of interface states

The enhancement of the interface states by gold impurity has been observed for the first time. Figure 12[7] shows the relationship between the interface state and the diffusion temperature of the Au impurity. The solid solubility of Au depends on the diffusion temperature. Figure 12 shows the number of interface states as a function of the density of the Au impurity. The energy distribution of the interface states has also been investigated. Figure 13[8] depicts the density of interface states as a function of energy from the mid gap of (the energy bandgap of) Si. The distribution is U-shaped; this phenomenon is well-established for the Si/SiO$_2$ interface.[9] The density of interface states increases proportionately with the diffusion temperature.

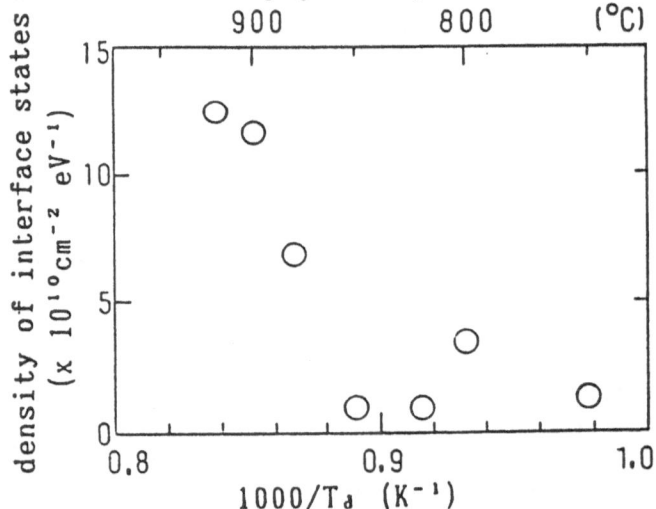

Fig. 12 : Plot depicting the increase in the number of interface states with an increase in diffusion temperature of gold impurity

86

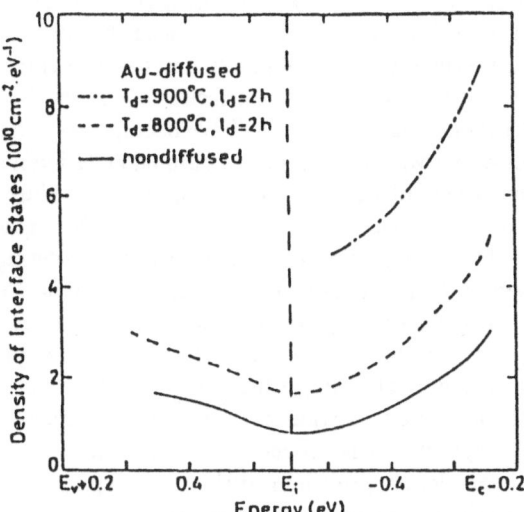

Fig. 13 : Distribution of interface states enhanced by gold diffusion within energy bandgap

Optimal gettering for VLSI process

Heavy metal impurities, harmful to the device characteristics, must be eliminated from the device area. Strong measures against the harmful impurities have been taken since the semiconductor device has been developed; these measures are called "gettering procedures". Many gettering procedures[10] have been proposed since the development of semiconductor devices and have been put into practice.

Of the many gettering techniques, backside damage gettering is adopted here as a representative technique. In this technique, a gettering region is positioned at the rear surface of the Si wafers as shown in Fig. 14. The gettering region is separated from the device site, normally located at the front surface of the wafer.

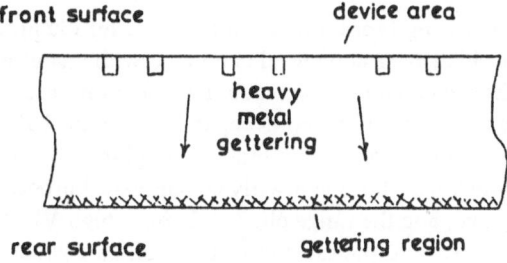

Fig. 14 : Schematic sectional view of Si wafer showing gettering procedure

The production yield can be increased through the elimination of harmful impurities from a device area with the use of a gettering procedure. This is the main advantage of the gettering procedure, but the procedure has a serious drawback of dislocation propagation from the gettering site region to the device area. This causes device failure through the adverse effects of dislocation. Another drawback is wafer warpage induced by dislocations that are introduced by the gettering procedure itself. Wafer warpage is harmful to VLSI production because optical lithography for micro-fabrication is adversely affected by surface unevenness. Surface unevenness is induced by dislocations. Such drawbacks have become very serious especially in newer VLSI processes, for which conditions of optimal gettering need to be secured.

Figure 15[6] shows a simulation result for a gettering procedure that is applied to the production of discrete devices. The discrete device yield is shown as a function of gettering capability. Dislocation is introduced in the gettering site region and contamination level of the production process is assumed to be 1×10^{15} cm^{-3}. As a result, the production yield becomes zero without the use of gettering procedure (see the case $D_d = 0$), as seen in Fig. 15. However, high production yields are achieved when the gettering capability D_d is larger than 10^4 cm^{-3}. This result shows that high device yield can be obtained over a wide range of gettering capability. The gettering procedure is readily applicable to the production of semiconductor devices. Gettering procedures have been already proven profitable for the production of LSI devices with low integration levels as well as discrete devices.

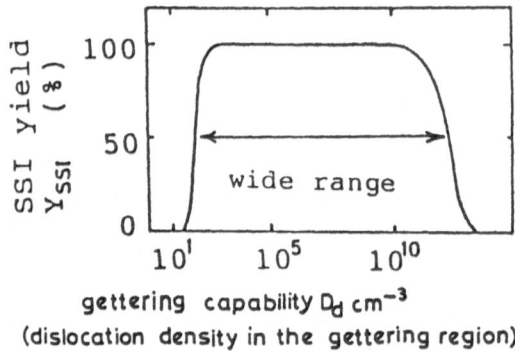

Fig. 15 : Plot of SSI yield versus gettering capability

However, the gettering procedure is not the same for the production of present VLSI devices with a high integration level compared with that of the discrete device. Figure 16 shows the similar simulation result for a VLSI process. The VLSI yield is similarly shown as a function of gettering capability. Figure 16(a) and (b) show the production process being heavily (10^{15} cm^{-3}) and lightly (10^{12} cm^{-3}) contaminated with heavy metal impurities. For the heavily contaminated process, a large gettering capability is required, and the range obtainable for a high VLSI yield shifts to the right side in Figure 16(a). At the same time, the profitable gettering range becomes narrower because of drawbacks in the gettering procedure, as seen in Fig. 16(a). This shows that the application of the gettering procedure in the process becomes complicated.

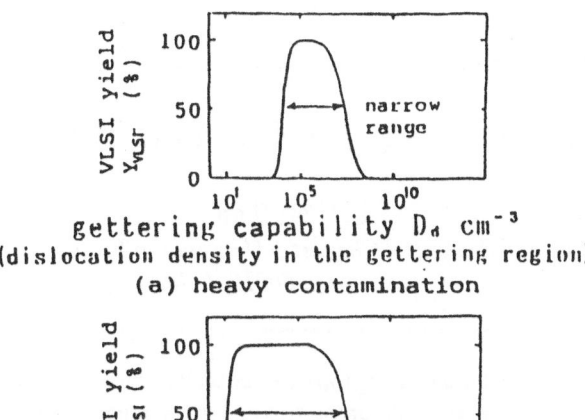

gettering capability D_d cm^{-3}
(dislocation density in the gettering region)
(a) heavy contamination

gettering capability D_d cm^{-3}
(b) light contamination

Fig. 16 : VLSI yield versus gettering capability for heavy and light contamination with heavy metal impurities

If the contamination level of the production process is low, a better result is expected with the same model even in the VLSI case, as seen in Fig. 16(b). A high VLSI yield could be obtained over a wide range of gettering capability, and the gettering procedure can be easily applied to the VLSI process. A soft gettering is more profitable for the VLSI process.

References

1. S.M. Sze, VLSI Technology, McGraw-Hill, 1983, pp. 1-7.
2. J.S. Kilby, IEEE Trans. on Electron Devices. ED-23, 1976, pp. 648-654.
3. P. Disessa and S. Stone, ASMC '91 Proceedings, IEEE, Inc., 1991, pp. 54-63.
4. L.C. Parriollo, VLSI Process Integration, In: *VLSI Technology*, ed., S.M. Sze, McGraw-Hill, 1983, pp. 445-505.
5. S. Kishino, Proc. 3rd. International Symposium on VLSI Science and Technology, 1985, pp. 399-418.
6. S. Kishino, H. Yoshida and H. Niu, *IEEE Trans. on Semiconductor Manufacturing*, 6 (1993) 251.
7. H. Yoshida, H. Niu and S. Kishino, *Jpn. J. Appl. Phys.*, 30 (1991) L1293.
8. H. Yoshida, H. Niu and S. Kishino, *J Appl. Phys.*, 73 (1993) 4457.
9. E.N. Nicollian and J.R. Brews, MOS Physics and Technology, John Wiley & Sons, New York, 1982, p. 794.
10. J.R. Monkowski, *Solid State Technology*, 24 (1981) 44.
11. T. Aoshima, Y. Kosaka and A. Yoshinaka, Proc. 6th International Symposium on Silicon Materials Science and Technology, In: *Semiconductor Silicon 1990*, ed., H.R. Huff, New Jersey, The Electrochem. Soc., 1990, pp. 724-733.

Main Group Elements and Their Compounds
V.G. Kumar Das (Ed)
Copyright © 1996 Narosa Publishing House, New Delhi, India

Properties of C_{60} and C_{70} in the Solid State

C.N.R. Rao

*Solid State and Structural Chemistry Unit, Indian Institute of Science
Bangalore 560012, India*

Fullerenes exhibit novel molecular properties as well as chemical reactivity. In the solid state, they exhibit interesting properties and phenomena. In this presentation, we discuss a few important aspects of solid C_{60} and C_{70}. The topics covered include phase transition associated with orientational order, effects of pressure on the structure and electronic properties, ferromagnetism in doped C_{60} and interaction of C_{60} and C_{70} films with metal clusters.

Phase transitions of C_{60} and C_{70}

Spherical molecules generally show orientational disorder in the solid state[1], and we would therefore expect that C_{60} and C_{70} show phase transitions associated with orientational order. Early structural studies on solid C_{60} at room temperature showed that large thermal factors are necessary to obtain reasonable fits to X-ray data[2]. Based on the narrow NMR linewidths due to short orientational correlation times at room temperature, solid C_{60} was considered to be orientationally disordered[3,4]. X-ray diffraction and DSC measurements showed that C_{60} undergoes a phase transition at around 250 K on cooling from an orientationally disordered fcc phase to a sc phase where molecular rotation persists but only along preferred axes[5]. NMR studies show that below 250 K molecules jump between preferred orientations with a barrier of *ca.* 3000 K[6]. Molecular Dynamics (MD) simulations [7,8] reproduce experimental observations quite well. It seems necessary to add a Coulombic term to the L-J potential in order to reproduce the charge-transfer effects. Charge-transfer is also indicated by neutron diffraction measurements at 5 K[9], which show the proximity of electron-rich double bonds to adjacent electron-poor pentagonal faces.

Raman spectra of C_{60} under pressure and related studies indicate the occurrence of an orientationally glassy state in C_{60}. Monte-Carlo studies[10] show that instantaneous cooling of the plastic-crystalline phase of C_{60} leads to an orientational glassy phase with Tg *ca.* 80 K (Fig. 1). A glassy state in C_{60} has been experimentally observed using dilatometry[11].

Ellipsoidal C_{70} shows a more complicated phase behaviour than C_{60}. Vaughan *et al.*[12] reported two phase transitions of C_{70} at 337 K and 276 K. However, the structures of these three phases are yet to be established unequivocally. At room temperature and above, both face-centred cubic (fcc) and hexagonal close-packed (hcp) structures seem to coexist as they are energetically similar. Single crystals of as-grown C_{70} are usually twinned and unsuitable for X-ray studies. Based on X-ray studies, Verheijen et al.[13] reported the occurrence of fcc, rhombohedral, hcp2, hcp1 and distorted hcp phases on going from high to low temperatures. We have carried out powder X-ray diffraction studies on C_{70} as a function of temperature. The high-temperature phase is clearly fcc and sluggishly transforms to a hcp phase. Thermal cycling increases the crystallinity of these materials as seen from X-ray line-widths. MD simulations[14] show that on cooling the high-temperature rotator phase, a transition to a phase with trigonal symmetry occurs followed by a transition to a monoclinic phase.

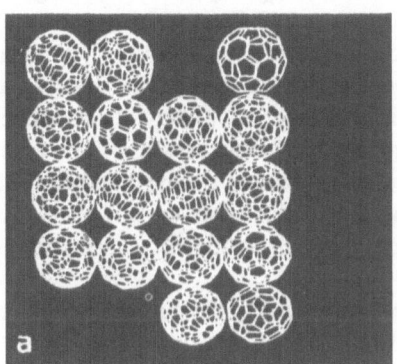 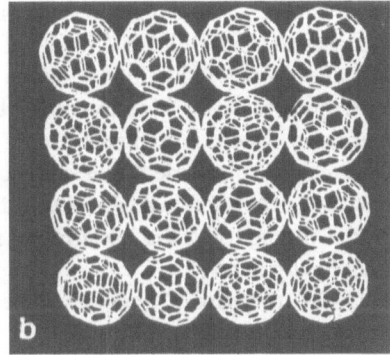

Fig.1 : Snapshot pictures showing the packing of molecules of C_{60} in (a) the plastic crystalline phase and (b) quenched glassy phase.[10]

Phase transitions of C_{70} have been investigated in our laboratory by both infrared[15] and Raman[16] spectroscopy. IR and Raman bands of sublimed C_{70} films show absorbance and line-width changes across the orientational phase transitions around 340 K and 280 K (Fig. 2).

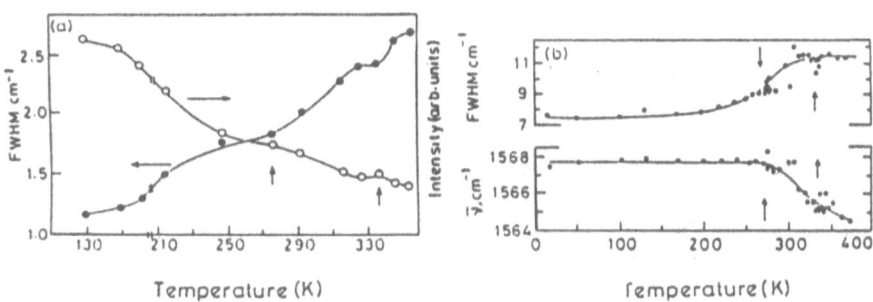

Fig. 2 : (a) Temperature variation of the full width at half maximum (FWHM) (●) and of the intensity (○) of the 643 cm⁻¹ band in the IR spectrum of C_{70}. (b) Temperature variation of the FWHM of the 1566 cm⁻¹ Raman band of C_{70} and of the phonon frequency. Transition temperature are indicated by arrows.[16]

The phonon frequencies harden on cooling across these transitions possibly due to a combination of decreasing unit cell volume and increased intermolecular interactions. The linewidths decrease on cooling due to the hindering of the molecular motion resulting in a decrease in the rotational density of states. The changes in Raman frequency and linewidth across the phase transitions are continuous. This also appears to be true of C_{60} contrary to the earlier Raman work of van Loosdrecht et al.[17] IR studies show evidence for the coexistence of phases in the 210-170 K region. Both IR and Raman spectra show that orientational disorder starts to freeze out below 150 K.

The effect of pressure on the phase transitions in C_{60} and C_{70} is interesting. In the case of C_{60}, the temperature of the transition at 250 K (at ambient pressure) increases at a rate of 10.3° kbar^{-1}. The DSC curve of Samara et al.[18] shows a shoulder beyond 6 kbar indicating the presence of two states of similar energies at ambient pressure. Measurements by Ramasesha et al.[19] show two distinct transitions. Raman investigations on C_{60} single crystals under pressure[20] show that the pentagonal pinch mode undergoes considerable softening around 3.5 kbar. The linewidth increases at higher pressures till around 130 kbar when the line shape has almost merged into the background indicating an orientationally glassy state as observed by others[11].

Under pressure, C_{70} shows three phase transitions[19], the highest transition temperature increasing with pressure at a rate of 6.8° kbar^{-1}, the lowest at 8.4° kbar^{-1} and the intermediate one at 5.3° kbar^{-1} (Fig. 3). At ambient pressure, two transitions are generally observed. A recent DSC study,[21] however, shows three distinct transitions at 280, 330 and 337 K.

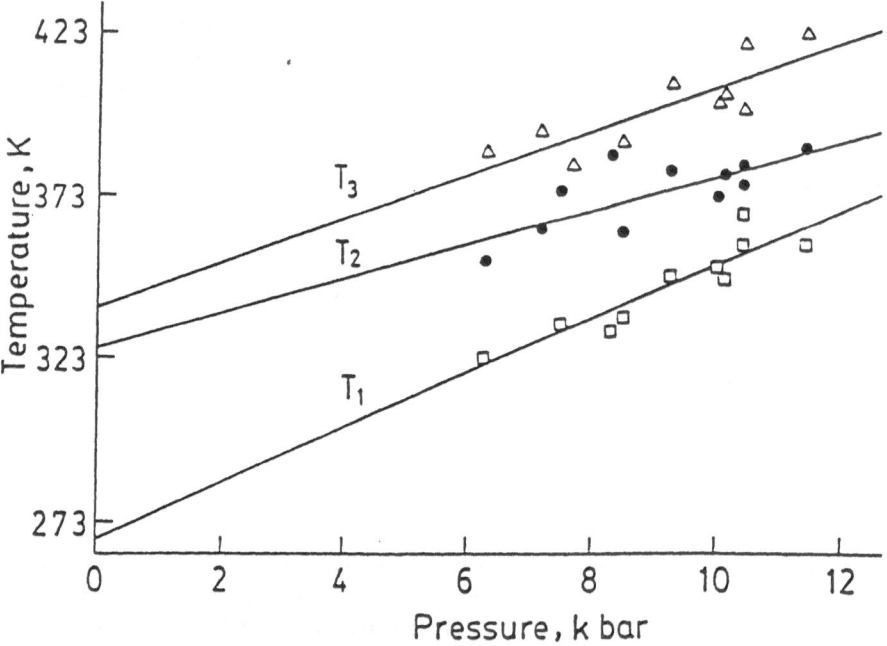

Fig. 3 : **Pressure dependence of the three transition temperatures of C_{70}.**[19]

C_{60} Photoluminescence under pressure

Early high pressure studies on C_{60} crystals from our laboratory[22] showed that the photoluminescence band of C_{60} is red-shifted continuously with increasing pressure until at 3.2 Gpa, the colour of the crystal changes from red to black, and the band disappears (Fig. 4). Since the C_{60} ball itself is largely incompressible, this closing of the PL band is interpreted as being due to a broadening of the valence and conduction bands arising from increasing interball resonance. The near disappearance of the optical gap is consistent with decreasing activation energy or conduction with increasing pressure as reported by a few workers. This study has certain implications for the strengths of electron-phonon coupling and therefore for superconductivity in alkali metal fullerides.

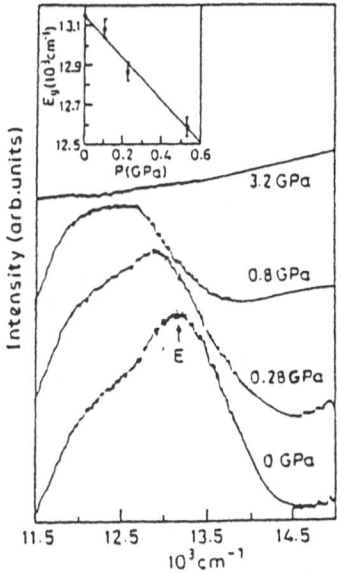

Fig. 4 : Photoluminescence band of C_{60} as a function of temperature.[22]
Ferromagnetism in $C_{60} \cdot$ TDAE

C_{60} is an excellent electron acceptor forming radical anions. With aromatic amines, it forms weak ground state complexes and exciplexes in solution[23]. A strong amine such as TDAE (tetrakis dimethylaminoethylene) forms a radical ion salt with C_{60}. $C_{60} \cdot$ TDAE is the best organic ferromagnet known to date[24] with a T_c of 16 K. The lattice structure has been established by X-ray studies[25]. The cell is monoclinic with a short C_{60}-C_{60} contact along the c-axis. ESR studies have shown that a single electron is doped from TDAE to C_{60}[26,27] The ESR linewidth decreases with decreasing temperature indicating the $C_{60} \cdot$ TDAE is metallic. Single-electron doping has been confirmed by our Raman studies by titrating the observed phonon frequencies against n[28], using the data for C_{60}^{n-} from the $A_x C_{60}$ phases. As with $A_x C_{60}$, electron doping results in softening of the phonons. This phonon softening is consistent with theoretical calculations. Raman studies also show the the A_g modes of C_{60} in $C_{60} \cdot$ TDAE are split due to the lower symmetry of the system. By fitting the low-temperature susceptibility data to a series expansion for the spin-1/2, 1-D Heisenberg system[29], we have found evidence for quasi-ID behaviour. The magnitude of the ferromagnetic coupling constant is 50 K. McConnell's model for ferromagnetic exchange has been sued to understand this system[27,30] (Fig. 5). Comparing $C_{60} \cdot TDAE$ with superconducting $A_3 C_{60}$, it would seem that the lower dimensionality and single electron doping are key features.

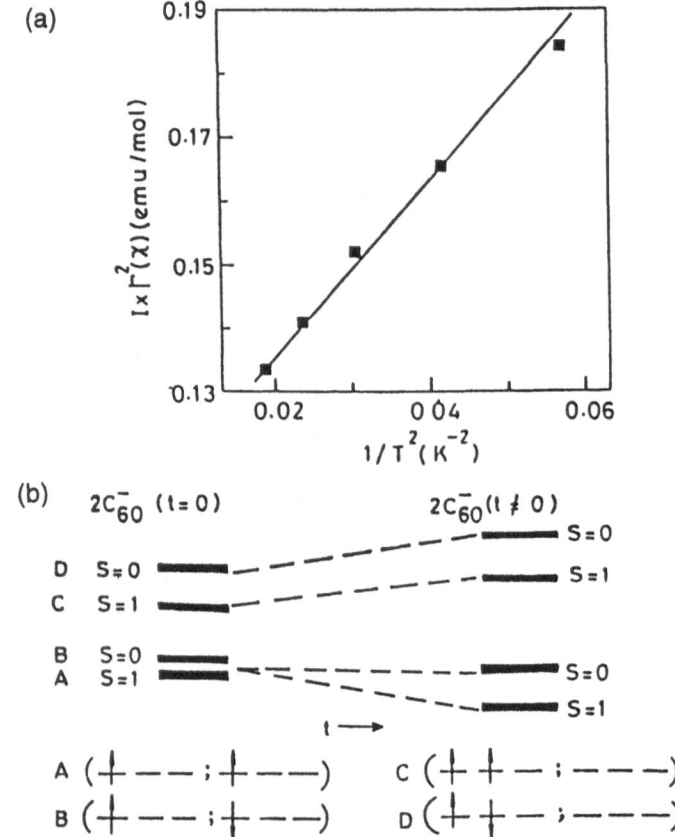

Fig. 5 : (a) Plot of the ESR susceptibility of $C_{60} \cdot TDAE$ against $1/T^2$.
(b) Configuration interaction in the ferromagnetic ground state.[27]

References

1. J.N. Sherwood (ed.), *The plastically Crystalline State*, John Wiley, London, 1979.
2. R.M. Fleming *et al.*, In: *Fullerenes: Synthesis, Properties and Chemistry of Large Carbon Clusters*, eds., G.S. Hammond and V.J. Kuck, ACS Symposium Series, **481** (1991) 25.
3. C.S. Yannoni *et al.*, *J. Phys. Chem.*, **95** (1991) 9.
4. P.A. Hainey *et al.*, *Phys. Rev. Lett.*, **66** (1991) 2911.
5. R. Sachidanandam and A.B. Harris, *Phys. Rev. Lett.*, **67** (1991) 1467.
6. R.D. Johnson *et al.*, *Science*, **225** (1992) 1235.
7. M. Sprik, A. Cheng and M.L. Klein, *J. Phys. Chem.*, **96** (1992) 2027.
8. J.P. Lu *et al.*, *Phys. Rev. Lett.*, **68** (1992) 1551.

9. W.I.F. David *et al.*, *Nature*, **353** (1991) 147.
10. A. Chakrabarti, S. Yashonath and C.N.R. Rao, unpublished results.
11. F. Gugenberger *et al.*, *Phys. Rev. Lett.*, **69** (1992) 3374.
12. G.B.M. Vaughan *et al.*, *Science*, **254** (1991) 1350.
13. M.A. Verheijen *et al.*, *Chem. Phys.*, **166** (1992) 287.
14. M. Sprik, A. Cheng and M.L. Klein, *Phys. Rev. Lett.*, **69** (1992) 1660.
15. V. Varma, R. Seshadri, A. Govindaraj, A.K. Sood and C.N.R. Rao, *Chem. Phys. Lett.*, *203* (1993) 545.
16. N. Chandrabhas, K. Jayaram, D.V.S. Muthu, A.K. Sood, R. Seshadri and C.N.R. Rao, *Rhys. Rev.*, **B47** (1993) 10963.
17. P.H.M. van Loosdrecht *et al.*, *Phys. Rev. Lett.*, **68** (1992) 1176; *ibid*, **69** (1992) 1147.
18. G.A. Samara *et al*, *Science*, **225** (1992) 1235.
19. S.K. Ramasesha, A.K. Singh, R. Seshadri, A.K. Sood and C.N.R. Rao, *Chem. Phys. Lett.*, in press.
20. N. Chandrabhas, M.N. Shashikala, D.V.S. Muthu, A.K. Sood and C.N.R. Rao, *Chem. Phys. Lett.*, **197** (1992) 319.
21. E. Grivei *et al.*, *Phys. Rev.*, **B47** (1993) 1705.
22. A.K. Sood, N. Chandrabhas, D.V.S. Muthu, A. Jayaraman, N. Kumar, H.R. Krishnamurthy, T. Pradeep and C.N.R. Rao, *Solid State Commun.*, **81** (1992) 89.
23. R. Seshadri, C.N.R. Rao, H. Pal, T. Mukherjee and J.P. Mittal, *Chem. Phys. Lett.*, **205** (1993) 395.
24. P.M. Allemand *et al.*, *Science*, **253** (1991) 301.
25. P.W. Stephens *et al.*, *Nature*, **355** (1992) 331.
26. K. Tanaka *et al.*, *Phys. Lett.*, **A164** (1992) 221.
27. R. Seshadri, A. Rastogi, S.V. Bhat, S. Ramasesha and C.N.R. Rao, *Solid State Commun.*, **85** (1993) 971.
28. D.V.S. Muthu, M.N. Shahsikala, A.K. Sood, R. Seshadri and C.N.R. Rao, *Chem. Phys. Lett.*, to be published.
29. M. Takahashi and M. Yamada, *J. Phys. Soc. Japan*, **54** (1985) 2808.
30. F. Wudl and J.D. Thomson, *J. Phys. Chem. Solids*, **53** (1992) 1449.

Main Group Elements and Their Compounds
V.G. Kumar Das (Ed)
Copyright © 1996 Narosa Publishing House, New Delhi, India

Adsorption Characteristics of Activated Carbon

P.M. Afenya

University of Technology, Lae, Papua New Guinea

Activated carbon has been used in a variety ways in industry for many years. Among the earliest applications are the removal of colour from sugar, odour and taste from water and objectional impurities from industrial wastes. In recent years activated carbon has attained new prominence as an important material for gold and silver adsorption from cyanide leach solutions. This new use has attracted many researchers to investigate the mechanism of aurocyanide adsorption onto activated carbon.

Activated carbon is similar in some respects to graphite. Whereas graphite has a hexagonal structure with perfect basal planes, activated carbon has mainly a distorted graphitic structure and a small proportion of microcrystallites composed of roughly parallel layers of hexagonally ordered atoms. It is characterised by well developed internal porosity and large specific surface area ranging from 300 to 2500 m²/g and has been widely employed as a gas-phase and liquid-phase adsorbent for the removal of impurities from gases and liquids.

A typical gas-phase activated carbon has a surface area of 1000 - 2000 m²/g and is made in larger particle sizes of greater strength and density than liguid-phase carbons. The small pores ($\leq$ 3nm) of gas-phase carbons provide high adsorptive capacity and selectivity for gases and vapours. To be a good gas adsorbent it is desirable that the activated carbon should have high adsorptive capacity per unit volume, high retentive capacity, high preferential adsorption of gases in the presence of moisture, high strength of breakage resistance and complete release of adsorbates at increased temperature and decreased pressure.

This paper reviews the chemical and physical characteristics of activated carbon and its mode of adsorption of organic and inorganic species from solutions. Recent use of X-ray photoelectron spectroscopy (XPS) for investigating the nature of the adsorbed aurocyanide species on activated carbon has been discussed. The results of XPS investigations have provided further insight into the mechanism of aurocyanide adsorption by activated carbon.

Application of gas-phase activated carbon

The largest single application for gas-phase carbon is in gasoline vapour emission control canisters on automobiles. The vapours emitted from both fuel tank and carburetor are adsorbed by activated carbon. It is also used in the purification and separation of many industrial gas streams. Odours are removed from air in air-conditioners by activated carbon and individuals are protected from toxic gases or vapours by gas masks containing activated carbon. Activated carbon is used also for removal of objectionable odours from process exhaust gases, removal of corrosive gases and vapours, for instance sulphur compounds from intake air to protect electrical switching equipment. It is also employed in the nuclear reactor systems to adsorb radioactive gases in carriers or coolant gases from air in reactor emergency exhaust systems. Its application is also found in industry for the recovery of volatile organic compounds from process air streams, in refrigerator deodorizers and vacuum equipment for removal of harmful contaminant vapours and in cigarette filters to adsorb some of the harmful components of tobacco smoke.

Activated carbon is also employed in the catalytic oxidation of many substances, for instance sodium arsenite, potassium nitrite, potassium ferrocyanide and ferrous sulphate in solution and for oxidizing H_2S in air to sulphur. It is used as a catalyst in the manufacture of phosgene from CO and Cl_2, and in the production of sulfuryl chloride by the reaction of SO_2 with Cl_2.

Liquid-phase activated carbon

Most of the activated carbon for liquid-phase application is produced in powdered form. The two principal uses of powdered activated carbon are for the (i) removal of colour, odour, taste or objectionable impurities such as substances causing foaming from solution or retarding recrystallization and (ii) concentration or recovery of a solute from solution.

In the first group, activated carbon is used widely to remove colour from sugar and glucose; to remove protein and hydroxymethyl furfural from syrup and render it colourless and stable. Although there are many chemical, physical and biological techniques of treating water contaminated with industrial and municipal wastes to produce safe and palatable drinking water none has the potential comparable to activated carbon. Activated carbon is used for removal of toxic or biorefractive substances from water. Herbicides, insecticides, chlorinated hydrocarbons and phenol, often present in many water supplies can be reduced to acceptable levels by activated carbons.

In the dry cleaning industry powdered and granular carbons are used for reclaiming liquid solvent that has been contaminated by dyes and rancid, oils and grease. The adsorption process is more effective than distillation in removing the solvent. In the food industry it is used in removing colour during the processing of fruit juices, honey, maple syrup, candy, soft drinks and alcoholic beverages. Its pharmaceutical uses include removal of pyrogens from solutions for injection, vitamin decolorising and deodouring and insulin purification.

In the second group, activated carbon is used for solution concentration and purification. Its use for gold and silver cyanide solution concentration and purification in the mineral industry is one of the recent innovations in the processing of precious metals.[1,2] One of the achievements of this process is that activated carbon so used can be reactivated and its adsorptive capacity restored and the carbon used many times.

Manufacture of activated carbon

Activated carbon has been made from materials of plant origin (hardwood, softwood, corncobs, rice hulls, nutshells, coffee beans etc.), and also from peat, lignite, soft and hard coals, tar, asphalt, petroleum residue and carbon black. However, for economic reasons, lignite, peat, coal, bones, wood, and paper mill waste are often used for the manufacture of liquid-phase or decolorizing carbon, and coconut shells, coal and petroleum residues are used for the manufacture of gas-phase and gold recovery carbons.[3]

Carbon activation results in the removal of hydrogen or hydrogen-rich fraction from the carbonaceous raw material to produce an open and porous product having large speficic surface area. The reaction is carried out in two stages.

In the first stage, the raw material is carbonized under controlled conditions between 400 °C and 500 °C to remove such volatile materials like CO, CO_2, acetic acid etc. from the feed. Tar-like residues may remain on the carbon. The carbonized carbon produced has micropores and increased surface area.

In the second stage, the carbonized carbon is heated to 700-1000 °C in an oxidizing atmosphere of steam, CO_2 and/or air to burn off the tar-like residues and develop internal pore structures. A distorted and broken graphitic structure is finally produced. The steam reaction with the carbonized carbon is exothermic and the chemical reaction that occurs in the process may be represented as,[4]

$$]C + H_2O \rightarrow]C(H_2O) (+ CO) \rightarrow]C(O) + H_2$$
$$\overset{\|}{CO}$$

where] is carbon surface.

Physical and chemical properties of activated carbon

Physical properties

The structure of activated carbon, as stated earlier, is to some extent similar to that of graphite. An important feature of activated carbon is its large surface area which may vary from 600 to 1500 m^2/g. The internal surface is large in contrast to the outer surface area which is relatively low. Activated carbon has a variety of pore sizes and shapes which are difficult to determine. Dubinin's generally accepted pore size classification[5] for activated carbon is shown in Table 1.

The type of pores in an activated carbon influences the adsorption kinetics of the adsorbates. For instance, coal-based activated carbon has greater number of mesopores than those of coconut shell carbon. It should therefore follow that coal-based carbon should have a greater rate of adsorption of aurocyanide than coconut shell carbon. However, in industry coconut shell carbon is preferred to coal-based activated carbon on the basis of the former's greater abrasive resistance.

Table 1 : Pore size classification for activated carbon[5]

Pore	size (nm)
Macropores	> 100 - 200
Mesopores	1.6 - 100 or 200
Micropores	< 1.6
Supermicropores	0.6 - 1.6

Chemical properties

The source of raw material, the presence of inorganic matter in the raw material, the conditions of carbonization and activation are responsible for the physical and chemical properties of the final carbon produced. The chemical characteristics of activated carbon although not fully understood are attributed to the following:[4]

1. The microcrystalline structure similar to graphitic structure and distorted and cross linked graphitic structure.

2. The presence of oxygen, hydrogen and inorganic matter in the carbon due to the source material. Oxygen, OH group and inorganics bonded to the carbon affect the type and degree of adsorption of adsorbates.

Activated carbons can either be H-carbons or L-carbons. H-carbons are activated at temperatures between 700 °C and 1000 °C. These carbons adsorb H^+ ions when immersed in water and increase the pH of water. L-carbons are activated normally between 300 °C and 400 °C and adsorb OH⁻ ions. Activated carbon used for gold adsorption are steam activated and belong to the H-carbon group.

Among the elements, oxygen combines most readily with activated carbon to form various functional groups at the edge surface of the structure. Functional groups proposed to be present on activated carbon include carboxyl, phenol, quinone, hydroxyl, ester, lactones, carboxylic anhydride and cyclic peroxides. Figure 1 shows some of these functional groups.

Activated carbon can also undergo catalytic oxidation to form anodic and cathodic sites most probably at the graphitic edges where electrostatic adsorption may occur. The basal planes also play an important role in the adsorption of neutral and nonpolar species.

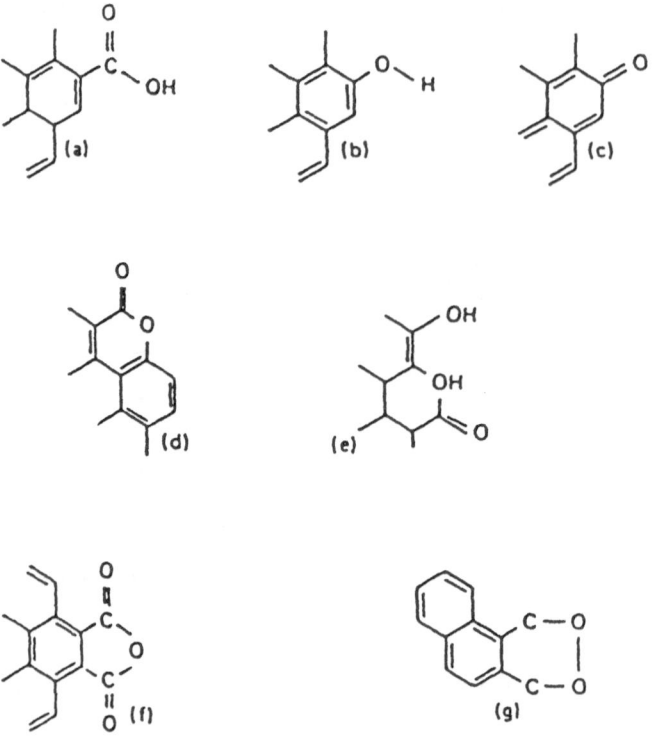

Fig. 1 : **Some of the functional groups proposed as being present on the edges of graphitic structure of activated carbon: (a) carboxyl groups, (b) phenolic hydroxyl groups, (c) quinone-type carbonyl groups, (d) normal lactones, (e) fluorescein-type lactones, (f) carboxylic acid anhydrides, and (g) cyclic peroxides[6]**

Adsorption characteristics

Adsorption of organics

Adsorption of neutral molecules whether in the gas-phase or liquid-phase is achieved by association with the basal planes of carbon through van der Waals forces of attraction. An example of such mode of adsorption is the abstraction of gasoline vapour emitted by vehicles by activated carbon. Neutral organic molecules are similarly adsorbed from liquid-phase onto the basal planes of carbon.

If the adsorbate is heteropolar (consisting of polar and nonpolar groups) the adsorption may be directed to both the basal planes and edge surfaces of the carbon with the rate of adsorption onto one of the two surfaces being greater. Where the organic adsorbate has polar and nonpolar groups as in the fatty acids and alcohols, the preferred adsorption sites are the edge surfaces containing functional groups. This may be followed by adsorption on the basal planes if the carbon chain of the nonpolar group is long. Organic species known to adsorb readily at the edge surfaces of carbon have -COOH, -CO, -OH, -COOC- functional groups.

Adsorption of inorganic compounds

Gold and other metals form a large number of complexes with ligands such as thiourea, thiocyanide, cyanide, chloride, iodide, bromide, sulphide and thiosulphate. Most of the published work on the interaction between activated carbon and metal complexes are reported for metal halides, cyanides, thiocyanates and thioureates.[4] Consequently this paper will focus on the adsorption of metal halides, metal cyanides and thioureates with special emphasis on gold complexes.

Adsorption of gold complexes

Although leaching of gold by the halogens, thiourea or thiosulphate is not commercially important currently, it is probable that treatment of gold by these reagents will attain importance in the future and therefore the abstraction of their gold complexes by activated carbon will be considered here. The capacity of these gold complexes for activated carbon is reported to be of the following order :[6,7]

$$Au\ halide > Au(CN)_2^- > Au(SCN)_2^- > Au(CS(NH_2))_2^+ > Au(S_2O_3)_2^{3-}$$

Besides gold complexes Ag, Hg, Fe, Cu, Zn, Ni, etc. complexes of these ligands also adsorb onto activated carbon at different rates and capacities. Most of the adsorption studies of the metal complexes have been carried out on gold halides, gold and silver cyanides and gold thiourea. Their adsorption mechanisms, rates and capacities under various conditions have been investigated.[8-10]

Adsorption of gold halide complexes

. Chloride leaching of gold was practised at the beginning of this century before the advent of cyanidation. Since then it has been replaced by cyanidation. In recent years, however, there has been renewed interest in chloride, bromide and iodide leaching of gold. The adsorption of gold halide onto activated carbon is accompanied by reduction of the halide to metallic gold. The reduction reactions of gold chloride at the cathodic sites of carbon are as follows:

$$AuCl_4^- + 3e \rightarrow Au + 4Cl^- \tag{1}$$

$$AuCl_2^- + e \rightarrow Au + 2Cl^- \tag{2}$$

$$AuCl_4^- + 2e \rightarrow AuCl_2^- + 2Cl^- \tag{3}$$

where the electrons for reduction are supplied by the activated carbon. At the anodic sites the reaction is as follows:

$$C + 2H_2O \rightarrow 4H^+ + CO_2 + 4e \qquad (4)$$

The dissolution reactions of gold with bromide in dibromo dimethyl hydantoin are expressed as:[11]

$$Br_2(DMH) + 2H_2O \rightarrow 2HOBr + H_2(DMH) \qquad (5)$$

$$2Au + 3HOBr + 3NaBr \rightarrow 2AuBr_3 + 2NaOH \qquad (6)$$

$$AuBr_3 + NaBr \rightarrow Na^+(AuBr_4)^- \qquad (7)$$

The reduction reaction is expressed as:

$$4(AuBr_4)^- + 3C + 6H_2O \rightarrow 4Au + 3CO_2 + 12H^+ + 16Br^- \qquad (8)$$

The basic thermodynamic data supporting the reduction mechanism of gold halides to metallic gold on activated carbon are shown in Table 2.

Table 2 : Reduction potentials of some gold complexes and carbon[10]

Species	$E°(V)$ Reduction Au^+/Au	Potential Au^{3+}/Au
H_2O	+1.73	+1.50
Cl^-	+1.15	+1.00
Br^-	+0.96	+0.86
I^-	+0.56	+0.57
SCN^-	+0.67	+0.64
NH_3	+0.56	+0.33
$S_2O_3^{2-}$	+0.20	+0.10
$(NH_2)_2C:S$	+0.40	+0.30
CN^-	-0.61	-0.50
carbon	-0.14	

Since the reduction potentials for gold halides are much more positive than that for carbon, it is most probable that carbon would reduce gold halides to metallic gold from their respective solutions. This was confirmed by various investigators.[10,11] On the other hand $Au(CN)_2^-$ having a reduction potential of -0.61V would most unlikely be reduced to metallic gold by activated carbon. The conclusion reached by many investigators is that metallic gold is not formed when aurocyanide is adsorbed by activated carbon.

Adsorption of gold thiourea

There are a number of publications on the leaching of gold by thiourea as an alternative to cyanidation[12] but not much has been published on gold thiourea adsorption by activated carbon. The reduction potential (E°) of gold thiourea is 0.40V indicating that $Au[CS(NH_2)_2]_2^+$ would be reduced to metallic gold when adsorbed on carbon. However, the work of Juarez and Oliveira [13] indicated that gold thiourea interacts with the anionic functional groups present on carbon surface. They found that increase of temperature from 15 to 55 °C increased gold adsorption. They did not indicate the nature of the adsorbed species. The adsorption of thiourea present in the solution, on the contrary, decreased as the temperature increased and thus suggesting that thiourea is physically adsorbed onto the graphitic planes.

Adsorption of gold and silver cyanide from solution

Efforts to understand the adsorption of gold cyanide onto activated carbon began early this century.[14,15] Since the publication of Green's work[14] in 1913 reporting the precipitation of metallic gold from cyanide solution onto charcoal, no other investigator has confirmed his finding. All subsequent workers[15-18] of his time on, the contrary, believed that $Au(CN)_2^-$, $M^{n+}[Au(CN)_2^-]_n$ or $AuCN$ was the species adsorbed onto carbon but not metallic gold. In 1927, Gross and Scott[18] proposed the mechanism of adsorption of $M^{n+}[Au(CN)_2^-]_n$ on carbon. At the time of publication of Gross and Scott's work the prevailing view was that activated carbon had positive and negative charged sites and that gold cyanide adsorption occurred at the positive sites. The cation for $M^{n+}[Au(CN)_2^-]_n$ was believed to be a natural component of the carbon. This was later disproved.

Further work on gold cyanide adsorption by activated carbon from the 1950s led to new theories. Garten and Weiss[19] proposed the formation of carbonium ion sites on carbon by oxidation of chromene groups existing on carbon surface to chromenol groups by oxygen as follows:

$$(9)$$

$$(10)$$

But Garten and Weiss[19] did not explain why gold cyanide adsorption is increased when the pH of the solution decreased.

Further work that followed included the study of the effects of various cations and anions species in cyanide solutions on gold cyanide adsorption, the mechanism and kinetics of adsorption on and desorption of gold cyanide from carbon.[6,8,9,20] During this period it was observed that Cl^-, I^- and ClO_4^- have no depressing effect on aurocyanide adsorption. Kuzminykh and Tjurin[21] argued that since Cl^- or I^- ions have no effect on aurocyanide adsorption but kerosene and octyl alcohol depress gold cyanide adsorption, the mechanism of adsorption was by neutral gold cyanide uptake. Under acidic conditions the neutral adsorbed species was $HAu(CN)_2$ but in neutral and alkaline solutions the species was $M^{n+}[Au(CN)_2^-]_n$. The neutral species is believed to be held to the surface by van der Waals type of interaction. (This type of interaction is expected to occur betweem the graphitic planes of carbon and neutral molecules).

In the work of Davidson[8] on the factors that influence the adsorption of $Au(CN)_2^-$ on activated coconut shell carbon, he found that reproducible results could only be obtained in a borate buffer solution. In that publication Davidson suggested that the mechanism of gold cyanide adsorption is by neutral $M^{n+}[Au(CN)_2^-]_n$ abstraction from solution onto carbon. But his main contribution was in recognizing that the alkaline metal ions, Ca^{2+} or Mg^{2+} ions enhance gold cyanide adsorption more than Na^+ or K^+ ions and he presented the order of adscrption enhancement as follows:

$$Ca^{2+} > Mg^{2+} > H^+ > Li^+ > Na^+ > K^+$$

The practical implication of his finding is that $CaCO_3$ which is often used for pH control enhances gold cyanide adsorption, while Na_2CO_3 can be used for desorpting gold cyanide from carbon surface by an ion exchange mechanism presented as,

$$Ca[Au(CN)_2]_2 + K_2CO_3 = 2K[Au(CN)_2] + CaCO_3$$

In practice a solution of NaOH and NaCN is used to elude gold cyanide from activated carbon whereas lime used to control solution pH enhances gold cyanide adsorption. Davidson suggested that $Ca^{2+}[Au(CN)_2^-]_2$ ion pair is more firmly bound to carbon than $Na^+[Au(CN)_2^-]$ ion pair.[8] Davidson[8] and McDougall et al.[24] observed that the amount of gold adsorbed on carbon at pH 4-7 is about twice that adsorbed between pH 8 and 11.

Other factors which influence gold adsorption are the presence of molecular oxygen and temperature. Oxygen in solution increases gold adsorption whilst the presence of nitrogen does not.[22] Dixon et al.[22] also showed that gold cyanide adsorption is inversely proportional to temperature.

Clauss and Weiss[23] observed that hydrogen peroxide enhances the activity of carbon, whereas reductants such as hydrazine and hydroquinone inhibit gold adsorption. They therefore suggested that the adsorption sites were probably quinone-type groups and discounted that the basal planes, carboxylic acid groups and basic oxides of carbon are involved in gold cyanide adsorption.

In the kinetic studies of the adsorption of gold and silver cyanides on activated coconut shell carbon and the factors influencing their adsorption, Cho et al.[20] found that the adsorption of silver cyanide (by analogy gold cyanide) was reversible and stated that the reaction was more of physisorption than chemisorption. The weakly hydrated silver cyanide was assumed to be specifically adsorbed onto the carbon surface whilst the strongly hydrated CN^- ions and Na^+ or Ca^{2+} were adsorbed into the electrical double layer. The Ca^{2+} cations on the carbon were said to be sites for further $Au(CN)_2^-$ adsorption and thus a multilayer adsorption of $Au(CN)_2^-$ was proposed. This mechanism is in agreement with the ionic solvation-energy theory developed by Anderson and Bockris.[24] This theory allows weakly hydrated large ionic species to be specifically adsorbed on surfaces in preference to strongly hydrated small ionic species. They presented the decreasing order of strength of the adsorption of gold and silver cyanide complexes and CN- ions onto carbon surface as,

$$Au(CN)_2^- > Ag(CN)_2^- > CN^-$$

in decreasing size of anions. This mechanism is unable to explain why large Cl^- or ClO_4^- ions, for instance, are not specifically adsorbed onto carbon.

McDougall et al.[6] also pointed out that the adsorption of gold cyanide by carbon could not be attributed to simple electrostatic interaction with the positive sites on carbon as proposed by many other workers.[19,20,21] By using x-ray photoelectron spectroscopy(XPS) to study gold cyanide loaded on carbons, they concluded that the oxidation state of the adsorbed gold was neither 1 as present in $Au(CN)_2^-$ or $AuCN$ nor 0, when metallic gold is present, but an intermediate value of 0.3. They consequently proposed that gold cyanide adsorption occurred in two steps. The first step was the adsorption of $M^{n+}[Au(CN)_2^-]_n$ onto carbon and the second step was the reduction of the neutral gold cyanide to unidentified species. They also observed that neutral $Hg(CN)_2$ in solution directly competes with gold cyanide for surface sites on carbon and that ionic or acidic strengths have no effect on $Hg(CN)_2$ adsorption.

Recently Adams et al.[25] and McDougall et al.[26] provided further indirect support for the adsorption of ion pair, $M^{n+}[Au(CN)_2^-]_n$ onto activated carbon. Adams[27] emphasized an ion pair adsorption mechanism in high ionic strength solutions similar to those found in gold processing plants. He showed that under low ionic strength conditions, however, gold cyanide is adsorbed by electrostatic interaction with the positive sites formed by oxidation of the carbon surface by molecular oxygen. This implies that the adsorption mechanism is dependent on cyanide solution conditions.

Klauber[28] and Jones et al.[29], by applying XPS to study gold cyanide adsorbed onto carbon concluded that $Au(CN)_2^-$ is adsorbed intact on the basal planes of carbon. They stated that bonding between $Au(CN)_2^-$ and carbon is accomplished through the interaction of π-electron of carbon and the central gold atom of $Au(CN)_2^-$. They indicated that similar bonding mechanism applies to the adsorption of tetramer anion, $Au_4(CN)_5^-$ formed by acid induced oligomerization onto carbon (Fig. 2).

The recent work of Ibrado and Fuerstenau[30] demonstrated that the adsorption of $KAu(CN)_2$ has strong correlation with the degree of graphitization of the carbon (Fig. 3). They therefore suggested that the adsorption of most of the gold cyanide is related to the graphitic nature of activated carbon.

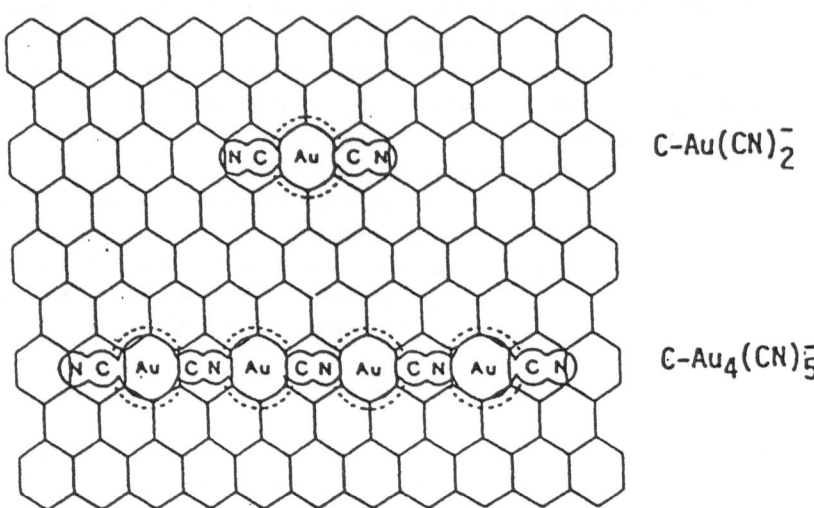

$C-Au(CN)_2^-$

$C-Au_4(CN)_5^-$

Fig. 2 : A model of the adsorption of gold cyanide on the graphitic plane of activated carbon, as $Au(CN)_2^-$ at basic pH values and as $Au_4(CN)_5^-$ at acidic pH values.[29]

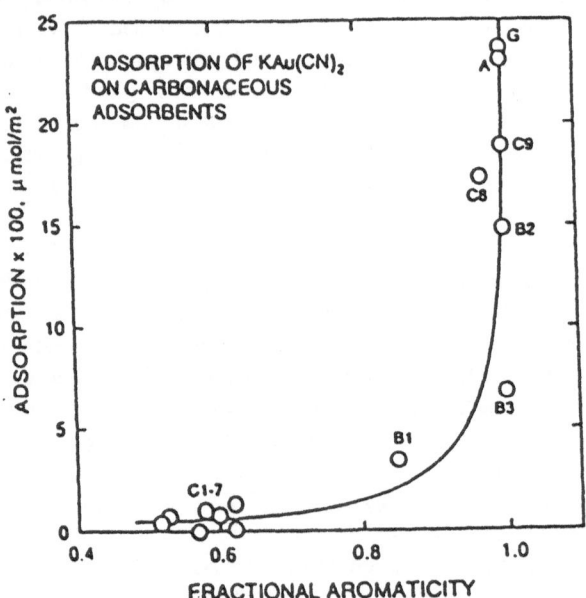

Fig. 3 : Adsorption of gold cyanide as a function of the fractional aromaticity of carbonaceous adsorbents.[29] A = coconut shell activated carbon, G = graphite, B1 = bone charcoal, B2, B3 = carbon blacks, C1 = subbutiminous coal, C7 = lignite.

Sibrell and Miller[31] using radiochemical, autoradiographic and XPS techniques showed that the adsorption density of gold cyanide is much higher at the edges than the basal plane of synthetic, highly-oriented pyrolytic graphite. Their results imply that site-specific adsorption is prevalent in adsorption of gold cyanide by graphitic carbons including activated carbon. In this respect the positive sites at the edge surfaces may most probably be the adsorption sites.

Adams[32] also showed a strong correlation between temperature of carbon activation and adsorption of aurocyanide and phenol. For the activated carbon used, those which showed no graphitic peaks had the lowest adsorption capacity for aurocyanide and phenol. His work also showed that the adsorption capacity for aurocyanide or phenol is not necessarily dependent on total surface area of the carbon.

The stage is now reached where many researchers have realised that the graphitic nature and the pores of activated carbon produced during activation are the dominant factors responsible for aurocyanide adsorption.

Kinetics of gold cyanide adsorption onto activated carbon

In the kinetic studies of gold cyanide adsorption onto activated carbon it was observed that the rate of adsorption was pore diffusion controlled .[20] Nicol et al.[33] proposed an adsorption rate controlled by mass transport across homogeneous boundary at carbon-solution interface. Recently Ahmed et al.[34] modelled the kinetics of the adsorption on homogeneous solid diffusion model (HSDM) which incorporates film mass transfer and intraparticle diffusion. Nicol et al.[33] also modelled the adsorption kinetics for steady state system, non-steady and batch systems. For the steady state system he proposed a kinetic model as follows:

$$C_o - C_{c,o} = kC_s t^n$$

where C_o = concentration of gold on carbon at time t,
$C_{c,o}$ = concentration of gold on carbon at time, $t = 0$
C_s = concentration of gold in solution at steady state,
k = rate constant, n = constant, t = time.

The adsorption isotherms at temperatures between 25 °C and 55 °C for gold cyanide concentration from 90 to 180 ppm approximate Freundlich equation which is expressed as,[20]

$$Q = bC^n$$

where Q = equilibrium quantity of gold adsorbed onto carbon from cyanide solution (μmol/l),
C = equilibrium gold concentration in solution in contact with carbon (μmol/l)
b and n are constants.

Factors affecting gold adsorption

Among the factors known to affect the adsorption of gold cyanide by activated carbon are temperature, gold concentration in solution, cyanide concentration, pH of solution, ionic strength of other species, concentration of other metals and dissolved oxygen.

Increase of temperature, cyanide concentration or solution pH decreases gold adsorption onto carbon. Simple cations in cyanide solution increase gold adsorption capacity in the order:

$$Ca^{2+} > Mg^{2+} > H^+ > Li^+ > Na^+ > K^+$$

and the anions decrease gold loading capacity in the order:

$$CN^- > S^{2-} > SCN^- > S_2O_3^{2-} > OH^- > Cl^- > NO_3^-$$

Adsorption of other metal cyanides

Cyanidation of gold under industrial conditions results in the dissolution of other metals to form metal cyanide complexes. Some of the metals often encountered in gold cyanidation are Ag, Cu, Ni, Zn, Fe, Pb and Hg. The cyanide complexes of these metals adsorb on carbon in competition with gold cyanide. Fortunately, however, activated carbon is most selective to gold cyanide with the exception of mercury cyanide. The general order of preference of adsorption for the metal complexes is as follows:[4]

$$Au(CN)_2^- > Hg(CN)_2 > Ag(CN)_2^- > Cu(CN)_3^{2-} > Zn(CN)_4^{2-}$$
$$> Ni(CN)_4^{2-} > Fe(CN)_6^{4-}$$

The mechanism of $Ag(CN)_2^-$ adsorption by carbon is similar to that of $Au(CN)_2^-$. Mercury cyanide complex directly competes with gold cyanide for surface sites on carbon. The adsorption of copper cyanide complexes is pH dependent and is of the order (from low to high pH):

$$Cu(CN)_2^- > Cu(CN)_3^{2-} > Cu(CN)_4^{3-}$$

In the absence of precious metals in cyanide solutions, activated carbon can be used to abstract nonprecious metal cyanide complexes.

Concluding remarks

The adsorption property of activated carbon is dependent on its structure. Unlike graphite which has a well developed hexagonal crystal system, activated carbon is a combination of well ordered elementary crystallites as in graphite and distorted, cross-linked hexagonal structures. Activated carbon has very large internal surface area because of the numerous pores produced during activation whilst graphite has low surface area.

The adoption of graphitic model for activated carbon requires that the adsorption of species onto activated carbon be considered in terms of the graphitic planes, edge surfaces having functional groups and electrochemically produced anodic and cathodic sites. Gases and organic species having polar and non-polar groups would adsorb onto either the basal planes or edge surfaces or both. In the case of cationic and anionic inorganic species and complexes it is expected that they adsorb on the edge surfaces having functional groups and the cathodic or anodic sites. Neutral inorganic molecules would adsorb on the basal planes through van der Waals forces of attraction. Some other adsorbates may adsorb on the graphitic planes by interacting with the π-electrons donated by the carbon.

It is relatively easy to rationalise the mode of adsorption of organic species and neutral molecules on activated carbon. For instance, organic species may adsorb onto the basal planes and the edge surfaces or both. With regard to the adsorption of gold halides the uptake is followed by reduction to metallic gold. The mechanism is assumed to be electrochemical reduction of gold halides to metallic gold. The reduction is likely to be most active at the edges of the graphitic structure.

Hitherto, the mechanism of gold cyanide adsorption by activated carbon is not fully understood. One of the obstacles to the consideration of degree of graphitization of activated carbon as vital to gold cyanide adsorption is the finding by the earlier researchers that graphite did not adsorb gold cyanide. The consequence was the rejection for a long time of the fact that gold cyanide adsorbs also on graphite. With respect to activated carbon, one of the mechanisms proposed by researchers is the adsorption of gold cyanide onto the cationic (carboniom ion) sites on carbon where $Au(CN)_2^-$ anions exchange with OH^- ions of the chromenol groups on activated carbon. This mechanism appeared to provide a satisfactory mode of adsorption of gold cyanide on carbon. In support of this mechanism Adams (1990) showed that in solutions of low ionic strength a limited amount of gold cyanide is abstracted by ion-exchange mechanism. Furthermore, since activated carbon has electrochemical property and is known to reduce gold chloride to metallic gold, it had been claimed to reduce gold cyanide in alkaline medium to metallic gold or AuCN. None of these products (metallic gold and AuCN) has been specifically identified on activated carbon in high alkaline solutions. The work of Sibrell and Miller[31] with synthetic graphite which showed that the adsorption density of gold cyanide at the edges is much higher than that on the basal plane is a strong evidence in favour of adsorption occurring at the edge sites of carbon.

The adsorption of $Au(CN)_2$ anions as such onto the graphitic planes of carbon has been suggested by Klauber.[28] He stated that $Au(CN)_2^-$ from $KAu(CH)_2$ solution adsorbs intact onto the basal plane of carbon through the interaction of the π-electrons donated by carbon with the central gold atom (unoccupied $Au\ 6p_{x,y}$ or $6s\ 6p$ hybrid orbital) of $Au(CN)_2^-$. The cation needed to neutralise the charge for $Au(CN)_2^-$ is said to be adsorbed on the carbon. If, however, a neutral molecule such as $M^{n+}[Au(CN)_2]_n$ is the adsorbate as suggested by many investigators then adsorption onto the graphitic plane of carbon appears most probable. In that case the direct competition of neutral

Hg(CN)$_2$ and phenol with gold cyanide provides a strong evidence supporting the basal planes being the adsorption sites for neutral KAu(CN)$_2$. Although Adams *et al.*[25] apparently provided evidence for the existence of neutral KAu(CN)$_2$ in solution, determination of the presence of this species on the graphitic plane by XPS has not been possible.

In conclusion the results of recent investigators strongly indicate that gold cyanide adsorbs on both the graphitic planes and the edge surfaces of activated carbon under industrial processing conditions of high ionic strength and pH. The ion pair, $M^{m+}[(Au(CN)_2^-]_n$ and $Au(CN)_2^-$ are most probably the adsorbed species. It appears that an ion pair, $M^{m+}[Au(CN)_2^-]_n$ is initially adsorbed on the basal plane and finally converted to $Au(CN)_2^-$. At the edge sites, however, the preferred adsorbed species is $Au(CN)_2^-$.

Acknowledgement

The assistance given by colleagues in the Department of Mining Engineering, Papua New Guinea Univesity of Technology in the preparation of this paper is gratefully acknowledged.

References

1. J.B. Zadra, A Process for the recovery of gold from activated carbon by leaching and electrolysis, U.S. Bureau of Mines Washington, D.C., R.I. No. 4672, 1950.
2. J.B. Zadra, A.K. Engel and H.J Heinen, A process for recovering gold and silver from activated carbon by leaching and electrolysis. U.S.B.M., Washington, D.C., R.I. No. 4843, 1952.
3. M. Šmisek and S. Cerny, Active Carbon, Elsevier, New York, 1970.
4. J. Marsden and I. House, The Chemistry of Gold Extraction. Ellis Horwood Ltd., Chichester, England, (1992), p.597.
5. M.M. Dubinin, Porous structure and adsorption properties of active carbons. Chemistry and Physics of Carbon, Vol.2, Marcel Dekker, 1966.
6. G.J. McDougall, *J. S. Afr. Inst. Min. Metall.*, September (1980) 344.
7. G.J. McDougall and R.D. Hancock, *Gold Bull.*, **14**(4) (1981) 138.
8. R.J. Davidson, *J.S. Afr. Inst. Min. Metall.*, Nov. (1974) 67.
9. C.A. Fleming and M.J. Nicol, 1984. *J. S. Afr. Inst. Min. Metall.*, April **84**(4) (1984).
10. J. Avraamides, G. Hefte and C. Budiselic, *Aus. Inst. Min. Metall.*, 7 (1985) 57.
11. R.K. Mensah-Biney, K.J. Reid and M.T. Hepworth, *Minerals and Metallurgical Processing*, **10**(1) (1993) 13.
12. C.L. Caldeira and V.S.T. Ciminelli, Thiourea leaching of gold ore. *XVIII International Mineral Processing Congress*, Vol.5, Sydney, 23-28 May, (1993), p.1123.
13. C.M. Juarez and J.F. Oliveira, J.F., 1993. Recovery of gold from solutions of thiourea by adsorption on activated carbon. *XVIII International Mineral*

Processing Congress Vol.5, Sydney, 23-28 May, pp.1425-1428.

14. M. Green, *Trans. Inst. Min. Metall.*, **23** (1913-1914) 65.
15. W.R. Feldtmann, *Trans. Inst. Metall.*, **24** (1914-1915) 329.
16. A.W. Allen, *Metall. Chem. Engng.* **18** (1918) 642.
17. L.B. William, *Min. Mag.*London, **23** (1923) 139.
18. J. Gross and J.W. Scott, Precipitation of gold and silver from cyanide solutions on charcoal. U.S. Bureau of Mines, Washington, D.C., Technical Paper No. 378, 1927.
19. V.A. Garten and D.E. Weiss, *Rev. Pure Appl. Chem.*, **7** (1957) 69.
20. E.H. Cho, S.N. Dixon and C.H. Pitt, *Metall. Trans. Bulletin*, **108** (1979) 185.
21. V.M. Kuzminykh and N.G. Tjurin, *Izv. Vyssh. Ucheb. Zaveb. Tsved. Metall.*, **11**(4) (1968) 65.
22. S. Dixon, E.H. Cho and C.H. Pitt, C.H., The interaction between gold cyanide and high surface area charcoal, AIChE Meeting, Chicago, IL., 26 Nov.-2 Dec, 1976.
23. C.R.A. Clauss and K. Weiss, K., Adsorption of aurocyanide on carbon. Council for Scientific and Industrial Research, Pretoria, Report of Investigation No. CENG, 206, September, 1977.
24. T.N. Anderson and J.O'M. Bockris, *Electrochim. Acta*, 9(4) (1918) 347.
25. M.D. Adams, G.J. McDougall and R.D. Hancock, *Hydrometallurgy*, **18** (1987) 139.
26. G.J. McDougall, M.D. Adams and R.D. Hancock, R, *Hydrometallurgy*, **18** (1987) 125.
27. M.D. Adams, *Hydrometallurgy*, **25** (1990) 171.
28. C. Klauber, C., *Surface Science*, **203** (1988) 118.
29. W.D. Jones, C. Klauber and H.G. Linge, Fundamental aspects of gold cyanide adsorption on activated carbon. World Gold '89, R.B. Bhappu and R.J. Harden, eds, SME, Littleton, CO, (1989) pp.278-281.
30. A.S. Ibrado and D.W. Fuertenau, *Hydrometallurgy*, **30** (1992) 243.
31. P.L. Sibrell and J.D. Miller, *Mineral and Metallurgical Processing*, November (1992) 189.
32. M.D. Adams, Influence of the chemistry and structure of activated carbon on the adsorption of aurocyanide. *XVIII International Mineral Processing Congress*, Vol.5, Sydney, 23-28 May (1993), pp.1175-1187.
33. M.J. Nicol, C.A. Fleming and G. Cromberge, *J. S. Afr. Inst. Metall.*, **84**(2) (1984) 50.
34. F.E. Ahmed, B.D. Young and A.W. Bryson, *Hydrometallurgy*, **30**(1-3) (1992) 257.

Main Group Elements and Their Compounds
V.G. Kumar Das (Ed)
Copyright © 1996 Narosa Publishing House, New Delhi, India

Hydrotalcite like Materials: Their possible Role in Environmental Control

C.S. Swamy, S. Kannan and S. Velu

Department of Chemistry, Indian Institute of Technology
Madras - 600 036, India

Clays, one of the most common minerals present in the Earth's crust, are indispensable to our existence. They find applications on a wide front, embracing ceramics, building materials, adsorbents, ion-exchanges, catalysts and decolourising agents.[1-3] Their adsorptive power and high water retention capacity are responsible for their wide applications.

Clays can be broadly classified into two categories, namely, cationic or smectite- type clays having a layered lattice structure in which two dimensional oxy anions are separated by layers of hydrated cations[4], and the anionic or brucite-type clays in which the charge on the layer and the gallery ion is reverse of that in the smectite clays.[5]

Hydrotalcite-like materials (HT) belong to the class of anionic clays (also referred to as layered double hydroxides (LDH)). They consist of brucite-like ($Mg(OH)_2$) networks wherein an isomorphous substitution of Mg^{2+} by a trivalent element M^{3+} occurs and the excess charge is compensated by gallery anions which, along with waters of hydration, occupy the interlayer space.[6,7]

Hydrotalcite, first discovered in Sweden around 1842, is a hydroxy carbonate of Mg and Al and occurs in nature in foliated and contorted plates. E. Manasse was the first to point out the correct molecular formula of HT and other isomorphous materials; he was also the first to recognise that carbonate ions were essential for this type of structure.[8] Aminoff and Broome[9] first recognised the existence of two polytypes of hydrotalcite with rhombohedral and hexagonal symmetry from X-ray investigations. In 1942 Feitknecht synthesised a large number of compounds having HT structure, to which he gave the name "doppelschichtstrukturen" (double sheet structure).[10] The structural features of these hydrotalcites were determined by Taylor and Allmann based on the X-ray analysis of single crystals.[11,12] They concluded that both the cations (M^{2+} and M^{3+}) are localised in the same layer and only carbonate and water are in the interlayer. Miyata and coworkers[13-15] have performed extensive studies on the synthesis and physico-chemical properties of these materials, especially on their anion exchange properties. However, the first documented study of the application of hydrotalcite compounds in catalysis was a 1970 patent which claimed that hydrotalcite material functioned as an effective catalytic precursor for hydrogenation reactions.[16]

The basic property of these materials was demonstrated by Nakatsuka *et al.*[17] in their study of base catalysed polymerisation of alkene oxides. Later in 1985, Reichle[18] utilised this property and carried out aldol condensation on these materials. Recently Laylock and coworkers have carried out the synthesis of polyether polyols and stereospecific polymerisation of propylene oxide over these materials.[19,20] The anion exchange property of these materials was efficiently used by Pinnavaia and coworkers[21-23] in their pioneering work on the intercalation of polyoxometallate which yielded pillared materials of improved catalytic activity. Thermal calcination of these materials results in the formation of non-stoichiometric mixed metal oxides which are used as active catalysts for many catalytic transformations like steam reforming, methanol and higher alcohol synthesis and nitrous oxide decomposition.[24-27] A recent review by Cavani et al.[28] gives an excellent account on the nature of hydrotalcite-like materials.

Structural features

The structures of these HT-like materials[11] are best visualised by starting with the structure of brucite. In brucite, Mg is surrounded by six hydroxyl groups in octahedral co-ordination, and these octahedra share edges to form infinite sheets. The infinite sheets are stacked upon one another to give a layered network held together by hydrogen bonding. If one of the Mg^{2+} is substituted by any trivalent metal like Al^{3+}, the positive charge density of the layer increases. To maintain electrical neutrality the anions occupy the interlayer positions which also accommodate the co-crystallizing water molecules. The brucite-like sheets can stack on each other in two different modes, namely, rhombohedral consisting of three sheets per unit cell ($c = 3c'$), and hexagonal consisting of two sheets per unit cell ($c = 2c'$). Pyroaurite and hydrotalcite crystallise in rhombohedral 3R symmetry, whereas sjogrenite and manessite (their corresponding polymorphs) crystallise with 2H symmetry. A schematic representation of the structure of brucite and hydrotalcite is given in Figure 1. It will be seen that the main features determining the hydrotalcite structure are the brucite-like sheets, the position of anions and water in the interlayer region, and the type of stacking of the brucite-like layers.

General molecular formula

These materials are characterised by the general formula

$$[M(II)_{1-x}M(III)_x(OH)_2)]^{x+} (A_{x/n}{}^{n-}).mH_2O$$

where M(II) and M(III) are various divalent and trivalent metal ions and A is the interlayer anion. The main criteria for inducing the metals to crystallise in this network are their ionic radius and their stereochemistry. Ions such as Be^{2+} are too small and ions such as Cd^{2+} are too large to be incorporated into the network. Table 1 gives the ionic radius of various divalent and trivalent metal ions. The single phase hydrotalcite occurs even after the proper selection of metals only for $0.2 < x < 0.33$. For x values outside this range, either the pure hydroxides or other compounds with different structures have been obtained.[29] However, a in recent report Schaper et al.[30] showed that a pure hydrotalcite with a Mg/Al atomic ratio of 10 ($x \approx 0.1$) could be prepared without any other phase formation. A wide range of compositions present in this class of materials provides the possibility of producing materials with specific properties, especially multicomponent catalysts.[31]

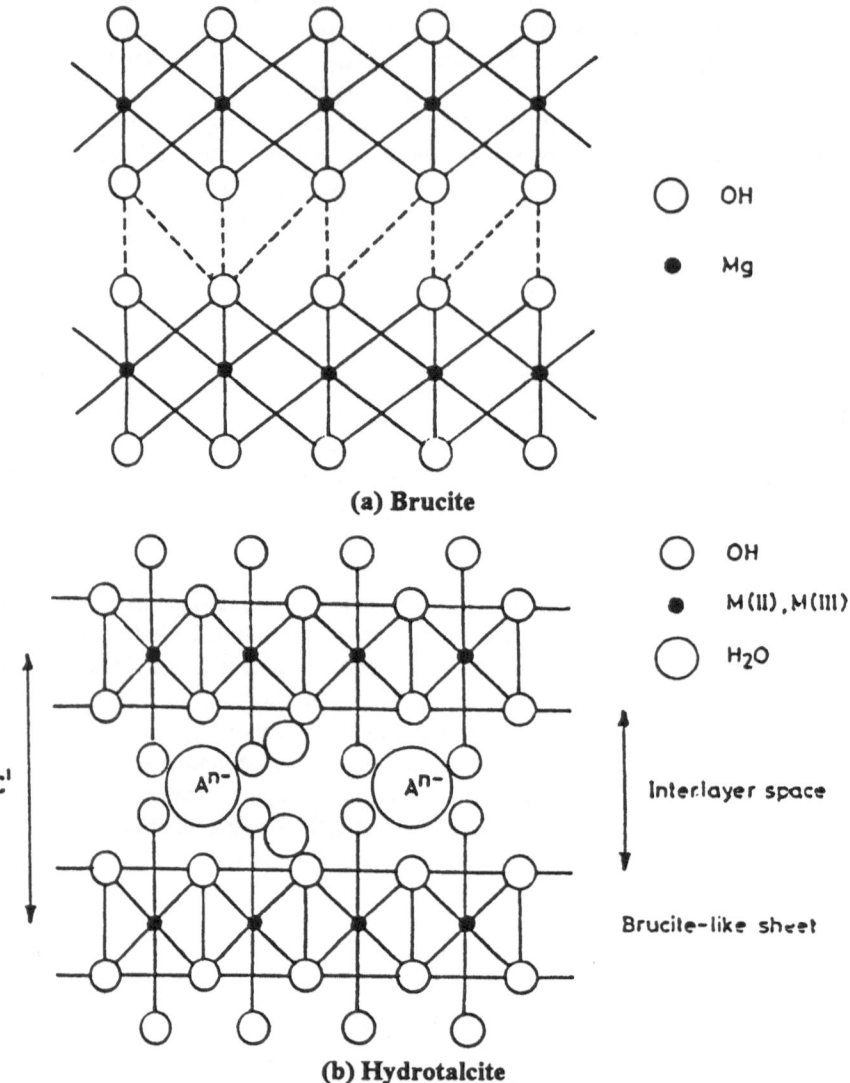

(a) Brucite

○ OH

● Mg

○ OH

● M(II), M(III)

○ H₂O

c¹

Aⁿ⁻   Aⁿ⁻

Interlayer space

Brucite-like sheet

(b) Hydrotalcite

Fig.1 :   Schematic representation of Brucite and Hydrotalcite

There is no limitation in respect of the anions that could be intercalated into the interlayer position of the hydrotalcite network. This is especially useful for the function of these materials. Among anionic types intercalated are the following:

1. Inorganic anions[15] e.g. Cl^-, Br^-, CO_3^{2-}, SO_4^{2-}

2. Heteropoly anions[32] e.g. $[V_{10}O_{28}]^{6-}$, $[Mo_7O_{24}]^{6-}$

3. Complex anions[33] e.g. $[Fe(CN)_6]^{3-}$, $[IrCl_6]^{2-}$

114

4. Organic acids[34] e.g. malonic, succinic and sebacic acids in their divalent ionic forms

5. Metallo macrocyclics[35], e.g. cobalt phthalocyanine

A detailed discussion on the anion exchange properties of these materials is given later.

Table 1 : Ionic radii of some divalent and trivalent cations

M(II)	Å	M(II)	Å	M(III)	Å	M(III)	Å
Be	0.30	Zn	0.74	Al	0.50	Mn	0.66
Mg	0.65	Fe	0.76	Ga	0.62	Cr	0.69
Cu	0.69	Mn	0.87	Ni	0.62	V	0.74
Ni	0.72	Cd	0.97	Co	0.63	Ti	0.76
Co	0.74	Ca	0.98	Fe	0.64	In	0.81

Preparation of Hydrotalcite-like materials

The hydrotalcite-like materials are synthesised generally by the coprecipitation technique[18] which allows homogenous precursors to be used as starting materials, where two or more elements are intimately mixed together and synergistic effects are favoured. The coprecipitation method involves a large number of unit operations, viz.: precipitation, aging and hydrothermal treatment, filtration, washing, drying, grinding and shaping. The preparation of hydrotalcite-like compounds can be carried out by any of the following routes:

a. sequential precipitation wherein the precipitants namely NaOH and/or Na_2CO_3 are added to the metal nitrate solutions with increasing pH

b. precipitation under low supersaturation wherein both the precipitants and the metal nitrates are added slowly and simultaneously holding the pH constant

c. precipitation under high supersaturation where the metal nitrates are added to the precipitants very quickly at constant pH.

Generally, precipitation under low supersaturation conditions yields precipitates with more crystallinity compared to those obtained under high supersaturation. This is because in the latter case the rate of nucleation is larger than the rate of crystal growth, resulting in a greater number of particles of smaller size.

The single phase formation of hydrotalcites and their crystallinity depend on various factors, namely, nature of M(II) and (III) ions, M(II)/M(III) atomic ratio, method of precipitation, temperature and aging, pH under which precipitation is carried out, and hydrothermal treatments.

To M^{2+} and M^{3+} metal nitrates of appropriate concentration, the precipitants NaOH (0.2 mol) and Na$_2$CO$_3$ (0.02 mol) dissolved in 50 ml of distilled water, are added over 30 min. with constant stirring at room temperature. The pH of the solution which increases during the addition was maintained at the final value of around 10. The precipitate formed is aged by heating the solution at 65 °C for 24 h. The aged precipitate is filtered, washed several times with distilled water and dried overnight in an air oven at 70-90 °C. A part of the sample was hydrothermally treated in a teflon coated autoclave at 110 °C for 2 days under autogenous conditions.

Physicochemical characterization

X-ray diffraction

X-ray diffraction (XRD) is one of the most versatile tools for the characterisation of hydrotalcite-like materials. Allmann[12] has performed single crystal diffraction studies on such materials and derived their general structures. As it is difficult to obtain single crystals of all hydrotalcites, powder diffraction studies are generally performed. The general XRD pattern of these materials, illustrated in Fig. 2, shows sharp and symmetric reflections for (003), (006), (110) and (113) planes and broad asymmetric reflections for (102), (105) and (108) planes which are characteristic of clay minerals possessing a layered structure.[13] The lattice parameters 'a' and 'c' are calculated by indexing the peaks under hexagonal crystal system using least square fitting of the peaks, mainly taking the reflections whose 2θ > 45°, the values for some of the hydrotalcites are given in Table 2. From Table 2 it is seen that the unit cell volume is dependent both on the nature of M(II) and M(III) ions and the M(II)/M(III) atomic ratio. In the case of M-Al-HT (M = Ni and Co), an increase in the M^{2+}/Al^{3+} atomic ratio results in increases in both the lattice parameters 'a' and 'c' with consequent increase in the unit cell volume. The increase in 'a' can be attributed to higher ionic radius of M^{2+} (Ni^{2+} = 0.72Å and Co^{2+} = 0.74Å) in comparison with Al^{3+} (0.50Å),[36] whereas the increase in the 'c' parameter is explicable in terms of electrostatic interaction between the layer and the interlayer. Increase in the Al^{3+} content increases the positive charge density of the brucite sheet, which in turn increases the electrostatic interaction between the layer and the interlayer, thereby resulting in a reduction in the thickness of the interlayer. Fig. 3 shows the XRD patterns of aged and hydrothermally treated Ni-Al-HT, indicating that the hydrothermally treated samples are well crystallised. This result could be corroborated with the particle size measurements determined by X-ray line broadening. Table 3 shows that the particle sizes of hydrothermally treated samples are larger than the aged samples. The crystallinity of the materials formed also depends on the method of precipitation. We have observed[37] that precipitation under low supersaturation yields better crystalline material in comparison with other methods. The results are evidenced from the XRD pattern of the samples given in Fig. 4. However, the lattice parameters calculated for these samples showed only a minimum variation.

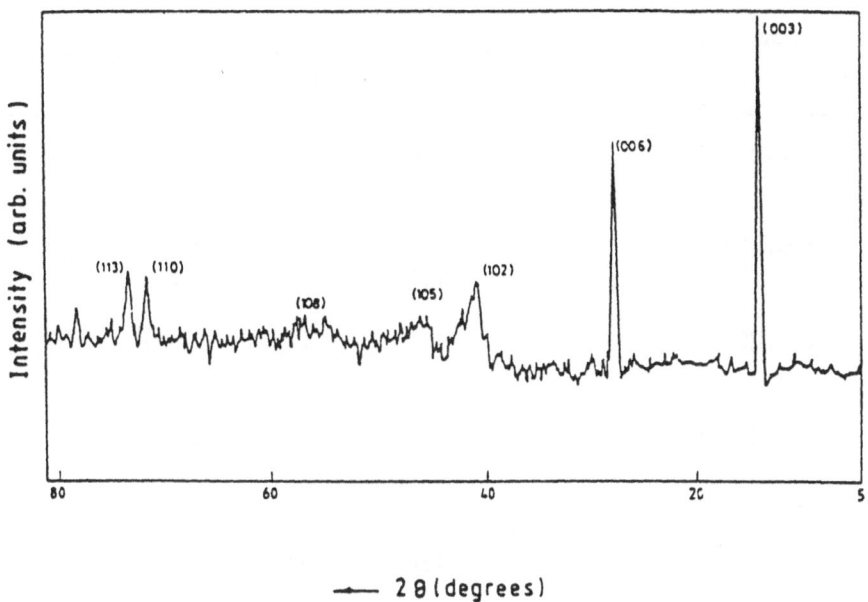

Fig.2 :   X-ray powder diffraction of Co-Al-CO$_3$-HT (Co/Al = 2.0)

Table 2 : Lattice parameters of some synthesised hydrotalcites

Sample	M^{2+}/M^{3+} (atomic ratio)	a (Å)	c (Å)	V (Å^3)	Surface area (m^2/g)
Co-Al2.0:1-A	2.0	3.052	22.144	178.6	63.3
Co-Al3.0:1-A	3.0	3.059	23.578	191.1	35.0
Ni-Al2.0:1-A	2.0	3.020	22.432	177.2	91.9
Ni-Al3.0:1-A	3.0	3.037	22.785	182.2	83.3
Cu-Al3.0:1-A	2.9	3.070	20.170	164.6	35.0

Table 3 : Particle sizes of some of the samples measured by X-ray line broadening

Compound	2θ (degrees)	hkl	β (degrees)	t (Å)	Average t(Å)
Co-Al2.5:1-A	27.657	006	0.500	225.0	237
	73.187	113	0.400	230.0	
Co-Al2.5:1-H	27.635	006	0.325	289.2	300
	73.187	113	0.375	303.2	
Co-Al3.0:1-A	27.902	006	0.425	221.3	209
	73.480	113	0.550	207.1	
Co-Al3.0:1-H	13.875	006	0.425	216.4	220
	73.102	113	0.525	216.4	
Ni-Al2.5:1-A	23.760	006	1.200	67.1	79
	61.260	110	0.975	91.2	
Ni-Al2.5:1-H	23.088	006	0.575	160.5	209
	61.260	110	0.400	257.4	
Ni-Al3.0:1-A	23.246	006	1.2	66.7	90
	59.870	110	0.800	113.1	
Ni-Al3.0:1-H	22.982	006	0.800	120.0	136
	60.820	110	0.575	151.6	

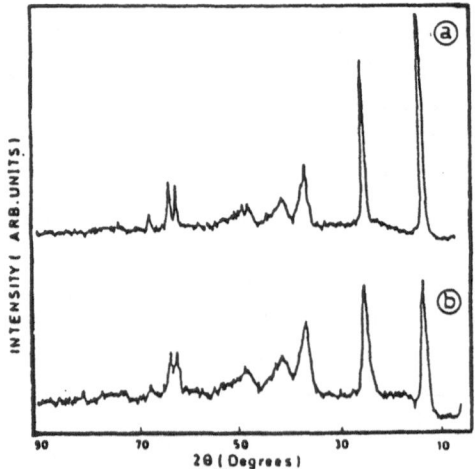

Fig. 3 : XRD patterns of Ni-Al-CO$_3$-HT (Ni/Al = 2.5)
(a) Aged sample (b) Hydrothermally treated sample

118

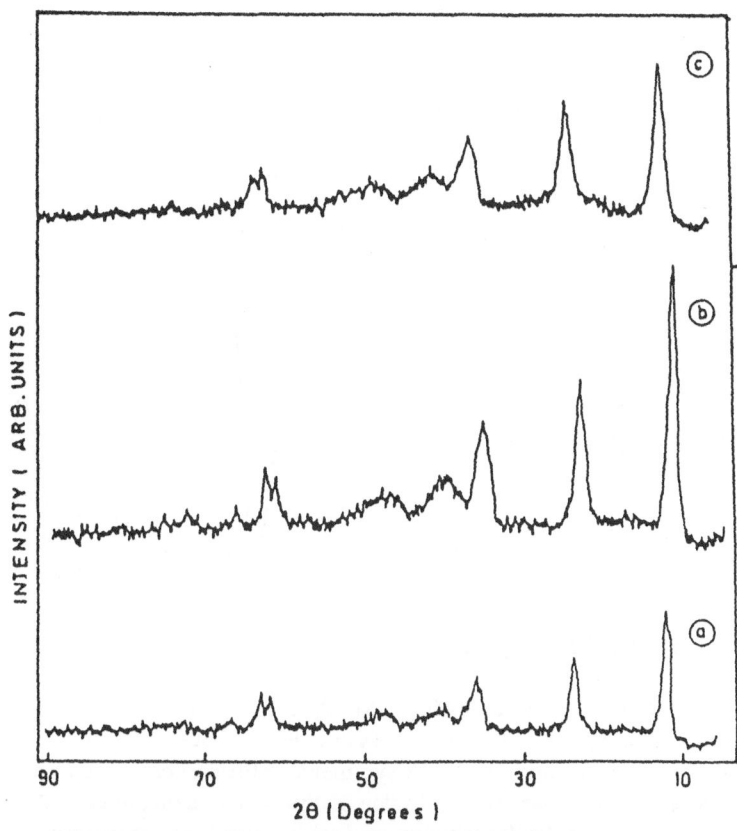

Fig.4 : XRD patterns of Ni-Al-CO$_3$-HT (Ni/Al = 3.0).
(a) Sequential precipitation; (b) Coprecipitation under low
supersaturation; (c) Reversel addition

IR studies

Although IR analysis is not a diagnostic tool for the characterisation of
hydrotalcites, it can be used to identify the presence of foreign anions like
polyoxometallates, chromates etc. in the interlayer and also provide information
regarding the type of bonds formed by the anions and their orientation in the
interlayer. Fig.5 depicts a typical FT-IR spectrum; the sample chosen for the
illustration is Co-Al-CO$_3$-HT (Co/Al 2.0), which may also be represented as Co-Al
2.0 : 1-A. The absorption band around 3400-3500 cm^{-1} is attributed to hydroxyl
stretching (v_{OH}), while the bands around 1630-1640 cm^{-1} and 1360-1380 cm^{-1},
respectively, are due to the deformation mode of OH and v_3 of CO$_3^{2-}$ stretching.[38]
The low value of v_{OH} in comparison with free hydroxyl group (3650 cm^{-1}) indicates
that all OH groups are hydrogen bonded. A weak absorption band around 2950
cm^{-1}- 3050 cm^{-1} is observed which is attributed to water molecules hydrogen bonded
to the carbonate ion present in the interlayer.[39] Bands observed around 800-400
cm^{-1} are assigned to the lattice vibrations like M^{2+}-M^{3+}-O stretching and bending and
M-O stretching.[40]

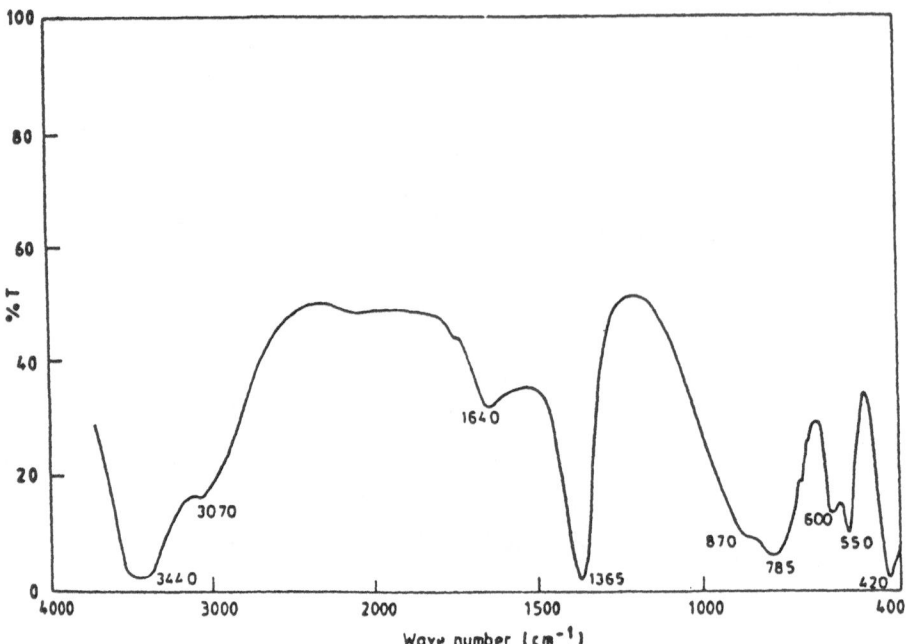

Fig.5 : FT-IR spectrum of Co-Al-CO₃-HT (Co/Al = 2.0)

For some of the aged samples (e.g. Ni-Al3.5:1-A), a sharp peak around 1380 cm⁻¹ with a shoulder around 1420 cm⁻¹ (v_3 mode of carbonate) is observed. This probably arises on account of a lower symmetry of the carbonate anion present in the interlayer. This can also cause activation of the v_1 mode (symmetric stretching) of carbonate at around 1030 cm⁻¹; this mode is inactive for a symmetric carbonate environment.[39] Upon hydrothermal treatment, the sample showed a sharp single band for v_3 stretching indicating that the treatment enhances the ordering in the interlayer space. In all these cases either the v_1 is completely absent or reduced in its intensity, confirming the above conclusion. However, there is no significant shift in the band position of the carbonate stretching upon hydrothermal treatment. For all the aged samples v_{OH} is broad and asymmetric in nature, but following hydrothermal treatment it is narrow and symmetric, indicating that better orderliness prevails in the hydrothermally treated samples.

A closer look at the v_{OH} and v_3 of values for aged samples indicates that as the M^{2+}/Al^{3+} atomic ratio increases both these vibrations suffer shifts to higher wave numbers. This can be explained on the basis of the increased electrostatic interaction between the layer and the interlayer with consequent influence on the hydrogen bonding between the hydroxyl group and carbonate.[40]

Thermal studies

The thermal behaviour of hydrotalcites is generally characterised by two transitions, which depend qualitatively and quantitatively on many factors. Fig. 6 shows the thermogravimetric and attendant differential curves for some samples synthesised by aging. All compounds showed two stages of weight loss except for some samples like Ni-Al3.5:1-A. The first weight loss occurring in the temperature range 150-250 °C is attributed to the removal of interlayer water molecules, which can also be seen from the XRD by the decrease in the interlayer spacing from 7.8Å to 6.7Å. The second weight loss occurring in the temperature range 250-350 °C is associated with the removal of water from the brucite sheet and CO_2 from the interlayer carbonate anion which leads to the destruction of the layered structure and the formation of mixed metal oxides.[41] In some cases as with Co-Al2.5:1-A the second transition occurs as two stages corresponding to the two above processes. Pesic et al.[42] have reported that initially CO_2 is lost followed by water molecules from the brucite sheet. Table 4 gives the net weight loss and the transition temperature of some of the samples synthesised. It is clearly seen from the data that as the M^{2+}/Al^{3+} atomic ratio decreases, the transition temperatures T_1 and T_2 for the corresponding weight losses increase. The net weight loss measured in the temperature range increases with increase in the aluminium content, since more carbonate and water molecules are required for charge compensation. The differential scanning calorimetry results, given in Fig.7 substantiate the thermogravimetric analysis results showing two endothermic peaks for the corresponding two weight losses. The DSC transition temperature also showed the same variation with the M^{2+}/M^{3+} atomic ratio. Comparison of the DSC curves of the aged and hydrothermally treated samples showed an interesting observation. The DSC curves are more intense and sharper for hydrothermally treated samples which is indicative of high crystallinity.[43] That no new peaks are found in the DSC of the hydrothermally treated samples indicates that no new phases are formed and that only the HT phase is present.

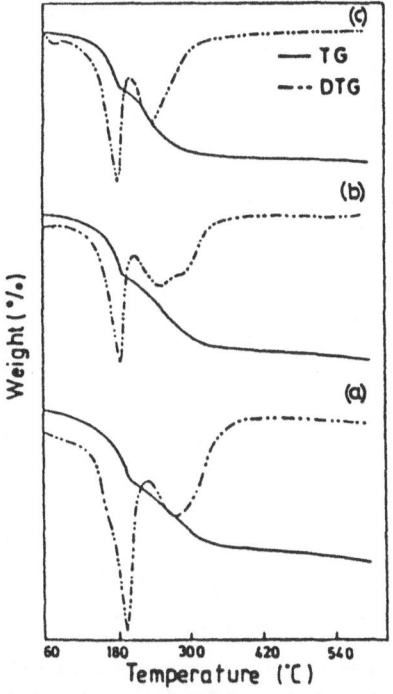

Fig.6 : **TG-DTG traces for hydrotalcites aged at 65 °C/30 min.**
(a). Co-Al2.0:1-A; (b). Co-Al2.6:1-A; (c). Co-Al3.0:1-A

121

Table 4 : TG transition temperatures and net weight loss for aged hydrotalcite samples

Compound	TG transition temperature (°C)		Net weight loss (%)
	T_1	T_2	
Co-Al2.5:1-A	184.7	253.4	32.4
		289.5	
Co-Al3.0:1-A	178.7	235.4	29.8
Ni-Al2.5:1-A	204.5	325.5	29.2
Ni-Al3.0:1-A	190.7	303.2	35.6
Cu-Al3.0:1-A	110.0	245.0	37.7

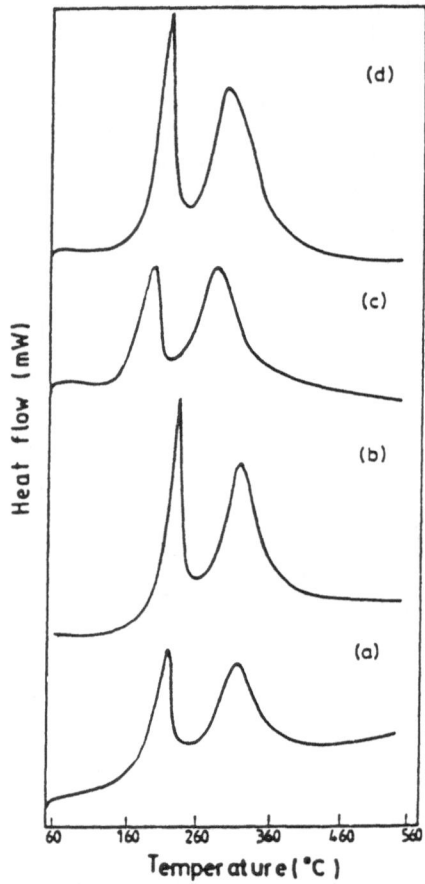

Fig.7 : DSC patterns of aged and hydrothermally treated hydrotalcites
(a) Co-Al2.0:1-A; (b) Co-Al2.0:1-H; (c) Co-Al3.0:1-A;
(d) Co-Al3.0:1-H

Other studies

Transmission electron microscopic studies have also been carried out on HT materials to understand their morphological features (Fig. 8). They show spherical to hexagonal platelets which are thin and wide. Solid-state Al^{27} MAS NMR was used to study the coordination of Al^{3+} present in the brucite sheet.[44] The study showed that Al^{3+} is present in octahedral coordination in hydrotalcite, whereas in the thermally calcined form it is present both as tetrahedral and octahedral sites. Mössbauer studies have also been carried out on Fe-containing hydrotalcites to assess the oxidation state and stereochemistry of Fe in the samples.[45]

Thermal calcination of the materials has been carried out in order to get a better insight into the nature of the thermally decomposed materials. This yielded non-stoichiometric, high surface area mixed metal oxides which are stable towards thermal sintering. In the case of Ni-Al-HT, upon thermal calcination NiO is obtained whose lattice parameters are less than those of the pure NiO indicating non-stoichiometry. The crystallinity of the material increases with increase in the calcination temperature, as evidenced from XRD showing the increased sharpness of the peaks. A similar result is observed for Co-Al-HT, given in Fig. 9, but the phase obtained is spinel.[46] This result correlates with studies on surface area measurements which showed that a decrease in surface area attends an increase in the calcination temperature. However, the extent of decrease in the surface area is less for Ni-containing HTs than for Co-containing HTs owing to the strong mutual interaction of NiO and Al_2O_3 which prevents the sintering of the particles. This result is also supported from the reduction behaviour of the NiO particles in the NiO/Al_2O_3 solid solution. Fig. 10 shows the variation of surface area with calcination temperature for Ni-Al-HT and Co-Al-HT (M/Al = 3.0).

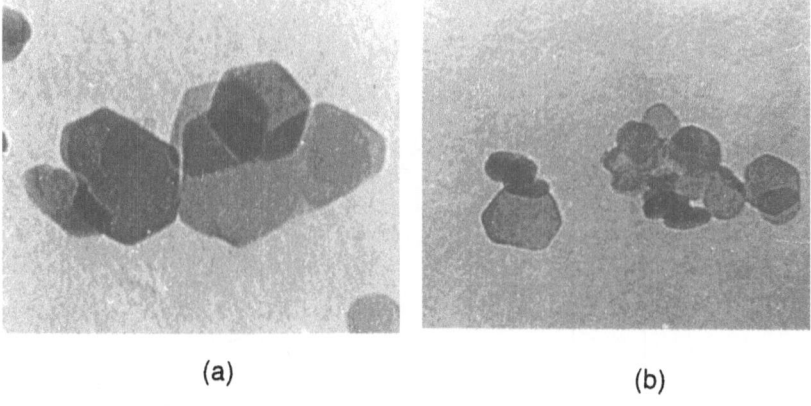

(a)   (b)

Fig. 8 : Transmission electron micrographs of
(a) Co-Al-CO$_3$-HT; (b) Mg-Al-CO$_3$-HT (M^{2+}/Al^{3+} = 2.0)

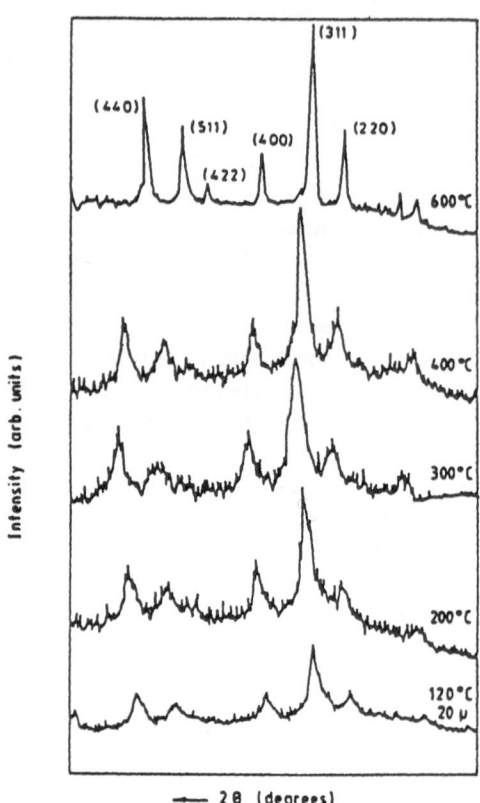

Fig. 9 : XRD patterns of Co-Al2.0:1-A calcined at different temperatures

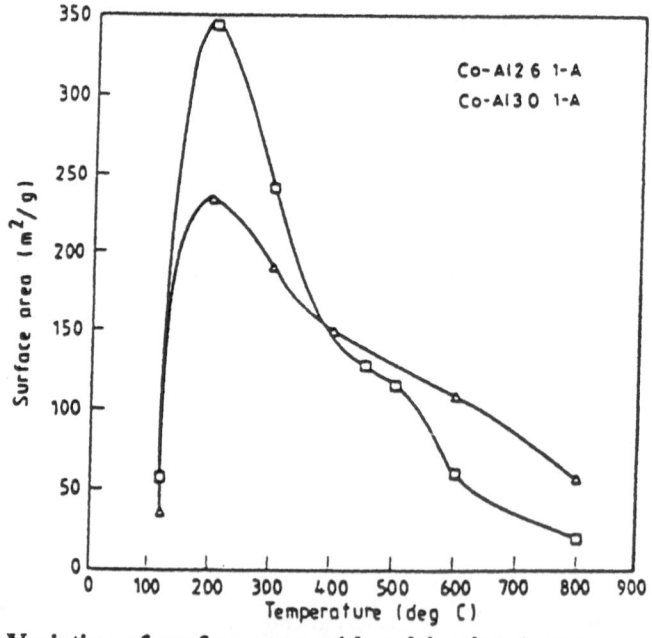

Fig. 10 : Variation of surface area with calcination temperature for Ni-Al3.0:1-A and Co-Al3.0:1-A

124

Catalytic reactions over hydrotalcites and their derived materials

Hydrotalcite-like compounds used in catalysis are mainly thermally calcined materials. These have the following properties:

1. High surface area

2. Basicity

3. Non-stoichiometry

4. Formation of homogeneous mixed metal oxides (solid solution of metal oxides) with very small crystal size stable to thermal treatments.

5. Memory effect- which allows the reconstruction of the layered HT structure when the product of the thermally treated materials is kept in contact with aqueous solutions containing various anions.

Base catalysis

Reichle has carried out the vapour phase aldol condensation of aldehydes and ketones on a number of thermally calcined hydrotalcite-like materials.[18] Mg-Al-CO_3-HT, calcined at 450 °C, showed the maximum activity and selectivity for isophorone formation. Using other hydrotalcites containing different M^{2+} and M^{3+} cations (M = Ni, Co, Zn, Al and Cr), resulted in a substantial decrease in isophorone production but a large increase in mesityl oxide production. Catalysts containing nickel yielded fairly large quantities of mesitylene. The catalytic activity is also influenced by the nature of anions present in the interlayer. Among the anions studied, carbonate showed the highest activity, whereas anions like SO_4^{2-}, X^- (X = Cl, Br) and CrO_4^{2-}, gave poor activities.

The basic property of the thermally calcined materials has been exploited in the catalytic synthesis of polyether polyols used in polyurethane industries.[19] Mg-Al-CO_3-HT showed a dual selectivity towards the production of both amorphous and crystalline polymers. The material calcined at 500 °C showed the maximum activity which is closely correlated with the maximum base strength of the catalyst. Narrower molecular weight distribution and lower levels of isotactic polymer, which are industrially desirable, can be realised by carrying out the polymerisation at very high catalyst concentrations (> 300 catalyst/mole initiator OH), thereby adjusting the statistical balance of the catalytic sites and monomer concentration.

Recently, Davis and coworkers[47] have studied the influence of the preparative method, calcination temperature, and Mg/Al atomic ratios on the base strength and consequent catalytic activity of the Mg-Al-CO_3-HT system for the steady state decomposition of 2-propanol. Relative to MgO and alumina, the catalyst showed both activity and selectivity to propanone formation in the order, MgO > calcined HT >> alumina, indicating the presence of catalytically active basic sites in the calcined hydrotalcite. No significant change in activity and selectivity was observed for the catalyst calcined in the temperature range 723-923K, whereas calcination at 1173K showed a considerable decrease in the activity (due to decrease in the surface area and $MgAl_2O_4$ spinel formation). Recently in our laboratory, we have carried out vapour phase alkylation of phenol with methanol on the thermally calcined Mg-

Al-CO$_3$-HT with various Mg/Al atomic ratios. The reaction was carried out in a tubular reactor with phenol:methanol ratio of 1:7 in the temperature range 200-450 °C. On all the catalysts, at low temperatures only anisole was obtained, while at higher temperatures 2,6-xylenol was obtained. The life span of these catalysts was investigated at 350 °C for 5 h on stream. The results indicated that catalysts with Mg/Al = 4.0 showed around 50% conversion with the activity levelling off without any deactivation. This was not the case with other Mg/Al ratios for which the catalytic activity deteriorated with time. The temperature dependence of the activity was also studied on these catalysts which showed that the same composition (Mg/Al = 4.0) exhibited 100% conversion with nearly 55% selectivity for 2,6-xylenol.

Synthesis of methanol and higher alcohols

Methanol is now synthesised from syngas employing a recent process operating at a lower temperature and pressure (around 250 °C and 5.0 MPa) based on Cu-containing mixed oxides. Cu-Zn-Al hydrotalcite on thermal calcination and subsequent reduction yields Cu/ZnO/Al$_2$O$_3$ with very small crystal size, stable under both reductive and catalytic conditions, which is thought to be the best catalyst available to date for the methanol synthesis.[31] The catalyst prepared by this route is characterized by higher stability and longer life span with respect to other conventionally prepared catalysts. However, the highest activity for methanol synthesis is observed for the catalyst whose precursor is a mixture of hydrotalcite and malachite-type phases. The influence of the second divalent cation in Cu-M-Cr-HT (or Al) (38:38:24) on the synthesis of methanol has been investigated.[48] The results showed that Cu-Zn-Cr(or Al)-HT manifested the maximum activity both on the basis of per kg of the catalyst as well as per kg of copper. Cu-Cr-HT and Cu-Mg-Cr-HT showed similar activities, although the latter compound crystallises in the HT network. This result implies that the nature of the divalent cation and the relative stability of the mixed metal oxide under reaction conditions are more important than the nature of precursors. Cobalt containing hydrotalcites showed an increase in activity, but with a high selectivity towards hydrocarbons. This high selectivity for the formation of hydrocarbons was explained on the basis of a synergic effect that exists between cobalt and copper, with formation of centres responsible for the synthesis of hydrocarbons. However, when cobalt is present in small amounts, a strong deactivation is observed, without change in selectivity, probably because of selective poisoning.

Catalytic decomposition of nitrous oxide

Issues such as the green house effect and ozone layer depletion have been the focus of recent global attention because of the dramatic and disastrous consequences they portend.[49] Nitrous oxide, is now considered an environmental pollutant, since it is involved in the catalytic stratospheric destruction of the ozone layer; it is also a greenhouse gas. This kind of radiative trapping by the so called greenhouse gases like N$_2$O, CH$_4$ and CO$_2$ causes the mean surface temperature of the earth to be 33K higher than it would be in their absence.[50] There are both natural and anthropogenic sources for N$_2$O. However, man-made sources are increasing at a much higher rate than natural ones. Nitrous oxide is produced in maximal quantities by using fossil fuels which involves both direct N_2O emissions and indirect emissions through NO_x. N_2O also arises as a co-product from some chemical processes, such as the use of circulating fluidised bed reactors for combustion and production of large amounts of adipic acid for Nylon 66.[51] The transportation sector and stationary sources are also significant sources of N_2O. Among the stationary sources, utility boilers are the largest N_2O producers. N_2O is also reported to be increasing at the rate of 0.2% per year in the Earth's atmosphere. Because N_2O has an atmospheric life time of 150 years, a strong effort is underway to identify the sources of N_2O and to limit its production or emission into the atmosphere.[49] For the protection of the environment, the N_2O formed should be decomposed, and it is here that catalysts can play a significant role. The stoichiometric reaction for the decomposition is as follows (equation 1):

$$N_2O \rightarrow N_2 + 1/2\ O_2 \tag{1}$$

The reaction is exothermic and the products are the stable and simple molecules, nitrogen and oxygen. The reaction can be easily followed by measuring the change in pressure as a function of time. The decomposition pathway follows the scheme indicated by equations (2) through (5).

$$N_2O + e^- \rightarrow N_2O^-_{ad} \tag{2}$$

$$N_2O^-_{ad} \rightarrow N_2 + O^-_{ad} \tag{3}$$

$$O^-_{ad} + O^-_{ad} \rightarrow O_2 + 2e^- \tag{4}$$

$$O^-_{ad} + N_2O \rightarrow O_2 + N_2 + e^- \tag{5}$$

The kinetic results for the decomposition of N_2O can obey any of the following rate expressions reported by Cimino et al.[52] corresponding to no, strong and weak inhibition by oxygen as given by equation (6), (7) and (8) respectively.

$$-dP_{N_2O}/dt = k_1 P_{N_2O} \tag{6}$$

$$-dP_{N_2O}/dt = k_2 P_{N_2O}/P_{O_2}^{1/2} \tag{7}$$

$$-dP_{N_2O}/dt = k_2 P_{N_2O}/1 + bP_{O_2}^{1/2} \tag{8}$$

If step 2, involving the adsorption of N_2O, is rate determining, then the kinetics of decomposition will obey equation (6). If step (4) or (5) involving desorption of oxygen is rate limiting, then the kinetics will follow equation (7). If adsorption of N_2O and desorption of oxygen are competing for the rate limiting step, then the kinetics will obey equation (8).

In our laboratory, we have carried out the decomposition of nitrous oxide on a wide variety of catalysts namely, metal oxides, mixed metal oxides like spinels, perovskites, K_2NiF_4-type materials, double perovskites, pyrochlores and

127

superconducting oxides.[53-56] Most of the studies are primarily concerned with the kinetics of decomposition rather than with their conversion levels. These catalysts showed considerable activity in the temperature range 300-500 °C. Hall and coworkers[57] studied the decomposition of N_2O on Fe-Y and Fe-Mordenite catalysts. Both catalysts yielded the same kinetics showing a steady state production of N_2 and O_2 in the stoichiometric ratio, 2:1. The reaction was more rapid over Fe-Mordenite than over Fe-Y (60 times higher on the basis of rate constants measured at 500 °C) in spite of the fact that the latter has a more accessible three dimensional pore system. Aparacio et al.[58] have studied the decomposition of N_2O on cation exchanged zeolites (Y-zeolites exchanged with both Fe and either Ca, La, Eu, Cr, Mn, Co and Ni or Cu). Fe,Eu-Y and Cu-Y zeolites showed the maximum conversion levels of 40% and 25% respectively, at 450 °C. Panov et al.[59] have carried out the same reaction and reported that Fe-ZSM-5 is the most active catalyst and the atomic activity of Fe in this zeolite is very high in comparison with the activity of Fe-Mordenite, Fe-Y and Fe_2O_3. Recently, Li et al.[60] have carried out the catalytic decomposition of N_2O on metal exchanged zeolites and have shown that Co-ZSM-5,Cu-ZSM-5 and Rh-ZSM-5 are very good catalysts based on the conversion levels. Under their experimental conditions (in a microcatalytic reactor in a steady state, plug flow mode using 990 ppm of N_2O in He) both Cu-ZSM-5 and Co-ZSM-5 showed appreciable conversions at 330 °C and reached 100% and 50% levels respectively, at 400 °C. Rh-ZSM-5 and Ru-ZSM-5 showed appreciable conversion levels even at 250 °C. These authors reported the interesting observation that all ZSM-5 forms of catalysts are much more active than Al_2O_3 supported ones, although all Al_2O_3 supported catalysts have higher metal loadings compared to the corresponding ZSM-5 forms. We have carried out the decomposition of N_2O on "in situ" generated thermally calcined hydrotalcites of general formula M-Al-CO_3-HT, where M = Ni, Co and Cu, in the temperature range 140 - 310 °C at 50 torr initial pressure of the gas in an all-glass recirculatory static reactor.[27] The catalysts showed first order dependence on N_2O, following equation (6), without any inhibition by oxygen. The integrated form of the equation is employed for the calculation of the rate constants given in Table 5. The data reveal the following order of catalytic activity: Ni-Al-CO_3-HT>Co-Al-CO_3-HT>Cu-Al-CO_3-HT. A typical kinetic plot for the decomposition is given in Fig.11. Arrhenius plots were constructed employing the rate constants measured at various temperatures and the corresponding kinetic parameters are also given in Table 5. For the evaluation of kinetic data the conversions were restricted to less than 20%. Fig. 12 depicts the catalytic efficiency as a function of temperature. Ni-Al-CO_3-HT and Co-Al-CO_3-HT showed appreciable activity even at 150 °C, whereas the Cu-containing catalysts showed appreciable activity around 300 °C. Ni-Al-CO_3-HT and Co-Al-CO_3-HT reached their 50% and 100% conversion levels at 190 °C and 250 °C respectively. Among the catalysts studied in our laboratory, under identical experimental conditions, these catalysts showed the maximum activity for the decomposition of N_2O. In comparing the catalysts reported in the literature under our experimental conditions, our catalysts 'light off' at the lowest temperature for the decomposition of N_2O.[61] The high activity of these materials could be attributed to the high surface area and non-stoichiometry of the mixed metal oxides generated upon thermal calcination.

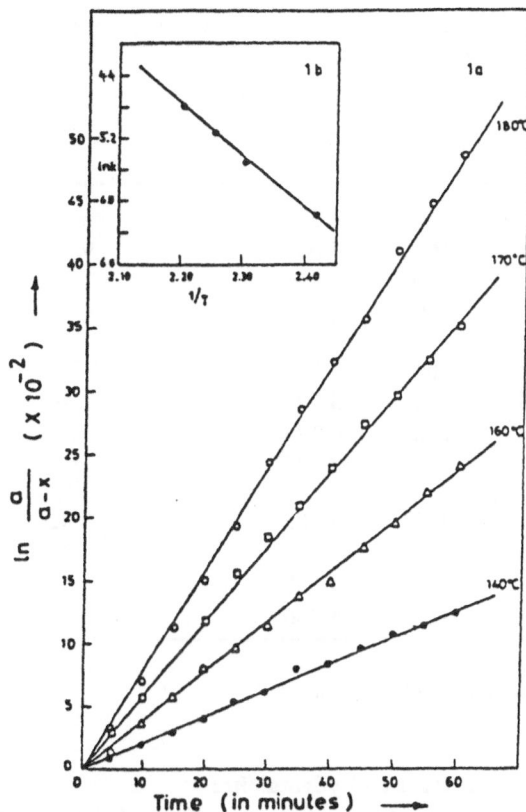

Fig.11 : (a). Kinetic plots for the decomposition of N_2O on thermally calcined
Ni-Al3.0:1-A at 50 torr; (b). Arrhenius plot for the decomposition of
N_2O

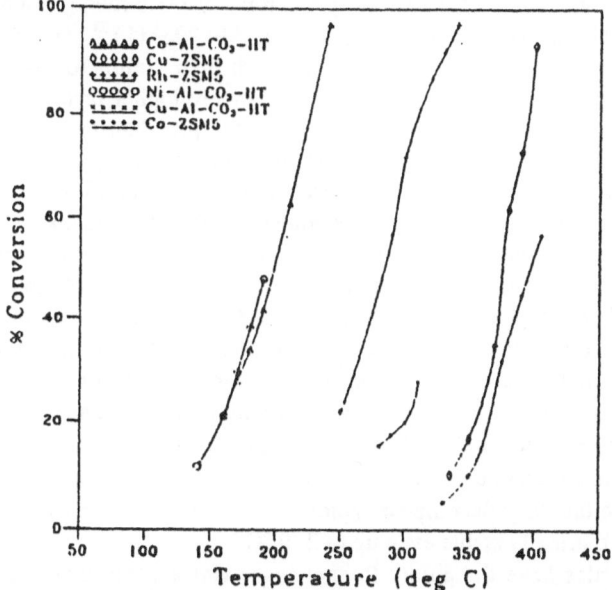

Fig.12 : N_2O conversion on various catalysts as a function of temperature

Table 5 : Kinetic parameters for the decomposition of N₂O at 500 torr on M-Al-CO₃-HT

Catalyst	Temp (°C)	k (x 10⁻³)ᵃ	E_a (kJ/mol)	ln A	% conversion at 30 min time intervals
	140	2.058			6.0
Ni-Al-CO₃-HT	160	4.000	54.7	9.73	11.1
	170	5.950			16.9
	180	8.333			21.3
	160	3.652			10.4
Co-Al-CO₃-HT	170	5.000	46.8	7.40	13.8
	180	6.473			17.0
	190	7.878			21.0
	280	2.702	53.9	5.82	7.7
Cu-Al-CO₃-HT	290	3.250			9.3
	300	4.137			10.4
	310	5.000			14.0

ᵃ Unit of k for no inhibition by oxygen : min⁻¹

Anion exchange reactions on hydrotalcites

The anion exchange property of the HT-like materials has been much projected in recent years owing to their potential impact on environmental control. These materials have an exchange capacity of approximately 1-1.5 meq/g (much less than the theoretical maximum of 3.3 meq/g for Mg-Al-CO₃-HT).[21] The low value of their ion exchange capacity in comparison with their complementary analogue, namely smectite type clays, is due to their relatively higher layer charge density (0.04 e⁺/Å² for LDH vs 0.01 e⁻/Å² for smectites). Despite their low anion exchange capacity, they find potential application as anion exchangers and adsorbents especially for the ecologically undesirable anions from dilute and possibly radioactive aqueous waste streams. Thus, ions like chromate, arsenate, cyanide and ferricyanide are strongly bound by these clays. Thermal treatment of these exchanged clays converts them into a metal-oxide matrix in which metal anion residue could be in water insoluble form e.g. Cr₂O₃ in the case of chromate, which could be easily disposed.[62] Radioactive metals could be then buried in this form to remove them from the environment. One of the major advantages of these materials over typical commercial anion exchangers namely, polystyrene-divinyl benzene quaternary ammonium ion exchange resin, lies in their thermal stability. The hydroxide forms of these commercial resins have severe temperature limitations in application in that they decompose rapidly above 60 °C, whereas the anionic clay minerals are thermally stable even up to 200 °C.

Hydrotalcites have the ability to exchange with a wide variety of anions as previously mentioned. The order of stability of the intercalating anions, reported by

130

Miyata et al.[15] is approximately $OH^- > F^- > Cl^-\ NO_3^- > I^-$ for monovalent anions, and $CO_3^{2-} > SO_4^{2-} > CrO_4^{2-}$ for divalent anions. However, there is a strong preference for the divalent anions, in particular for carbonate, as it is found in the naturally occurring minerals more than monovalent anions.

Hydrotalcites with non-carbonate anions can be prepared by different methods. These include:

1. Synthesis under rigorous exclusion of CO_2
2. Treatment of CO_3^{2-} form with dilute acid
3. Heating the hydrotalcite to 450-500 °C, followed by rehydration and readsorption of anions
4. Anion exchange by dispersing the compound in an unbuffered solution containing the exchange anions

Bish[63] has synthesised pure NO_3^-, SO_4^{2-} and ClO_4^- containing HT's by carefully removing CO_3^{2-} in the interlayer by adding dilute acid. Organic anions like oxalate, succinate, sebacate etc., were also intercalated with Zn-Al-Cl-HT by direct anion exchange reactions. From the 'c' lattice parameter it was presumed that these long chain acid anions are perpendicular to the inorganic planes, connecting them as pillars[34]. Since the gallery anions are exchangeable, it is possible, in principle, to intercalate robust anions like polyoxometallates with high negative charge. Such anions should impart large gallery heights and large lateral anion spacing (due to anion-anion repulsion) thereby providing access to the intracrystalline gallery surfaces. These type of clays interlayered by polyoxometallate (POM) lead to the new class of pillared materials for selective adsorption and catalysis.[21] Decavanadate anion $V_{10}O_{28}^{6-}$ was intercalated in Zn-Al-Cl-HT by the coprecipitation method, and photo-oxidation of 2-propanol was carried out on the intercalated hydrotalcite.[64] The intercalated material was three times more active for acetone formation in comparison with the reaction carried out under homogenous conditions. Drezdzon[32] has described a promising route to POM intercalated basic LDH, wherein the exchange takes place between desired POM anions such as $[V_{10}O_{28}]^{6-}$ and large anions like p-toluene sulphonate or terephthalate intercalated hydrotalcite. In this way it was possible to prepare pillared forms of highly basic Mg-Al-HT with acidic POM pillars. Yamagishi et al.[65] have investigated the removal and fixation of harmful oxometallate ions, by intercalating them in hydrotalcites. Chromate and permanganate anions were removed from aqueous solution by the hydration of an Mg-Al oxide solid solution, which was prepared by the thermal calcination of Mg-Al-CO_3-HT. They found that the amount of divalent anion adsorbed was larger than that of monovalent anion in the wide region of equilibrium concentrations. Mg-Al-CO_3-HT, interlayered with Co(II) phthalocyanine tetrasulphate was prepared in 92% yield in aqueous solution.[35] The intercalated complex was tested for the catalytic auto-oxidation of 1-decane thiol. Thiols, even in small amounts are environmentally undesirable contaminants in certain petroleum distillates and industrial effluents. The intercalated catalyst was twice as active in comparison with the homogenous complex under equivalent reaction conditions. Suzuki et al.[66] have utilised the anion exchange properties of synthetic HTs in the catalysis of halide exchange reactions between alkyl halides. Hydrotalcites with various M^{2+} and M^{3+} (M^{2+} = Mg, Ni, Co, Mn and Zn ; M^{3+} = Al

131

and Cr), and containing Cl^- as the interlayer anion, were used as the halide exchanger between alkyl bromide and benzyl chloride. The alkyl bromide undergoes halide substitution by the interlayer Cl^- ions, leaving Br^- in the interlayer. These intercalated Br^- ions would in turn attack benzyl chloride to yield benzyl bromide, leaving Cl^- in the interlayer. In our laboratory, we have studied the conversion of benzyl chloride to benzyl cyanide using HTs containing Cl^- as the interlayer anion. A reaction of benzyl chloride with NaCN and 1g of Mg-Al-Cl-HT has been carried out which yielded 100% conversion with 76% selectivity towards benzyl cyanide and the remaining towards benzyl alcohol.[67] In the absence of the catalyst, only about 40% conversion could be obtained. A detailed investigation including kinetics of the reaction is under progress.

A hydrotalcite-like mineral containing CH_3O^- and Cl^- anions catalysed the disproportionation of $(CH_3O)_3SiH$ to $(CH_3O)_4Si$ and SiH_4.[68] The production of SiH_4 is very important since its thermal decomposition gives pure silicon used in the semiconductor industry. The disproportionation catalysed by CH_3O^- intercalated HT gave higher activity (in comparison with Cl^-) with 95% conversion and 95% selectivity after 9 h.

Similar to zeolites, HT-like materials also possess molecular sieving properties despite their two dimensional layered structure. The sieving effect primarily depends on the interlayer thickness which is mainly controlled by the size of the anion. Miyata et al.[69] have studied the adsorption of N_2, O_2, CO_2 and H_2 on the dehydrated HT containing $[Fe(CN)_6]^{4-}$ as the interlayer anion. The specific surface area calculated for the sample dehydrated at 100-200°C was 4-5 times higher than that of untreated samples indicating the adsorption of N_2 in the interlayer. In order to ascertain the presence or absence of the molecular sieve effect, adsorption of CO_2 was carried out on Mg-Al-X-HT ($X= CO_3^{2-}$, SO_4^{2-}, ClO_4^- and $[Fe(CN)_6]^{4-}$. The results showed that the CO_3^{2-} type compound having the interlayer thickness of 2.87Å does not permit adsorption of CO_2 whereas the other anions (interlayer thickness of more than 3.8Å) permit significant adsorption. The amount of CO_2 adsorbed increases with increase in the interlayer thickness of these hydrotalcites.

Both HT and thermally treated HT are capable of capturing anions and are therefore used as halogen scavengers for polyolefins produced by Zeigler catalysts[70]. By utilising their characteristic ion selectivity, HTs are expected to find application in the removal of acid dyes, environmentally undesirable anions in waste water, as well as in the neutralisation and thermal stabilisation of halogen-containing polymers. A broad application in the environmental area is anticipated on account of their pronounced anion exchange capacity for both inorganic and organic anions. The molecular sieving and catalytic properties of these novel materials promise to be an exciting area for future research.

References

1. R.M. Barrer, *Zeolites and Clay Minerals as Sorbents and Molecular Sieves*, Academic press, New York, 1978, p.407.
2. T.J. Pinnavaia, *Science*, **220** (1983) 365.
3. F. Figueras, *Catal. Rev. Sci. Engg.*, **30** (1988) 457.

4.   T.J. Pinnavaia and H. Kim, *Zeolite Microporous Solids: Synthesis, Structure and Reactivity,* 1992, p.79.

5.   W.T. Reichle, *Solid State Ionics,* **22** (1986) 135.

6.   K.A. Corrado, A. Kostapapas and S.L. Suid, *Solid State Ionics,* **26** (1988) 77.

7.   W.T. Reichle, *Chemtech.,* **16** (1986) 58.

8.   E. Manasse, *Atti. Soc.Toscana Sc. Nat., Proc. Verb.,* **24** (1915) 92.

9.   G. Aminoff and B. Broome, *Kungl.Sven. Vetensk. Handl. 9,* **3** (1930)23.

10. W. Fietknecht, *Helv. Chim. Acta.,* **25** (1942) 131.

11. H.F.W. Taylor, *Miner. Mag.,* **39** (1973) 377.

12. R. Allmann, *Acta Crystallogr.,* **B24** (1968) 972.

13. S. Miyata, *Clays Clay Miner.,* **23** (1975) 369.

14. S. Miyata, *Clays Clay Miner.,* **28** (1980) 50.

15. S. Miyata, *Clays Clay Miner.,* **31** (1983) 305.

16. F.J. Brocker and L. Kainer, *German Patent* 2,024,282 (1970)

17. T. Nakatsuka, H. Kawasaki, S. Yamashita and S. Kohijiya, *Bull. Chem. Soc. Jpn.,* **52** (1979) 2449.

18. W.T. Reichle, *J. Catal.,* **94** (1985) 547.

19. D.E. Laylock, R.J. Collacott, D.A. Skelton and M.F. Tchir, *J. Catal.,* **130** (1991) 354.

20. D.E. Laylock and R.A. Newman, In: *Proc. 12th Canadian Symp. Catal.,* Alberta, Canada, eds., K.J. Smith and E.C. Sanford, Elsevier, Amsterdam, 1992, p.269.

21. T.J. Pinnavaia, T. Kwon and S.K. Yun, *Zeolite Microporous Solids: Synthesis, Structure and Reactivity,* 1992, p.91.

22. T. Kwon and T.J. Pinnavaia, *J. Mol. Catal.,* **74** (1992) 23.

23. T. Kwon and T.J. Pinnavaia, *Chem. Mater.,* **1** (1989) 381.

24. E.C. Kruissink, L.L. Van Reijen and J.R.H. Ross, *J. Chem. Soc., Faraday Trans. I,* **77** (1981) 665.

25. G. Fornasari, S.Gusi, F. Trifiro and A. Vaccari, *Ind. Eng. Chem. Res.,* **26** (1987) 1500.

26. A.J. Marchi, J.I. Di Cosimo and C.R. Apestiguia, *Catal. Today,* **15** (1992) 383.

27. S. Kannan and C.S. Swamy, *Appl. Catal. B* (in press)

28. F. Cavani, F. Trifiro and A. Vaccari, *Catal. Today,* **11** (1991) 173.

29. G.W. Brindly and S. Kikkawa, *Amer. Miner.,* **64** (1979) 836.

30. H. Schaper, J.J. Berg-Slot and W.H.J. Stork, *Appl. Catal.,* **54** (1989) 79.

31. A. Vaccari, *La Chimica & L'Industria,* **74** (1992) 174.

32. M.A. Drezdzon, *Inorg. Chem.,* **227** (1988) 4628.

33. A.P. Cavancanti, A. Schutz and P. Biloen, In: *Prep. Catal. IV,* eds., B. Delmon, P. Grange, P.A. Jabos and G. Poncelet, Elsevier, Amsterdam, 1987, p.165.

34. S. Miyata and T. Kumura, *Chem. Lett.,* (1973) 843.

35. M. Elena Perez-Bernal, R. Ruano-Casero and T.J. Pinnavaia, *Catal. Lett.,* **11** (1991) 55.

36. R.D. Shannon and C.T. Prewitt, *Acta Crystallogr.,* **B25** (1969) 925.

37. S. Kannan and C.S. Swamy, In: *INDO-US workshop on Perspectives in New Materials,* New Delhi, India, Mar. 23-24, 1992, Abs. p.75.

133

38. R.M. Taylor, *Clay Miner.*, **15** (1980) 369.
39. F.M. Labojos, V. Rives and M.A. Ulibarri, *J. Mat. Sci.*, **27** (1992) 1546.

40. M.J. Hernandez-Moreno, M.A. Ulibarri, J.L. Rendon and C.J. Serna, Phys. *Chem. Miner.*, **12** (1985) 34.
41. M.J. Hernandez, M.A. Ulibarri, J.L. Rendon, C.J. Serna, *Therm. Acta*, **81** (1984) 311.
42. L. Pesic, L. Salipurovic, V. Markovic, D. Vucelic, W. Kaguny and W. Jones, *J. Mater. Chem.*, **2** (1992) 1069.
43. A.J. Marchi, J.I. Di Cosimo, C.R. Apestigua, In: *Proc. 9th Int. Congress on Catal*, Chemical Institute of Canada, Ottawa, eds., M.J. Philips and M. Ternan **2** (1988) 529.
44. W.T. Reichle, S.Y. Kang, and D.S. Everhardt, *J. Catal.*, **101** (1986) 352.
45. E. Murad and R.M. Taylor, *Clay Miner.*, **19** (1984) 77.
46. S. Kannan and C.S. Swamy, *J. Mat. Sci. Lett.*, **11** (1992) 1585.
47. A.L. McKenzie, C.T. Fishel, and R.J. Davis, *J. Catal.*, **138** (1992) 547.
48. S. Gusi, F. Pizzoli, F. Trifiro, A. Vaccari and G. Del Piero, In: *Prep of Catalysts IV*, eds., B. Delmon, P. Grange, P.A. Jacobs and G. Poncelet, Elsevier, Amsterdam, 1987, p.723.
49. B. Delmon, *Appl. Catal. B*, **1** (1992) 139.
50. A.N. Hayhurst and A.D. Lawrence, *Prog. Energy Combust. Sci.*, **18** (1992) 529.
51. M.H. Thiemens and W.C. Trogler, *Science*, **251** (1991) 932.
52. A. Cimino and M. Schiavello, *J. Catal.*, **20** (1971) 202.
53. S. Louis Raj, B. Viswanathan and V. Srinivasan, *J. Catal.*, **75** (1982) 185.
54. K.V. Ramanujachary, N. Kameshwari and C.S. Swamy, *J. Catal.*, **86** (1984) 121.
55. J. Christopher and C.S. Swamy, *J. Mol. Catal.*, **62** (1990) 69.
56. J. Christopher and C.S. Swamy, *Catal. Rev. Sci. Engg.*, **34** (1992) 409.
57. J. Leglise, J.O. Petunchi and W.K. Hall, *J. Catal.*, **86** (1984) 392.
58. L.M. Aparacio, M.A. Ulla, W.S. Millman, J.A. Dumesic, *J. Catal.*, **110** (1988) 330.
59. G.I. Panov, V.I. Sobolev and A.S. Kharitonov, *J. Mol. Catal.*, **61** (1990) 85.
60. Y. Li and J.N. Armor, *Appl. Catal. B*, **1** (1992) L21.
61. S.Kannan, C.S. Swamy, Y. Li, J.N. Armor and T.A. Braymer, *U.S. Patent* (filed) (1993).
62. H.C.B. Hansen and R.M. Taylor, *Clay Miner.*, **25** (1990) 161.
63. D.L. Bish, *Bull. Miner.*, **103** (1980) 170.
64. T. Kwon, G.A. Tsigdions and T.J. Pinnavaia, *J. Amer. Chem. Soc.*, **110** (1988) 3653.
65. T. Yamagishi, *Nippon Kakagu Kaishi*, **4** (1993) 329. (Chem. Abs. No: 118, 260431t)
66. E. Suzuki, M. Okamoto and Y. Ono, *J. Mol. Catal.*, **61** (1990) 283.
67. S. Velu and C.S. Swamy (unpublished results).
68. E. Suzuki, M. Okamoto and Y.Ono, *Chem. Lett.*, (1989) 1485.
69. S. Miyata and T. Hirose, *Clays Clay Miner.*, **26** (1978) 441.
70. S. Miyata and M. Kuroda, *U.S. Patent* 4,284,762.

Main Group Elements and Their Compounds
V.G. Kumar Das (Ed)
Copyright © 1996 Narosa Publishing House, New Delhi, India

Microwave Ovens in the Laboratory: Recent Advances

Frank E. Smith

Department of Chemistry, Laurentian University, Sudbury
Ontario P3E 2C6, Canada

Laurentian University in Northern Ontario was one of the first centres in the world to carry out research into the laboratory use of microwave ovens for the rapid digestion of mineral samples prior to chemical analysis, and for speeding up reactions in chemical synthesis. The first publications describing the use of sealed teflon vessels for both microwave-assisted digestions and chemical synthesis came out of Laurentian University.[1,2] It is interesting to briefly review the circumstances which led up to these developments.

Laurentian University is located in Sudbury, in the heart of Canada's mining and smelting industry. Efficient exploitation of minerals in the earth's crust requires extensive knowledge of the detailed geology, mineralogy and geochemistry of the various deposits, as well as a comprehensive understanding of the behaviour of the various ores and minerals at different stages of the mining, milling, smelting and refining processes. Analytical chemistry is fundamental to the acquisition of this knowledge, and the mining industry houses some of the most sophisticated and advanced analytical chemistry laboratories to be found as well as supports a wide range of associated research and development activities. Over the past decade, the mining and smelting industry has come under increasing pressure on two main fronts - economic and environmental. On the economic front, inflation along with weak metal prices, increasing competition and lower quality ore bodies have forced companies to improve efficiency and productivity. At the same time, society has become increasingly concerned with the environmental effects of the industry. This has led to a demand for more stringent controls on emissions and effluents, and for the revegetation of previously devastated areas. These influences have in turn made efficient management more and more dependent on reliable chemical analysis at all stages of the mining and smelting process. Analytical chemistry has never had a more important role to play. And the signs are that this role will become even more crucial in the future. As its significance in these and other industries has increased, analytical chemistry has become more sophisticated. Improved instruments with the capability to detect ever-smaller quantities of more and more elements and compounds, more rapidly than before, are constantly being developed and typically incorporate powerful computational capabilities. However, many of these advanced techniques require the sample to be in solution form prior to analysis and yet dissolution methods had hardly changed in the sixty years prior to the time the first pressurised microwave digestions were carried out in Sudbury. The inspiration for these studies came from a U.S. Bureau of Mines report,[3] which described how rapid dissolution of some mineral samples had been achieved by using a microwave oven to heat samples and an acid mixture in polycarbonate bottles. When this experiment was repeated, the method showed great potential, but the vessels proved unsuitable for more rigorous conditions. A great deal of effort went into the search for better reaction vessels - a range of commercially-available Teflon® bottles was tried, but these deformed or burst during the heating process. Some vessels were machined out of solid Teflon® blocks, but it was found that the screw threads in the cap deformed during heating, allowing the cap to explode off. Finally, a vessel was constructed from Teflon® PFA tubing plus end caps purchased from the Savillex Corporation,[4] and this proved successful. Arrangements were then made with the manufacturer to make vessels from the tubing. These were marketed and eventually became the vessel of choice for microwave experimentation. The identification of this more suitable vessel was a very important step forward and led to the development of a variety of digestion protocols,[5-8] and to a range of organic and organometallic syntheses,[9-12] all extremely fast compared with the traditional methods they were intended to replace.

The discussion now continues along three separate themes:

    i)   microwave methods in analytical chemistry
    ii)   microwave methods in chemical synthesis
    iii)   new technology

Microwave methods in analytical chemistry

Microwave ovens have been used in chemical laboratories for moisture analysis and wet ashing procedures of biological and geological materials for many years.[13,14] The pressure dissolution of geological specimens in Parr bombs heated in water baths has also been exploited for a long time.[5] However, it is only recently that the two techniques have been combined to provide rapid pressure dissolutions of analytical samples prior to AA or ICP analysis.[5-8]

A variety of protocols for digesting analytical samples has now been developed, but here, two examples are presented to illustrate the microwave dissolution method:

Determination of nickel in blister copper (INCO Ltd. Copper Cliff, Ontario)

This analysis which is performed on a daily basis for many samples, was previously carried out by an open beaker digestion of 2.5 g samples in sulphuric acid. The samples were heated for 16 hr to dryness and then redissolved in distilled water prior to analysis.

The current method involves placing a 1.0 g sample in a microwave digestion vessel with a mixture of nitric and hydrochloric acids (3:5) and then heating the vessel in the microwave oven for 2 min at a pressure setting of 70 psi. The resulting solution is diluted to volume prior to analysis. The analytical results obtained using the rapid microwave technique are comparable in terms of accuracy and precision with those from the open beaker technique. Adoption of this technique has enabled INCO to release one technician for work elsewhere in the plant.

Determination of selenium in ore samples [7] (Kidd Creek Mines Ltd., South Porcupine, Ontario)

The open beaker technique involved a slow digestion of the sample in a mixture of nitric and perchloric acids, followed by an extraction into toluene containing 1% *o*-phenylene diamine and then a colorimetric determination of the selenium. Two analysts were needed to process 50 samples for results the next day.

The microwave procedure (for a tray of 18 vessels) consists of placing a 300 mg sample into each vessel with 3 mL of concentrated nitric acid and letting the mixture react for 2 min., capping the vessels and microwaving for 3 min. at 280 W. Next, the vessels are cooled in ice-water, uncapped and 5 mL of concentrated hydrochloric acid added. The vessels are then recapped and microwaved for 3 min. at 280W. The final step involves cooling and uncapping the vessels once more, adding 22 mL of distilled water to each, and then recapping and microwaving for 5 min. at 600 W. The resulting solutions are ready for Zeeman Flame Atomic Absorption (ZFAA) analysis. Each of the steps in the microwave dissolution procedure outlined above is performed robotically. Using this technique, one analyst can process 50 samples for results the same day. In addition, there is a marked reduction in the loss of selenium compared to the open beaker method, and the solutions produced may be used for the determination of other analytes such as arsenic, copper and zinc.

This major breakthrough in sample preparation methodology has particular significance for analytical laboratories in the mineral processing industry for a number of reasons:

a)   Analytical results can be available within a matter of minutes if necessary. Tedious open beaker digestions which often took many hours can be replaced by pressure dissolutions which take just a few minutes. This means that managers may respond to changes in plant operating conditions more quickly, leading to better process control and efficiency.

b)   Loss of volatile analyte is eliminated in the closed system. Both fusion and open beaker techniques can lead to volatilisation if an element that forms volatile compounds is present in the system.

c)   The closed microwave digesion system reduces the possibility of sample contamination or loss through vigorous boiling.

d)   Microwave dissolution techniques lend themselves readily to automation and robotics.[9] This can mean the elimination of some of the more repetitive laboratory tasks, freeing analysts for more creative work.

e)   The potential economic advantages of the microwave techniques are considerable, and encompass savings in labour, power, chemicals and the need for fume hoods. In addition come the advantages of a cleaner and healthier laboratory environment.

The microwave technique has become standard in analytical laboratories around the world. A variety of complete computer controlled microwave dissolution systems are now on the market, some of which allow for pressure and temperature monitoring and control throughout the heating cycle. The digestion vessels have also been dramatically improved, and operating pressures up to 1600 psi can be safely attained, enabling rapid dissolution of very difficult samples.

Microwave methods in chemical synthesis

The use of microwave ovens for the rapid synthesis of organic compounds was first reported by a team from Laurentian University.[2] It was observed in our laboratories that organic reactions were dramatically accelerated by irradiating the reaction mixture in a sealed Teflon container with microwaves. This could be exploited in the rapid synthesis of a wide variety of compounds. Some examples of these reactions included the hydrolysis of amides and esters to carboxylic acids, the esterification of carboxylic acids with alcohols and the conversion of alkyl halides to ethers.[2] Shortly after our initial communication, other workers reported substantial reductions in reaction times in syntheses for the Diels-Alder, Claisen and ene reactions using a microwave oven.[11]

This rapid and convenient method has since attracted a great deal of attention and a number of synthetic and analytical applications of microwave heating of organic reactions have been reported during the last several years.[9,12,16-25] The field has recently been reviewed.[26,27] Some examples of the Laurentian University results in this field are now presented.

A number of different types of organic reactions were carried out both in sealed Teflon containers in microwave ovens and under conventional reflux conditions. Substantial reductions in reaction times were observed in the microwave reactions and rate enhancements varying from 6 times in the hydrolysis of benzamide, to 240 times in the S_N2 rection of 4-cyanophenoxide ion with benzyl chloride in methanol, were achieved using the normal technique. A rate enhancement of 1200 times was found when the latter reaction was carried out in a 50 mL Teflon bomb.[9,13] The results of some representative reactions are shown in Table 1.

Table 1. Reaction times and yields in representative syntheses using microwave and reflux techniques.

Compound synthesised	Procedure followed	Reaction time	Average yield (%)	Solvent	Rate enhancement
Hydrolysis of benzamide					
C_6H_5COOH	Microwave	10 min	99	H_2O	6
C_6H_5COOH	Reflux	1 hr	90		
Hydrolysis of methyl benzoate					
C_6H_5COOH	Microwave	2.5 min	84	H_2O	24
C_6H_5COOH	Reflux	1 hr	95		
Esterification of benzoic acid with methanol					
$C_6H_5COOCH_3$	Microwave	5 min	76	CH_3OH	96
$C_6H_5COOCH_3$	Reflux	8 hr	74		
Conversion of benzophenone to oxime					
$(C_6H_5)_2C{:}NOH$	Microwave	2 min	71	C_2H_5OH	60
$(C_6H_5)_2C{:}NOH$	Reflux	2 hr	68	pyridine	
S_N2 reaction of 4-cyanophenoxide ion with benzyl chloride					
$NCC_6H_5OCH_2C_6H_5$	Microwave	4 min	89	CH_3OH	240
$NCC_6H_5OCH_2C_6H_5$	Reflux	16 hr	93		

For safety reasons, the microwave method was only used for preparing relatively small amounts (up to 2-3 grams) of products, since our early attempts to use larger reaction volumes resulted in very high pressures being produced. On one occasion, a violent explosion occurred when approximately 40 mL of a mixture containing toluene, $KMnO_4$ and aqueous KOH was heated in a 50 mL Teflon bomb. Microwave heating was found to be particularly useful for preparing derivatives for the identification of organic compounds[10] where only small amounts of products are required. For example, oximes of aldehydes and ketones could be prepared in good yield in only 2 min, compared with a reflux time of 2 hr.[28]

Once it had been shown that microwave heating could be useful in the rapid synthesis of a wide range of compounds, attention was turned to a study of the factors which affected the rates of the reactions as well as approaches to improve the safety of the method.

Solvent effects

An important factor which affects the rate of a reaction carried out in a microwave oven is the polarity of the solvent. It has been known for some time that polar molecules absorb microwaves, resulting in increased rotation and this causes heating of the substance through frictional effects.[29,30] It would therefore be expected that the rate at which a solvent absorbs microwave irradiation depends on its polarity, that is, its dielectric constant. An experiment was performed to determine the increase in temperature when equal amounts (0.5 mol) of a number of liquids were heated for 15 seconds in the microwave oven. It was found that strongly polar compounds such as water, alcohols and ketones heated up the most rapidly whereas less polar liquids, e.g., ethers, amines and alkyl halides showed smaller temperature increases. As expected, little or no heating occurred with non-polar liquids, e.g., carbon tetrachloride and hydrocarbons. The data in Table 2 show that the temperature increase is related to the dielectric constant of the liquid.

Thus microwave heating would be expected to produce the most dramatic rate increases in solvents of high dielectric constant. The results in Table 1 show that the greatest rate enhancements were obtained for reactions in methanol. Reactions in water and higher boiling alcohols, on the other hand, showed smaller rate increases. The results in Table 3 show that in the esterification of benzoic acid the rate enhancement is greatest for methanol and decreases as the chain length of the alcohol increases. The dielectric constants of alcohols decrease only gradually as the chain length increases. Thus, the heating rates for all these alcohols in the microwave oven are similar. A more important factor affecting the rate enhancements is the fact that the rate of the reflux reaction increases with the boiling point of the alcohol. These results might suggest that the microwave technique produces significant rate increases only in reactions in low boiling solvents. However, the rate of reactions in higher boiling solvents can be increased significantly by using a higher power level in the microwave oven. For example, the rate enhancement for the esterification of benzoic acid with 1-pentanol was increased from 1.3 to 6.1 by a small increase in power level from 560 to 630 watts.

Table 2. The increase in temperature found when 0.5 mol of liquids are heated for 15 sec at 560 watts in an open vessel in a microwave oven

Compound	Temp. increase after 15 sec	Boiling point °C	Dielectric constant
1,4-dioxane	11	101	2.2
Tripropylamine	10	156	2.4
Propanoic acid	19	141	3.3
Chloroform	24	62	4.8
n-Propyl acetate	28	102	5.6
Ethyl acetate	29	77	6.0
Acetic acid	38	118	6.2
1-Hexanol	45	158	13.8
1-Pentanol	51	137	13.9
2-Pentanone	49	102	15.4
1-Butanol	56	117	17.8
2-Butanone	41	80	18.5
1-Propanol	62	97	20.1
Water	44	100	78.5

Table 3. Reaction times and rate enhancements for the esterification of benzoic acid with different alcohols under conventional and microwave heating[a]

Alcohol	Approx. reaction temp. °C	Reaction Time	Yield %	Rate Enhancement
Methanol	65	8 hr (reflux)	74	
	134	5 min (M-wave)	76	96
1-Propanol	97	4 hr (reflux)	78	
	135	6 min (M-wave)	79	40
1-Butanol	117	1 hr (reflux)	82	
	135	7.5 min (M-Wave)	79	8
1-Pentanol	137	10 min (reflux)	83	
	137	7.5 min (M-wave)	79	1.3
1- Pentanol at 630 watts	162	1.5 min (M-wave)	77	6.1

[a] All reactions were carried out using 10 mL of the alcohol in a 300 mL Berghof bottle. A ten-fold molar excess of the alcohol to benzoic acid was used. With the exception of the 1-pentanol reaction at 630 watts, all the reactions were carried out at 560 watts.

Pressure effects

In our earlier experiments it was observed that very high pressures were generated in microwave heated reactions if low boiling solvents, such as methanol, were used. In some of these reactions, deformations of the Teflon containers or even explosions occurred. The increased rates of reaction are believed to be the result of the rapid superheating of the solvent. This leads to the substantial increase in pressure that is observed. Since the effect of pressure appeared to play a significant role in the rate enhancements, the effect of pressure on the rate of the S$_N$2 reaction of 4-cyanophenoxide ion with benzyl chloride in methanol was investigated. The rate enhancements of this reaction when identical amounts of reagents were reacted to 65% completion in 50mL, 120 mL and 300 mL Teflon vessels were determined. The results in Table 4 show that the rate of reaction increased markedly when the volume of the Teflon container is reduced. A plot of 1/volume of the vessel *vs.* 1/reaction time was linear and thus the rate of reaction varies inversely with the volume of the vessel, and hence is proportional to the pressure developed.

Table 4. Time required to obtain 65% yield of benzyl-4-cyanophenyl ether using equal amounts of sodium 4-cyanophenoxide, benzyl chloride and methanol in vessels of different sizes

Volume of the Teflon vessel (mL)	Time for 65% reaction	1/Volume (1/L)	1/Time (1/min)	Rate Enchancement
50	35 sec	20.0	1.71	1240
120	1.3 min	8.3	0.75	540
300	3.0 min	3.3	0.33	240

Initially the microwave energy is used to bring the reaction mixture to the boil. Any extra energy will presumably be used to convert solvent molecules to the vapour state and the pressure inside the vessel will increase. If a smaller vessel is used, higher pressures will result. Since the boiling point of the solvent is higher at higher pressures, the reaction occurs at a higher temperature and therefore faster in smaller reaction vessels.

Superheating

The effect of pressure on the boiling point of a solvent is well known and widely understood, but it has recently been reported that many organic solvents, when heated in a microwave oven at atmospheric pressure, may boil at temperatures 13-26°C above their normal boiling points.[31] The explanation for this probably lies in the fact that conventional boiling occurs via heat transfer from the walls of a vessel where there are large numbers of nucleation sites. In microwave heating, however, heat transfer takes place far from the walls where the number of nucleation sites around which bubbles may readily form is very limited.

Specific "microwave" or athermal effects

A few reports have appeared which suggest that, in certain instances, reactions carried out by microwave heating give different products from those carried out by conventional heating [32,33] This question is of great concern to the food industry, in view of the widespread use of domestic microwave ovens, and so Unilever and Nestle sponsored a special meeting on this issue during the summer of 1993, in Switzerland. It was generally agreed that all the anomalous results reported could be explained on the basis of localised superheating effects at different hotspots in the reaction mixtures, and that there were no athermal effects. This conclusion is not surprising in view of energy considerations. Microwave radiation is of low energy, about 1 joule per mole of photons. When this is compared to bond energies (H-bonds ~ 20 kJ/mole, the C-H bond energy ~ 350 kJ/mole), it seems unlikely that microwave energy could preferentially rupture any chemical bonds.

Inorganic, organometallic and solid state reactions

Recent reports have described the use of microwave energy in the synthesis of mixed metal oxides (including high-temperature superconductors), metal binary compounds such as borides, carbides, chalcogenides, halides and nitrides and metal cluster compounds.[27]

Radioisotope studies - microwave syntheses of radioactively-labelled compounds for tracer studies in medicine

The isotope of carbon, ^{11}C, which is a positron emitter, has a half-life of just 20 minutes. This means that if ^{11}C-labelled compounds are used in PET (positron emission tomography) studies, at the end of the testing period, residual radiation will quickly decay, reducing the risk of radiation damage to the patient. But, as the half-life is so short, speedy synthesis of labelled compounds is essential. This is an area where microwave syntheses can be used very effectively.[34]

New Technology

Lined reaction vessels

The latest reaction vessels available for digestions or syntheses incorporate a Teflon® PFA liner, cover and rupture membrane enclosed in a casing and cap of Ultem® polyetherimide. These are safe at pressures of up to 200 psig (1380 kPa) and temperatures up to 250°C. Special vessels that can operate at up to 600 psig (4140 kPa), or even 1600 psig (11,040 kPa) are also available. The vessels are also available fitted with ports to accommodate temperature and pressure probes[35-39]

Temperature probes

Temperature is a crucial factor in all chemical reactions and digestions, but some difficulties arise if conventional thermocouples are used in the reaction mixture inside the microwave cavity. This problem has been resolved by the use of fibre optic probes. Several types are used, depending on the manufacturer.[35-38]

Digestion systems with both temperature and pressure monitors and settings

Four systems are now commercially available which incorporate simultaneous pressure and temperature monitoring.[35-38] The reaction/digestion may be pre-set to run at either a given temperature, or a given pressure. The temperature or pressure may be programmed to vary in a step-wise fashion over a pre-determined time period. The systems are microprocessor-controlled, and printouts of reaction conditions may be obtained.

Protein hydrolysis apparatus

A specialised reactor for vapour phase protein hydrolysis, which fits inside a microwave oven, is available. Up to ten samples may be hydrolysed in each batch, and the time taken is less than ten minutes per batch. This is becoming widely used for amino acid analysis.

Flow-through systems

A flow-through system for chemical synthesis was announced several years ago, but has been plagued with development problems and is not yet on the market. However, it is anticipated that development will soon be completed and the apparatus will be commercially available in the near future.[40]

A flow-through digestion system for chemical analysis is now on the market.[36] The system is fully-automated, and incorporates a cleaning step between samples to eliminate carry-over. Before delivery into the final auto-sampler, all samples are cooled and filtered, and ready for final analysis by ICP, AA or other techniques.

Acknowledgments

I would like to express my gratitude to the following researchers with whom the work described was carried out :

i)   Richard Gedye, Ken Westaway and John Huang from Laurentian University with the help of many Laurentian students.

ii) John Bozic and members of his team from INCO Ltd. (Copper Cliff)

iii) John Labrecque from Kidd Creek Mines Ltd. (Timmins)

References

1. F. E. Smith, B.G. Cousins, J. Bozic and W. Flora, *Anal. Chim. Acta*, **177** (1985) 243.

2. R.N. Gedye, F.E. Smith, K.C. Westaway, H. Ali, L. Baldisera, L. Laberge, and J. Rousell, *Tetrahedron Letters*, **27** (1986) , 279.

3. S. A. Matthes, R. F. Farrell and A. J. Mackie, Tech. Progr. Rep. U.S. Bureau of Mines, Number 120 (1983).

4. The Savillex Corporation, 5325 Hwy. 101, Minnetonka, Minnesota 53345, USA.

5. F.E. Smith, B.G. Cousins and J.-Y. Maillet, *Education in Chemistry*, **24** (1987) 13.

6. J. Huang, D. Goltz and F.E. Smith, *Talanta*, **35** (1988) 907.

7. J.M. Labrecque, *Proceedings of the 17th Canadian Mineral Analysts Conference*, September 1985, pp17-38

8. H.M. Kingston and L.B. Jassie (eds), *Introduction to microwave sample preparation*, American Chemical Society, Washington DC, 1988, pp 203-230.

9. R.N. Gedye, F.E. Smith and K.C. Westaway, *Canadian Journal of Chemistry*, **66** (1988) 17.

10. R.N. Gedye, F.E. Smith, K.C. Westaway, *Education in Chemistry*, **25** (1988) 55.

11. R.J. Giguere, T.L. Bray, S.M. Duncan and G. Majetich, *Tetrahedron Letters*, **27** (1986) 4945.

12. R.J. Giguere, A.M. Naman, B.O. Lopez, A Arepally and D.E. Ramos, *Tetrahedron Letters*, **28** (1987) 6553.

13. J. A. Hesek and R. C. Wilson, *Analyt. Chem.*, **46** (1974) 1160.

14. A. Abu-Samra, J. S. Morris and S. R. Koirtyohann, *Analyt. Chem.*, **47** (1975) 1475.

15. R. T. Rantala and D. H. Loring, *Atomic Absorption Newsletter*, **14** (1975) 117.

16. H.M. Yu, S.T. Chen, S.H. Chiou and K.T. Wang, *Journal of Chromatography*, **456** (1988) 357.

17. M. Jie and C. Yankit, *Lipids*, **23** (1988) 367.

18. J. Linders, J.P. Kotje, M. Overhand, T.S. Lie and L. Maat, *Recueil des Travaux Chimiques des Pays-Bas*, **107** (1988) 449.

19. A.J.J. Straathof, H. Vanbekkum and A.P.G. Kieboom, *Recueil des Travaux Chimiques des Pays-Bas*, **107** (1988) 647.

20. E. Gutierrez, A. Loupy, G. Bram and E. Ruiz-Hitzky, *Tetrahedron Letters*, **30** (1989) 945.

21. A.B. Alloum, B. Labiad and D. Villemin, *Journal of the Chemical Society, Chemical Communications*, (1989) 386.

22. W.G. Stroop and D.C. Shaefer, *Analytical Biochemistry*, **182** (1989) 222.

23. S. Agod, I. Gyoker, G. Bene, A. Puskas, J. Vari, A. Pinter and S. Takacs, *Hungarian Patent HU*, **47** (1989) 613.

24. S. Takano, A. Kijima, T. Sugihara, S. Satoh and K. Ogasawara, *Chemistry Letters*, **1** (1989) 87.

25. S.H. Chiou, *Journal of the Chinese Chemical Society*, **36** (1989) 435.

26. R.A. Giguere, *Organic Synthesis: Theory and Application*, Vol. 1, p 103-172, JAI Press Inc., 1989.
27. D.R. Baghurst and M.P. Mingos, *Chem. Soc. Rev.*, **20** (1991) 1.
28. R.L. Shriner, R.C. Fuson, D.Y. Curtin and T.C. Morill, *The Systematic Identification of Organic Compounds* (6th Edition), p 181 , John Wiley, New York, 1980.
29. D.A. Copson, *Microwave Heating*, (2nd Edition), p 8-18, AVI Publishing Company, Westport, CT, 1975.
30. K.W. Watkins, *Journal of Chemical Education*, **60** (1983) 1043.
31. D.R. Baghurst and D.M.P. Mingos, *Journal of the Chemical Society, Chemical Communications*, (1992) 674.
32. J. Berlan, P. Giboreau, S. Lefeuvre, and C. Marchand, *Tetrahedron Letters*, **32** (1991) 2363.
33. J. Berlan, K. Cann-Pailler, J.L. Imbert and R. Tessier, French Patent 9114720, 28 Nov. 1991.
34. J.O.Thorell, Ph.D. Diss., Karolinska Hospital and Inst., Stockholm, Sweden 1993.
35. Questron Corporation, P.O. Box 2387, Princeton, NJ 08543, USA.
36. CEM Corporation, P.O. Box 200, Matthews, NC 28106, USA.
37. Floyd Associates Inc., Lake Wyle, SC, USA.
38. MILESTONE s.r.l., via Fatebenefratelli 1/5, 24010 Sorisole (Bergamo), Italy.
39. Parr Instument Company, 211 Fifty-Third St., Moline, Illinois 61265 USA.
40. Dr Chris Strauss, CSIRO Division of Chemicals and Polymers, Private Bag 10, Clayton, Victoria 3168, Australia

Main Group Elements and Their Compounds
V.G. Kumar Das (Ed)
Copyright © 1996 Narosa Publishing House, New Delhi, India

Raman Spectroscopy of Semiconducting Materials

S. Radhakrishna

Institute of Advanced Studies, University of Malaya
59100 Kuala Lumpur, Malaysia

Raman spectroscopy has attained a new significance with the advent of modern instrumentation utilising fibre optics which allow the possibility of *in situ* Raman measurements. This feature has taken Raman spectroscopy to the forefront of the semiconductor industry where it is now possible to scan samples to a spatial resolution of 1 μm allowing a complete surface scanning of *in situ* semiconductor samples as they are being processed. This, coupled with the fact that we can get tunable laser sources practically in the complete range 2000-15000 Å, has given Raman spectroscopy a new lease of life. Indeed, it is now possible to observe many features which could not be studied a few years ago. With these new sources and ultra high sensitive CCD (charge coupled devices) detectors, the limits of resolution and sensitivity have been pushed to the extreme.

Raman spectroscopy deals with the appearance of additional frequencies which are characteristic of the material under investigation. The number of Raman lines or the new frequencies observed depends on the nature of the vibrational properties of the sample.[1-3] Because of this a study of the Raman spectrum of an as-grown semiconductor material will provide information about the vibrational characteristics.

A light beam incident on a sample causes an induced polarisability, P, given by the expression

$$P = [\alpha_o + \alpha_{vib} \cos (2\pi \nu_{vib} t)].[E_o \cos (2\pi \nu_o t)]$$

comprising the three terms,

$\alpha_o E_o \cos 2\pi \nu_o t$ (Rayleigh scattering), $\alpha_{vib} E_o \cos [2\pi(\nu_o - \nu_{vib})t]$ (Stokes scattering) and $\alpha_o E_o \cos [2\pi \nu_o + \nu_{vib})t]$ (Anti-stokes scattering)

Two of these terms will appear only because of the vibrational (ν_{vib}) frequency of the sample. A homopolar crystal like silicon will have only one characteristic vibrational frequency (at 520 cm^{-1}), while more complex crystals with more vibrating atoms of different kinds will have more vibrational frequencies. However Raman scattering is a very weak phenomenon; typically for every 30000 photons incident on a sample of 1 m thickness, only one Raman scattered photon is observed. This means that the incident light must be as intense as possible. For the reason that the intensity of the scattered light falls off very rapidly with increase in the wavelength of the incident light, Raman spectrometers were traditionally equipped with high pressure gas ultraviolet sources, but the recent advent of powerful coherent laser sources has ensured that the "weak" scattered light is strong enough to be observed. As the Rayleigh scattered light is about 10^{12} times the intensity of the scattered Raman light, the rejection of the Rayleigh frequencies is as important as the gathering of the weak Raman scattering. The use of the holographic notch filter for rejection of the Rayleigh scattered light and the use of powerful CCD detection techniques are among features that modern Raman instruments such as the Renishaw[4] offer present day researchers.

Experimental

The experimental setup used in the experiments described here (see Fig. 1) consists of a commercially available Renishaw Raman imaging microscope consisting of a low power air cooled HeNe Laser source(the laser power delivered at the sample was 5mW), an Olympus microscope,a single spectrograph for spectroscopy mode, a set of band pass filters for the imaging mode, a peltier cooled CCD detector and a suitable recorder for recording the Raman spectra. Single crystals of silicon grown from melt and cut into wafers were used in the present experiments.

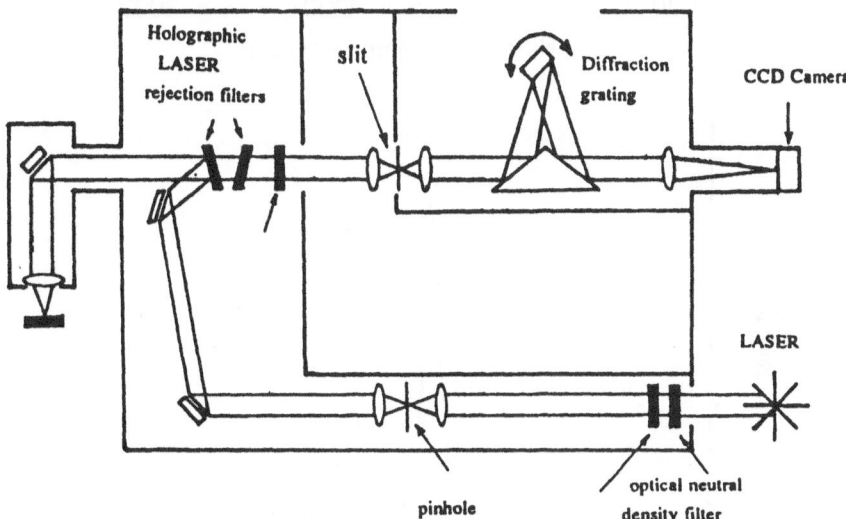

Figure 1 : Schematic diagram of a modern imaging Raman microscope

Results and Discussion

Figure 2 shows the Raman spectrum obtained with a typical silicon sample. Fig 2a shows the Raman spectrum recorded for the sample in the high order spectral range;[5] Fig 2b shows the very good signal to noise ratio and quality of the spectrum that can be obtained. One of the important applications to which the high spatial resolution has been put is in respect of material contamination. By covering a typical 6 inch silicon wafer it is possible to look for impurity contamination at the 1 μm level both in the spectroscopic mode as well as in the Raman image mode. By fixing the microscope arrangement at 519 cm^{-1} it is possible to quickly see if there are any white patches on the sample. A white patch in an otherwise dark background will quickly indicate the presence of an impurity, and the identification

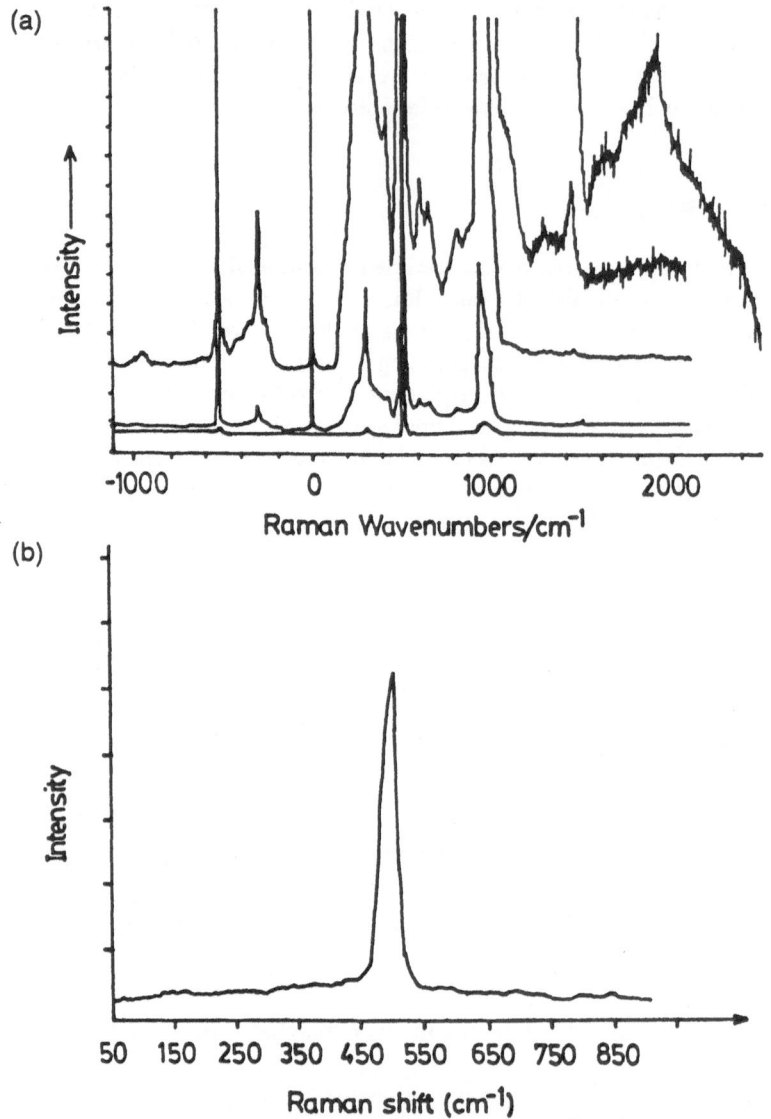

(a)

(b)

Figure 2 : a) **Raman spectrum of a silicon single crystal using a 5 mW laser and 633 cm⁻¹ excitation showing the higher order Raman lines.[5]**
b) **Raman spectrum of a silicon single crystal showing the excellent signal to noise ratio that is attainable.**

In Raman spectroscopic imaging spatial information can be obtained by a Raman mapping procedure in which the focus spot is moved in a two dimensional surface of the sample. This rapid analysis of the sample *in situ* has made enormous contributions to the semiconductor industry. The use of Raman spectroscopy for studying impurities in crystals has been established in a number of cases.[6-8] Raman spectroscopy has also been useful in the study of silica fibers doped with Nd^{3+} and Er^{3+} ions.[9]

148

Another important application of Raman spectroscopy has been in the micrometer sized dot imaging of GaAs samples.[10] This has established Raman spectroscopy as a technique for nondestructive and quantitative microanalysis of semiconductor microstructures. Since a typical device element in many VLSI applications is about 1 μm in dimensions, the ability of the Raman microprobe to look at spatial resolutions of 1 μm has proved invaluable. GaAs samples give two Raman lines at 292 and 269 cm^{-1} representing LO and TO phonon modes of GaAs crystals. The presence of additional bands at 242 cm^{-1} has been interpreted as being due to structural damage which makes it possible to observe a forbidden zone edge Raman line. Figure 3 shows the Raman spectrum of GaAs samples with spatial resolution corresponding to the dot size. Curve (a) shows the Raman spectrum of undamaged GaAs and this is obtained for the most part of the sample. However there are regions which have been damaged and these regions show the additional peak at 242 cm^{-1}.

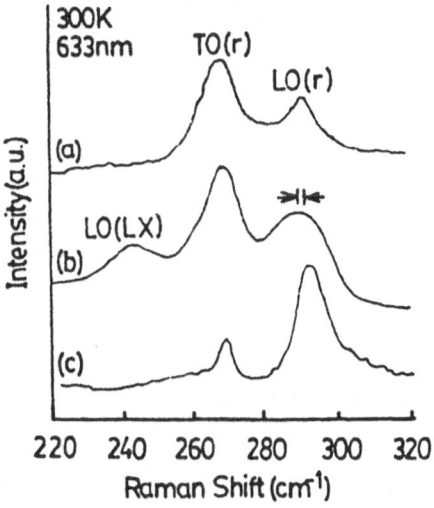

Figure 3 : Raman spectrum of a GaAs dot shown in (a) and (b) shows Raman spectrum with a damaged GaAs,[4] and (c) a typical spectrum of substrate.

Another important area of application of the Raman spectroscopy in material characterisation is the determination of the quality of diamond films. A diamond crystal shows a strong and sharp Raman line at 1332 cm^{-1}. Figure 4 shows three Raman spectra obtained for the case of a) a diamond crystal, b) good quality diamond films, and c) bad quality diamond films. The spectra were obtained using a low power 25 mW HeNe Laser.[5] It should be noted that the vertical axis is not to scale and the diamond Raman line shown in curve a is much more intense than the lines shown in curves b and c. What is important to note is that as the quality of the diamond film deteriorates the Raman line broadens considerably on account of the presence of graphite and consequent photoluminescence effects. This observation is very significant because the appearance of a broad Raman peak will clearly indicate the presence of graphite and the growth process can then be altered to bring about good quality diamond films. The important thing is the rapidity with which this scan can be observed.

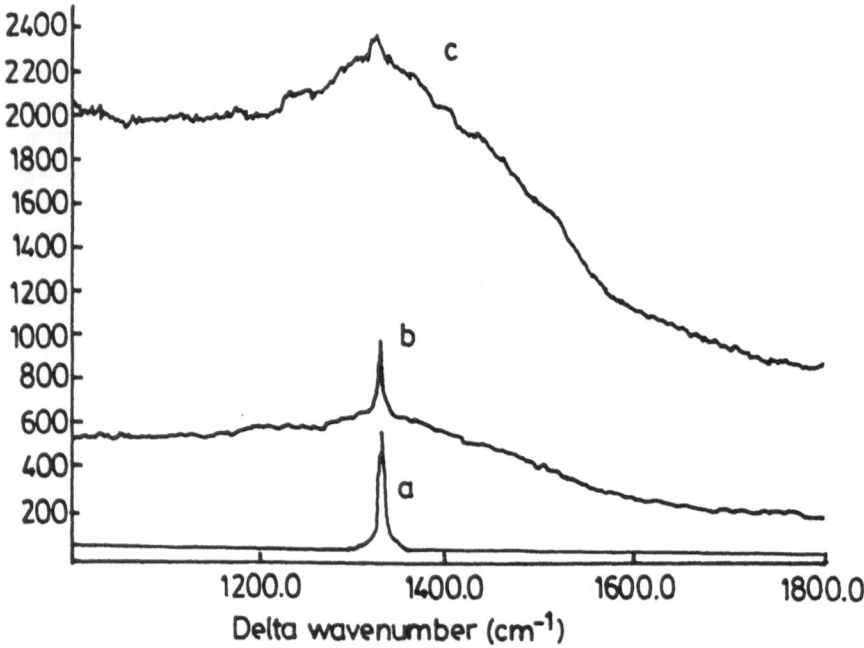

Figure 4 : Raman spectrum of (a) a diamond crystal (b) good quality diamond films and (c) bad quality diamond films.[5]

References

1. T.R.Gilson and P.J.Hendra, *Laser Raman Spectroscopy*, Wiley, New York 1970

2. M.Cardona (ed), *Light Scattering in Solids*, vol. 1-6, Springer, Berlin 1983-1992.

3. G.J.Rosasco, in *Advances in Infra red and Raman spectroscopy*, R.J.H.Clark and R.E.Hester (eds), Wiley Heydon, Chichestor, 1980.

4. K.P.J.Williams, "Recent Trends in Raman Spectroscopy", In: *Optical Techniques and Materials*, eds., S.Radhakrishna and Harith Ahmad, Nova Science Publishers, New York, 1995.

5. K.P.J.Williams, In: *Renishaw Raman Technical literature*, 1994.

6. S. Radhakrishna, *Phy. Rev.*, **48** (1971) 1382.

7. S. Radhakrishna and B.D. Sharma, *J. Phy. Soc. Jpn.*, **31** (1971) 184.

8. S. Radhakrishna, K.P. Jain and A.K. Prabhakaran, *Phy. Rev.*, 5 (1972) 2325.

9. F.J. Perera, H.B. Ahmad and S. Radhakrishna, *Materials Science and Engineering* (1995), in press.

10. P.D.Wang, C.Cheng, C.M.Sotomayor Torres and D.N.Batchelder, *J. Appl. Phy.*, **74** (1993) 5907.

Main Group Elements and Their Compounds
V.G. Kumar Das (Ed)

Use of an Interfaced GC-AA Instrument with Full Chromatographic Integrity and Results for the Analysis of Organometallic Species with confirmation by Interfaced GC-MS

Peter J. Craig[a], Stuart H. Laurie[a] and Darren Mennie[b]

*[a]Department of Chemistry, De Montfort University Leicester
The Gateway, Leicester, LE1 9BH, UK.*

*[b] Ministry of Agriculture, Fisheries and Food, Torrey Research Station
135 Abbey Road, PO Box 31, Torrey, Aberdeen, AB9 8DG, UK.*

Gas chromatography is frequently used in the identification of compounds from the natural environment. However, many of the usual methods of detection are non-selective, detecting all or most of the chemical species separated by the column (e.g. with flame ionization or electron capture detection). Interfacing a gas chromatograph (GC) with an element specific detector (e.g. an atomic absorption spectrophotometer (AA) or a Mass Spectrometer (MS) renders the GC technique much more selective. Both of the above interfacing methods have been used by our group over a number of years.[1-6] A typical interfaced GC-AA quartz furnace apparatus is shown in Figure 1. Used for the analysis of organotin compounds from the environment (e.g. derivatized $((C_4H_9)_3Sn)_2O$, TBTO), this apparatus will separate the various tin and other compounds present in the analysis, and it will detect only tin-containing species. One drawback of this apparatus is that the heated transfer line itself did not previously contain chromatographic packing material (stationary phase), leading to incomplete chromatography from injector to detector with subsequent peak broadening. The development of the apparatus in Figure 1 so as to incorporate chromatographic separation ability throughout is described. Detection is via the quartz furnace atomic absorption method using a hollow cathode lamp.[5-7] Chromatographic separation is by use of an 30 m SE30 capillary column. Full chromatographic separation is achieved by uncoiling the final portion of the capillary column and threading it through the one metre transfer line directly to an interface with the quartz cell.

The work described here involves organometallic compounds of a type often found in the aqueous or sedimentary natural environment, viz butyl tins or methyl mercury. These compounds are converted (derivatized) to their corresponding hydrides or ethyl derivatives according to equations 1 and 2 so as to allow them to become sufficiently volatile for chromatography and analysis.

$$CH_3HgCl \xrightarrow{\quad NaB(C_2H_5)_4 \quad} CH_3HgC_2H_5 \qquad (1)$$

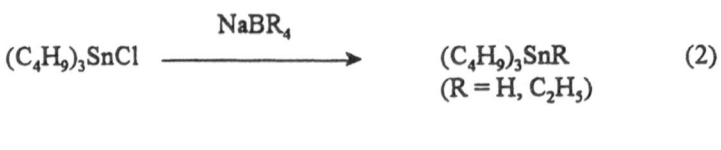

$$(C_4H_9)_3SnCl \xrightarrow{\text{NaBR}_4} \begin{array}{c} (C_4H_9)_3SnR \\ (R = H, C_2H_5) \end{array} \qquad (2)$$

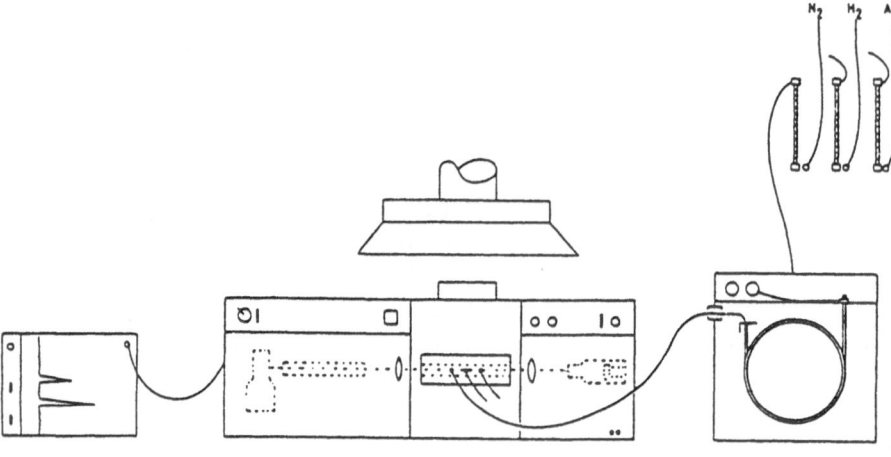

Fig. 1 : General arrangement of interfaced GC-AA apparatus
(Reprinted from ref. 5. Copyright 1988 - John Wiley & Sons)

Experimental

Derivatization

Derivatization methodology is now described. The method used here is ethylation of ethanolic butyltin and octyltin solutions with 1% $NaB(C_2H_5)_4$ in ethanol for 15 min. This time was arrived at after optimization studies. The extraction procedure from sediments is similar to that described previously.[6]

2 g of accurately weighed sediment (containing butyltin) was suspended in 20 cm³ of distilled water and 5 cm³ of concentrated hydrochloric acid was added carefully. 1 cm³ of a 1ppm solution of tripropyltin chloride in ethanol was added as the internal standard. The suspension was then left overnight. The suspension was then extracted with a solution (10 cm³) of 0.05% tropolone in dichloromethane for ten minutes in darkness with magnetic stirring. Uncomplexed tropolone was then removed using 10 cm³ of saturated iron (II) sulphate solution and the sediment was separated by vacuum filtration. The organic layer was then removed and evaporated to dryness with a gentle stream of nitrogen. The dried extract was then resuspended in 0.5 cm³ of ethanol and derivatized using 0.2 cm³ of a 1% ethanolic $NaB(C_2H_5)_4$ solution for 15 min. 1 or 10 μL was then injected into the GC-AA for analysis. Detection was by use of a tin hollow cathode lamp at 286.6 nm. The temperature of the quartz furnace was 1000°C. The furnace walls were pretreated with hydrofluoric acid (Care!).

152

For mercury compounds derivatization took place in 20 cm³ headspace vials sealed with PTFE faced butyl rubber septa within an aluminium crimped top. These contained 10 cm³ of mixed aqueous solutions of $HgCl_2$ and CH_3HgCl at various concentrations. The solutions were derivatized by injection of 1 cm³ of a 1% aqueous solution of $NaB(C_2H_5)_4$ solution. For analysis 1 cm³ of headspace vapour was injected using a gas tight syringe (Hamilton). The solutions were left for 30 min. after derivatization prior to analysis. Here the quartz furnace was at 600 °C. The pH of the mercury containing solution was equally efficient between 4 and 7. Treatment of the cell walls with hydrofluoric acid seemed to have little effect for mercury compounds. The AA line used for mercury was 253.6 nm.

The mercury compounds were either extracts in toluene or present in fish tissue. Extraction of methylmercury from toluene was carried out as follows: 10 cm³ of the toluene containing methyl mercury was extracted with two freshly prepared aliquots (5 cm³) of 0.005 M sodium thiosulphate solution using a magnetic stirrer for 10 min. each. Separation of the organic and aqueous phases was carried out using Whatman phase separating filter papers. The two aqueous aliquots were combined in a vial as described above and then extracted.

The analysis was also carried out on 5 g of dried and powdered tuna fish muscle, spiked with CH_3HgCl to a 1 ppm level in the fish. 2 g of this fish muscle was weighed into a 100 cm³ beaker and suspended in 15 cm³ of 10% sulphuric acid and left overnight. This was then stirred for 10 min. using a magnetic stirrer. The solid material was removed by vacuum filtration. The aqueous phase was extracted with 10 cm³ of toluene. Phase separating filter paper was used to separate the organic from the aqueous phase (the filler papers were able to separate the organic/aqueous phase emulsions resulting from stirring). The toluene extract was then re-extracted using two 7.5 cm³ aliquots of freshly prepared 0.005 M sodium thiosulphate solution. Again phase separating paper was used to separate the organic and aqueous phases (including emulsions). The combined 15 cm³ aqueous solution was placed in a crimp topped vial and derivatized as above.

The analysis of mercury from sediment matrices was carried out as follows: the sediments were spiked to the 10 ppm level in methyl mercury. 5 g of sediment was suspended in 15 cm³ of 10% sulphuric acid and left overnight. The suspension was then stirred vigorously for 10 min. with a magnetic stirrer and the mixture then separated using vacuum filtration. 10 cm³ of toluene was used to extract the aqueous phase as for the fish sample described above. The toluene extract was then extracted by magnetic stirring for 10 min. with two separate 7.5 cm³ aliquots of freshly prepared 0.005 M sodium thiosulphate solution. The aqueous layer was then separated using phase separating paper as above and the aqueous phases combined and derivatized as above.

GC-AA Apparatus

An Analytical Instruments Model 92 GC was used. Peaks were recorded on a Spectra Physics Chromjet Integrator. A Perkin Elmer 3100 AA with quartz furnace was used with a hollow cathode lamp operated at 286.6 nm (or 253.6 nm) with a slit width of 0.7 nm. The GC used was an Analytical Instruments Model 92 used with a 30 m SE30 capillary column (Aldrich). Initially Injector temperature was 150 °C, transfer line 200 °C and GC column ramped from 100° to 210 °C over 5

153

minutes (for tin work).

The transfer line is now described (inner sections first); SE30 capillary column, 1.09 mm internal diameter stainless steel capillary tube, 1.59 mm internal diameter PTFE tube, 28 SWG Nichrome heating wire, PTFE tape, 4.65 mm internal diameter PTFE tube, wound asbestos rope insulation, PTFE tape. Heating is required for organotin species which are relatively non-volatile; organomercury compounds are sufficiently volatile so as to require little or no heating of the transfer line. Initially the interface at the quartz furnace, which was held at 1000 °C by electrical heating, was not deliberately heated although it achieved a temperature of 50 °C by conduction from the quartz cell.

The inner portion of the transfer line is stainless steel capillary tubing, 1.59 mm outside diameter, 1.09 mm internal diameter. The internal diameter of 1.09 mm permitted the capillary column to be threaded down the stainless steel capillary tube, whilst fitting closely inside it, thereby ensuring good thermal contact between the tubes. The stainless steel tube was sleeved with 1 m of 3 mm outside diameter and 1.59 internal diameter PTFE tubing which served as electrical insulation. Around the PTFE, 10.1 m of 28 SWG nichrome wire was wound to serve as a heating element. The nichrome wire was held in place using PTFE tape. The entire transfer line as described was then sleeved with 1 m of 6.35 mm outside diameter, 4.65 mm internal diameter PTFE tubing to electrically isolate the nichrome wire.

The transfer line was then insulated using a double thickness of asbestos rope (Jencons) held in place with PTFE tape. Power was then supplied using an independent variable transformer (24 V) which gave a temperature (thermocouple) inside the transfer line of 200 °C. The transfer line was connected directly to the quartz furnace using a 6.25 to 1.59 mm Swagelock reducing union, fitted with a 6.25 mm PTFE ferrule and a 1.59 mm graphite ferrule. The furnace has been described previously.[5]

The GC-AA interface as described was used for experimental work, but with derivatised butyltin compounds, poor peak shapes were obtained (Fig. 2(A)). This was the result of the interface and transfer line temperature being too low, leading to substantial condensation of the analyte in the transfer line and thus to poor peak shape. The temperature of the interface with the quartz furnace was 50 °C due to conduction from the quartz furnace, from its 1000 °C centre to its exposed, (non insulated) edges. The temperature of this interface was too low, as it should be 20-50 °C higher than that of the transfer line itself, i.e. 220-250 °C[8]. This also contributed to failure to produce Gaussian peaks with butyltin species.

The above interface with the quartz cell was particularly successful with mercury compounds, due to the greater volatility of these compounds. For tin species it is likely that the unheated interface of the capillary column to the quartz furnace created a "cold spot" of stationary phase, for which the analyte species had a high affinity, and on which it therefore condensed owing to the decrease in temperature from 200 °C in the transfer line to 50 °C in the interface.

The interface was therefore rebuilt to incorporate a method of heating. However, the existing transfer line incorporated PTFE which has a maximum upper working temperature of 250 °C. Above this temperature the material decomposes liberating a variety of noxious substances.

154

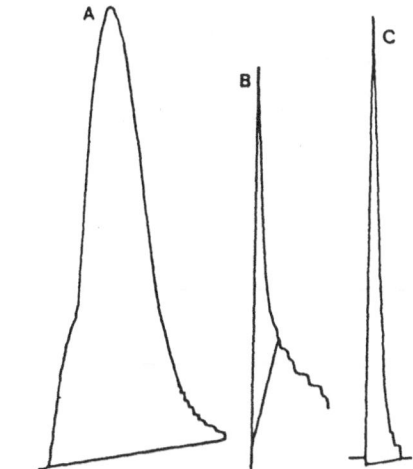

Fig. 2 : Chromatograms of $(C_4H_9)_3SnC_2H_5$.
A is without direct GC-AA interface heating and without make up gas; B is with
interface heating and without make up gas; C is with both heating and make up
gas. Retention time = 8.3 min.

Consequently it was decided to remove the PTFE insulation from the transfer
line. Ceramic beads (Electrothermal Elements Hinckley, Leicestershire, UK)
appeared to be a good choice to replace PTFE as an insulator as they are
inexpensive, easily available and are excellent insulators. The transfer line was then
reconstructed as now described.[8] 3 m of the 28 SWG Nichrome wire was sleeved
with ceramic beads. These were then wrapped in an even helix over 0.25 m of 3
mm outside diameter 1.59 mm internal diameter stainless steel tubing (Phase Sep).
The beads were held in place with fire clay. A convenient method of encasing the
beads in fire clay was by the lengthways splitting of a piece of 10 mm internal
diameter rubber tube, which was then filled with fireclay. The stainless steel tube
and the beads could then be introduced into the fireclay via the slit, and the fireclay
cured overnight in an oven at 100 °C and the rubber removed. A further piece of
stainless steel, 1.59 mm outside diameter, 1.09 mm internal diameter was threaded
down the fireclay encased steel tube to serve as a thermal contact between the tube
and the capillary column.

Finally, the column was threaded down the transfer line and connected directly
to the quartz furnace using a 6.25 to 1.59 mm Swagelock reducing union. Electrical
connections were made using ceramic connectors. The transfer line was heated
using an independently variable transformer.

To summarize, the transfer line consisted, from inner to outer sections, the SE
30 capillary column, a 1.09 mm internal diameter stainless steel tube, a 1.59 mm
internal diameter stainless steel tube, ceramic insulating beads containing Nichrome
wire, and then the fire clay outer covering.

This rebuilt transfer line was capable of a maximum temperature of 450 °C (and
was capable of analysing octyltin species). To generate peaks of sufficient quality,
it is still necessary to heat the interface between the quartz furnace and the transfer
line. This was achieved by wrapping excess of the ceramic bead sleeved 28 SWG

155

wire around the 6.25 to 1.59 mm Swagelock reducing union.

Considering Figure 3, we can see that Port A is the nitrogen and analyte eluent from the GC (ie the transfer line), Port B is the hydrogen inlet and Port C is the air inlet. Respective flow rates were initially 5 cm^3 min^{-1} (A), 250 cm^3 min^{-1} (B) and 13 cm^3 min^{-1} (C). Use of Port A produces the best peak shape due to its higher temperature compared to Ports B and C. However, if point X (Fig. 3) is considered, it becomes clear that one is trying to force 5 cm^3 min^{-1} of carrier gas against a combined gas flow of 263 cm^3 min^{-1} from Ports B and C. This is not an ideal situation (Fig. 2(B)).

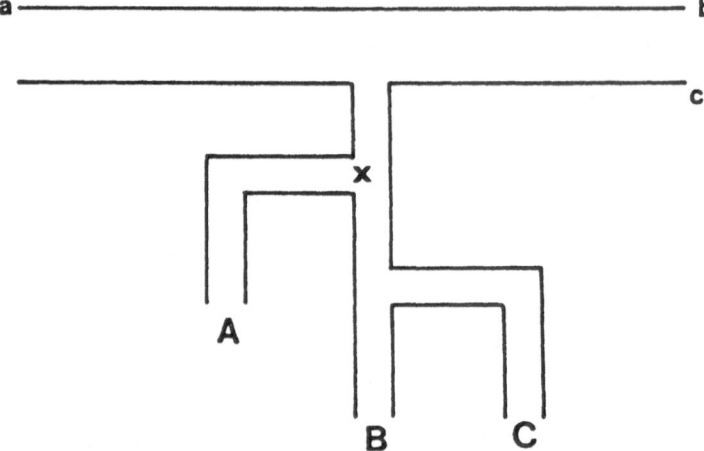

Fig. 3 : Arrangement of quartz furnace and supply of gases and analyte
A - port for entry of transfer line and analyte/nitrogen; B - port for entry of hydrogen gas; C - port for entry of air; ab = 165 mm; bc = 14 mm.

Hence it was decided to introduce an additional supply of nitrogen (make up gas) at Port (A). Many chromatographic detectors when operated with a capillary column utilise a make up gas[8]. Here the extra gas supply was introduced into a modified Swagelock reducing union at Port A. Full details of the construction of this are given elsewhere.[9] By this means total nitrogen gas flow rate was raised to 60 cm^3 min^{-1} and good chromatography was obtained (Fig. 2(C)). A schematic diagram of the arrangement is given in Figure 4. GC-AA conditions for the various species analysed are given in Table 1.

GC-MS Apparatus

The above GC-AA technique was complemented also by using a similarly integrated GC-MS system.

Here the GC used was a Hewlett Packard 5890 fitted with a 12 m SE54 capillary column (Altech) interfaced to a VG Trio 3 triple quadrupole MS instrument. Helium gas pressure was 5 x 10^4 Nm^{-2}. GC conditions for octytins here were; injector temperature 200 °C, oven temperature 100 °C for 2 min. rising at 20 °C per min. to a maximum of 280 °C which was held for 5 min. For butyltins injector temperature was 220 °C, oven temperature 100 °C for 3 min. rising at 20 °C per min. to a maximum of 220 °C. For mercury compounds oven temperature was

100 °C, the injection temperature was 110 °C, and the transfer line temperature 180 °C.

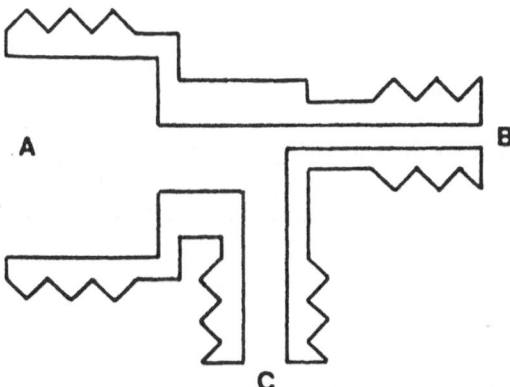

Fig. 4 : Detail of the union between the transfer line (and capillary column) and the quartz furnace: introduction of nitrogen make up gas.
AB is a 6.25 mm to 1.59 mm Swagelock reducing union modified to include a 3.175 mm union (C). To C is attached stainless steel tubing for nitrogen make up gas. A is attached to the quartz furnace and B is attached to the capillary column. The whole is heated by ceramic beaded 28 SWG nichrome wire. A 3.175 mm PTFE tube is attached to the stainless steel tube at C via a 3.175 mm Swagelock fitting. The flow of nitrogen make up gas to C was controlled by a needle flow valve (Phase Sep). The capillary column at B terminates at exactly the end of the 1.59 mm graphite ferrule and does not project further along the AB axis. This reduces the extent of cooling as the analyte enters the union at a port with good thermal contact with the heating wire.

Table 1 : Conditions Used for GC-AA Analysis

Compound	Flow Rates (cm³ min⁻¹)			Temperatures (°C)			Retention Time (min)
	Air	H_2	N_2	Injector	Oven	Transfer Line	
Bu_3SnEt	13	250	60	150	100-220	200	8.3
Bu_2SnEt_2	13	250	60	150	100-220	200	6.1
$Bu Sn Et_3$	13	250	60	150	100-220	200	2.9
Oct_2SnEt_2	13	250	60	200	100-280	280	11.0
$Oct Sn Et_3$	13	250	60	200	100-280	280	8.0
Et_2Hg	5	0	1.5	110	100	180	1.9
$MeHgEt$	5	0	1.15	110	100	180	1.2

$Me = CH_3$; $Et = C_2H_5$; $Bu = n\text{-}C_4H_9$; $Oct = n\text{-}C_8H_{17}$

Mass spectra to confirm the identity of butyltin derivatives are discussed; those for relevant mercury and octyltin compounds are given elsewhere.[1,10,11]

Results and Discussion

Figures 3 and 5 show the chromatographic peaks obtainable with the capillary column present up to the interface with the quartz furnace and with interface heating. Derivatizations were as carried out in the Experimental section.

Fig. 5 : **Chromatograms of $(C_2H_5)_2Hg$ and $CH_3HgC_2H_5$ obtained with the capillary GC-AA system.**
$A = CH_3HgC_2H_5$, retention time = 1.2 min; $B = (C_2H_5)_2Hg$, retention time = 1.9 min

Use of this apparatus at Leicester for the analysis of five aliquots of a sediment doped with $(C_4H_9)_3SnOAc$ and sent as an unknown to numerous laboratories for analysis by the Bureau of Community Reference (BCR) of the European Community, Brussels are now given: concentrations found of $(C_4H_9)_3SnOAc$ were 2.85, 3.10, 3.00, 2.85 and 3.05 ppm. The BCR later gave the original concentration to be 3.3 ppm. This apparatus was also able to separate and detect similar concentrations of mixed tri-, di- and monobutyltin species after derivatization. Retention time for $(C_4H_9)_3SnC_2H_5$ was 8.3 min, for $(C_4H_9)_2Sn(C_2H_5)_2$ 6.1 min, and for $C_4H_9Sn(C_2H_5)_3$ 2.9 min. There was no evidence of dismutation.

The analysis of methyl mercury from fish and sediment matrices described here is simpler than many previous extraction methods described and the method is either element (GC-AA) or compound (GC-MS) specific. The limit of detection for both matrices is 0.01 ppm if 10 g of sample is used. Recovery is 93% from sediment and

158

90% from tuna fish based on average values from three analyses each from samples of known concentration. Improved chromatography over the method without full chromatography in the transfer line was obtained. With regard to the mercury work, it should be noted that derivatization yield with $NaB(C_2H_5)_4$ for aqueous solutions approaches 100%. However, derivatization yields from aqueous solutions containing sodium thiosulphate is a constant 50% both from standard and sediment/fish extract solutions. This is allowed for in the calculations. It should be noted that derivatization of separate or mixed inorganic mercury/methyl mercury solutions showed no evidence of dismutation (i.e. we did not see Hg(O) or $(CH_3)_2Hg$ from these derivatizations).

With regard to the butyltin work, identities were confirmed with GC-MS and typical spectra are shown in Figure 6. Similar confirmation was achieved with other analytes.[10,11]

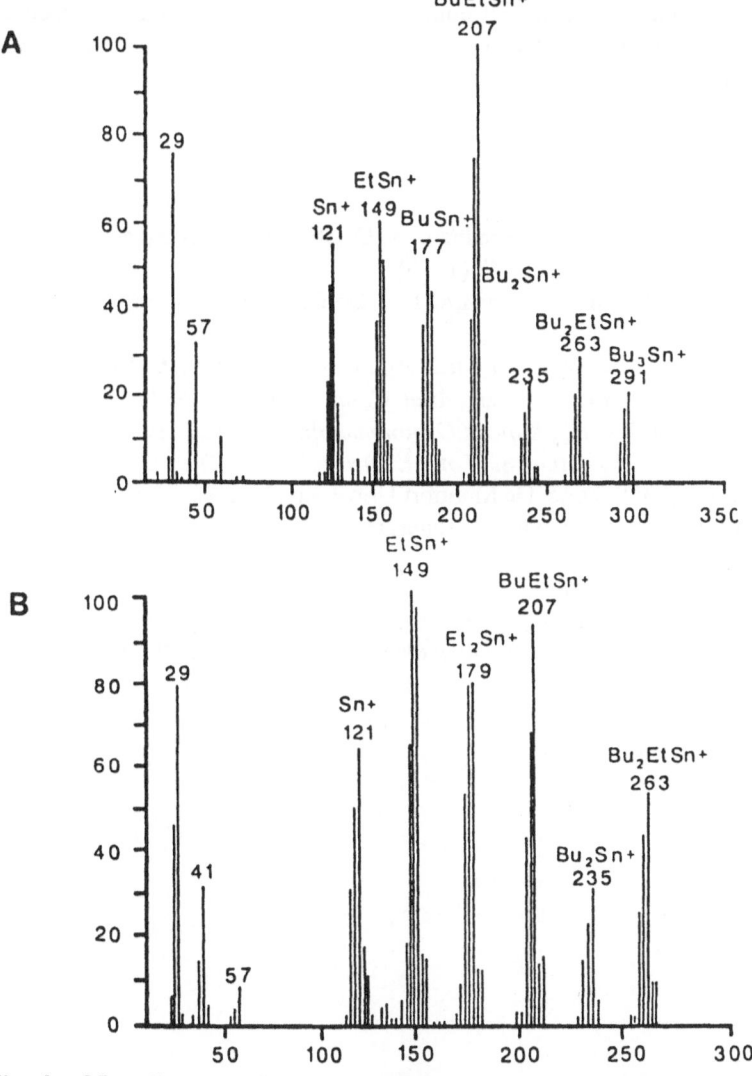

Fig. 6 : Mass Spectra of (A) $(C_4H_9)_3SnC_2H_5$ and (B) $(C_4H_9)_2Sn(C_2H_5)_2$

159

Conclusion

An element specific interfaced GC-AA detector with full chromatographic separation ability using a capillary column and capillary transfer line has been constructed. This apparatus has been used to separate and detect mixed butyltin, mixed octyltin, and mercury species from environmental matrices and at concentrations relevant to the natural environment. Results were confirmed by GC-MS.

Acknowledgements

PJC and DM acknowledge a SERC earmarked quota studentship to DM from the Science and Engineering Research Council, UK. Support from the British Council under the Alliance (France) and A.R.C. (Germany), which allowed valuable collaborative work in Europe, is also gratefully acknowledged.

References

1. P.J. Craig, D. Mennie, M. Needham, N. Ostah, O.F.X. Donard and F. Martin, *J. Organometallic Chem.*, **447** (1993) 5.
2. P.J. Craig, D.Mennie, N. Ostah, O.F.X. Donard and F. Martin, *Analyst*, **117** (1992) 823.
3. J.R. Ashby and P.J. Craig, *Applied Organometallic Chem.*, **5** (1991) 173.
4. S. Rapsomanikis and P.J. Craig, *Anal. Chim. Acta*, **248** (1991) 563.
5. S. Clark and P.J. Craig, *Applied Organometallic Chem.*, **2** (1988) 33.
6. J.R. Ashby and P.J. Craig, *Sci Total Environ.*, **78** (1989) 219.
7. D. Mennie, PhD Thesis, De Montfort University, Leicester. UK, 1993.
8. R.D. Braun, Introduction to Instrumental Analysis, McGraw Hill, New York, 1986.
9. Reference 7, pp 75-77.
10. D. Mennie and P.J. Craig, *Main Group Metal Chemistry*, **17** (1993) 453.
11. P.J. Craig and D. Mennie, *Applied Organometallic Chem.*, **8** (1994) 1.

Main Group Elements and Their Compounds
V.G. Kumar Das (Ed)
Copyright © 1996 Narosa Publishing House, New Delhi, India

Identification and Determination of Selenium Compounds in Samples of Environmental Importance

Gottfried Kölbl, Kurt Kalcher and Kurt J. Irgolic
Institut für Analytische Chemie, Karl-Franzens Universität
Universitätsplatz 1, A-8010 Graz, Austria

Selenium has been recognized as an essential nutrient almost four decades ago.[1] A sufficient intake of selenium has been reported to reduce the risks of heart disease and the occurrence of cancer.[2] In animals selenium is known to reduce the toxic effects of heavy metals such as arsenic and mercury.[3] Selenium is also an integral part of several enzymes and other proteins; the function of some of these proteins is still unknown.[4] However, selenium is toxic at higher doses: diets which contain more than 5 mg Se kg^{-1} are considered to be poisonous to man and animals.[5] The difference between the required and the toxic dose is small.[6] Toxicity is not a property of "selenium", but of selenium compounds. Selenium compounds do not all show the same toxicity.[7] Because of this dichotomy of metabolic action (beneficial and detrimental effects vary within a very narrow range of concentration) selenium was called an "element with two faces".[8]

Selenium can be present in the environment and in organisms in inorganic form and in organic forms with direct Se-C bonds. The naturally-occurring inorganic selenium compounds are selenite and selenate, which are the most common selenium compounds in the aquatic environment,[9] the insoluble elemental selenium, metal selenides (HgSe, PbSe, CdSe) and metal sulfides with selenium replacing some of the sulfur atoms. The volatile hydrogen selenide is believed to be an intermediate in the conversion of selenite and selenate to organic selenium compounds by organisms.[10] Dimethyl selenide and dimethyl diselenide have been identified as volatile excretory products of selenium in breath and sweat of animals. The trimethylselenonium ion was found in urine. Several selenoamino acids (*e.g.*, selenocystine and selenomethionine) were shown to be building blocks of selenium-containing enzymes and other selenoproteins. In most cases these selenoamino acids are part of the active sites of the enzymes. Selenium-modified nucleosides occur in tRNAs from several bacterial, mammalian and plant species. The precise biochemical role of selenium in these compounds is yet unknown. Some evidence exists also for the occurrence of selenium-containing carbohydrates and selenolipids.

All of these selenium compounds are part of a natural selenium cycle which links the inorganic selenium species with the more complex organic selenium compounds. However, most chemical and biochemical pathways of this cycle are still unknown. A complete understanding of the biochemical transformation of selenium compounds in nature and of the action of these compounds on organisms requires the identification and quantification of all naturally-occurring selenium compounds.

Only a few methods have been worked out for the identification and quantification of selenium compounds in biological and environmental samples. Some information can be obtained from *sequential extraction procedures* which are able to separate different groups of compounds such as inorganic compounds, lipids/lipoproteins, amino acids, organic acids, humic acids, carbohydrates, and proteins. In this manner the association of selenium with certain groups of compounds can be established. However, hardly any conclusions can be drawn about the bonding (covalent, ionic, adsorptive) of selenium to these compounds. The identification and determination of a definite selenium compound is not possible with this method. *Gas chromatography* is a valuable tool for the determination of the few naturally-occurring, volatile selenium compounds (dimethyl selenide, dimethyl diselenide). But most selenium compounds are not volatile and have to be converted to volatile derivatives before they can be subjected to gas-chromatographic analysis. These derivatisation procedures are in most cases time-consuming and are restricted to selenium compounds that react with the derivatisation reagents. Additionally, the derivatisation process might destroy information about the chemical nature and origin of the compounds to be determined. *Conventional liquid chromatography* (using normal pressure) can be used for the separation and determination of soluble selenium compounds. But column chromatographic methods at atmospheric pressure tend to have rather long analysis times and require large volumes of mobile phase. The resulting dilution of the sample complicates the determination of selenium compounds at trace levels.

Substantial improvements can be achieved by using high performance liquid chromatography (HPLC).[11] This technique can be employed for the separation and determination of a variety of compounds. Ionic species can be separated with ion chromatography (cation or anion exchange stationary phase) or with ion-pair partitioning reversed-phase chromatography (lipophilic stationary phase); neutral molecules can be separated by reversed-phase chromatography. HPLC provides high separating efficiency at reasonable short analysis times. The method also offers low detection limits in the $\mu g \ L^{-1}$ range if appropriate detectors are used. The detectors can be divided into three groups:

General detectors register the physico-chemical change of the eluate (response to "bulk properties"). Such detectors (*e.g.*, differential refractive index detector) are not specific for the substances to be determined. In biological or environmental samples, which often contain a large excess of compounds besides the compounds to be determined, the signals from co-eluting substances can overlap with the signals from the analytes or obscure them entirely.

Selective detectors use one or more properties of the eluted compounds for detection (response to "solute properties"). These detectors are based on UV-, IR-absorption, fluorometry, conductivity, radiometry, or electrochemistry. Detectors of this kind can provide some selectivity but their response may also be affected by co-eluting substances.

"Element-specific" detectors respond specifically to the presence of a substance containing a specific element (response to "property of one element"). Detectors of this type respond to atomic transitions (atomic absorption or atomic emission spectrometers) or register different ions or ionized fragments of the analytes (mass spectrometric devices). Element-specific detectors can drastically simplify the

chromatographic procedure, because

- only the species containing the element to be detected need to be separated;
- even large excesses of co-eluting substances interfere little or not at all with the determination;
- the stability of the baseline is better (no negative signals are possible) than with other detectors;
- no system peaks can occur in the chromatogram; and
- low detection limits are achievable.

In our work we investigated the advantages and disadvantages of three selenium-specific detectors (graphite furnace atomic absorption spectrometer, flame atomic absorption spectrometer, and inductively coupled plasma mass spectrometer) for the ion-chromatographic separation and determination of selenite and selenate.

Experimental

Reagents

Selenium stock standard solutions were prepared from anhydrous sodium selenate (Fluka), sodium selenate pentahydrate (Merck), selenocystine, selenomethionine, and selenoethionine (all from Sigma). Trimethylselenonium iodide was prepared from dimethyl selenide (Alfa) and methyl iodide.[12] A modified version[13] of the procedure of Tanaka and Soda [14] was used for the preparation of selenohomocystine. Mobile phases contained potassium hydrogen phthalate, nitric acid, 2-propanol, tetrabutylammonium phosphate, and methanol. Potassium hydrogen phthalate solutions (3 mM) were chemically modified by stirring 2.0 L overnight with an excess (5 g) of nickel hydroxide. Eluents with lower nickel concentrations were prepared by mixing the 3 mM phthalate solution saturated with nickel hydroxide with nickel-free 3 mM phthalate solution. The pH of the phthalate-containing mobile phases was adjusted with potassium hydroxide. All mobile phases were filtered through 0.2 μm cellulose nitrate filters (Sartorius) prior to use. Water was first purified by deionization and then distilled twice in a quartz still.

Instrumentation

The chromatographic system consisted of a double head pump (Waters 600 E multisolvent delivery system) with a Waters U6K injector, a precolumn and an analytical column mounted in a thermostated chamber held at 25 °C. Separations were performed on a strongly basic anion-exchange column (ESA Anion III), a strongly acidic cation-exchange column (Hamilton PRP-X200), and a polymer-based reversed-phase column (Hamilton PRP-1). All columns had dimensions of 250 x 4 mm. The stationary phases had particle sizes of 10 μm. Solutions (100 μL) to be chromatographed were injected onto the column with a microliter syringe (Hamilton).

A simultaneous multi-element atomic absorption spectrometer (Hitachi Z-9000) was operated with graphite tubes from Ringsdorff-Werke (RWO 521) and a selenium hollow-cathode lamp at 12.5 mA. Argon at 0.280 L min^{-1} was employed as carrier gas. Volumes of 20 μL were transferred to the furnace, dried at 80-120 °C for 5 s and at 200 °C for 5 s, ashed at 390 °C for 20 s, and finally atomized at 2600 °C for 5 s. Signals from the GFAA spectrometer were processed and quantified as described.[15] The chromatographic system and the GFAAS detector were connected with a laboratory-built flow-through cell[16] made of Teflon with a dead volume of 30 μL.

A Zeeman-effect flame atomic absorption spectrometer (Hitachi Z-6100) was operated with an air-acetylene flame (fuel pressure 15 kPa, oxidant pressure 160 kPa, optimum burner height 7.5 mm, Se hollow-cathode lamp at 12 mA). The HPLC system was connected to the inlet of the atomizer with a short piece of Teflon tubing. The spectrometer signals were read into a personal computer via the Hitachi recorder interface (Hitachi 171-9124) after analogue-to-digital conversion. The resulting data were treated by a modified version of the program for the GFAAS.

An inductively-coupled plasma mass spectrometer (VG Elemental PQ2 Turbo Plus) was operated with an argon plasma (forward power: 1380 W, reflected power: <1 W). Argon at 13.5 L min^{-1} was used as cooling, at 1.1 L min^{-1} as auxiliary, and at 0.88 L min^{-1} as nebulizer gas. The sample solution was aspirated with a Meinhard concentric glass nebulizer (Tr-30-A3) with an uptake rate of 0.92 mL min^{-1}. The ions from the plasma were extracted into the mass spectrometric unit through nickel cones (sampling cone: 1.00 mm orifice, skimmer cone: 0.75 mm orifice). The vacuum in the mass spectrometer was 1.6 mbar at the expansion, 1.0×10^{-4} mbar at the intermediate, and 2.1×10^{-6} mbar at the analyzer stage. The HPLC system was connected to the ICP-MS instrument via a short piece of Teflon tubing. The eluent was monitored for the ^{82}Se-isotope with factory-supplied software in the single-ion monitoring mode.

Results & Discussion

GFAAS as detector for HPLC

Graphite furnace atomic absorption spectrometry (GFAAS) offers low detection limits for the determination of selenium (0.5 to 50 μg Se L^{-1} with 20 mL sample volume).[17] However, its use as detector for chromatographic experiments is difficult. Chromatographic separations are performed continuously, while GFAAS is a discontinuously working technique: the sample solution has to be injected into the furnace, dried, ashed, and atomized at a temperature between 2000 and 3000 °C; after the furnace has cooled down to room temperature the next injection can occur. This entire procedure requires one to two minutes. Therefore, a HPLC/GFAAS chromatogram can consist only of a limited number of signals. Very narrow peaks in the chromatogram might be missed by the detector entirely. We designed a system in which the action of the different modules of the AAS device (autosampler and power unit) can be controlled independently by a personal computer via a home-built interface.[15] With this system it was possible to optimize the whole analysis sequence to a minimal repetitive cycle time, because the autosampler of the AAS system can start to take the aliquot from the chromatographic eluate while the graphite furnace is still active. With such a system the time between two consecutive signals in a chromatogram can be lowered to 30-45 s. The signals from the spectrometer are transferred to the personal computer after analogue-to-digital conversion and can be stored and treated by routines of the computer program. A self-designed Gaussian algorithm which reconstructs the chromatographic peaks from the GFAAS data improved the quantification of the chromatograms.[15]

GFAAS needs only small sample volumes. Therefore, the flow-rate of the mobile phase can be kept low. Because GFAAS works discontinuously, the chromatographic peaks should be broad to obtain sufficient numbers of GFAAS signals for each selenium compound. However, the time for the total analysis should be as short as possible to avoid unnecessary consumption of graphite cuvettes and argon gas. Two fully resolved peaks for selenite and selenate were observed on the anion exchange column when a mobile phase of 3 mM potassium hydrogen phthalate at flow rates between 0.3 and 0.5 mL min^{-1} was used (Fig. 1).

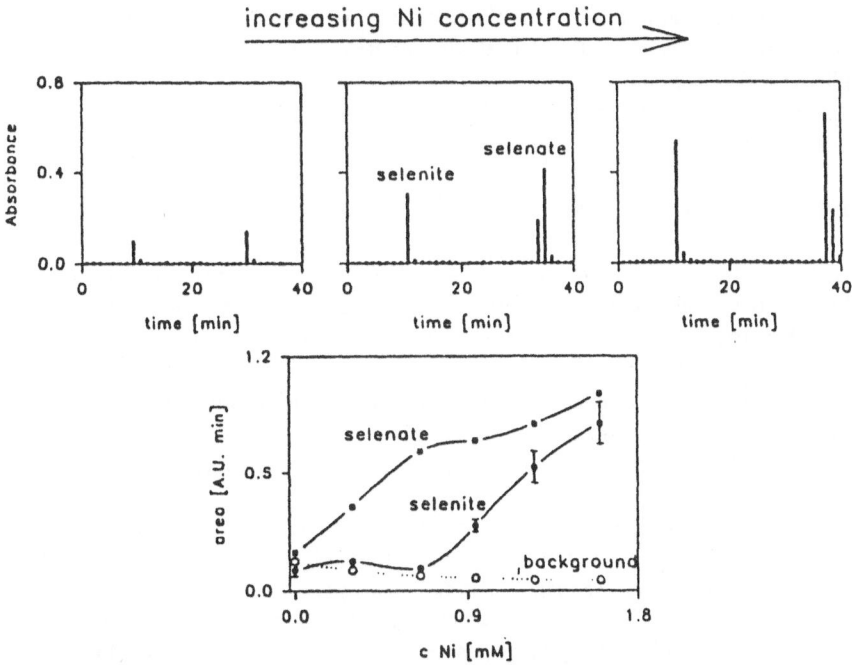

Fig. 1 : **Influence of nickel in the mobile phase on selenite and selenate signals. Column: ESA Anion III; mobile phase: 3 mM potassium hydrogen phthalate, pH 7; flow rate: 0.4 mL min^{-1}; injection volume: 100 µL onto column; detector: GFAAS, 20 µL sample volume; 300 ng Se as selenite, 300 ng Se as selenate (corresponding to 3 mg Se L^{-1} as selenite and 3 mg Se L^{-1} as selenate).**

GFAAS may also suffer from various interferences such as background absorption from molecular species, light scattering, and detrimental reactions during the atomization process. Matrix modification (addition of a certain salt mixture to the analyte solution) can prevent these interferences, increase the signal/noise ratio, and increase the signal intensity. For the determination of selenium with GFAAS,

165

nickel, copper, silver, molybdenum, and platinum salts have been used as modifiers. The ion-chromatographic separation of selenite and selenate on the anion exchange column was performed with solutions of potassium hydrogen phthalate which is a common eluent for anions in single-column ion chromatography. These solutions were saturated with the slightly soluble nickel hydroxide. The direct modification of the mobile phase has the advantage that the sample solution is not diluted unnecessarily by the addition of the modifier solution and the modification process is also performed without additional effort or time during the analysis. The nickel-matrix-modified mobile phases showed the expected increase of the selenium signals; additionally the background absorption is lowered to 50% of the value characteristic for an unmodified solution (Fig. 1). The alteration of the retention times of selenite and selenate can be explained in terms of the formation of a nickel:phthalate 1:1 compound[18] which lowers the concentration of "free" phthalate in the mobile phase necessary to elute the anions. Absolute detection limits (3 σ of noise) with the matrix-modified mobile phase were found to be 1 ng Se for selenite and 0.6 ng Se for selenate; these are the lowest detection limits for selenium with an HPLC/GFAAS system reported so far. Therefore, HPLC/GFAAS seems to be an appropriate method for the identification of selenium compounds in complex matrices of biological and environmental samples.

FAAS as detector for HPLC

Flame atomic absorption spectrometry (FAAS) is a well established technique for the determination of several elements. In FAAS the sample solution is nebulized and then introduced into and atomized in a flame. However, detection limits for selenium are rather high (>> 0.1 mg Se L^{-1}) with an conventional air/acetylene flame. The advantage of FAAS lies in the continuous mode of operation. Thus, even very narrow peaks in the chromatogram can be detected easily which allows high flow rates of the eluent to be used. In this manner the total analysis time can be kept short.

A problem is the high sample-uptake rate (about 6 mL min^{-1}) by common FAAS nebulizers. Such high flow rates are not achievable with conventional analytical HPLC columns. The practical, maximal back pressure of 200 bar restricts the maximal flow rate to 2-3 mL min^{-1} (a value which is dependent on stationary phase material and mobile phase). The missmatch between the uptake rate of the FAAS nebulizer and the flow rate of the chromatographic system causes an increase of the detection limits. Detection limits for selenium with HPLC/FAAS are usually one order of magnitude higher than with an HPLC/GFAAS system.

The high stability of the flame atomization and the continuous registration of the signal by the FAAS detector make quantifications of the chromatograms very easy and lead to more reliable results than possible with the GFAAS detector. The FAAS detector was used for the optimization of chromatographic conditions and can be applied to the determination of selenium compounds at higher concentrations (see Table 1) in complex matrices .[19] The separation of selenite and selenate with an HPLC/FAAS system is shown in Fig. 2.

Table 1 : Characteristics of GFAAS, FAAS and ICP-MS as selenium-specific detectors for HPLC

HPLC-detector	GFAAS	FAAS	ICP-MS
HPLC flow rates (mL min^{-1})	0.2 - 0.5	2 - 3	1
Total analysis time (min)	30 - 60	3 - 10	5 - 20
Absolute detection limit (ng Se)	1	10	< 0.1
Concentration detection limit for 100 µL injection (µg Se L^{-1})	10	100	< 1
Linear calibration graphs for 100 µL injection (mg Se L^{-1})	0.2 - 5.0	10 - 100	0.01 - 10

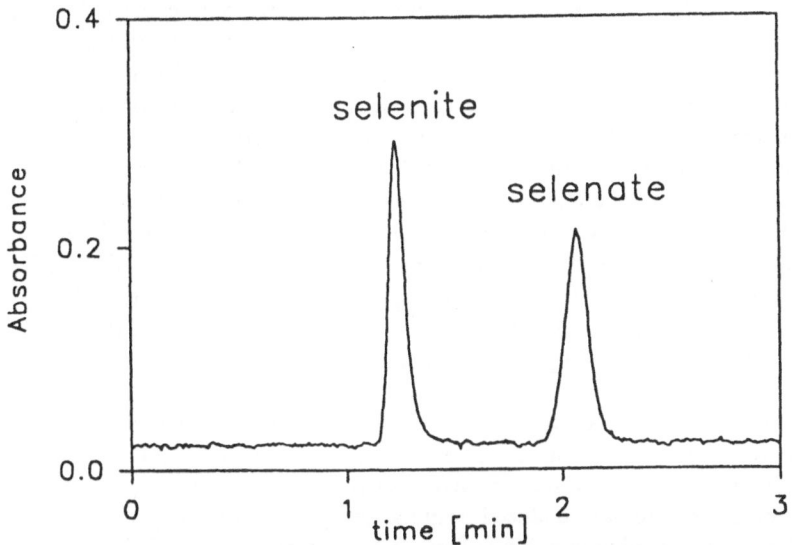

Fig. 2 : Separation of selenite and selenate using HPLC with FAAS detection. Column: ESA Anion III; mobile phase: 10 mM potassium hydrogen phthalate, pH 7; flow rate: 2.5 µL min^{-1}; injection volume: 100 µL onto column; 10 µg Se as selenite, 10 µg Se as selenate (corresponding to 100 mg Se L^{-1} as selenite and 100 mg Se L^{-1} as selenate).

ICP-MS as detector for HPLC

Inductively coupled plasma mass spectrometry (ICP-MS) is a very young technique; the first commercial instruments became available in 1983[20]. For ICP-MS measurements the sample solution is aspirated and the resulting aerosol is introduced into an argon plasma kept at temperatures up to 10000 ° C. At this temperature compounds are decomposed to atoms which are ionized. In this way, the plasma serves as the ion source for the quadrupole mass spectrometer.

ICP-MS combines the advantages of GFAAS and FAAS: the instrument does not only provide excellent detection limits ($>> 1$ µg Se L^{-1}) for the determination of nearly all elements, but also continuous registration of the signals. Linearity between concentration of the analytes and signal response can usually be obtained for a range of 7 orders of magnitude. Several elements can be monitored simultaneously without additional effort or analysis time. Observation of different isotopes of one element is also possible which makes the instrument suitable for metabolic studies with stable isotopes.

A flow rate matching the normal uptake rate of an ICP-MS nebulizer (1 mL min^{-1}) is easily achievable with conventional HPLC columns. Therefore, interfacing an HPLC system with an ICP-MS instrument requires only the connection of the end of the chromatographic column to the inlet of the ICP nebulizer via a short piece of Teflon tubing.

Problems in instrument performance might arise from the long-time signal drift of the ICP-MS. This drift can be corrected with isotope dilution analysis or with the use of internal standard solutions. The internal standard solution has to be supplied to the nebulizer at a constant rate as an additive to the effluent from the chromatographic system. The signal from the element in the internal standard solution (an element not present in the sample) is then used to correct the signals from the analytes. Additional problems arise when organic solvents (methanol, acetonitrile, tetrahydrofuran) are used as mobile phase components in the chromatographic separation process. High concentrations of organic solvents can extinguish the plasma, or cause signal depression and interferences. Thompson and Houk[21] stated that the maximal concentration of methanol that could be tolerated before extinguishing the plasma is about 20%. Results in our laboratory showed that 10% (v/v) methanol depressed the selenium signals to 10-40% of the signal intensity of an aqueous solution (Fig. 3). The dependence of the signal intensity for different selenium isotopes showed a maximum at a concentration of 2% (v/v) methanol.

This enhancement effect was explained by volatilization effects in the plasma[22] and was found to be dependent on the ionization energy of the element to be determined.[23] However, a conclusive explanation of this effect is not given in the literature and cannot be deduced from the data given. These experiments suggest, that chromatographic conditions have to be chosen very carefully to match the requirements of the ICP-MS detector. Some of the interferences at higher organic solvent concentrations might be avoided by aerosol desolvation performed in modern nebulizers. The addition of a few percent (up to 10%) of oxygen to the argon gas of the plasma can also eliminate interference problems.

ICP-MS is the most powerful element-specific detector for HPLC at the present time. Selenium compounds (and other trace element compounds) can be determined at very low concentrations even in complex matrices. The detection limits for the determination of selenium compounds are lower than 0.1 ng Se absolute. Additionally, several "element-specific channels" can be observed simultaneously, a very useful feature for investigation of different elements present in the same trace element compounds. Fig. 4 shows an HPLC/ICP-MS chromatogram of selenite and selenate. Table 1 summarizes the important characteristics of GFAAS, FAAS, and ICP-MS as selenium-specific detectors for HPLC.

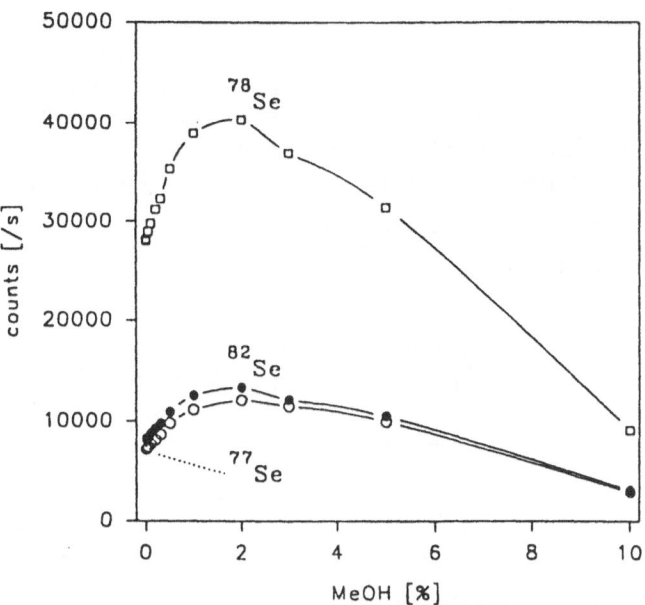

Fig. 3 : Influence of the methanol concentration in the sample solution on ICP-MS selenium signals. Measurement conditions are given in the experimental section.

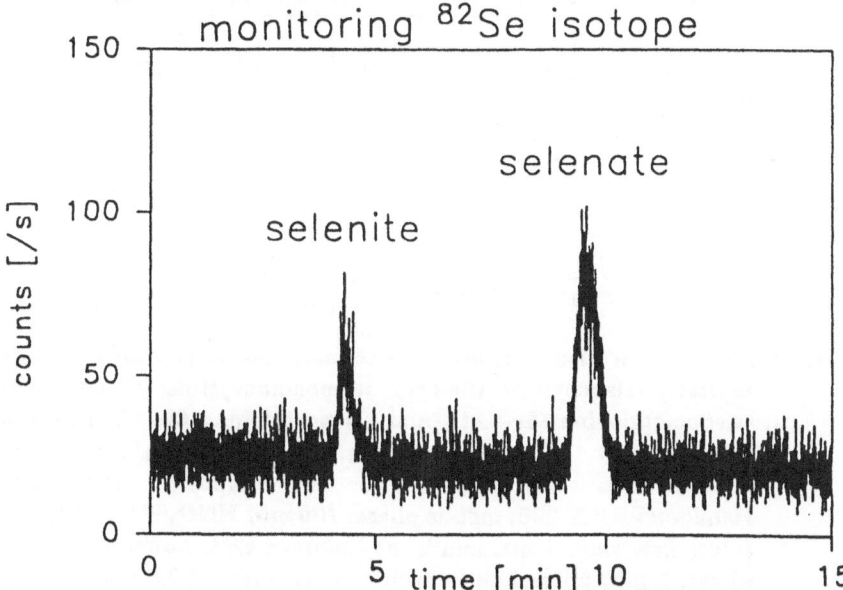

Fig. 4 : Separation of selenite and selenate using HPLC with ICP-MS detection. Column: ESA Anion III; mobile phase: 5 mM potassium hydrogen phthalate, pH 7; flow rate: 0.92 mL min^{-1}; injection volume: 100 µL onto column; 0.1 ng Se as selenite, 0.1 ng Se as selenate (corresponding to 1 µg Se L^{-1} as selenite and 1 µg Se L^{-1} as selenate).

169

Conclusion

Most research work in the past concentrated on the determination of total selenium concentrations in biological and environmental samples. Various techniques are available for this purpose.[11] Some efforts were made to determine the inorganic species selenite and selenate. Although methods for the identification and quantification of trace element compounds using hyphenated techniques (mainly HPLC with element-specific detection) have been known for two decades,[24-27]

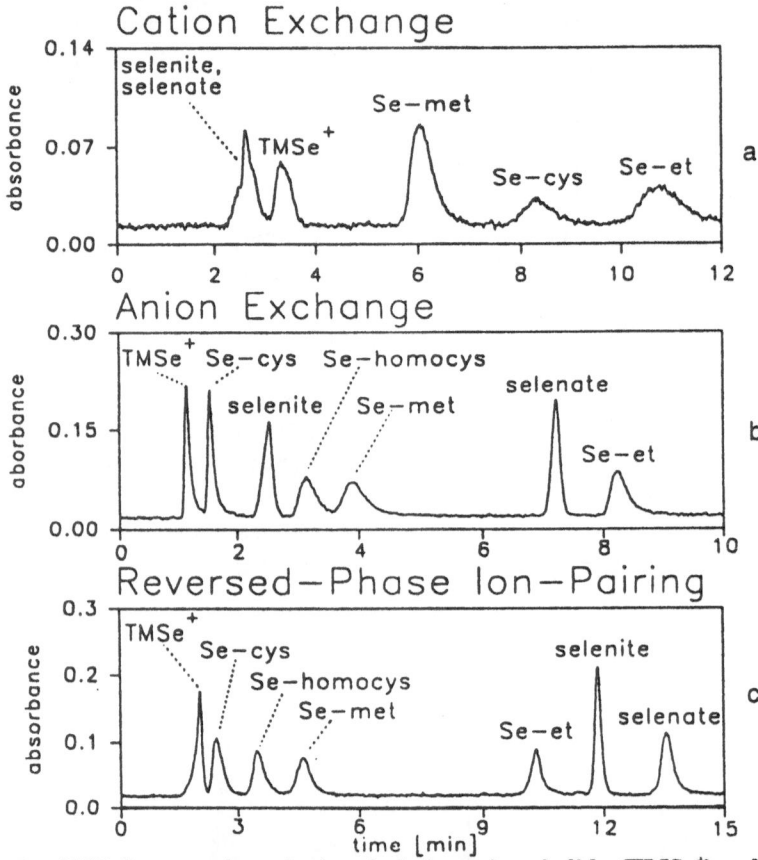

Fig. 5 : **HPLC separation of trimethylselenonium iodide (TMSe$^+$), selenite, selenate, selenocystine (Se-cys), selenohomocystine (Se-homocys), selenomethionine (Se-met), and selenoethionine (Se-et). Injection volume: 100 μL onto column; detector: FAAS; selenium concentration is 100 mg Se L^{-1} for each compound. a) Column: Hamilton PRP X-200; mobile phase: 100 mM HNO$_3$/9% 2-propanol (v/v); flow rate: 1 mL min^{-1}. b) Column: ESA Anion III; mobile phase: 2 mM potassium hydrogen phthalate, pH 9.0 _after 2.5 min_ 12 mM potassium hydrogen phthalate, pH 9.0/20% methanol (v/v); flow rate: 2 mL min^{-1}. c) Column: Hamilton PRP-1; mobile phase: 2 mM tetrabutylammonium phosphate _after 2 min_ 2 mM tetrabutylammonium phosphate/50% methanol (v/v); flow rate: 1 mL min^{-1}.**

knowledge about the occurrence and the concentrations of organic selenium compounds is not very extensive. Since the introduction of these hyphenated techniques, the instrumentation has improved, the procedures simplified, and the detection limits lowered. HPLC seems to be the method of choice for the separation of selenium compounds.[11]

Our present work concentrates on the elaboration of chromatographic procedures for the separation of organic selenium compounds, such as trimethylselenonium ion, selenoamino acids, and of the inorganic species selenite and selenate in single chromatographic runs. Preliminary results with seven selenium compounds separated on cation exchange, anion exchange, and on reversed-phase columns are very promising (Fig. 5). These methods will be applied to the determination of these compounds in biological and environmental samples in the near future.

Acknowledgement

This work is part of the PhD thesis of G.K. under the supervision of K.J.I.

References

1. K. Schwarz and C.M. Foltz, *J. Amer. Chem. Soc.*, **79** (1957) 3292.
2. M.L. Jackson, *Appl. Geochem.*, **1** (1986) 175.
3. O.A. Levander, *Curr. Top. Nutr. Dis. (Clin. Biochem., Nutr. Aspects Trace Elem.)*, **6** (1982) 345.
4. A. Wendel, *Phosphorus, Sulfur, Silicon, Relat. Elem.*, **67** (1992) 405.
5. L.P. Gough, H.T. Shacklette and A.A. Case, *U.S. Geol. Surv. Bull.*, No. **1466** (1979) 42.
6. K. Forchhammer and A. Böck, *Naturwissenschaften*, **78** (1991) 497.
7. R.J. Shamberger, *Biochemistry of Selenium*, Plenum Press, New York, 1983.
8. J.E. Oldfield, *J. Nutr.*, **117** (1987) 2002.
9. J.M. McNeal and L.S. Balistrieri, *Soil Sci. Soc. Am., Spec. Publ. (Selenium Agric. Environ.)*, **23** (1989) 1.
10. W. Maher, S. Baldwin, M. Deaker and M. Irving, *Appl. Organomet. Chem.*, **6** (1992) 103.
11. G. Kölbl, K. Kalcher, K.J. Irgolic and R.J. Magee, *Appl. Organomet. Chem.*, in press.
12. J.L. Hoffman, *J. Chromatogr.*, **588** (1991) 211.
13. K. Schachl, *Diploma Thesis*, Karl-Franzens Universität Graz, 1993.
14. H. Tanaka and K. Soda, *Methods Enzymol.*, **143** (1987) 240.
15. G. Kölbl, K. Kalcher and K.J. Irgolic, *J. Autom. Chem.*, **15** (1993) 37.
16. F.E. Brinckman, W.R. Blair, K.L. Jewett and W.P. Iverson, *J. Chromatogr. Sci.*, **15** (1977) 493.
17. M. Verlinden, H. Deelstra and E. Adriaenssens, *Talanta*, **28** (1981) 637.
18. I.R. Desai and V.S.K. Nair, *J. Chem. Soc.*, (1962) 2360.
19. G. Kölbl, K. Kalcher and K.J. Irgolic, *Anal. Chim. Acta*, in press.

20. K.E. Jarvis, A.L. Gray and R.S. Houk, *Handbook of Inductively Coupled Plasma Mass Spectrometry*, Blackie, Glasgow and London, 1992.
21. J.J. Thompson and R.S. Houk, *Anal. Chem.*, **58** (1986) 2541.
22. J. Goossens, F. Vanhaecke, L. Moens and R. Dams, *Anal. Chim. Acta*, **280** (1993) 137.
23. P. Allain, L. Jaunault, Y. Mauras, J.-M. Mermet and T. Delaporte, *Anal. Chem.*, **63** (1991) 1497.
24. L. Ebdon, S. Hill and R.W. Ward, *Analyst*, **112** (1987) 1.
25. S.J. Hill, M.J. Bloxham and P.J. Worsfold, *J. Anal. At. Spectrom.*, **8** (1993) 499.
26. N.P. Vela, L.K. Olson and J.A. Caruso, *Anal. Chem.*, **65** (1993) 585A.
27. N.P. Vela and J.A. Caruso, *J. Anal. At. Spectrom.*, **8** (1993) 787.

Main Group Elements and Their Compounds
V.G. Kumar Das (Ed)
Copyright © 1996 Narosa Publishing House, New Delhi, India

Polysiloxane-bound Ether-phosphines and Ruthenium Complexes : Characterization by Solid-state NMR Spectroscopy and Catalytic Properties

Ekkehard Lindner, Martin Kemmler, Herman A. Mayer
and Peter Wegner

Institut für Anorganische Chemie der Universität
Auf der Morgenstelle 18, D-7400 Tübingen 1, Germany

Over the past few years considerable progress has been made in the preparation and application of immobilized reagents.[1] Inorganic oxides as base matrices, especially synthetic silicates, have gained prominence due to their diversity in physical properties, their purity, chemical resistance and the ease of functionalization through silanol groups.[2] In contrast to surface-modified inorganic materials,[3-5] simultaneous co-condensation of organo-functionalized silanes $RSi(OMe)_3$ with $Si(OEt)_4$ (sol gel process[6]) yields novel networks (equation 1) with preferred mechanical and chemical properties, namely, a high degree of flexibility and high ligand densities as well as a reduced leaching of functional groups:[7,8]

$$x\ RSi(OMe)_3 + y\ Si(OEt)_4 + (3x+4y)/2\ H_2O \rightarrow [RSiO_{3/2}]_x \cdot [SiO_{4/2}]_y$$
$$\text{T groups} \qquad \text{Q groups}$$

$$+\ 3x\ MeOH + 4y\ EtOH \qquad (1)$$

When R contains donor atoms (such as P, O, N or S), trimethoxysilyl-(T-) functionalized organometallic complexes can be synthesized and copolymerized to yield three dimensional networks.[7-10,11] Our approach in respect of this goal was to generate T-functionalized ligands containing oxygen and phosphorus as donor atoms. This type of potentially bidentate ligand has recently found application in coordination chemistry and homogeneous catalysis.[12] A flexible α,ω-alkanediyl spacer was used to link the ether-phosphine moiety to the support. These versatile building blocks allow us to tailor and tune the properties of the materials to various applications.

Heteronuclear cross-polarization magic-angle spinning (CP-MAS) solid-state NMR spectroscopy has been shown to be a powerful tool for investigating insoluble and non-crystalline solids.[2,13,14] The phosphorus nucleus has proved to be a very suitable probe for investigating supported phosphines and the stereochemistry of immobilized complexes.[7-9,15-17] Carbon-13 CP-MAS NMR spectroscopy has been used to verify the integrity of the ligand framework.[9,11,15,16] In addition, solid-state ^{29}Si NMR spectroscopy is a unique method to characterize the support of sol gel materials.[7-9,11,14,18] Furthermore, scanning probe microscopy combined with the BET method has been used to characterize the surfaces of sol gel products.[18]

Although much synthetic work has been reported, detailed structural information about such materials is still lacking. In this article we report, on the preparation and characterization of sol gel matrices of materials of this category polymerized under various conditions and with a range of complex densities. The dynamic behaviour of the polysiloxane-bound ether-phosphine ligands and their ruthenium complexes is also discussed. In addition, we correlate sol gel routes, crosslinking, particle sizes, and surface areas with catalytic activities in hydrogenation experiments.

Results and Discussion

Preparation of the monomeric complexes 3a-c

In the case of trimethoxysilyl-(T-) modified phosphines as 1a-c, the general synthetic methodology complexes of the type $HRuCl(CO)(PR_3)_3$ from $RuCl_3$ and an excess of the phosphine and aqueous formaldehyde in alcohol[19] could not be applied. This is on account of hydrolysis of the T-functionality, undesirable at this stage, and a possible redox reaction between $RuCl_3$ and the phosphines. Therefore, the phosphines in $HRuCl(CO)(PPh_3)_3$ (2) were exchanged by the more basic ether-phosphines 1a-c by treating 2 with 1a-c in toluene (Scheme 1).

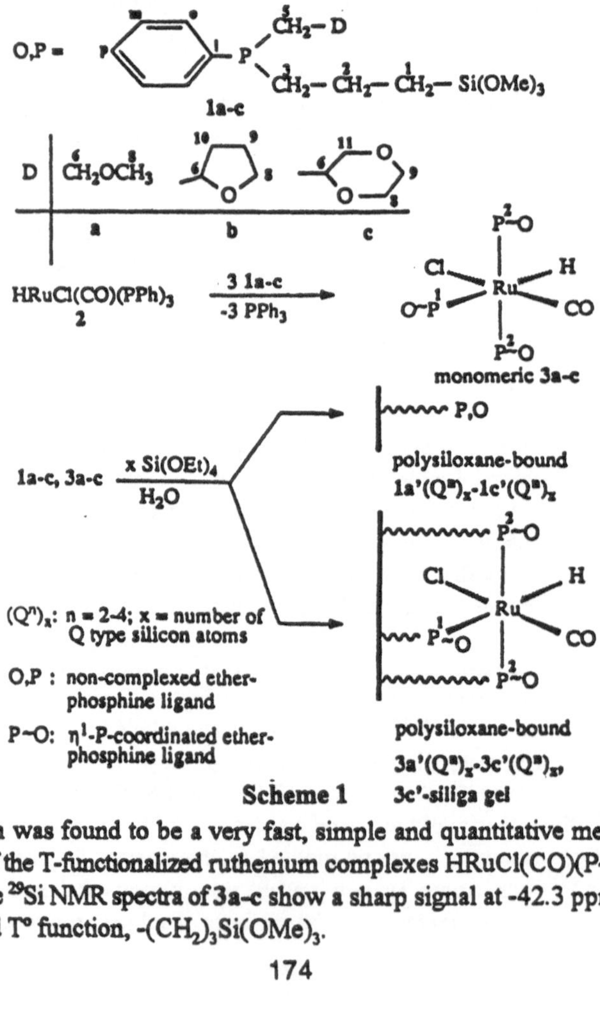

Scheme 1

This reaction was found to be a very fast, simple and quantitative method for the preparation of the T-functionalized ruthenium complexes $HRuCl(CO)(P\sim O)_3$ (3a-c). The solid-state ^{29}Si NMR spectra of 3a-c show a sharp signal at -42.3 ppm due to the unhydrolysed T^0 function, $-(CH_2)_3Si(OMe)_3$.

174

In previous studies, the polycondensation of T-functionalized ether-phosphine complexes of ruthenium(II) and palladium(II) was attempted in the absence of Si(OEt)$_4$.[3,9] These polymers show a low degree of condensation[20] (59-77%) and consist of branched polyorganosiloxane chains, which are connected via ruthenium and palladium complexes to form relatively open three dimensional networks. The retention of the stereochemistry of the transition metal complexes and the Si-C bond during the gelation process has been established by ^{31}P, ^{13}C and ^{29}Si solid-state NMR spectroscopy.[9,11] However, for many applications, e.g., catalysis or chromatography, it is favourable to design two- or three-dimensional polysiloxane matrices in a crosslinking copolymerization of Si(OEt)$_4$ with T-functions.[6,8,10] In this process, it is expected that the T species are concurrently incorporated together with Q-species or that the matrix, which is formed, contains domains of either T or Q groups. The sol gel process can be described by the following reaction scheme.[2]

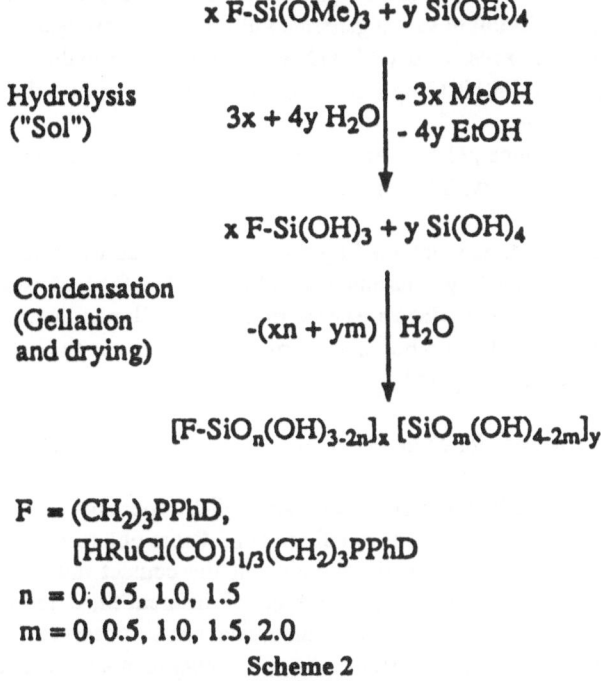

$$x\ F\text{-}Si(OMe)_3 + y\ Si(OEt)_4$$

Hydrolysis ("Sol")

$$3x + 4y\ H_2O \quad \begin{array}{l} -3x\ MeOH \\ -4y\ EtOH \end{array}$$

$$x\ F\text{-}Si(OH)_3 + y\ Si(OH)_4$$

Condensation (Gellation and drying)

$$-(xn + ym)\ H_2O$$

$$[F\text{-}SiO_n(OH)_{3-2n}]_x\ [SiO_m(OH)_{4-2m}]_y$$

$$F = (CH_2)_3PPhD,$$
$$[HRuCl(CO)]_{1/3}(CH_2)_3PPhD$$
$$n = 0,\ 0.5,\ 1.0,\ 1.5$$
$$m = 0,\ 0.5,\ 1.0,\ 1.5,\ 2.0$$

Scheme 2

Suitable monomer precursors are first hydrolysed and the resulting colloidal solutions subsequently undergo polymerization. The hydrated, highly swollen gells are then dried to obtain further condensation and densification.

When organometallic complexes such as **3a-c** are incorporated by sol gel processing, they have to be chemically inert towards their T-functionalized phosphines, water, Si(OEt)$_4$, and the polycondensation catalyst. Thus the sol gel process has to be adjusted to individual properties of the complexes. Several pathways of sol gel processing have been applied which can be considered as examples of various possibilities of creating functionalized polysiloxane materials.

In **method 1**, the components were homogenized with a minimum amount of methanol and a catalytic amount of (n-Bu)$_2$Sn(OAc)$_2$. Hydrolysis and condensation steps increased slowly the viscosity of the warmed mixture and eventually solidification led to a highly swollen gel. When the supporting solvent was removed in vacuo, the activities in the polymer network collapsed with a multiple loss of its volume. Since the resulting gel did not undergo swelling in alcohols to its original volume, this was taken as indication that simultaneous condensation of silanol groups had occurred. The hardened, brittle gels were ground mechanically into fine powders. Subsequent separation of PPh_3 by-product from the powdered material could be conveniently achieved by stirring the material in n-hexane ('solvent processing').

The precipitation of the hydrolysed complexes 3a-c and $Si(OEt)_4$ was carried out by *method 2a*. At low concentrations of the components in methanol [0.2 M solution of $Si(OEt)_4$], no gel transition was observed. After the addition of aqueous NH_4HCO_3, coagulation and condensation occurred, leading to fine, cloudy precipitates. The gel was isolated and stirred with dry acetone to facilitate condensation steps and to remove physically adsorbed water. In contrast to *method 1*, further drying in vacuo generated ultrafine gel particles.

In *method 2b*, excess of dry $MgSO_4$ was added to the hydrolysed complexes instead of aqueous NH_4HCO_3 to remove the water and to initiate the condensation and the gel transition.

In *method 3*, silica gel was prepared from $Si(OEt)_4$ alone and subjected to a second sol gel reaction with the monomer 3c to give the immobilized complex 3c'-silica gel.

The different materials obtained by the above procedures (methods 1-3) were thoroughly characterized by heteronuclear solid-state NMR spectroscopy, scanning electron microscopy, and surface area determinations, following which the catalytic behaviour of the polysiloxane-bound ruthenium complexes in the hydrogenation of butyraldehyde was investigated.

Characterization by Solid-State NMR

One-dimensional heteronuclear solid-state NMR spectroscopy can be used to yield a great variety of complementary information. Cross polarization constants T_{XH}, which can be calculated from the variation of the contact time, and spin-lattice relaxation times T_{1X} (X = Si, C, P) are typical of different Si, C and P nuclei in the materials.[21] Thus, local environments of the nuclei can be observed. The relaxation time of the protons in the rotating frame, $T_{1\rho H}$, however, is not specific to the various nuclei, but rather, it is a quantity which is averaged within domains of 1-2 nm in diameter, at least in those cases where there is sufficient dipolar coupling among the protons to provide spin diffusion.[22-23] If these dynamic aspects are taken into account, valuable quantitative information can be obtained. Thus, in the case of quantitative ^{29}Si CP-MAS NMR spectroscopy, the correct proportion of the silyl species, T:Q, the degree of condensation,[20] and the distribution of the T-functionalities through the polysiloxane matrix can be described.

^{29}Si **Solid-State NMR Spectroscopy**: In Figure 1, ^{29}Si CP-MAS NMR spectra of the polysiloxane-bound complex $HRuCl(CO)(P{\sim}O)_3$, $3b'(Q^n)_6$, obtained at two different contact times T_C, are depicted. This complex has been prepared according to **method 1** by cocondensation with six equivalents of $Si(OEt)_4$ (T:Q = 1:2, Schemes 1 and 2). The silyl species in Figure 1 and the corresponding ^{29}Si chemical shifts[4,5,24] are summarized in Table I.

Table 1. Silyl species, $\delta^{29}Si$, T_{SiH}, relative $[I(T_c)]$, and corrected relative values I_o of the silyl moieties in $3b'(Q^n)_6$.

notation	structural type	$\delta^{29}Si$ (ppm)	T_{SiH} (ms)	I_o $[I(4\ ms)]$	I_o $[I(5\ ms)]$	I_o $[I(8\ ms)]$	I_o $[I(10\ ms)]$	MAS (SPE)
T^1	$O_{1/2}Si(OH)_2F^a$	-49.9	-	0.8(0.8)	0.5(0.5)	0.4(0.3)	0.5(0.4)	0.6
T^2	$O_{2/2}Si(OH)F$	-59.2	1.45	3.1(3.2)	3.4(3.4)	2.8(2.7)	3.0(2.8)	3.0
T^3	$O_{3/2}SiF$	-66.5	1.95	6.1(6.0)	6.2(6.1)	6.8(7.0)	6.5(6.8)	6.4
Q^2	$O_{2/2}Si(OH)_2$	-91.6	1.46	2.1(2.2)	1.8(1.7)	2.0(1.9)	2.2(2.1)	2.2
Q^3	$O_{3/2}Si(OH)$	-100.5	2.04	$-^b$	$-^b$	9.6(10.0)	9.1(9.6)	9.2
Q^4	$O_{4/2}Si$	-108.6	3.43	$-^b$	$-^b$	9.1(10.1)	8.2(10.2)	8.4

^{a}F $= (CH_2)_3PPhD(HRuClCO)_{1/3}$, ^{b}Contact time too short for cross polarization of Q^3 and Q^4 species.

As is evident from Fig. 1, various contact times T_C result in different peak intensities in the ^{29}Si solid-state NMR spectra. Due to the strong dependence of the cross polarization efficiency on the internuclear distances between protons and the various silicon atoms, the relative peak intensities cannot be used to portray the correct population in the sample.[21,25,26] However, the observed signal intensities $I(T_C)$ can be corrected by performing variable contact time studies under conditions of the Hartmann-Hahn match. The amplitude $I(T_C)$ of each signal observed in the ^{29}Si spectrum at the contact time T_C has been described by the following `classical' equation:[26]

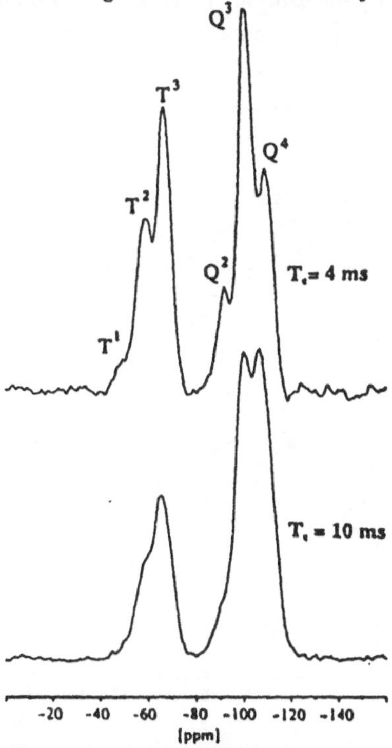

Figure 1. Solid-state ^{29}Si CP-MAS NMR spectra of $3b'(Q^n)_6$ prepared by method 1 at different contact times Tc.

177

$$I(T_c) = \frac{I_o}{1 - \dfrac{T_{SiH}}{T_{1\rho H}}} \left(e^{\frac{T_C}{T_{1\rho H}}} - e^{\frac{-T_C}{T_{SiH}}} \right) \qquad (2)$$

The value of I_o represents the relative amount of each of the silyl species present in the sample.

First of all, cross polarization constants T_{SiH} (Table 1) were determined by a fit of equation 2 to the intensities of the signal $I(T_c)$ as a function of the contact time T_C. Silicon atoms with an equal number of proton-containing substituents have similar T_{SiH} constants (T^2/Q^2 and T^3/Q^3). The fully condensed Q^4 species with rather long ^{29}Si-^1H distances magnetize the slowest. The T_{SiH} constants for the Q species of the polysiloxane-bound complexes (Table 1) are in the range of those determined for silica gel prepared by **method 3**: 1.07 ms (Q^2), 1.86 ms (Q^3) and 3.90 ms (Q^4). It is evident that the T_{SiH} constants, especially those of the silyl species in the T/Q copolymers, must be considered as an average value of all possible local environments of the particular silicon species.

Secondly, the loss of proton magnetization in the rotating frame, described by $T_{1\rho H}$,[27] has to be investigated to obtain the quantitative ratios I_o of the various T and Q species (equation 2) and to describe the distribution of the T-anchored ruthenium complexes throughout the material.

As shown in Fig. 2A by the sample $3b'(Q^n)_6$, $T_{1\rho H}$ is constant (5.4 ± 0.2 ms) for all protons near the T and Q silicon atoms with no dependence on the applied contact time T_C (4 and 10 ms) being observed.

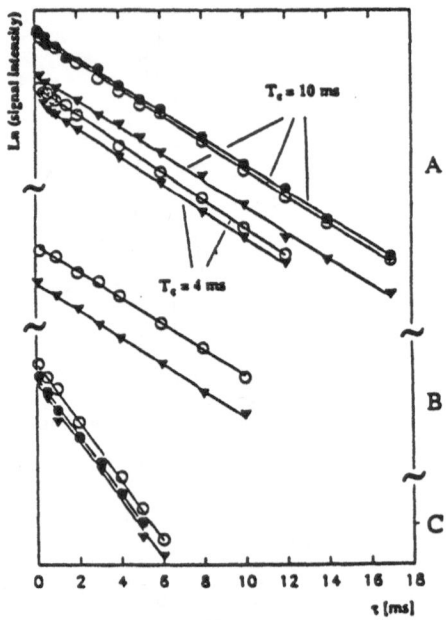

Figure 2. $T_{1\rho H}$ via ^{29}Si: A, $3b'(Q^n)_6$; B, $3a'(Q^n)_6$; C, $1a'(Q^n)_2$ (∇ T^3, $\circ Q^3$, $\bullet Q^4$).

Obviously, T_{1pH} is a uniform quantity in the sample due to efficient dipolar coupling of the protons. This indicates that the ligand-containing T species and the Q groups are covalently bound and mixed in the gel. Hence, the ruthenium complexes, which are fixed by the T groups via the ether-phosphine ligand, are distributed homogeneously across the whole T/Q copolymer. Surprisingly low, single exponential T_{1pH} values for the protons of the T and Q silicon species have been found for the ether-phosphine ligands $1a'(Q^n)_2$, $1c'(Q^n)_2$, $1c'(Q^n)_{10}$ (Table 2 and Fig. 2C).

Table 2 : Quantification of materials prepared by methods 1, 2a,b, 3 by ^{29}Si Solid-State NMR spectroscopy

compound	T_{1pH}(ms)	T¹	T²	T³	Q²	Q³	Q⁴	Found T:Q	degree of condensation (%) T	Q	degree of hydrolysis (%) T	Q
Method 1												
$1a'(Q^n)_2$[b]	2.47	-	1.9	8.1	1.6	8.7	7.9	1:1.82	94	84	100	-
$1a'(Q^n)_{10}$[b]	3.61	-	2.1	7.9	6.8	46.1	40.7	1:9.36	93	84	98	-
$1b'(Q^n)_2$[b]	3.72	-	2.8	7.2	2.2	10.8	8.5	1:2.15	91	82	100	98
$1c'(Q^n)_2$[c]	3.96	-	2.2	7.8	1.9	8.9	8.6	1:1.94	93	84	100	97
$3a'(Q^n)_6$[c]	5.5	0.6	3.0	6.4	2.1	9.3	8.3	1:1.97	86	83	96	-
$3b'(Q^n)_6$[c]	5.4	0.5	3.1	6.4	2.0	9.4	8.7	1:2.01	86	83	100	100
$3c'(Q^n)_6$[c]	5.4	1.0	3.1	5.9	2.7	9.6	7.6	1:1.99	83	81	96	99
$3c'(Q^n)_{10}$[c]	5.4	1.5	2.5	6.0	4.2	17.8	11.9	1:3.39	82	81	97	98
$3c'(Q^n)_{30}$[c]	5.4 - 20[d]	0.7	2.8	6.5	15	43.5	29.6[e,f]	1:8.83[c,f]	86	82	100	99
Method 2a												
$3a'(Q^n)_{3.4}$[c]	5.5	1.7	3.0	5.3	1.7	5.2	4.4	1:1.13	79	81	98	
$3b'(Q^n)_{4.4}$[c]	5.5	1.1	3.4	6.5	1.9	6.5	5.5	1:1.46	81	81	100	100
$3c'(Q^n)_{4.2}$[c]	5.4	1.2	3.8	5.0	1.9	6.9	4.8	1:1.41	79	80	98	100
Method 2b												
Mg^{2+}-$3a'(Q^n)_3$[c]	5.4	1.5	3.8	5.5	1.5[g]	4.6	3.1	1:0.99	78	76	89	-
Mg^{2+}-$3a'(Q^n)_6$[c]	5.4	1.4	3.6	5.0	4.3[h]	7.6	5.2	1:1.94	79	73	92	100
Method 3												
silica gel[c]	40.5	-	-	-	0.6	4.3	6.3[e]	-	-	88	-	100
3c'-silica gel[c]	5.5-32[d]	2.2[i]	5.3	1.0	0.9	4.3	8.0[e,k]	1:1.32[c,k]	53	88	89	100

[a] Determined via ^{29}Si. [b] Values from deconvoluted ^{29}Si MAS NMR (SPE) spectra. [c] Values determined by the cross polarization method. [d] At least two components of T_{1pH}. [e] Q⁴ moieties maximal four bonds from protons. [f] Value determined by a ^{29}Si MAS NMR (SPE) spectrum: $I_a = 43.9$ (T:Q = 1 : 10.03). [g] Additional signal: Q^z ($I_a = 1.5$). [h] Additional signals: $Q^{1'}$ ($I_a = 0.3$), Q^z ($I_a = 2.0$). [i] Additional signal T⁰ ($I_a = 1.5$). [k] Value determined by a ^{29}Si MAS NMR (SPE) spectrum: ($I_a = 14.4$ (T:Q = 1 : 1.96)).

Upon coordination of three immobilized phosphine ligands to ruthenium to form the complexes $3a'(Q^n)_x$-$3c'(Q^n)_x$ (x = 3.4-10), T_{1pH} increases to a typical value of 5.4

-5.5 ms (Table 2 and Figs. 2A, B). The fact that all protons in the polysiloxane matrix, even those of the Q species, change their T_{1pH} quantity equally upon complexation clearly demonstrates that the T functions of the ruthenium complexes are thoroughly mixed with the Q moieties. In the case of domain structures, which may be formed from either T or Q species, Q^4 silicon atoms located too far from protons (typically 4 bonds away) would not contribute significantly to the intensity of the Q^4 signal in the spectrum. Such 'problematic' Q^4 silicons are present in silica gels and silicon glasses to a considerable extent.[2,28,29] Therefore, the number of non-detectable 'bulk' species depends on the ratio of T:Q. At low Q contents (x = 3.4-10, see above), and if the T and Q silicon are totally disordered in the polysiloxane matrix,the total amount of Q^4 species has to contribute to the intensity of the signal according to equation 2. However, for a quantitative determination, the condition $T_{1pH} \gg T_{SiH}$ must be satisfied.[23] The Q^4 groups in the polysiloxane-bound ether-phosphine ligands $1a'(Q^n)_2$-$1c'(Q^n)_2$ and $1a'(Q^n)_{10}$ do not meet this condition because cross polarization and proton relaxation in the rotating frame occur at similar rates (e.g. $1a'(Q^n)_2$: $T_{SiH} = 2.77$ ms, $T_{1pH} = 2.47$ ms, see Table 2). In the case of the polysiloxane-bound complexes ($T_{1pH} = 5.4 - 5.5$ ms, see Table 2), the determination of I_o of the silyl species by CP-MAS NMR spectroscopy is well detailed in Table 1 for $3b'(Q^n)_6$. The 'calculated' real amounts I_o have been compared to the absolute areas of a ^{29}Si MAS NMR spectrum (SPE) in Table 1. Although the accuracy of quantitative ^{29}Si CP-MAS NMR has been generally estimated to be approximately $\pm$ 10%,[18,26] both methods are in acceptable agreement. Since the expected ratio of T:Q = 1:2 of the polysiloxane matrix has been verified by both methods, domains related to silica gel with 'bulk' Q^4 species seem to be absent. Moreover, the discrepancy in the I_o values of the T and Q type silicon species between the CP- and the SPE-method is within this margin of error. Therefore, the T-functionalized (ether-phosphine)ruthenium moieties and the Q species in $3b'(Q^n)_6$ cannot be phase separated or located in domains of either T or Q groups but have to be totally mixed in the material. Fyfe and coworkers have recently shown by a two dimensional ^{1}H/^{2}Si correlation experiment,that the Q species in a methyl-substituted, D_2O treated gel have been directly cross polarized from the methyl group of the T species to a considerable amount.[14] Thus the components are mixed in the gel and this is in excellent agreement with our studies.

In Table 2, various samples prepared by the methods 1, 2a, b and 3 have been listed with their relative content of silyl moieties I_o. In the ^{29}Si solid-state NMR spectra of the polysiloxane-bound ether-phosphine ligands, no T^1 and only low amounts of T^2 species have been detected. This is reflected in the high degree of condensation[20] for the T groups (91-94%). T moieties of the complexes prepared by methods 1 and 2a also show a high degree of condensation (79-86%), which is one basic requirement for application in catalysis without leaching of the T species and the ruthenium complexes, respectively. The degree of condensation of the Q silicon species in samples prepared by the methods 1 and 2a (Table 2) shows negligible tolerances ($\pm$ 2%). In each of the polysiloxane-bound complexes prepared by method 1, the anticipated ratio of T:Q has been found by quantification with CP excitation, except for sample $3b'(Q^n)_{30}$. In this case, just 67% of the Q^4 silicon species, found by ^{29}Si MAS NMR spectroscopy (SPE), were determined by the cross polarization method. The presence of a considerable amount of 'bulk' Q^4 groups in this case indicates the formation of domains related to silica gel, which are at least 1-2 nm in diameter.

Ether-phosphine complexes prepared by **method 2a** show a lower content of Q silicons. 25-45% of the total amount of Q groups were removed as uncondensed silicic acid or oligomer during the precipitation and washing process. The degree of condensation is similar for T and Q species and somewhat lower compared to complexes obtained by **method 1**. The T_{1pH} data of the polysiloxane matrix in the samples prepared by **method 2a** also show a single exponential behaviour and are uniform for the T and Q silicons, too. The values are comparable to those found for the polysiloxane-bound complexes prepared according to **method 1** (Table 2). Thus, in principle, the polymer network obtained by both methods should not differ. For the samples prepared by the **methods 1** and **2a** (Table 2) the degree of hydrolysis[30] as determined by the residual $-OCH_3$ and $-OCH_2-$ signals in the ^{13}C CP-MAS NMR spectra (50.3 and 59.8 ppm, respectively), ranges between 96 and 100%. Hence, in these cases nearly all the uncondensed Si functions are polar SiOH groups.

In the sample **3c'-silica gel**, which was obtained by 'heterogeneous polycondensation' according to **method 3** (T:Q $\simeq$ 1.2), the condensation reaction of the T-functionalized ruthenium complexes during the sol gel process was less efficient. Even uncondensed T^0 species at -43.2 ppm have been observed in the ^{29}Si CP-MAS NMR spectrum (Table 2). The low degree of condensation of 53% can be explained in terms of the lack of suitable Q-copolymerization partners and isolated T^0 functions[9,11] in the expected T-T/Q-Q material. These experiments demonstrate that a high degree of condensation of the T-functionalities can solely be obtained by a simultaneous cocondensation of T with Q moieties starting from the monomers according to the **methods 1** and **2**. This is necessary for a good fixing of the ruthenium complexes in the gel matrices in view of a successful application as a stable, heterogeneous catalyst.

As a result of the quantification, the average structures of the polysiloxane matrix of the ether-phosphines and their ruthenium complexes prepared by the **methods 1** and **2a** (Table 2, except sample $3c'(Q^n)_{30}$] can be visualized. 50-65% of the matrix consists of mixed T^3 and Q^3 species, which form a two-dimensional layer structure. Incorporated Q species (25-30%), as elements of the three-dimensional network, crosslink these layers. A smaller amount of T^2 and Q^2 species are either random elements of the disordered layers or are included as chain segments. In the case of the complexes, three T-functionalized (ether-phosphine) moieties are additionally linked by a ruthenium atom. Thus, the main part of the silicon building blocks in the polysiloxane matrices (70-75%) are one- and two-dimensional elements.

^{31}P and ^{13}C CP-MAS NMR Spectroscopy

In the ^{31}P MAS NMR spectra of the polysiloxane-bound ether-phosphines $1a'(Q^n)_2-1c'(Q^n)_2$, one signal has been observed. The chemical shifts are comparable to those observed by $^{31}P\{^1H\}$ NMR spectroscopy of the monomers in solution. In Table 3, the ^{31}P signals of the polysiloxane-bound O,P-ligands $1a'(Q^n)_2-1c'(Q^n)_2$ are listed together with their corresponding P-coordinated complexes $3a'(Q^n)_6-3c'(Q^n)_6$. No free ether-phosphine ligands, but the signals of *trans*-P^2 and *cis*-P^1 (Scheme 1) with the relative intensities of 2:1 are observed in the spectra. Since all ether-phosphine ligands are P-coordinated to ruthenium, each T moiety of the ligand contains 1/3 of the ruthenium complex fragment [HRuClCO]. The ^{31}P and ^{13}C chemicals shifts and the C≡O absorption in the IR spectra of the polysiloxane-bound complexes are in agreement with those of their monomeric counterparts. The $^{31}PC-$

MAS NMR spectra of the monomeric complex **3c** (Fig. 3A) and the polysiloxane-bound complex **3c'(Q^n)$_6$** (Fig. 3B) show that sol gel processing with Si(OEt)$_4$ does

Table 3 : δ ^{31}P, T$_{pH}$, T$_{1pH}$ and T$_{1P}$ data of **1a'(Q^n)$_2$-1c'(Q^n)$_2$** and **3a'(Q^n)$_6$-3c'(Q^n)$_6$**

compound	δ ^{31}P (ppm)	T_{pH} (ms)	T_{1pH} (ms)	T_{1P}(s)
1a'(Q^n)$_2$	-32.9	0.58	2.93	5.2
1b'(Q^n)$_2$	-32.5	0.23	4.42	6.9
1c'(Q^n)$_2$	-34.9	0.24	4.90	10.5
3a'(Q^n)$_6$	17.3 (P^2), -0.3 (P)	0.21 (P^2), 0.22 (P^1)	5.83	7.0 (P^2), 7.0 (P^1)
3b'(Q^n)$_6$	18.5 (P^2) -1.3 (P^1)	0.16 (P^2), 0.17 (P^1)	6.05	13.9 (P^2), 13.6 (P^1)
3c'(Q^n)$_6$	20.1 (P^2), 1.9 (P^1)	0.17 (P^2), 0.19 (P^1)	6.17	11.6 (P^2), 11.4 (P^1)

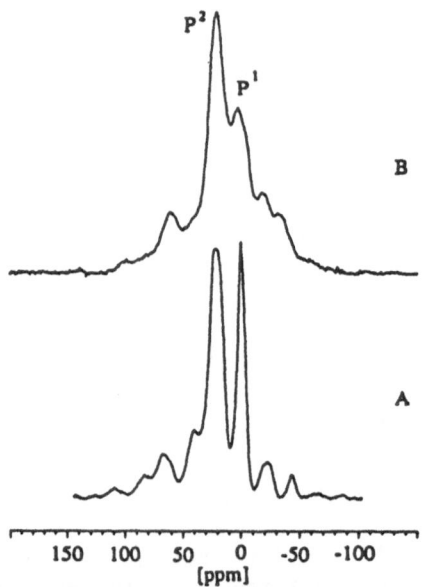

Fig 3 : Solid-state ^{31}P CP-MAS NMR spectra of **3c** (A, 3.5 kHz) and after sol gel processing with Si(OEt)$_4$ [B, **3c'(Q^n)$_6$**, 3.2 kHz]. The broadened signal of P^2 in A is due to diastereomers.

not change the complexes, but increases the linewidths of the signals of P^2 and P^1 due to an enhanced dispersion of the chemical shifts of the phosphorus nuclei. The mobilities of the ether-phosphine and their ruthenium complexes in the polysiloxane framework have been studied by ^{31}P CP-MAS NMR spectroscopy. The free ether-phosphine ligands **1b'(Q^n)$_2$** and **1c'(Q^n)$_2$**, containing tetrahydrofuryl- and 1,4-

dioxanyl groups, magnetize faster (lower T_{pH} constants) than methoxyethyl-substituted $1a'(Q^n)_2$ (see Table 3). The linewidths of the ^{31}P signals of the samples $1a'(Q^n)_2$, $1b'(Q^n)_2$, and $1c'(Q^n)_2$ are 640, 870 and 1160 Hz, respectively. Moreover, T_{1pH} and T_{1P} relaxation times are increased in this series. Thus, decreasing mobilities in the free ether-phosphine ligands from $1a'(Q^n)_2$ to $1b'(Q^n)_2$ and $1c'(Q^n)_2$ are evident. When the phosphine ligands are bound to ruthenium, the phosphorus nuclei are cross-polarized faster (T_{Ph} constants decrease), while the relaxation times T_{1pH} and T_{1P} (Table 3) and the chemical shift anisotropy increase. This is due to the expected reduced mobilities of the polysiloxane-bound ligands upon coordination to the ruthenium atom. The magnetization in the T_{1P} experiment of the (ether-phosphine)ruthenium complexes $3a'(Q^n)_6$-$3c'(Q^n)_6$ showed a multi-exponential decrease of the signal intensities, which was approximated by a bi-exponential function. The amount of the 'two' components has been taken into account with their T_{1P} times to give one averaged T_{1P} value (Table 3). In contrast, the magnetization in the T_{1P} experiments of each of the free ether-phosphine $1a'(Q^n)_2$-$1c'(Q^n)_2$ showed a single-exponential behaviour. This is due to the mobilities of the free ether-phosphine ligands, which lead to an averaging of the different local environments of the polysiloxane materials.

In Figure 4, T_{1pH} and T_{1P} of the ether-phosphine ligands $1a'(Q^n)_2$-$1c'(Q^n)_2$ (Fig. 4A-C) and of the complex $3a'(Q^n)_6$ (Figures 4D, 4E) are shown as a function of the temperature. In the very mobile, free ligand $1a'(Q^n)_2$, the increase in temperature is correlated with the increasing T_{1pH} values (Fig. 4A). However, in the case of the bulky, free ether-phosphines $1b'(Q^n)_2$, decreasing T_{1pH} values were determined by increasing the temperature, until minima at 337 K (1.83 ms) and 357 K (2.19 ms), respectively, were reached (Fig. 4A). In all samples, $1a'(Q^n)_2$-$1c'(Q^n)_2$, the T_{1P} times (Fig. 4B) and the linewidths of the peaks in the ^{31}P CP-MAS NMR spectra (Fig. 4C) decreased to a range of similar values (2.5 - 3.1 s and 234 - 374 Hz, respectively). The low T_{1P} data of the free ether-phosphines above 357 K are close to those of phosphines in liquids at ambient temperature. These studies clearly demonstrate the temperature dependent transition of the ether-phosphines in the gel to very mobile ligands. The high mobility behaviour enables the reversible exchange of P^1 in the polysiloxane-bound ruthenium complexes $HRuCl(CO)(P{\sim}O)_3$ with a CO ligand above 60 - 70 °C.[9] In the complex $3a'(Q^n)_6$, the temperature dependences of T_{1pH} and T_{1P} are completely different, the effects being much smaller (Fig. 4D and 4E, respectively). First of all, T_{1pH} and T_{1P} decreased slightly. Above 320 K, increasing T_{1pH} values of P^1 and P^2 in the complex were observed (Fig. 4D). The T_{1P} curve showed a sudden drop to lower values between 320 and 340 K (Fig. 4E) and the linewidth of P^2 in the ^{31}P CP-MAS NMR spectra of the complex decreased reversibly from 1400 to 1200 Hz. The ^{31}P CP-MAS NMR spectra of $3a'(Q^n)_6$ at 367 K and 397 K are shown in Fig. 5.

Above 347 K the magnetization decay in the T_{1P} experiment of the P nuclei in the ruthenium complex showed a single-exponential behaviour. This indicates that the ruthenium complex is fixed non-rigidly by its three carbon spacers to the matrices within the gel. At room temperature, the complex yields a ^{31}P CP-MAS NMR spectrum similar to that recorded at 367 K (Fig. 5A) after it was treated with a small amount of methanol to induce swelling. Both the change of temperature and the adsorption of methanol leads to a smaller dispersion of the chemical shifts of the phosphorus nuclei due to the non-rigid behaviour of the polysiloxane systems. The chemical shift anisotropy in the complex remained unchanged, which is consistent with a limited effect of temperature on the relaxation time $T_{1\rho}$ (Fig. 4E) and strong, rigid Ru-P bonds. Thus the transition in the $T_{1\rho}$ curves of the dry sample is caused by a thermal change in the material which might be related to the glass transition of organic polymers.[23]

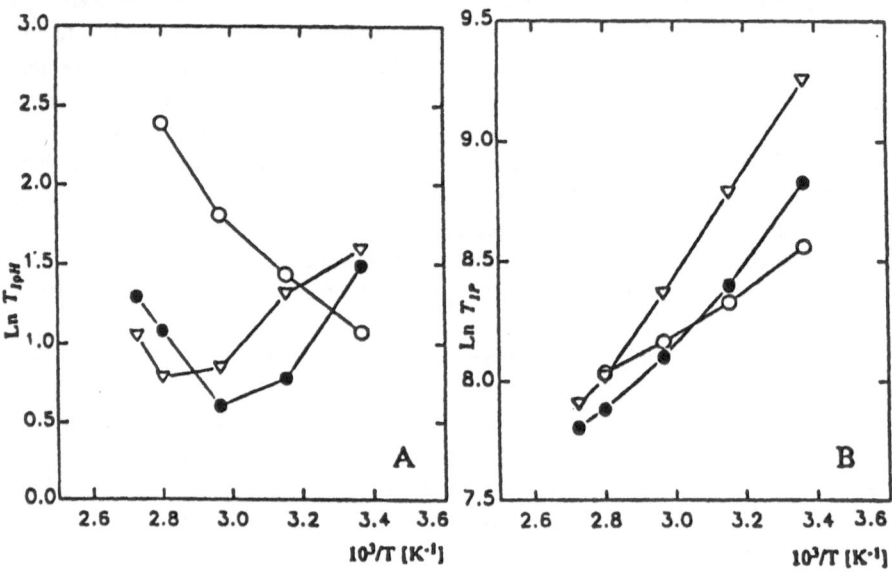

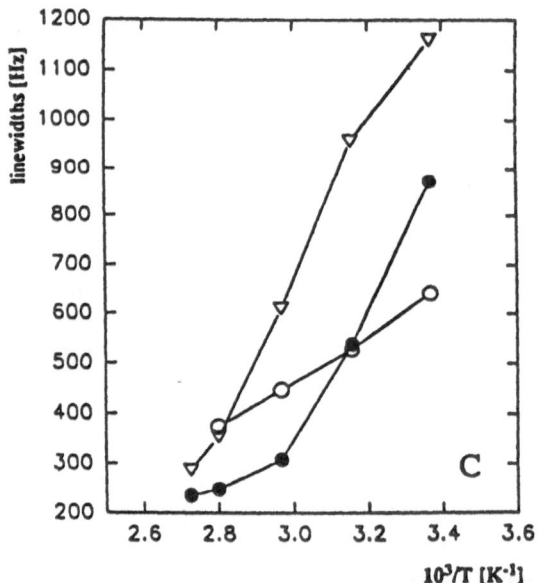

Fig 4A-C : $T_{1\rho H}$ (A), $T_{1\rho}$ (B), and linewidths (C) of the polysiloxane-bound ether-phosphine ligands $1a'(Q^n)_2$ o, $1b'(Q^n)_2$ •, and $1c'(Q^n)_2$ ▽ as a function of the temperature.

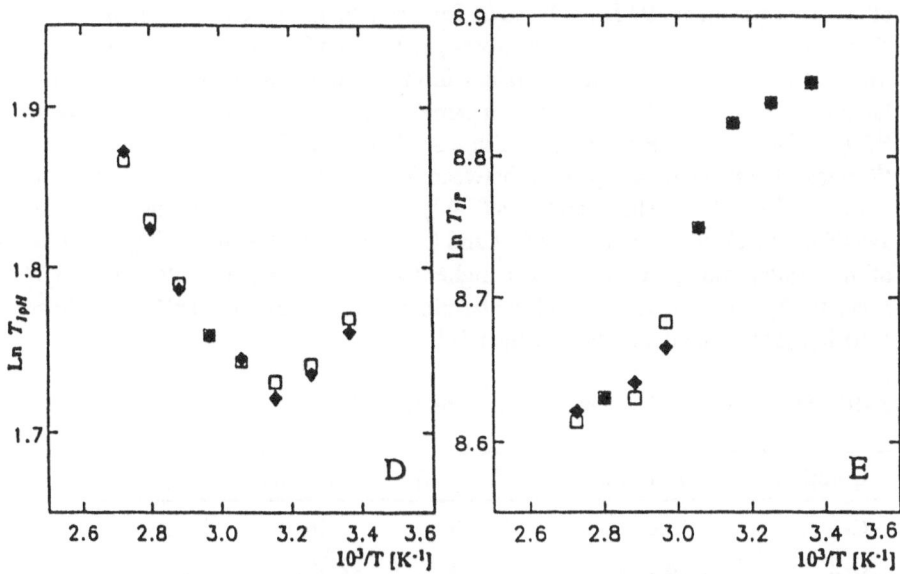

Fig 4D,E : $T_{1\rho H}$ (D) and $T_{1\rho}$ (E) of the P-coordinated ether-phosphine ligand in $3a'(Q^n)_2$ as a function of the temperature ($\square P^2$, $\blacklozenge P^1$).

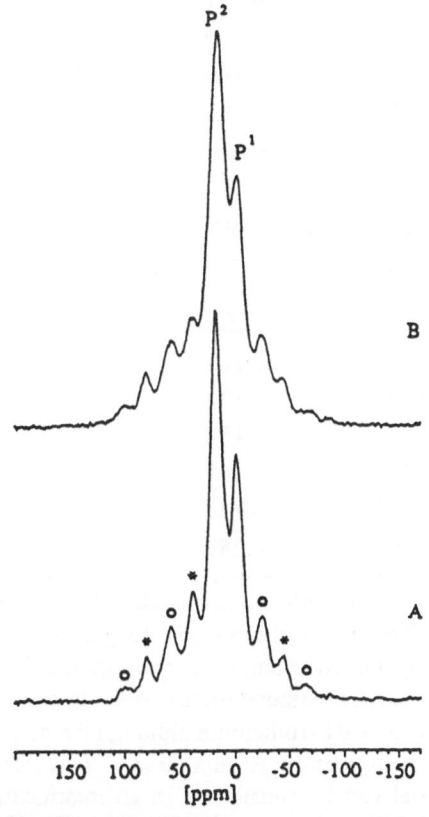

Fig 5 : Solid-state ^{31}P CP-MAS NMR spectra (3.4 kHz) of $3a'(Q^n)_6$ at 367 K (A) and 297 K (B)

185

The corresponding dynamic parameters obtained from ^{13}C CP-MAS NMR spectra of the same system at 297 K are listed in Tables 4a,b. In agreement with the results obtained by ^{31}P solid-state NMR spectroscopy, T_{1pH} and T_{1C} evaluated from ^{13}C CP-MAS NMR spectra show higher values in the complex $3a'(Q^n)_6$ than in the free ligand $1a'(Q^n)_2$. The T_{1pH} data in both samples are in the range determined via the ^{29}Si (see Table 2) and ^{31}P probes (see Table 3). Due to the low natural abundance of ^{13}C nuclei, the spin diffusion process between them is minimal and different carbons exhibit different T_{1C} relaxation times (Tables 4a,b).[22] Thus decreasing T_{1C} data in $1a'(Q^n)_2$ were obtained from C-1 to C-2 and C-3,5 showing the increasing mobilities of the carbons along the spacer. The higher value for the T_{1C} relaxation time (1.76 s) and the T_{CH} constant (350 µs) of the terminal methoxy group in $1a'(Q^n)_2$ indicate a methyl group rotation in the free ligand.[31]

Table 4a. $\delta\ ^{31}P$, T_{CH}, T_{1pH} and T_{1C} data of $1a'(Q^n)_2$

Compound	$\delta\ ^{31}P$ (ppm)	T_{CH} (ms)	T_{1pH} (ms)	T_{1C} (s)
C-1	13.9	0.189	2.46	1.25
C-2	18.3	0.114	2.55	1.14
C-3,5	27.2	0.165	2.68	0.54
C-6	70.0	0.251	2.51	0.86
C-8	58.2	0.350	2.38	1.76
C-i	139.5	0.140	2.44	2.64
C-o	132.1	0.203	2.43	1.09
C-m,p	128.7	0.179	2.42	0.84

Table 4b. $\delta\ ^{31}P$, T_{CH}, T_{1pH} and T_{1C} data of $3a'(Q^n)_6$

Compound	$\delta\ ^{31}P$ (ppm)	T_{CH} (ms)	T_{1pH} (ms)	T_{1C} (s)
C-1	14.1	0.182	5.35	3.80
C-2	17.9	0.182	5.97	4.70
C-3,5	27.3	0.158	5.85	4.81
C-6	68.5	0.165	-	3.61
C-8	57.9	0.725	5.48	3.75
C-o,m,p	129.4	0.289	5.63	6.36

The solid state NMR investigations elucidate the thermal properties and the mobility behaviour of the functional centers in the polysiloxane material with a T:Q ratio of 1:2. All polysiloxane-bound ether-phosphine drastically increase their mobility in the gel when the temperature is raised. In contrast, coordinated ether-phosphines are strongly bound to ruthenium although the materials show a non-rigid character when the temperature is increased. Therefore this tailoring of a functionalized material can be considered as an intermediate system between a solution and a solid and should combine the advantages of homogeneous and heterogeneous catalysis.

Relationship to silica gel attached systems

Silica gels normally show much higher T_{1pH} values (typically 20-50 ms)[24,32] than those reported in this work for functionalized polysiloxanes (Tables 1 and 2). Thus, if more Si(OEt)$_4$ is used for the cocondensation, it should be possible to find out the transition when silica gel domains are formed due to an inhomogeneous molecular mixing between T and Q functionalities. For the sample silica gel (Fig. 6A, **method 3**), the Q species show a T_{1pH} value of 40.5 ms. After the cocondensation with **3c** to form **3c'-silica gel** (T:Q 1:2, **method 3**), the Q species show the expected two components of T_{1pH} (Fig. 6B). The lower value is comparable with that calculated from the single-exponential magnetization decay of the signals of the T groups (T^1 and T^2) in this sample (5.5 ms); the larger one is in the range of 19-33 ms. The majority of the protons (70%) of the Q^3 species in **3c'-silica gel** show the low T_{1pH} value; hence they are dipolar-coupled with the protons of the T domains. Q^3 protons with longer T_{1pH} values are located in the bulk of the silica gel domains, separated and less influenced by 'surface' T and Q protons (see Table 2). Therefore, this sample consists of domains formed by the T groups of the ruthenium complexes **3c'** and a core of rigid silica gel, as might be expected from the preparation according to **method 3**.

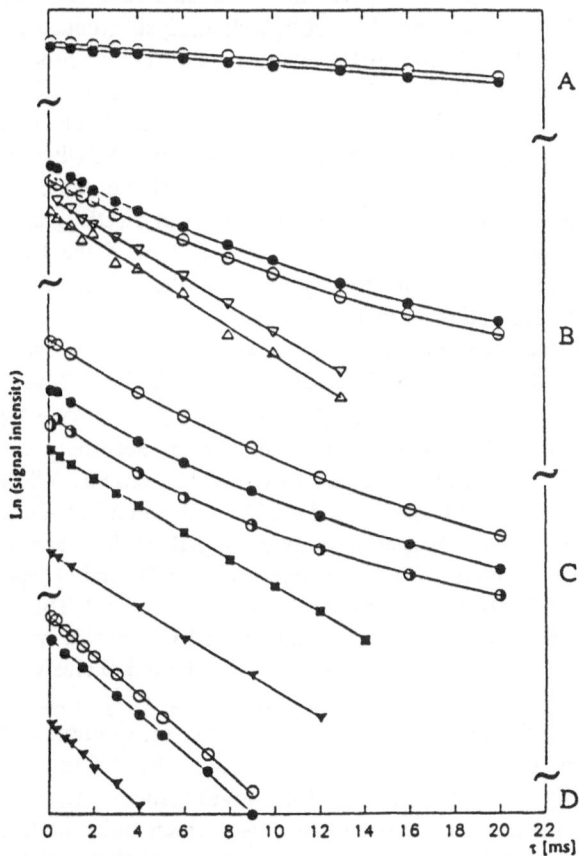

Fig 6 : T_{1pH} via ^{29}Si(T_C = 5 ms) and ^{31}P (T_C = 1 ms): A, silica gel; B, 3c'-silica gel; C, 3c'(Q^n)$_{30}$; D, 1a'(Q^n)$_{10}$ (^{29}Si:△ T^1, ▽ T^2, ▼ T^3, ○ Q^2, ○ Q^3, ● Q^4, ^{31}P:■ P^2).

187

In contrast to **3c'-silica gel**, the samples **3c'(Q^n)$_{30}$** and **1a'(Q^n)$_{10}$** were prepared by homogeneous copolymerization (T:Q = 1:10, **method 1**). Fig. 6C shows the T_{1pH} behaviour of **3c'(Q^n)$_{30}$**, which is comparable to that described in the previous sample. Those protons which are located in close proximity to the complexes show one low T_{1pH} component of 5.4 ms. This is demonstrated by a measure of T_{1pH} via the sensitive ^{31}P probe and the ^{29}Si nuclei of the T^3 species in **3c'(Q^n)$_{30}$**. Hence, 70-80% of the protons bound to Q species are dipolar-coupled with these fast relaxing protons, but 20-30% show components between 18 and 20 ms. However, the T_{1pH} behaviour of the protons in the polysiloxane matrix of the ether-phosphine ligand **1a'(Q^n)$_{10}$** (T:Q = 1:10 as above) is single-exponential and similar for both T and Q species (3.61 ms, Fig. 6D).

We conclude that the sample **3c'(Q^n)$_{30}$** (Fig. 6C) consists of domains, with high and low densities of the T-functionalized ruthenium complexes, whereas the distribution of the T functionalities in **1a'(Q^n)$_{10}$** is homogeneous, making the dipolar coupling of the protons efficient across the whole sample. This indicates that hydrolysis and condensation steps in the complex **3c**, which has been equipped with three T functions in one unit, is preferred compared to Si(OEt)$_4$ or **1a-c** containing one Q and one T unit, respectively. This can be explained by the formation of gels of **3a-c** in alcohols at much lower concentrations than observed with Si(OEt)$_4$ and **1a-c**. Hence, the concentration of Si(OEt)$_4$ or hydrolysis products increase leading to the formation of Q-domains related to silica gel during the process of the sol gel reaction. This is in agreement with experimental observations. Thus, when the gel process is stopped after two hours, the T-functionalized ruthenium complex is immobilized quantitatively, but a considerable amount of Si(OEt)$_4$ is washed out.

Summing up, these investigation show the transition of the matrix to silica gel related domains when more than 10-30 equivalents of Si(OEt)$_4$ are involved in the sol gel process with **3a-c** according to method 1 or if the two-steps process (**method 3**) is applied. Therefore, those matrices of the ruthenium complexes listed in Table 2 with T:Q ratios up to 1:3.3, prepared by the **methods 1 and 2a,b** can generally be defined as uniform systems, whereas this is not the case for the samples **3c'(Q^n)$_{30}$** (T:Q = 1:10, **method 1**) and **3c'-silica gel** (T:Q $\approx$ 1:2, **method 3**).

Polysiloxane matrices doped with Mg^{2+} cations: The ^{29}Si MAS NMR spectrum of **3c'(Q^n)$_6$**, polycondensated by the addition of MgSO$_4$, according to **method 2b**, is shown in Fig. 7. The material contained 1.04 wt % Mg^{2+} cations. While the chemical shifts of the T, Q^2, Q^3 and Q^4 species and the T_{1pH} behaviour in the sample remained similar to the Mg^{2+}-free counterparts, additional signals at -81.0 and -86.5 ppm have been found in the spectrum. They can be interpreted as Q$^{1'}$ and Q$^{2'}$ species, in which one SiOH proton has been exchanged by a Mg^{2+} cations.[33,34] This is in agreement with the correspondence of the sum of the Q$^{1'}$ and Q$^{2'}$ species with the amount of Mg^{2+} cations per unit. The Mg^{2+}-ions, which are bound exclusively to Q species, somewhat lower the degree of condensation of the Q type silicon species (see Table 2). Thus, the polysiloxane matrix and the distribution of the ruthenium complexes therein can be compared to the respective Mg^{2+}-free samples.

Summarizing the solid state NMR investigations, novel, uniform and homogeneous materials have been classified, which unify both 'inorganic' and 'organic' properties, if the copolymerization of the T and Q species has been carried out simultaneously starting from monomers with low amount of Si(OEt)$_4$. The T_{SiH} constants of the Q groups can be well compared to those of native silica gels, whereas their extremely low, uniform T_{1pH} values and the temperature behaviour, the quantities of T_{1pH} via ^{13}C and ^{31}P, T_{1C}, T_{CH} and T_{1P} are typical of organic polymers or related systems. These properties can be explained by the fact that the T moieties of the ruthenium complexes and the ether-phosphine ligands comprise the substantial building blocks of the support, which are interpenetrated with Q species to form a suitable, flexible organic-inorganic hybrid catalyst system. If more equivalents of Si(OEt)$_4$ or native silica gel are involved in the sol gel process with the ruthenium complexes 3a-c, the resulting materials are not uniform and should therefore differ in their catalytic properties.

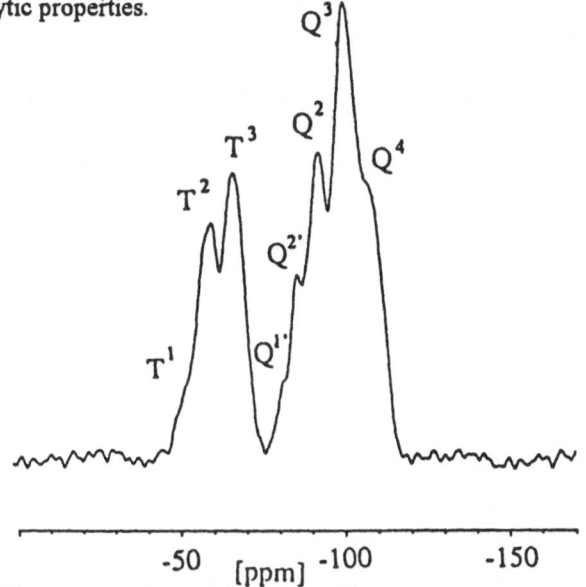

Fig 7 : Solid-state ^{29}Si CP-MAS NMR spectrum of Mg^{2+} doped 3c'(Q^n)$_6$ prepared by method 2b.

Surface structure of the materials

The surface areas (BET) of the samples prepared according to the **methods 1, 2a** and **2b** lie between 1 and 14 m^2/g. These low values are in the range of the external surface of the gel particles. This agrees with STM micrographs which show a non-porous surface of the gels. The materials prepared by **method 1** (e.g. **3c'(Q^n)$_6$** in Fig. 8A) are aggregates of irregular shapes (lumps) and show crushed gel particles with size distributions in the range of 5-30 μm. However, precipitation according to **method 2a** resulted in more uniform, tiny and rounded particles [e.g. **3c'(Q^n)$_{4.2}$** in Fig. 8B].

Samples prepared by **method 2b**, which contained low amounts of Mg^{2+}, showed a more structured, but non-porous surface with small particle sizes of 0.5 - 5 μm.

The surface area of silica gel (**method 3**) was determined to 180 m^2/g and the STM micrographs showed a porous surface as expected (particle size distribution 10-30 μm). After the sol gel process of silica gel with 3c to form **3c'-silica gel** (T:Q ≈ 1:2) the surface area decreased and huge, non-porous particles (STM: 30-50 μm) were formed. This coating of silica gel by a shell of 3c' without Q species is in agreement with the low degree of condensation of the T-functionalized ruthenium

189

complex moieties, the amount of 'bulk' Q^4 species (see Table 2), and the T_{1pH} behaviour (Fig. 6).

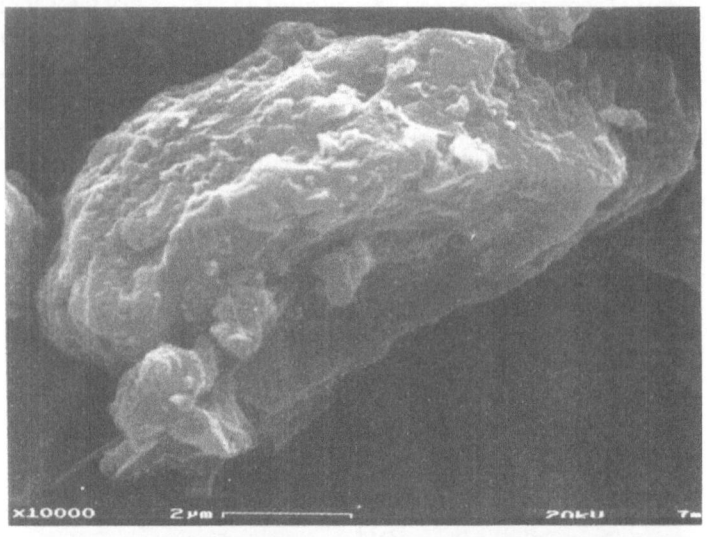

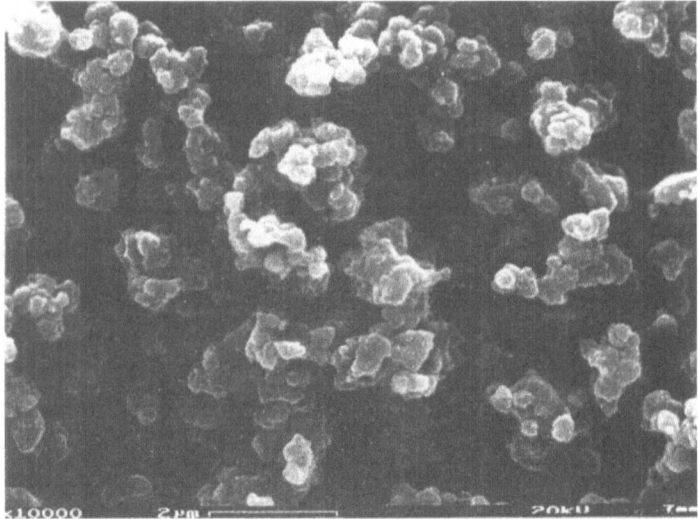

Fig 8. Scanning tunnelling micrographs of $3c'(Q^n)_6$ sol gel processed by method 1 (A) and $3c'(Q^n)_{4.2}$ precipitated according to method 2a (B).

For the sample $3b'(Q^n)_6$ with a total degree of hydrolysis of 100%, the molecular water, being adsorbed at 'surface' silanol groups, has been determined to 3.11% by the Karl Fischer Method.[35] This corresponds to 0.278 water molecules per silicon atom - approximately half of the concentration of silanol groups in the matrix (average 0.58 OH/Si) - and indicates that two of the silanol groups are associated to one water molecule. Thermogravimetric and Karl Fischer analyses showed that the adsorbed water was lost after treatment at 100 °C, while no further condensation of the matrix was detectable by ^{29}Si CP-MAS NMR spectroscopy. Although molecular water had been removed, the IR spectra of $3b'(Q^n)_6$ showed no free SiOH vibrations at 3700 - 3750 cm^{-1},[36] but a broad absorption band of associated SiOH...(H)OSi hydroxyl groups[36] at 3200 - 3600 cm^{-1}. Subsequently the sample was exposed to methanol vapour. Considerable swelling was induced and a sharp signal at 49.5 ppm (linewidth 28 Hz) appeared in the ^{13}C CP-MAS NMR spectrum (TOSS) due to methanol molecules being adsorbed at 'surface' SiOH groups. The amount of adsorbed methanol in **3b'(Q^n)$_6$** was determined to 0.54 molecules per silicon atom, which is in the range of the concentration of silanol groups (0.58 OH/Si). This indicates that each of the silanol functions is associated with one methanol molecule. The amount of adsorbed organic molecules decreased with their polarity (MeOH 0.54/Si, EtOH 0.39/Si, acetone 0.28/Si, THF 0.26/Si, Et$_2$O 0.08/Si, alkanes 0.03/Si). These studies indicate that the silanol groups, even in the bulk of the polymer network, are accessible to polar organic molecules like methanol. However, swelling was hardly induced by exposure of the materials to non-polar molecules because of the strongly bridging hydrogen bonds between the polar silanol groups. This is in agreement with a "convertible surface" for the materials. This point of view can be explained in terms of the quantitative description of the materials with a T:Q ratio in the range of 1:1 - 1:2 by ^{29}Si solid-state NMR spectroscopy: 70-75% of these matrices consist of flexible chain or folded two-dimensional layer elements, which are anchored by the Q^4 species or the homogeneously divided ruthenium centers in the gel.

The sample **3c'(Q^n)$_{30}$** hardly swelled in methanol or other solvent vapours (0.31 MeOH/Si compared to the total concentration of silanol groups per silicon of 0.69 OH/Si). This result can be interpreted in terms of inflexibility of the silica gel domains, which partially occlude the silanol groups. The increasing rigidity of the matrices is also evident from decreasing T_{SiH} values of the Q^3 species in the system **3c'(Q^n)$_x$** (amounts of Q species x = 6, 10 and 30; T_{SiH} = 2.01, 1.66 and 1.42 ms, respectively).

Catalytic activity of the materials

The different polysiloxane-bound ruthenium complexes described in this study displayed various interesting properties in hydrogenation of *n*-butyraldehyde to *n*-butanol. In accordance with the literature,[37] the catalytic activity of complexes of the type HRuCl(CO)(P ~ O)$_3$ is initiated by cleavage of the ligand P^1 *trans* to H and the coordination of the aldehyde to ruthenium. Following treatment with *n*-butyraldehyde, the ether-phosphine ligand **1a'(Q^n)$_2$**, which is dissociated from **3a'(Q^n)$_6$** could be observed in the ^{31} CP-MAS NMR spectrum.

The ether functions in ether-phosphine transition metal complexes can be considered as an intramolecular solvent.[12] They stabilize low coordinated species due to the chelate effect and control the catalytic cycle. This concept has been successfully applied in homogeneous catalysis,[12] and is extended here to polysiloxane materials. Indeed, we have already demonstrated that the polysiloxane-bound ether-phosphines show a behaviour similar to liquids at higher temperatures. The system should therefore be able to imitate a homogeneous chemical reaction at higher temperatures.

Catalytic experiments on complexes prepared according to **method 1** and the Mg^{2+}-doped sample **3c'(Q^n)$_6$** (**method 2b**) were carried out isothermally (H$_2$-pressure at the beginning of the reaction 50 bar; ruthenium:butyraldehyde = 1:1000,

191

see Table 5a and Fig. 9A). In contrast to $3a'(Q^n)_6$, the samples $3b'(Q^n)_6$ and $3c'(Q^n)_6$ exhibited no significant catalytic activity at 100 °C or below. This might be due to the higher mobility of the anchored ether-phosphine $1a'(Q^n)_2$ in contrast to $1b'(Q^n)_2$ or $1c'(Q^n)_2$ which was established by ^{31}P CP-MAS NMR relaxometry. At 150 °C, the turnover frequency[38] of the system $3c'(Q^n)_x$ (method 1) decreased with an increasing particle size and a higher content of cocondensated Q silicon moieties x, although the BET values (external surfaces) do not vary significantly. This can be explained by the occlusion of the ruthenium complexes into rigid, non-swelling silica gel related domains, which make them unavailable for catalysis. However, the catalysts $3c'(Q^n)_x$ (x = 10, 30) could be recovered quantitatively by centrifugation from the reaction medium. In the Mg^{2+}-doped sample the significantly higher turnover frequency can be traced back to a smaller particle size and a lower degree of condensation (see Tables 5a and 2).

Due to their high catalytic activity, hydrogenation experiments with some samples prepared according to the **methods 2a, b** could be performed isothermally under the same conditions (Table 5b). This is explained in terms of tiny particles (0.5 - 2 μm), a low content of Q silicon groups (T:Q = 1:1 - 1:1.5), and a somewhat lower degree of condensation compared to materials prepared by **method 1**. The aldehyde was quantitatively converted into n-butanol within several minutes (range of temperature 80-140 °C). Between 70-80 °C the decrease of the H_2-pressure correlated with an exothermic reaction, which was detected by the thermocouple introduced into the reaction medium [p/T-curve from $3c'(Q^n)_{4.2}$, Fig. 9B].

After opening the autoclave, the samples prepared by **methods 1 and 2a** with a T:Q ratio up to 1:2 were present in the form of a highly dispersed, swollen gel, in cases where the aldehyde was converted in a considerable amount. The gels were precipitated from n-butanol by the addition of alkanes (yields 70-80%).

This reaction behaviour can be explained by the formation of polar n-butanol within the gel, followed by swelling and a substantial increase of the gel surface during the course of the reaction. The resulting concentration gradient between the aldehyde on the outside and n-butanol on the inside of the gel results in a rapid diffussion of the reactants to and from the catalytic sites. Thus the differences in catalytic activities depicted in Figures 9A and 9B can be qualitatively interpreted in terms of a diffusion controlled reaction model.

The hydrogenation behaviour (p/T curve) of the sample 3c'-silica gel (**method 3**, Table 5b) is identical with that of the monomeric ruthenium complex 3c. The material was decomposed into pure silica gel and non-separable, polymeric ruthenium complexes 3c', on which was carried out the homogeneous hydrogenation. This leaching is the result of the low degree of condensation of the domains formed by the T-functionalized ruthenium complexes (53%).

We have shown that samples with a high amount of Q species can be well separated from the reaction medium, but suffer from lower catalytic activity due to the non-flexibility of the matrices and occlusion of the complexes. However, samples with a low content of Q silicons and small particle sizes are highly active, but separation is only possible by the addition of non-polar solvents. The Mg^{2+} doped catalysts $3c'(Q^n)_6$ and especially $3a'(Q^n)_3$ displayed a high catalytic activity and were completely recovered by centrifugation after the hydrogenation experiment due to the small content of ionic, insoluble magnesium silicates. This design of the polysiloxane matrix therefore unifies the properties of a heterogeneous catalyst with those of a homogeneous catalyst.

Table 5a. Isothermal hydrogenation of n-butyraldehyde[a]

catalyst	surface area (m^2/g)	particle size (μm)	temp. (°C)	reaction time[b] (min)	conversion (%)	turnover frequency (min^{-1})
Method 1						
$3a'(Q^n)_6$[c]	5.9	5-15	50	60	16.5	2.8
			80	60	62.2	10.4[d]
			100	60	81.1	13.5
			150	45	99.5	22.1
$3b'(Q^n)_6$[e]	1-10	10-20	50	60	2.4	0.04
			100	60	16.5	2.8
			150	55	98.5	18.0[f]
$3c'(Q^n)_6$	1-12	5-12	100	60	32.5	3.9
			150	45	100	22.2[g]
$3c'(Q^n)_{10}$	7-11	10-30	150	60	91.8	15.3
$3c'(Q^n)_{30}$	14	20-30	150	60	74.9	12.5
Method 2b						
Mg^{2+}-$3c'(Q^n)_6$[h]	3-8	0.5-5	150	15	100	66.6[i]

[a]Starting H_2-pressure 50 bar/296 K, Ru : n-butyraldehyde = 1:1000, selectivity 100%. [b]Counted from the point, when the reaction mixture in the autoclave has reached the stated temperatures. [c]Ru 7.55%. [d]Second run 10.0 min^{-1}, Ru 7.76%. [e]Ru 6.83%. [f]Second run 16.9 min^{-1}, Ru 7.18%; third run 17.0 min^{-1}, Ru 6.74%. [g]See Figure 9A. [h]Ru 5.95%. [i]Second run 65.9 min^{-1}, Ru 6.4%.

Table 5b. Non-Isothermal hydrogenation of n-butyraldehyde[a]

catalyst	surface area (m^2/g)	particle size (μm)	reaction time[b] (min)
Method 2a			
$3a'(Q^n)_{3.4}$	13.5	1 - 2	4 - 5
$3b'(Q^n)_{4.4}$	18.5	1 - 1.5	4 - 5
$3c'(Q^n)_{4.2}$	10.0	0.5 - 2	4 - 5[c]
Method 2b			
Mg^{2+}-$3c'(Q^n)_6$[h]	11.0	1 - 4	6 - 7[d]
Method 2c			
3c'-silica gel	9 - 15	30 - 50	1 - 2.5[e]

[a]Starting H_2-pressure 50 bar/296 K, Ru : n-butyraldehyde = 1:1000, selectivity 100%. [b]Estimated from the beginning of the exothermic reaction. [c]See Fig. 9B. [d]Reproducible data and constant C, H, Cl and Ru values during the catalytic runs 1-3. [e]Decomposing to a sol of oligomeric ruthenium complexes 3c' and pure silica gel.

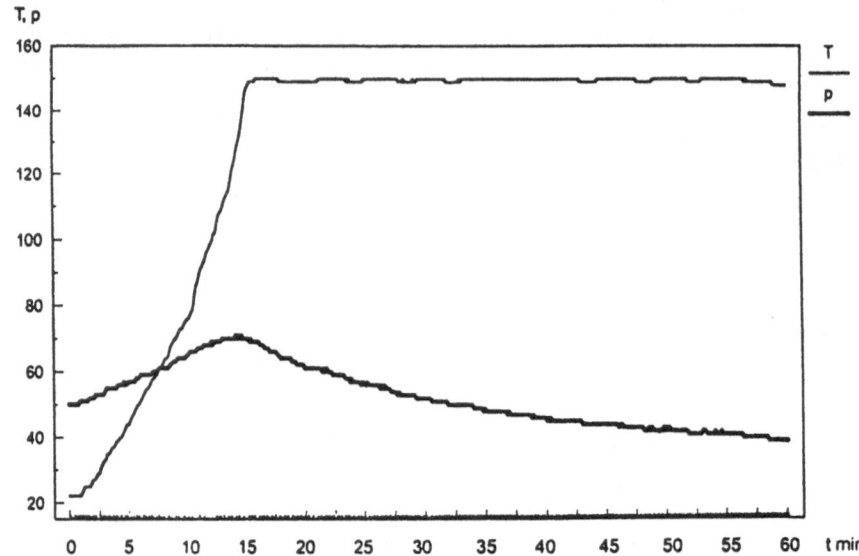

Fig 9A. Hydrogenaton of *n*-butyraldehyde with polysiloxane-bound $3c'(Q^n)_6$ prepared by method 1.

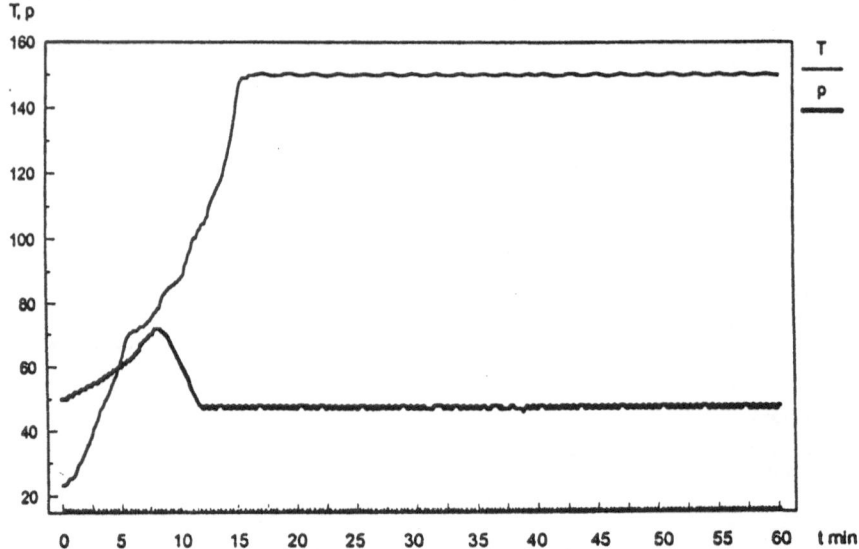

Fig 9B : Hydrogenation of *n*-butyraldehyde with polysiloxane-bound $3c'(Q^n)_{4.2}$ prepared by method 2a

The polysiloxane-bound catalysts described in this study showed no measurable leaching of ruthenium during the hydrogenation of *n*-butyraldehyde (except **3c'-silica gel**, as described above). Constant analytical values of the materials after several catalytic runs have been found (see Tables 5a,b) and no loss of catalytic activity has been observed with Mg^{2+}-doped $3a'(Q^n)_3$ or $3c'(Q^n)_6$. It is most surprising that the degree of condensation in the polysiloxane matrices has even been increased during the hydrogenation reaction. As an example, the degree of condensation of the T species in catalyst $3c'(Q^n)_6$ increased from 83 to 92%, and in the Mg^{2+} derivative from 79% to 86% during one catalytic run. The T:Q ratios remained constant as also the analytical values.

Conclusion

In this study, we have demonstrated that sol gel processing opens up the tantalizing prospect of designing novel catalyst systems with tailored properties, which can be investigated unequivocally by the combination of heteronuclear solid-state NMR spectroscopy, surface characterization and catalysis. The convenient and mild conditions offered by the sol gel process for the preparation of modified inorganic networks have paved the way for the synthesis of a new catalyst system. On the nanometer scale, the degree of interpenetration of the organometallic T moieties with the 'inorganic' Q species in the sol gel routes 1, 2a,b depends on the ratio of these components. At a low content of inorganic Q building blocks, the domain sizes are reduced to a level such that uniform molecular T/Q composites are formed. As a result of this intimate mixing, these hybrid catalysts show unique properties. The flexibility of the carbon pacers and the tailored polysiloxane matrices of the gels allow a ready imitation of chemical reactions of homogeneous catalysts.

References and Notes

1. U. Deschler, P. Kleinschmit and P. Panster, *Angew. Chem.*, **98** (1986) 237; *Angew. Chem. Int. Ed. Engl.*, **25** (1986) 236.
2. H. Eckert, *Prog. Nucl. Magn. Reson. Spectrosc.*, **24** (1992) 159.
3. G.E. Maciel, D.W. Sindorf and V. Bartuska, *J. Chromatogr.*, **205** (1981) 438.
4. D.W. Sindorf and G.E. Marciel, *J. Am. Chem. Soc.*, **105** (1983) 3767.
5. E. Bayer, K. Albert, J. Reiners, M. Nieder and D. Muller, *J. Chromatogr.*, **264** (1983) 197.
6. I.S. Khatib and R.V. Parish, *J. Organomet. Chem.*, **369** (1989) 9-16.
7. E. Linder, A. Bader and H.A. Mayer, *Z. Anorg. Allg. Chem.*, **598/599** (1991) 235.
8. E. Linder, A. Bader and H.A. Mayer, *Inorg. Chem.*, **30** (1991) 3783.
9. (a) E. Linder, M. Kemmler and H.A. Mayer, *Chem. Ber.*, **125** (1992) 2385.
    (b) E. Lindner, M. Kemmler and H.A. Mayer, *Z. Anorg. Allg. Chem.*, in press.
10. U. Schubert, C. Egger, K. Rose and C. Alt, *J. Mol. Catal.*, **55** (1989) 330.
11. E. Linder, R. Schreiber, M.Kemmler, H.A. Mayer, R. Fawzi and M. Steinmann, "Supported Organometallic Complexes. 5." Part 4, *Z. Anorg. Allg.Chem.*, **619** (1993) 202.
12. A. Bader and E. Lindner, *Coord. Chem. Rev.*, **108** (1991) 27 and references cited therein.
13. C.A. Fyfe, *Solid State NMR for Chemists* 1984, CRC Press, Guelph, Ontario.
14. C.A. Fyfe, Y. Zhang and P. Aroca, *J. Am. Chem. Soc.*, **114** (1992) 3252.
15. D.K. Liu, M.S. Wrighton, D.R. McKay and G.E. Maciel, *Inorg. Chem.*, **23** (1984) 212.

16. R.A. Komoroski, A.J. Magistro and P.P. Nicholas, *Inorg. Chem.*, **25** (1986) 3917.

17. L. Bemi, H.C. Clark, J.A. Davies, C.A. Fyfe and R.E. Wasylishen, *J. Am. Chem. Soc.*, **104** (1982) 438.

18. K.J. Shea, D.A. Loy and O. Webster, *J. Am. Chem.Soc.*, **114** (1992) 6700.

19. N. Ahmad, J.J. Levison, S.D. Robinson and M.F. Uttley, *Inorg. Synth.*, **15** (1974) 48.

20. Degree of condensation of T species = $100(T^1 + 2T^2 + 3T^3)/[3(T^1 + T^2 + T^3)]$; degree of condensation of Q species = $100(2Q^2 + 3Q^3 + 4Q^4)/[4(Q^2 + Q^3 + Q^4)]$; T^1, T^2, T^3, Q^2, Q^3 and Q^4: relative amounts of silyl species present in the sample (Table 3).

21. R. Voelkel, *Angew. Chem.*, **100** (1988) 1525; *Angew. Chem. Int. Ed. Engl.*, **27** (1988) 1468.

22. J.L. Koenig and M. Andreis, *Solid State NMR of Polymers*; L.J. Mathias (ed), Plenum Press, New York, 1991, pp 201-213.

23. R.A. Komoroski, *High Resolution NMR Spectroscopy of Synthetic Polymers in Bulk*; VCH Publishers, Inc.; Deerfield Beach, FL, 1986; pp 63-226, 247-306.

24. B. Pfleiderer, K. Albert, E. Bayer, L. van den Ven, J. de Haan and C. Cramers, *J. Phys. Chem.*, **94** (1990) 4189.

25. R.K. Harris, *Analyst*, **110** (1985) 649.

26. M. Mehring, In: *Principles of High Resolution NMR in Solids*, 2nd ed. Springer-Verlag, Berlin, Heidelberg, New York, 1983; pp 143-155.

27. J. Schaefer, E.O. Stejskal and R. Buchdahl, *Macromoelcules*, **10** (1977) 384.

28. S. Leonardelli, L. Facchini, C. Fretigny, P. Tougne and A.P. Legrand, *J. Am. Chem. Soc.*, **114** (1992) 6412.

29. K.L. Walther, A. Wokaun and A. Baiker, *Mol. Phys.*, **71** (1990) 769.

30. Degree of hydrolysis of T species = $(1 - n/3)100$, degree of hydrolysis of Q species = $(1 - n/4)100$; n = number of residual OCH_3 (T species) or OCH_2 groups (Q species) per silicon atom.

31. D.W. Sindorf and G.E. Maciel, *J. Am. Chem. Soc.*, **105** (1983) 1848.

32. G.E. Maciel and D.W. Sindorf, *J. Am. Chem. Soc.*, **102** (1980) 7607.

33. M. Lippmaa, M. Magi, A. Samoson, G. Engelhardt and A.R. Grimmer, *J. Am. Chem. Soc.*, (1980) **102** 4889.

34. J. Rocha, M.D. Welch and J. Klinowski, *J. Am. Chem. Soc.*, **113** (1991) 7100.

35. R.P.W. Scott and S. Traiman, *J. Chromatogr.*, **196** (1980) 193.

36. H.E. Fischer, S.A. King, J.B. Miller, J.Y. Ying, J.B. Benziger and J. Schwartz, *Inorg. Chem.*, **30** (1991) 4403.

37. R.A. Sanchez-Delgado, N. Valencia, R.-L. Marquez-Silva, A. Andriollo and M. Medina, *Inorg. Chem.*, **25** (1986) 1106.

38. Turnover frequency = mol product/(mol catalyst. 60 min).

39. E. Linder, A. Bader, E. Glaser, B. Pfleiderer, W. Schumann and E. Bayer, *J. Organomet. Chem.*, **355** (1988) 45.

40. E. Lippmaa and A. Samoson, *Bruker Rep.*, **1** (1982) 6.

41. W.T. Dixon, J. Schaefer, M.D. Sefcik, E.O. Stejskal and R.A. McKay, *J. Magn. Reson.*, **49** (1982) 341.

42. A.D. Torchia, *J. Magn. Reson.*, **30** (1978) 613.

Main Group Elements and Their Compounds
V.G. Kumar Das (Ed)
Copyright © 1996 Narosa Publishing House, New Delhi, India

Latent Organotin Catalysts : Mechanism of Activation

Bernard Jousseaume

Laboratoire de Chimie Organique et Organométallique, URA 35 CNRS,
Université Bordeaux 1, 351, cours de la Libération, 33405 Talence, France

---

Diorganotin compounds are mainly used as PVC stabilizers in the chemical industry. They also have important applications as catalysts in other areas, such as silicone curing, polyurethane preparation and esterification reactions because of their low cost, high efficiency and moderate toxicity.[1] For silicones, the curing of linear polymer chains necessary to develop the required elastomeric properties can be obtained by a condensation reaction of the terminal silicon-hydroxyl group with a curing agent, either a tetraalkoxysilane SiOH/SiOR condensation, or a hydrogenopolysiloxane SiOH/SiH condensation. Diorganotin dicarboxylates catalyze both processes.[2] Diorganotin dicarboxylates are also often employed as catalysts in the polyurethane industry.[3] However, for either silicones or polyurethanes, the high efficiency of organotin catalysts may cause some disadvantage when long pot lives are desired.[4] Usually condensation reactions start when reactants and catalysts are mixed together which leads to a rapid decrease in fluidity of the mixtures. To avoid this drawback, a study concerning the design and the development of new organotin catalysts with long pot lives and ability to be activated at will was undertaken.

Results and Discussion

The strategy that we have employed is one of designing suitable organotin pro-catalysts which are inactive at room temperature, but which upon brief heating are readily transformed into the catalytically active dicarboxylates. We have in this connection chosen tetraorganotins as the pro-catalyst structures and examined the scope of β-eliminations[5] as the required transformation to secure the active catalysts. β-Elimination was preferred because it is an intramolecular reaction which would not be subject, unlike a bimolecular process, to experimental uncertainties on account of the complexity of industrial mixtures and the low concentration of the catalytic species. As esters of β-hydroxyalkyl triorganotins have been shown to be thermally unstable and to decompose into olefin and triorganotin carboxylate, the corresponding disubstituted diorganotins were considered to be good precursors for diorganotin dicarboxylates. These have been prepared by the reaction of diorganotin dihydrides and vinyl esters. Hydrostannation of alkenes with dialkyltin dihydrides has not often been used in organotin chemistry and it has been shown to give somewhat unexpected results.[7] In this case, under UV irradiation, high yields of the

197

adducts were obtained in high purity (Eq. 1).

$$R^1_2SnH_2 + 2\ CH_2{=}CR^2OCOR^3 \xrightarrow[\text{irradiation}]{\text{UV}} R^1_2Sn(CH_2CHR^2OCOR^3)_2 \quad (1)$$

Upon heating they were indeed quantitatively transformed into the expected diorganotin dicarboxylates (Eq. 2).[8]

$$R^1_2Sn(CH_2CHR^2OCOR^3)_2 \xrightarrow{\Delta} 2\ CH_2{=}CHR^2 + R^1_2Sn(OCOR^3)_2 \quad (2)$$

Heating times necessary for the complete decomposition of some disubstituted diorganotins are given in Table 1.

Table 1 : Thermal decomposition of bis[2-(acyloxy)alkyl]dialkyltins

R^1	R^2	R^3	Temp. °C	Time[a]	$k*10^4$ (s^{-1})
Bu	H	H	(b)		
Bu	CH_3	H	110	3	2.5
Bu	$n{-}C_{11}H_{23}$	H	140	3	0.24
Bu	$CH(C_2H_5)C_4H_9$	H	140	5	0.07
Bu	$t{-}C_9H_{19}$	H	140	6.5	
Bu	$4{-}CF_3C_6H_4$	H			163
Bu	C_6H_5	H	120	2	25
Bu	$4{-}CH_3C_6H_4$	H			4
Bu	CH_3	C_6H_5	(b)		
Bu	CH_3	CH_3	80	0.3	435
Bu	CH_3	CF_3			0.01
Oct	CH_3	H	110	3	

(a) time necessary for complete decomposition; (b) unstable compound at RT

Decomposition temperatures were in the 80-150 °C range and longer decomposition times did not exceed 5 h. The nature of the substituent in β-position and the nature of the carboxylate group have a great influence on the decomposition conditions, whereas the change of the butyl group on tin by octyl has no effect on the stability of the adducts. As the decomposition conditions were highly dependent on the nature of the substituents, a systematic study was carried out by measuring the parameters of the reaction by differential scanning calorimetry. It was found that

(a) the decomposition is monomolecular or pseudo-monomolecular, (b) the longer and more branched the carboxylate is, the more difficult is the decomposition, and (c) the more electron withdrawing the β-substituent is, the more difficult is the decomposition. Thus, it appears that the rate of the reaction does not correspond to a case of one involving nucleophilic attack of the carbonyl on the tin atom. In fact, it depends on the ability of the β-substituent to stabilize a partially developing positive charge on the carbon at the β-position and also on the ability of the carboxylate to stabilize a partially developing negative charge on the ester moiety. The crucial factor is thus the cleavage of the β-carbon-oxygen bond. It is generally well established that the pyrolysis of esters is a syn-elimination corresponding to a cyclic, six-membered transition state. A stereochemical study of the thermal decomposition was thus undertaken to determine whether it was syn-elimination, as in esters,[9] or an anti-elimination, as in the acid-catalyzed decomposition of β-hydroxyalkyltriorganotins (Eq. 3).[10]

$$
\left[
\begin{array}{c}
\delta^+ \ \ R^2 \\
CH_2 \!\cdots\! CH \\
\diagdown Sn \quad\quad O^{\delta^-} \\
O \cdots C \\
\quad R^3
\end{array}
\right]^{\neq}
\quad or \quad
\left[
\begin{array}{c}
\delta^- \\
OCOR^3 \\
\delta^+ \\
CH_2 - CH \\
\diagdown Sn \quad R^2
\end{array}
\right]^{\neq}
\tag{3}
$$

The requisite *threo-* and *erythro-*[2-(acyloxyl)alkyl]triorganostannanes were prepared by esterification of the corresponding alcohols obtained from the disatereoselective opening of oxiranes by triorganostannyllithiums.[11] They were then subjected to thermal decomposition at 50 °C for a few hours until the starting material was completely consumed. The reactions were quantitative with respect to the yields of the olefin and triorganotin acetates (Eq. 4).

$$
\begin{array}{c}
R_3Sn \quad H \\
\quad\quad Me \\
H \quad\quad \\
Me \quad OAc \\
\textit{threo}
\end{array}
\xrightarrow{\ anti\ }
\underset{(Z)}{\diagup\!=\!\diagdown}
\qquad
\begin{array}{c}
R_3Sn \quad Me \\
\quad\quad H \\
H \quad\quad \\
Me \quad OAc \\
\textit{erythro}
\end{array}
\xrightarrow{\ anti\ }
\underset{(E)}{\diagup\!=\!\diagdown}
\tag{4}
$$

As *threo-*[2-(acetyloxy)butyl]tributylstannane gave (Z)-2-butene and *erythro-*[2-(acetyloxy)butyl]tributylstannane gave (E)-2-butene, it was concluded that the thermolysis was an anti-elimination proceeding through an open transition state,[12] and not a cyclic one as observed for the pyrolysis of esters[9] or postulated for the pyrolysis of β-silylated esters (Eq. 5).

$$
\left[
\begin{array}{c}
\delta^- \\
OCOR^3 \\
\delta^+ \\
CH_2 - CH \\
\diagdown Sn \quad R^2
\end{array}
\right]^{\neq}
\tag{5}
$$

The models used were tetraorganostannanes which only account for the first step of the thermolysis of bis[2-(acyloxy)alkyl]diorganostannanes. The second step involves (acyloxyl)[2-(acyloxy)alkyl]dialkylstannanes where the acyloxy group borne by the tin atom can render the metal sufficiently electrophilic to increase its coordination from four to five by internal complexation (Eq. 6).[14]

$$Bu_2Sn(CH_2CH_2OCOR)_2 \rightarrow Bu_2(RCO_2)SnCH_2CH_2OCOR + H_2C=CH_2$$

$$Bu_2(RCO_2)SnCH_2CH_2OCOR \rightarrow Bu_2Sn(O_2CR)_2 + H_2C=CH_2$$

(6)

As this intramolecular chelation could conceivably favour a syn elimination, a study has been undertaken on model compounds with such complexation possibilities. Although crystallographic evidence for intramolecular coordination is lacking in a triorganostannane with a carbonyl moiety in the δ-position, the presence of the carbonyl side-chain can exert a strong effect on the reactivity of the stannane.[15,16] Internal coordination has been demonstrated in cases where a five-membered ring can be formed intramolecularly with an ester as ligand.[17] Indirect evidence for the presence of such coordination in (acyloxyl)[2-(acyloxy)alkyl]dialkylstannanes comes from ^{119}Sn and ^{13}C NMR data. In ^{119}Sn NMR, an upfield shift of 60 ppm was measured for (acetoxy)[2-(acetoxy)ethyl]dibutylstannane from tributylacetoxy-stannane. By way of comparison, a 100 ppm shift is recorded for organotin halides with higher than four coordination.[18] ^{13}C NMR spectral measurements of ^1J (^{13}C-^{119}Sn) for iodo[2-(pivaloyloxy)but-3-yl]dibutylstannane yield values of 449 Hz and 401 Hz, respectively. These values are in the range for pentacoordinate and tetracoordinate tins, respectively. Model compounds were prepared by electrophilic cleavage of the phenyl group in [2-(hydroxy)but-3-yl]dibutylstannanes by iodine followed by esterification. The resulting *threo-* and *erythro-*iodo[2-(pivaloyloxy)but-3-yl]dibutylstannanes were then thermolyzed. *Anti-*elimination was observed as the *threo* compound led to (Z)-2-butene and the *erythro* compound led to (E)-2-butene.[20] Thus, the internal chelation did not decrease or reverse the selectivity observed for [2-(acyloxyl)-alkyl]triorganostannanes (Eq. 7).

(7)

Studies have also been conducted on bis(2-hydroxyalkyl)dialkylstannanes. These are potential latent catalysts as their decomposition leads to bis(hydroxy)dialkylstannanes where the tin bears two oxygen atoms, as in the usual catalysts (Eq. 8).

$$R^1{}_2Sn(CH_2CH_2OH)_2 \xrightarrow[-2\ H_2C=CH_2]{\Delta} [R^1{}_2Sn(OH)_2] \longrightarrow R^1{}_2SnO + H_2O \qquad (8)$$

For this purpose, *threo*-3-tributylstannyl-2-butanol and *erythro*-3-tributylstannyl-2-butanol, were chosen as models. When thermolyzed at 100 °C for 6 h, the *threo* alcohol yielded (E)-2-butene, contaminated with less than 10% of the (Z)-2-butene. This reaction was thus stereoselective and resulted from an *anti*-elimination (eq. 9).

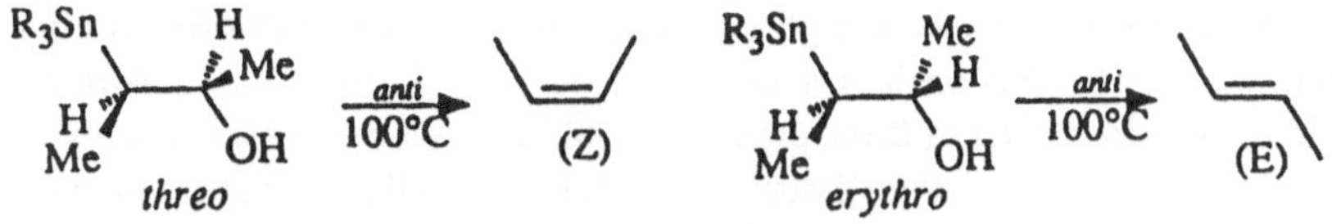

$$(9)$$

However, when the thermolysis of the alcohol was carried out at 160 °C, the stereoselectivity was reversed. Thus *threo*-3-tributylstannyl-2-butanol gave mainly (Z)-2-butene (Table 2).

The thermolysis at 160 °C thus appears to be proceeding via a *syn* elimination. This constitutes the first reported example where an inversion of stereoselectivity in a thermally induced decomposition has been observed in respect of an alcohol (Eq. 10).[21]

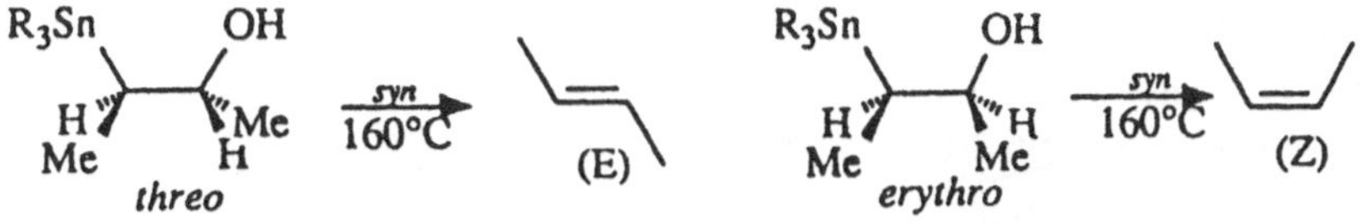

$$(10)$$

Table 2 : Decompositions of 3-tributylstannyl-2-butanols

Compound	Conditions	1-Butene (%)	2-Butene (%)	
			(E)	(Z)
erythro-Bu₃SnCHMeCH(OH)Me	100 °C, 6h	2	89	9
	160 °C, 20min	8	31	61
threo-Bu₃SnCHMeCH(OH)Me	100 °C, 6h	-	10	90
	160 °C, 20min	2	78	20

To account for these results the following explanation is proposed. The observed *anti*-eliminations are quite analogous to these reported for the solvolysis of β-stannylated[22] or β-silylated alcohols,[23] or for the abstraction of a β-hydrogen by trityl cation in tetraorganostannanes.[24] For these latter, an antiperiplanar geometry of the metal and of the leaving group is a feature of the transition state. Important also is the requirement that the developing positive charge on the β-carbon be stabilized by the metal.[25] The thermolyses reported herein were conducted in a neutral medium, less polar than that used for the solvolyses of β-metalated alcohols or for the abstraction of a β-hydrogen. Consequently, charge development may be postulated to be somewhat attenuated in the thermolysis reactions. Nevertheless, metal stabilization effects and geometrical requirements for the same are important. That the rates of decomposition of (acyloxyl)[2-(acyloxy)alkyl]dialkylstannanes are higher than those of bis[2-acyloxy)alkyl]dialkylstannanes may be explained in terms of a better stabilizing effect by the heteroatom-substituted metal, as the positive charge can be delocalized on the oxygen atoms.[26]

The opposite selectivity observed at high temperature when β-stannylated alcohols are thermolyzed has a different explanation. It is suspected that bis(tributyltin)oxide formed during the elimination by intramolecular dehydration of tributyltin hydroxide could act as a base to give β-stannylated tributyltin alkoxides. The formation of these products would be favoured by the removal of water. Thus, threo- and erythro-β-stannylated tributyltin alkoxides were prepared by reaction of the corresponding alcohols with a stoichiometric amount of dimethylaminotributylstannane, and subjected to thermolysis. At 100 or 160 °C, the process is rapid and furnished the syn-elimination products (eq. 11).

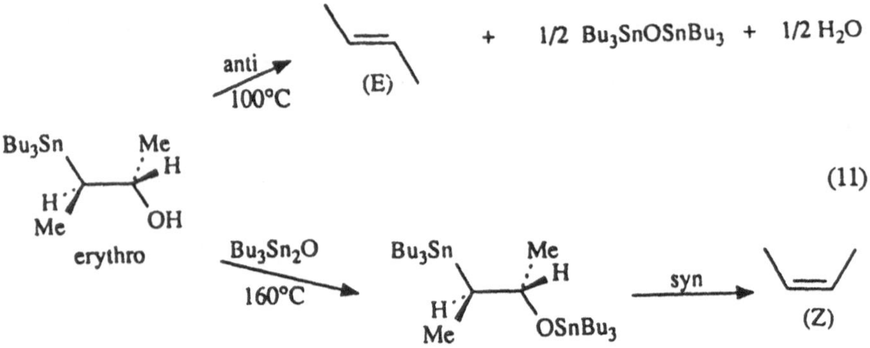

$$(11)$$

The crucial intervention of water was demonstrated by the following results. At 100 °C, under vacuum, threo-β-stannylated alcohol gave (Z)-2-butene indicating a syn-elimination. Thus, at the same temperature, it was possible to drive the reaction from an anti-elimination to syn-elimination just by displacing water. This reversal of selectivity, as in Peterson olefination,[23] can be explained by the change in the nucelophilic properties of the oxygen atom and in the leaving group ability induced by the presence of a metal on the oxygen atom.[26]

References

1. C.J. Evans and S. Karpel, *J. Organomet. Chem. Lib.* **16** (1985); C.J. Evans, *Chemistry of Tin*, ed., P.G. Harrison, Blakie, Glasgow, 1989, p 421.
2. S. Karpel, *Tin and Its Uses*, **6** (1984) 142.
3. F.W. Van Der Weij, *Makromol. Chem.*, **181** (1980) 2541; S.L. Axelrood, C.W. Hamilton and K.C. Frisch, *Ind. Eng. Chem.*, **53** (1961) 889.
4. R.P. Eckberg, *High Sol. Coat.*, (1983) 14.
5. P.J. Davidson, M.F. Lappert and R. Pearce, *Chem. Rev.*, **76** (1976) 216;

J. Kochi, *Organometallic Mechanisms and Catalysis*, Academic press, 1978, p 249.

6. G.J.M. Van der Kerk, J.G. Noltes and J.G.A. Luijten, *J. Appl. Chem.*, **7** (1957) 356; J.G. Noltes and G.J.M. Van der Kerk, *Functionally Substituted Organotin Compounds*, Tin Research Institute, Greenford, 1958, p 59.

7. B.R. Laliberté, W. Davidsohn and M.C. Henry, *J. Organomet. Chem.*, **5** (1966) 527.

8. B. Jousseaume, V. Gouron, B. Maillard, M. Pereyre and J-M. Francés, *Organometallics*, **9** (1990) 1330; B. Jouseaume, V. Gouron, M. Pereyre and J-M. Francés, *Appl. Organomet. Chem.*, **5** (1991) 135; J-M. Francés, V. Gouron, B. Jousseaume and M. Pereyre, EP 338947, 343086 and 421895.

9. C.H. DePuy and R.W. King, *Chem. Rev.*, **60** (1960) 431; R. Taylor, G.G. Smith and W.H. Wetzel, *J. Amer. Chem. Soc.*, **84** (1962) 4817; G.G. Smith, D.A.K. Jones and D.F. Brown, *J. Org. Chem.*, **28** (1963) 403; See also for an alternate mechanism, D.H. Wert and N.L. Allinger, *J. Org. Chem.*, **42** (1977) 698.

10. T. Kauffmann, R. Kriegesmann and A.Hamsen, *Chem. Ber.*, **115** (1982) 1818; T. Kauffmann, *Angew. Chem. Int. Ed. Engl.*, **21** (1982) 410.

11. A. Mordini, M. Taddei and G. Seconi, *Gazz. Chim. It.*, **116** (1986) 239.

12. B. Jousseaume, N. Noiret, M. Pereyre and J-M. Francés, *Organometallics*, **11** (1992) 3910.

13. C. Eaborn, F.M.S. Mahmond and R. Taylor, *J. Chem. Soc., Perkin Trans. II*, (1982) 1313.

14. I. Omae, *J. Organometal. Chem. Lib.*, **18** (1986) 189.

15. H.G. Kuivila, J.E. Dixon, P.L. Maxfield, N.M. Scarpa, T.M. Topka, K.H. Tsai and K.R. Wursthorn, *J. Organometal. Chem.*, **86** (1975) 89.

16. B. Jousseaume and P. Villeneuve, *J. Chem. Soc. Chem. Commun.*, (1987) 517.

17. B. Jousseaume, P. Villeneuve, M. Draeger and J.M. Chezeau, *J. Organometal. Chem.*, **349** (1988) C1; U. Kolb, M. Draeger and B. Jousseaume, *Organometallics*, **10** (1991) 2737.

18. B. Wrackmeyer, *Ann. Rept. NMR Spectrosc.*, **16** (1985) 73.

19. T.M. Mitchell, *J. Organometal. Chem.*, **59** (1973) 189.

20. B. Jousseaume, V. Gouron, N. Noiret, M. Pereyre and J.-M. Francés, *J. Organometal. Chem.*, **450** (1993) 97.

21. B. Joussaeume, N. Noiret, M. Pereyre and J-M. Francés, *J. Chem. Soc. Chem. Commun.*, (1992) 739.

22. D.D. Davis and C.F. Fray, *J. Org. Chem.*, **35** (1970) 1303.

23. E. Colvin, In: *Silicon in Organic Synthesis*, Butterworths, London, 1980, p 142; A.R. Bassindale and P.G. Taylor, In: *The Chemistry of Organic Silicon Compounds*", eds., S. Patai and Z. Rappoport, John Wiley, London, 1989, p 893.

24. W. Hanstein, H.J. Berwin and T.G. Traylor, *J. Amer. Chem. Soc.*, **92** (1970) 829; S.J. Hannon and T.G. Traylor, *J. Org. Chem.*, **46** (1981) 3645.

25. J. March, *Advanced Organic Chemistry*, John Wiley, New York, 1985, p 146 and references cited therein.

26. C. Trindle, J.T. Wang and F.A. Carey, *J. Org. Chem.*, **38** (1973) 2664.

Main Group Elements and Their Compounds
V.G. Kumar Das (Ed)
Copyright © 1996 Narosa Publishing House, New Delhi, India

Optimization and Characterization of Piezo-Ceramic Sensors used in Smart Systems

V.R. Singh and Ranvir Singh

National Physical Laboratory, New Delhi - 110012, India

Sensors hold the key to the further advancement of micro-electronics in a number of diverse fields .[1-11] In concert with the growth of large-scale integration in microelectronic circuits and the rapid developments in the field of high-precision mechanics, sensors have found an increasing application in a host-of technologies and, indeed, have contributed to the improved safety of many products.

Most new developments in the field of ceramic sensors have taken advantage of the availability of high-quality materials and the well established sophisticated methods of fabrication that are used in the production of integrated circuits. This trend has been further enhanced by the concept of smart materials and of their use to make smart sensors. Smart sensors are used in various smart systems to enhance the performance and efficiency of such systems. One of the qualities that distinguishes smart materials from inanimate materials is the ability of smart materials to adapt to changes in the environment. Smart materials, such as, for example, poled piezoelectric ceramics, have the ability to perform both sensing and actuating functions.

Smart materials can be divided into passively smart materials that respond to external changes without any assistance, and actively smart materials that utilize a feedback loop enabling them to both recognize the change and initiate an appropriate response through an actuator. Materials with a built-in learning function are smarter than those without it. A very smart material senses a change in the environment and responds by altering one or more of its property coefficients. In this paper we focus on the development and characterization of piezo-ceramic materials and also on the optimisation of various characteristics such as sensitivity, thermal stability, repeatability and accuracy for the sensors derived from these.

Fabrication Process

High-performance smart piezoelectro-ceramics with high mechanical strength and very high homogenous composition were prepared by using a "semi-chemical process". This process combines two methods, firstly the "multistage coprecipitation process" for synthesizing the ceramic powder, and secondly the "hydrothermal method." Here, only one constituent of the target compound is chemically synthesized, while the remainder is synthesized by conventional mixing and hydrothermal methods. The semi-chemical process was first applied to the synthesis of Pb(ZrTi)O$_3$. The ZrTiO$_2$ component was prepared by the conventional mixing or presintering method, while the Pb containing component was synthetically accessed through the oxalate salt. This approach proved to be rewarding in producing a material with high component activites.[2,5-9]

Smart Materials

Smart materials have the capability of sensing a change in the environment and responding to this in a useful way. Different types of sensing and actuating functions are summarized in Fig. 1.

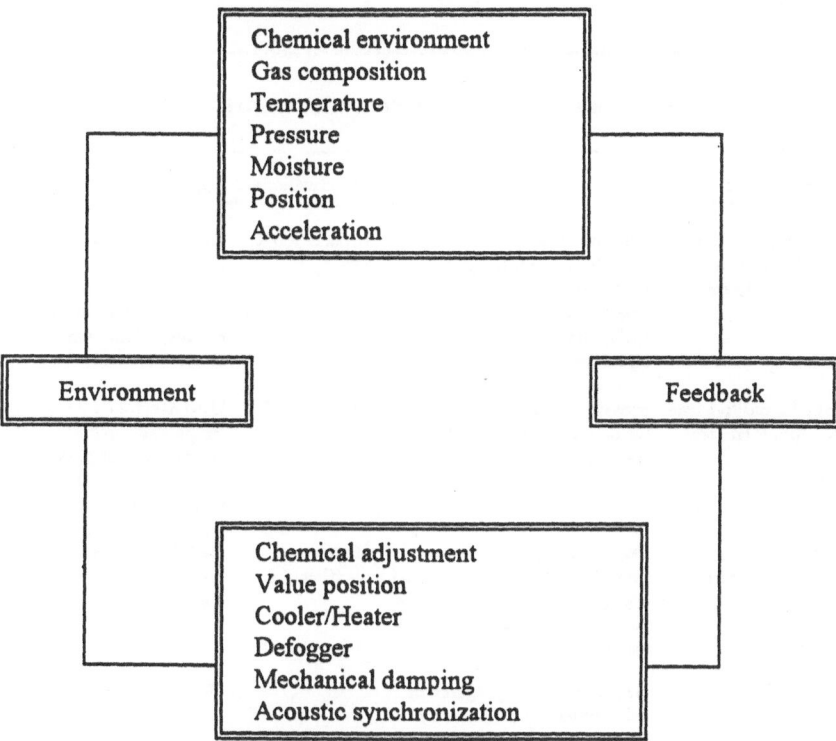

Fig. 1 : **Sensing and actuating functions of smart materials (some examples)**

As noted above, smart materials can be either passively or actively smart, based on the feedback loop used to trigger the actuation. A passively smart material responds automatically to a change in the environment. On the other hand, materials with external feedback loops, such as the electromechanical transducers used in tunable elastic compliance systems, are "actively" smart. Features of a "passive" smart sensor include selectivity, self-diagnosis, self-tuning, sensitivity, shapeability, self-recovery, simplicity, self-repair, stability and multi-stability, standby phenomena, survivability, and switchability.

Some examples of actuator materials are listed in Table 1.

205

Table 1 : Actuator materials

No.	Type	Property	Material
1.	Ferro-electric	Piezoelectric	$Pb(Zr.Ti)O_3$
		Electrostrictive	$Pb(Mg_{1.3}\ Nb_{2.3})O_3$
2.	Ferromagnetic	Magnetostrictive	Terfenol
3.	Ferroelastic	Shape Memory	Nitinol

Table 2 : Some examples of smart piezo-ceramic sensors

Type of sensor	Material	Specifications	Applications
Temperature Sensors (a) NTC negative temp. coefficient	Ceramic Semiconductors are made from transition metal oxides by doping with alio-valent cations Li_2O_3 - doped NiO is used.	Around 300°C	Oxide thermistors used as intake air temperature sensors, water temperature sensors and fuel-level sensors, exhaust gas temperature sensors.
(b) PTC positive temperature coefficient	Prepared by donor-doping a ferroelectric material like $BaTiO_3$	About 300°C	Used in automobiles as temperature and liquid level sensors; in control systems, Air-fuel ratio control is monitored in a vehicle.
Ceramic Piezoresistive Sensors	Doped $BaTiO_3$ type PTC ceramics possess larger uniaxial and hydrostatic piezoresistance coefficients than simple silicon sensors.	High temp. coefficient	Useful in pressure, torque, vibration and acceleration sensors.
Oxygen Sensors	i) ZrO_2 ceramic is prepared in cubic or tetragonal phase by adding CaO or Y_2O_3		Used in air-fuel mixture
	ii) Oxygen-sensitive porous TiO_2 ceramic conductivity depends upon temperature.		As thermistor
Humidity Sensors	Porous Al_2O_3, $MgCr_2O_4$ - TiO_2, and $CoAl_2O_4$ spinels, SeO_2,-ZnO, $ZnCr_2O_4$ - $LiZnVO_4$; K^+ doped ZrO_2 and aluminium and zinc phosphates.		Relative humidity sensors
Piezo-electric Sensors	Ferroelectric ceramics such as $Pb(ZrTi)O_3$ viz. PZT to generate ultra-dampness.		Roughness sensor, Sonar system

Composite Smart Materials

Increased sensitivity and selectivity demands on sensors require them to be structurally more complex, with their geometrics so designed as to permit a desired field concentration or composite symmetry to be acheived through the control of piezoelectric coefficients.[4,9] However, in the case of ceramic composites featuring SiC whiskers and SiC continuous fibres as matrix reinforcements, problems of oxidation present themselves at high temperatures which lead to a consequent reduction in mechanical properties. Coatings have ben offered as a solution to this problem.[9]

Types of Ceramic Sensors

Different types of piezo-ceramic smart sensors are given in Table 2 along with their compositional characteristics and possible applications.

Optimization of Sensor Material

Ceramic-sensor materials are classified in the three ways according to i) the physical properties of the grain itself (bulk properties), ii) the properties of the grain boundary, and iii) the surface effects (see Table 3).

The characteristics govern the basis for the generation and transportation of the charge carrier. The generation of the charge carrier is triggered by a gas molecule (or atom), such as water vapour or a hydroxide ion, interacting with the material. As most of the ceramic sensors are polycrystalline, the main characteristics depend upon their behaviour (bulk, grain, boundary, or surface).[8]

Table 3 : Types of ceramic sensor materials

Classification/ Property	Application	Functional Property	Material
Bulk properties	i)Temperature sensors	NTC thermistor	NiO, Fe_2O_3
	ii)Oxygen sensor	Solid electrolyte	ZrO_2
	iii)Oxygen sensor	Semiconductor	
	iv)Pressure sensor	Piezoelectric	TiO_2, $SrTiO_2$
	v)Radiation sensor	Pyroelectric	Pb(ZrTi)PZT type
Grain boundary properties	i)Temperature sensor	PTC thermistor	Doped $BaTiO_3$
	ii)Gas sensor	Semiconducting varistor	ZnO
Surface effects	i)Humidity sensor	Humidity measurement	ZnO-Cr_2O_3 $MgCr_2O_4$
	ii)Electronic sensor	Varistor cat.	$BaTiO_3$

Bulk properties depend upon the actual material structure. In thermally sensitive resistors, such as the NTC thermistor, the resistance of the material decreases as the temperature increases (Fig. 2) and this is related to the atomic packing patterns. PTC thermistors, with reverse characteristics, are examples of ceramic materials governed by the grain boundary behaviour rather than by the polycrystalline grains themselves (bulk properties). Barium titanate ($BaTiO_3$) and solid solutions exhibit a PTC thermistor behaviour (Fig. 3) which results from the presence of a space-charge barrier layer on the surface of every ceramic grain in the material.

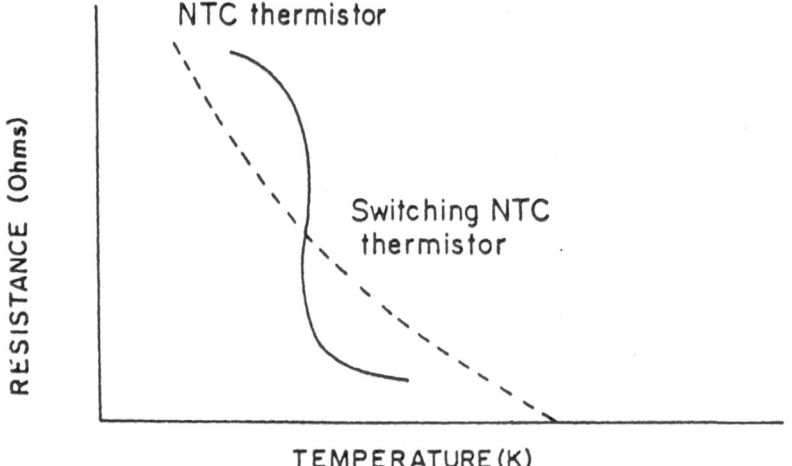

Fig. 2 : Resistance-temperature characteristics of a negative-temperature resistor (NTC)

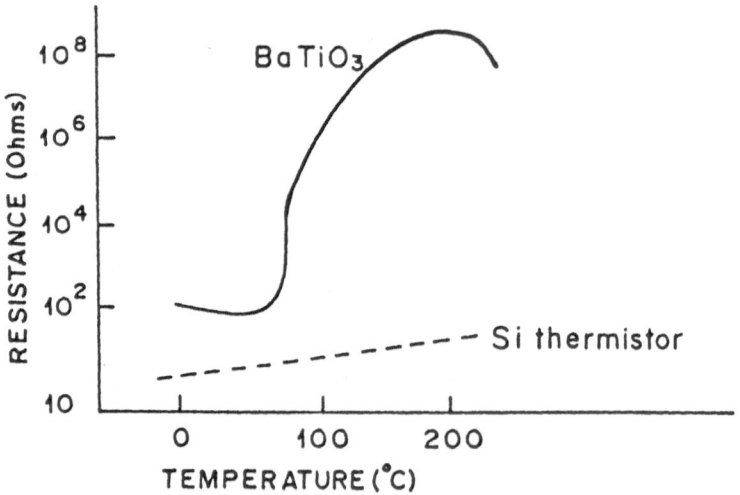

Fig. 3 : Resistance-temperature characteristics of a positive temperature resistor (PTC)

Optimisation of Sensor Performance Characteristics

The following characteristics of smart sensors require to be optimised for better performance:

Electrical Properties

For optimal output from a sensor, the following properties should be optimised:

i) Resonant frequency

The fundamental resonant frequency, in thickness mode, is given by

$$f_r = 1/2t \sqrt{(c/\rho)}$$

where "t" is the thickness dimension, "c" the elastic stiffness and "ρ" the density.

ii) Acoustic Impedance

The acoustic impedance, "Z" is given by

$$|Z| = \sqrt{(\rho c)}$$

iii) Mechanical damping coefficient (Q)

The mechanical damping coefficient Q is defined by

$$Q = \pi/\lambda\alpha$$

where "λ" in the acoustic wavelength and "α" the damping coefficient. Thus, Q is governed by "α" and is important because it controls ringing in the transducer.

iv) Electro-mechanical coupling coefficient (K)

K is defined as

$$K = d\sqrt{(c/\epsilon)}$$

where "d" is the piezoelectric charge coefficient and "ϵ" the electric permitivity. Thus K is controlled by the piezoelectric coefficient which, in turn, is controlled and fine-built using relaxor ferro-electric with large electrostatic effects.

v) Electrical impedance (Z_e)

The electrical impedance (Z_e) is given by

$$Z_e = t/\omega$$

where "ω" is the angular frequency and A the electrode area. The dielectric constant of relascor ferro-electric depends mainly on direct-current bias fields, allowing the electrical impedance to be tuned over a wide range.

Reliability

Reliability is the measure of the probability that the sensor will operate for a specified length of time. Reliable and smart sensors can be obtained by (i) identification by means of testing the functional and reliable portions of a batch of sensors, (ii) performing failure analysis and eliminating, through understanding of the mechanism of failure, subsequent failures. Failure analysis in smart sensor fabrication technology is of various types (i) catastrophic early life failures (ii) short-term drifts and failures.

Repeatability

The ability of a sensor to give the same output when the same measure and value is applied is called the repeatability of a sensor, and this requires to be maintained for optimum performance.

Drift

This is the undesired slow change of the output but not a function of the measurand, and is required to be controlled.

Frequency Response

Frequency response of a sensor is the change with frequency of the output/measurand amplitude ratio for a sinusoidally varying measurand.

Sensitivity

The ratio between the change in the sensor output and the value of the measurand is the sensitivity of a sensor. Theoretically, sensitivity, "S" is given as

$$S = \frac{\delta x}{\delta y}$$

where "δy" and "δx" are changes in the output and input respectively. Thus, sensitivity is the level of performance of a sensor and is important to be maintained.

Response Time

This is the time required for the sensor output to reach the final value and is optimised accordingly.

Resonant Frequency

This is the frequency at which the sensor shows a maximum output. The sensor optimum output is achieved at this frequency only.

Error-Free Measurement

The difference between the indicated value and the true value is called 'error'. In the case of a smart sensor, systematic uncertainties are mimimised by calibration, measurement and computation as these are reproducible. Also, use of an integral microcomputer may minimize random uncertainties.

Sensor Calibration

Automatic calibration is made for smart sensors equipped with EPROM. Sensors to be calibrated are placed in a controlled environment where these are exercised by changing the value of the input parameter and the environmental parameters. The values read out by each uncalibrated sensor are compared by an external computer with the indicated values of these parameters, as determined by transfer standards. EPROM is programmed with software coefficients that minimize the uncertainty in the measurement and the results are checked.

Accuracy

The accuracy of a sensor is the ratio of the error to the full-scale output expressed as a percentage. This is achieved from the optimum design of the sensor.

Stability

Control of mechanical, chemical and electrical factors is important since these influence the performance and stability of sensors (Fig. 4).

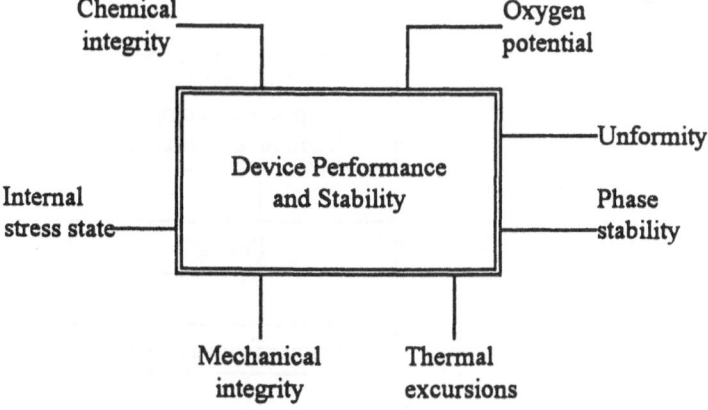

Fig. 4 : Factors affecting sensor performance stability and reliability

Developmental Studies on Sensors at the National Physical Laboratory

Electro-ceramic transducers and sensors are developed in our laboratories for various scientific and biomedical applications. For making PZT discs, the procedure as outlined in Fig.5 is followed. Raw material (ceramic powder) is subjected to grinding with the help of dry ball milling for about 24 h. The powder is then mixed with a few drops of polyvinyl acetate (PVA) and compressed into circular discs with diameter of 22.5 mm and thickness 2.5 mm. The discs are kept dry for 15 days, then electroded and poled with firing on silver by a vacuum process as well as by using air-dry paste on either side with a d.c. electric field of 2.5 kV/cm - 10 Kv/cm at elevated temperatures (60 - 80°C). Measurements are then taken on poled and unpoled samples.

Work is now in progress on the development of smart electro-ceramics as well as on composites. Smart semiconductor sensors have also been developed for measurement of blood pressure and other physical parameters.[10]

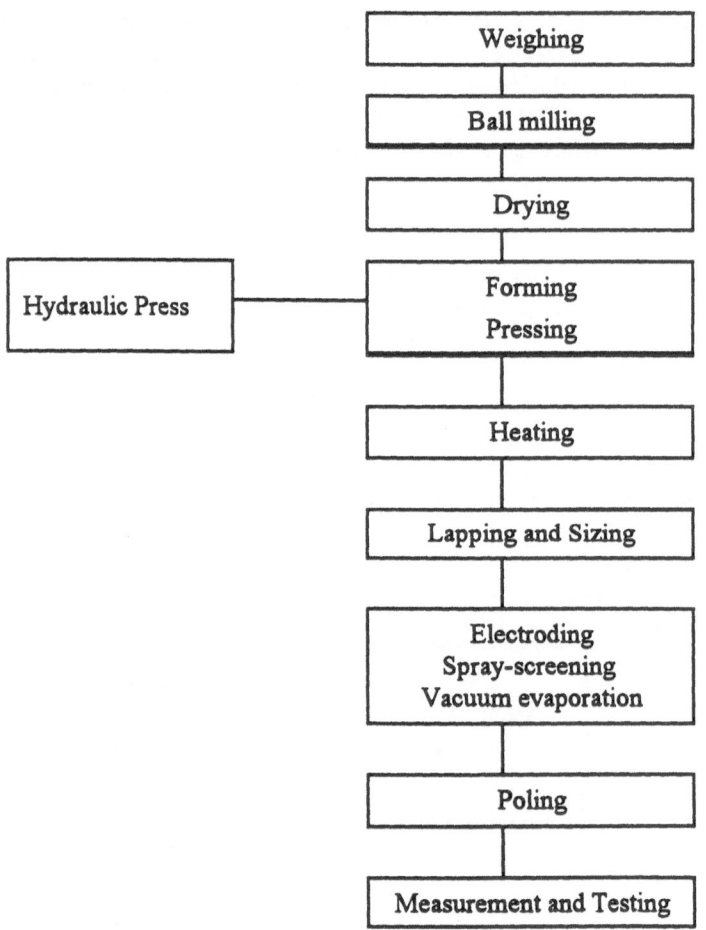

Fig. 5 : Process flow chart for making piezo-ceramic sensor elements

References

1. W.G. Wolber, Smart Sensors, *Acta IMEKO* - 1982, pp. 3-20
2. D.C. Hill and H.L. Tuller, Ceramic Sensors : Theory and Practice, In: *Ceramic Materials for Electroceramics*, ed., R. Budhanan, Marcel Dekker, N.Y., 1986.
3. R.E. Newnham, *M.R.S. Bulletin*, April, 1993, pp. 24-26.
4. S. Torolier-McKinstry and R.E. Newnham, *M.R.S. Bull.*, April, 1993, pp. 27-33.
5. R.E. Newnham and G.R. Ruschan, *J. Am. Ceram. Soc.*, **74**(3) (1991) 463.
6. R.E. Newnham, Q.C. Yu, S. Kumar and L.E. Cross, *Ferroelectrics*, **102** (1990) 1.
7. T. Yamamoto, *Ceramic Bull.*, **71**(6) (1992) 978.
8. L. Ketron, *Ceramic Bull.*, **68**(4) (1989) 860.
9. L.M. Sheppard, *Ceramic Bull.*, **71**(4) (1992) 617.
10. V.R. Singh, S. Bhatnagar, S. Verma and Ranvir Singh, 'A smart semiconductor silicon pressure sensor', Proc. Int. Workshop on Physics of Semiconductors, New Delhi, Dec 14-18, 1993.
11. S. Middelhock and S.A. Audet, "Silicon Sensors", Academic Press, New York, 1989.

'Main Group Elements and Their Compounds
V.G. Kumar Das (Ed)
Copyright © 1996 Narosa Publishing House, New Delhi, India

Optimization of the concentration of Lead in $Bi_{2-x}Pb_xSr_2Ca_2Cu_3O_{10+\delta}$ Superconducting Ceramics

**B. Madhu Sudhana, B. Srinivasulu Naidu, B.J. Reddy
and P.J. Reddy**

Department of Physics, S.V. University, Tirupati-517502, India

The discovery of superconductivity in the Ba-La-Cu-O system by Bednorz and Müller in 1986[1] spurred enormous interest in 'ceramic' superconductors, particularly in oxide ceramics of the general formula $(A_{1-x}B_x)M_bX_c$, where A is La, Y, etc., B is Ba, Sr, etc., M is Cu and X is 0.[2,3] This was soon followed by studies on the non-lanthanide systems, Bi-Sr-Ca-Cu-O and Tl-Ba-Ca-Cu-O where even higher critical temperatures (T_c above 120 K) could be reached.[4,5] In the Bi-Sr-Ca-Cu-O system, three superconducting phases[6-13] have been identified: the (2201)-phase ($T_c \sim 10$ K), the (2212)-phase ($T_c \sim 85$ K) and the (2223)-phase ($T_c \sim 110$ K). These three homogolous series are represented by the general formula $Bi_2Sr_2Ca_{n-1}Cu_nO_{2n+4}$ with n = 1, 2 and 3, respectively. It has been reported[14,15] that the zero resistance temperature above 100 K is associated with the phase dominance of the n = 3 material.

Following the discovery of Bi-Sr-Ca-Cu-O system, much effort has been expended in producing the monophasic 2223 compound which shows the highest T_c of 110 K. It has been reported[16] that the volume fraction of the high T_c phase is very sensitive to heat treatment and strongly depends upon the composition of the starting material. This renders the synthesis of the 105-115 K T_c single phase superconductor from the $Bi_2Sr_2Ca_2Cu_3O_{10+\delta}$ material of nominal composition extremely difficult. Hence, several procedures have been employed to increase the volume fraction of the (2223) phase, among which are the following:

i)   annealing at a very precise temperature[10]
ii) selecting a particular starting composition[15,17]
iii) providing a long sintering period[18]
iv) cationic substitution[19]

Of these, the cationic substitution method, especially involvling Pb, has proved to be the most efficient, with zero resistivity temperature $T_c(0)$ above 100 K being attained in a number of cases.[20-24] Pb substitution for Bi was reported to accelerate the formation of the 2223-phase and also to enhance the T_c of Bi-Sr-Ca-Cu-O system.[7,25,26] This was followed by a large number of reports[19,25,27-38] on Bi(Pb)-Sr-Ca-Cu-O system with $T_c \sim 110$ K, describing varying compositions and heat treatment conditions that tend to raise the volume fraction of the high T_c phase. In all these studies the samples usually contained, besides the 2223 phase, impurity phases like 2212, 2201, Ca_2PbO_4 and Ca_2CuO_3 or unreacted oxides. Whether Pb is incorporated into the Bi_2O_2 layer or simply acts as a flux is presently not clear. However, the results of band structure calculations suggest that Pb substitution may have some positive influence on the superconducting properties of the Bi-Sr-Ca-Cu-O system.[39] It was also observed that when the sintering temperature increased above 1148 K, the superconducting properties suddenly deteriorated. For the specimens treated at 1153 K, only a weak diamagnetic signal could be detected at 80 K.[40]

Even though the substitution of Bi with Pb enhances the formation of Bi-2223 phase and stabilizes it, an optimum concentration of Pb is essential in order to obtain a single phase 2223 compound. In the case of low Pb concentration, the formation of the 2223 phase is incomplete, whereas higher Pb content leads to the formation of impurity phases like Ca_2PbO_4. In this paper, we report the synthesis of $Bi_{2-x}Pb_xSr_2Ca_2Cu_3O_{10+\delta}$ ($0 < x < 0.5$) with differing Pb concentrations as well as their structural properties and superconducting behaviours.[41,42]

Experimental

A series of samples with starting composition of $Bi_{2-x}Pb_xSr_2Cu_3O_{10+\delta}$ ($x = 0, 0.1, 0.2, 0.25, 03.75$ and 0.45) were prepared by the solid state reaction method at normal atmosphere using 4N pure Bi_2O_3, PbO, $SrCO_3$, $CaCO_3$ and CuO mixed in the desired ratio using an agate mortar and pestle. Fine homogeneous powders are required to obtain dense uniformly sintered bodies, as the densely sintered body probably has good properties as a superconductor.[41,42]

The samples were prepared with different Pb concentrations under different heat treatment conditions. The mixed powders were thoroughly ground and heated at 1128 K in a silica crucible for 100 h with intermittent grinding for every 25 h of heat treatment and finally quenched in air. The air quenched samples were ground to a particle size of 100 mm and pressed to form pellets of 112 mm diameter and 1.5 mm thickness under a pressure of 75 kN. At least two pellets were made for each composition. The pellets were sintered at normal atmosphere for 72 h at $1118 + 10$ K. The samples were either air quenched or slowly furnace cooled. The samples were black in colour and stable towards exposure to atmosphere.

X-ray diffraction patterns were recorded using a Philips X-ray diffractometer with Cu-K_α radiation. Resistivity measurements were performed on the sintered samples using the standard four probe technique in the temperature range 303-70 K. The temperature dependence (303-60 K) of ac susceptibility was studied at 3117 Hz and at an ac field of 0.05 Oe.

Results and Discussion

In the compounds with the general formula $Bi_{2-x}Pb_xSr_2Ca_2Cu_3O_{10+\delta}$ ($0 < x < 0.5$) the presence of lead (Pb) results in the stabilization of the T_c (2223) phase.[20] The high T_c 110 K phase is reported to be most stable for Pb substitutions ($x = 0.2$-0.3). With higher x values impurity phases are formed.[43,44] It was also reported that as sintering times increased, increasing Pb loss was observed and the 2223 phase was converted into 2212 and 2201 low-T_c phases.[45-47]

In the present investigation the samples with higher concentrations of Pb were found to exhibit low T_c values and the sample with x = 0.225 showed the highest T_c at 105 K. The variation of $T_c(0)$ with the concentration of Pb is shown in Table 1. As shown in the table, the samples with 0.2 < x < 0.3 give $T_c(0)$ values > 100 K. The $Bi_{1.75}Pb_{0.25}Sr_2Ca_2Cu_3O_{10+\delta}$ samples were subjected to different sintering temperatures and periods and the results are summarized in Table 2. It is seen from the table that a sintering temperature of 1118 K and a sintering period of 72 h are essential to obtain $T_c(0)$ of 105 K.

Table 1 :   Variation of the transition temperature with the concentration of Pb is $Bi_{2-x}Pb_xSr_2Ca_2Cu_2O_{10+\delta}$ sintered at 1118 K for 72 hours

Sample	Zero resistivity temp, $T_c(0)$ (K)
$Bi_2Sr_2Ca_2Cu_2O_{10+\delta}$	85
$Bi_{1.9}Pb_{0.1}Sr_2Ca_2Cu_3O_{10+\delta}$	98
$Bi_{1.8}Pb_{0.2}Sr_2Ca_2Cu_3O_{10+\delta}$	103
$Bi_{1.75}Pb_{0.25}Sr_2Ca_2Cu_3O_{10+\delta}$	105
$Bi_{1.625}Pb_{0.375}Sr_2Ca_2Cu_3O_{10+\delta}$	100
$Bi_{1.55}Pb_{0.45}Sr_2Ca_2Cu_3O_{10+\delta}$	90

Table 2 :   Effect of sintering temperatures and sintering periods on $T_c(0)$ of $Bi_{1.7}Pb_{0.25}Sr_2Ca_2Cu_3O_{10+\delta}$ ceramics

Sintering temperature (K)	Sintering period (h)	$T_c(0)$ (K)
1098	24	73
	48	82
	72	100
	96	95
1118	24	70
	48	92
	72	105
	96	100
1138	24	85
	48	90
	72	94
	96	90

The X-ray diffraction pattern of the sample with x = 0.25, sintered at 1118 K for 72 h is shown in Fig. 1. The majority of the observed reflections were characteristic of an orthorhombic structure with lattice parameters;[48] a = 0.3866 nm, b = 0.3753 nm and $c = 3.6566$ nm. The samples with other Pb concentrations had the same orthorhombic structure with the same lattice parameters. From the XRD pattern it is seen that the majority of the samples contain the high T_c(2223)-phase with minor percentages of the (2212)- and (2201)-phases.

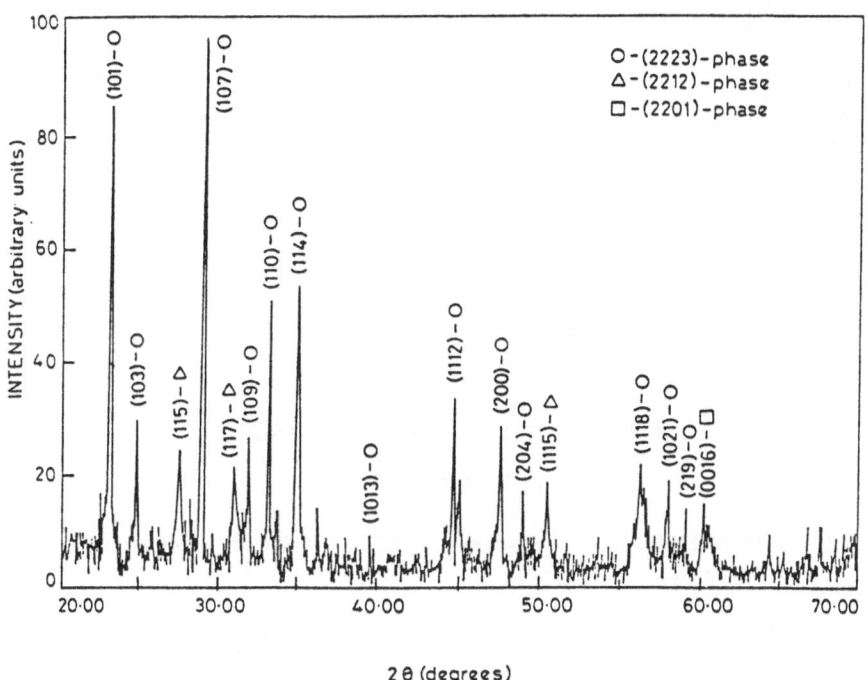

Fig. 1 : **X-ray powder diffraction pattern of the bulk sample of**
$Bi_{1.75}Pb_{0.25}Sr_2Ca_2Cu_3O_{10+\delta}$

Voltage and current leads were attached to the pellets using silver epoxy. The temperature dependance of electrical resistance for a typical sample (x = 0.25) is shown in Fig. 2. The behaviour is metallic between 303 and 110 K and the electric resistance starts dropping at about 115 K and becomes zero at 105 K.

Temperature dependence of ac susceptibility of the sample (x = 0.25) is shown in Fig. 3. From the figure it is seen that at room temperature the specimen is Pauli paramagnetic but changes to a diamagnetic state upon cooling to 115 K. The diamagnetic signal becomes saturated at 105 K. Samples with other concentrations of Pb showed zero resistivity temperatures in the range 85-103 K.

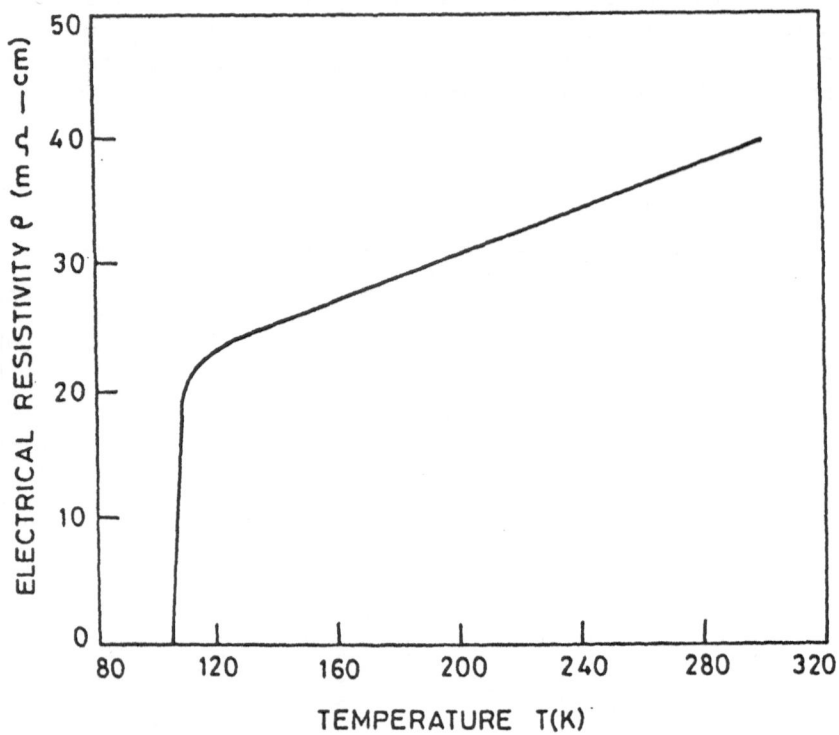

Fig. 2 : Temperature dependence of electrical resistivity for $Bi_{1.75}Pb_{0.25}Sr_2Ca_2Cu_3O_{10+\delta}$

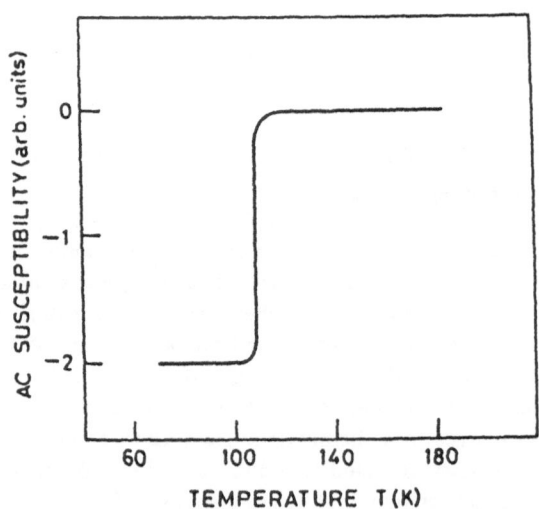

Fig. 3 : Temperature dependence of ac susceptibility for $Bi_{1.75}Pb_{0.25}Sr_2Ca_2Cu_3O_{10+\delta}$

References

1. J.G. Bednorz and K.A. Muller, *Z. Phys.*, **B64** (1986) 189.
2. S. Uchida, H. Takagi, K. Kitazawa and S. Tanaka, *Jpn. J. Appl. Phys.*, **26** (1987) L1.
3. H. Takagi, S. Uchida, K. Kitazawa and Z. Tanaka, *Jpn. J. Appl. Phys.*, **26** (1987) L123.
4. H. Maeda, Y. Tanaka, M. Fukutomi and T. Asano, *Jpn. J. Appl. Phys.*, **27** (1988) L209.
5. Z.Z. Sheng and A.M. Hermann, *Nature*, **332** (1988) 138.
6. C. Michel, M. Herrien, M.M. Borel, A. Grandin, F. Deslandes, J. Provost and B. Raveau, *Z. Phys.*, **B68** (1987) 421.
7. S.A. Sunshine, T. Sigrist, L.F. Schneemayer, D.W. Murphy, R.J. Cava, B. Batlogg, R.B. Van Dover, R.M. Flemming, S.H. Glarum, S. Nakaharo, R. Farrow, J.J. Krajewski, S.M. Zahurak, J.V. Waszczak, J.H. Marshall, P. Marsh, L.W. Rupp, Jr. and W.F. Peck, *Phys. Rev. B*, **38** (1988) 893.
8. H.W. Zandbergen, Y.K. Huang, M.J.V. Menken, J.N. Li, K. Kadowaki, A.A. Menovsky, G. Van Tendeloo and S. Amelincky, *Nature*, **333** (1988) 620.
9. H.W. Zandbergen, P. Goren, G. Van Tendeloo, J. Van Landuyt and S. Amelinckx, *Solid State Commun.*, **66** (1988) 397.
10. J.M. Tarascon, Y. le Page, P.Barboux, B.G. Bagley, L.H. Green, W.R. McKinnon, G.W. Hull, M. Girondi and D.M. Hwang, *Phys. Rev. B*, **28** (1988) 885.
11. R.M. Hazen, C.T. Prewitt, R.G. Angel, N.L. Roy, L.W. Finger, C.G. Hadidiacos, D.R. Vablen, P.J. Heaney, P.H. Hor, R.L. Meng, Y.Y. Sun, Y.Q. Wang, Y.Y. Xua, Z.J. Huang, L. Gao, J. Bechtold and C.W. Chu, *Phys. Rev. Lett*, **60** (1988) 1174
12. J.B. Torrance, Y. Tokura, S.J. Laplaca, T.C. Huang, R.J. Savoy and A.I. Nazzal, *Solid State Commun.*, **66** (1988) 703.
13. A. Sumiayama, T. Yoshitomi, H. Endo, J. Tsuchiya, N. Kijima, M. Miznno and Y. Oguri, *Jpn. J. Appl. Phys.*, **27** (1988) 542.
14. L. Ganapathi, J. Narayan and Ashok Kumar, *Appl. Phys. Lett.*, **55** (1989) 1460.
15. B.N. Veal, H. Claus, J.W. Downey, A.P. Paulikas, K.G. Vander Voort, J.S. Pan and D.J. Lam, *Physica C*, **156** (1988) 635.
16. Y. Tanaka, M. Fukutomi, T. Asano and H. Maeda, *Jpn. J. Appl. Phys.*, **27** (1988) L548.
17. W. Herkert, H.W. Neumueller and Wilhelm, *Solid State Commun.*, **69** (1989) 183.
18. N. Kijima, H. Endo, J. Tsuchiya, A. Sumiyama, M. Mizumo and Y. Oguri, *Jpn. J. Appl. Phys.*, **27** (1988) L821.
19. M. Takano, Y. Takada, K. Oda, H. Kitaguchi, V.Miura, Y. Ikeda, Y. Tommi and H. Mazaki, *Jpn. J. Appl. Phys.*, **27** (1988) L1041.
20. S.M. Green, C. Jiang, Y. Mei, H.L. Luo and C. Politis, *Phys. Rev. B*, **38** (1988) 5016.
21. R. Ramesh, G. Thomas, S.M. Green, C. Jiang, Y. Mei, M.L. Rudee and H.L. Luo, *Phys. Rev. B*, **38** (1988) 7070.
22. R. Ramesh, K. Remsching and J.M. Tarascon, *J. Mater. Res.*, **6** (1991) 278.
23. Y. Idemoto, S. Fujiwara and K. Fueki, *Physica C*, **179** (1991) 96.

24. S. Losch, M. Schlichenmaier, S. Kemmeler-Sack, Scholl, S. Dottinger, W. Forkel, D. Kolle, R. Gross and R.P. Huebener, *J. of Less-Common Metals*, **159** (1990) 261.

25. B.W. Statt, Z. Wang, M.J.G. Lee, J.V. Yakhmi, P.C. de Camergo and J.F. Rutter, *Physica C*, **156** (1988) 251.

26. P.V.P.S.S. Sastry, I.K. Gopalakrishnan, J.V. Yakhmi and R.M. Iyer, *Physica C*, **157** (1989) 491.

27. A. Oota, Y. Sasaki and A. Kirihigashi, *Jpn. J. Appl. Phys.*, 27 (1988) L1445.

28. Y. Tanaka, T. Asano, K. Jikihara, M. Fukutomi, J. Machida and H. Maeda, *Jpn. J. Appl. Phys.*, **27** (1988) L1655.

29. R. Retoux, F. Studer, C. Michel and B. Raveau, *Phys. Rev. B*, **41** (1990) 193.

30. M. Mizuno, H. Endo, J. Tsuchiya, N. Kijima, A. Sumiyama and Y.Oguri, *Jpn. J. Appl. Phys.*, **27** (1988) L1225.

31. A. Oota, A. Kirihigashi, Y. Sasaki and K. Ohba, *Jpn. J. Appl. Phys.*, **27** (1988) L2289.

32. Y. Yamada and S. Murase, *Jpn. J. Appl. Phys.*, **27** (1988) L996.

33. T. Ishida, T. Sakuma, T. Sasaki and Y. Kawada, *Jpn. J. Appl. Phys.*, **28** (1989) L559.

34. A. Maeda, K. Noda, K. Uchinokura and S. Tanaka, *Jpn. J. Appl. Phys.*, **28** (1989) L576.

35. E. Chavira, R. Escudero, D. Rios Jara and L.M. Leon, *Phys. Rev. B*, **38** (1988) 9272.

36. S. Koyama, U. Endo and T. Kawai, *Jpn. J. Appl. Phys.*, **27** (1988) L1861.

37. C.N.R. Rao, R. Vijayaraghavan, L. Ganapathi and S.V. Bhat, *J. Solid State Chem.*, **79** (1989) 177.

38. L. Ganapathi, S. Krishna, K. Murthy, R. Vijayaragavan and C.N.R. Rao, *Solid State Commun.*, **67** (1988) 967.

39. W.N. Wang, M.F. Tai, H.C. Ku, M.J. Shich, T.Y. Lin, Y.F. Wang, H.B. Lu, P.C. Yao, S.J. Yang and S.E. Hsu, *Supercond. Sci. Technol.*, **2** (1989) 55.

40. S. Kobayashi, Y. Saito and S. Wada, *Jpn. J. Appl. Phys.*, **28** (1989) L772.

41. R. Yamamoto, T. Furusawa, K. Park, M.L. Green, K. Kuwahara, N. Ooba, T. Hasegawa, K. Kishio and K. Kitazawa, *Jpn. J. Appl. Phys.*, **35** (1988) 841.

42. M. Awano, K. Kani, Y. Kodama, H. Takagi and Y. Kuwahara, *Jpn. J. Appl. Phys.*, **29** (1990) L254.

43. M. Pissas, D. Niarchos, C. Christides and M. Anagnostou, *Supercond. Sci. Technol.*, **3** (1990) 128.

44. A. Hammer and G. Srinivasan, *Jpn. J. Appl. Phys.*, **69** (1991) 4899.

45. J.L. Tallon, R.G. Buckley, P.W. Gilberd and M.R. Presland, *Physica C*, **158** (1989) 247.

46. B. Ramesh Tripathi and D.W. Johnson Jr., *J. Am. Ceram. Soc.*, **74** (1991) 247.

47. A. Hammer and G. Srinivasan, *Supercond. Sci. Technol.*, **3** (1990) 306.

48. F. Shi, T.S. Rong, S.Z. Zhou, X.F. Wu, J. Du, Z.H. Shi and N.C. Shi, *Phys. Rev. B*, **41** (1990) 6541.

Main Group Elements and Their Compounds
V.G. Kumar Das (Ed)
Copyright © 1996 Narosa Publishing House, New Delhi, India

Correlation between Ionization Potential and Superconductivity of Oxides

Wei Chen and Ruzhang Mai

*Department of Materials Physics, University of Science and
Technology Beijing, Beijing 100083, P.R. China*

Since the discovery of high Tc superconductivity in 1986 by Bednorz and Muller,[1] the superconductivity of oxides has been studied extensively. Although several models[2] have been proposed to explain this fascinating phenomenon, the mechanism of high Tc superconductivity has not been revealed unambiguously. It is important to have a guided approach to superconductivity in the search for new and higher Tc superconductors. Because some semi-empirical rules can provide the right direction for the search of high Tc superconducting materials, much work has been done to secure these through correlating the properties of superconductors that relate intimately to the effect.

In the early stages, the investigators paid attention to known superconducting elements. On the basis of electron theory of atomic structure the investigators deduced that a criterion for superconductivity of an element necessarily has something to do with the distribution of electrons: known superconducting elements all possess 2-5 valence electrons. Many empirical formulae[3] have been suggested for superconductivity, e.g. Pine's criterion[4] gave a Z*-rs curve line, above or below which lay superconducting or non-superconducting elements. But the use of all such semi-empirical criterions is rather limited. In recent years, electronegativity has been found to be a useful quantity for understanding the nature of superconducting materials.[5,6] Luo and Wang[5] reported a new criterion for superconductivity of all the elements in the periodic table. The criterion shows that the values of electronegativity of all superconducting elements concentrate in a narrow range from 1.3 - 1.9 and that elements with values outside this region would be non-superconductive. More recently Jayaprakash and Shanker[6] investigated the correlation between electronegativity and high temperature superconductivity. They found that the electronegativity parameters concentrate in a narrow range for about 60 superconducting oxides. In 1988 Villars and Phillips[7] used the golden coordinates to delineate the sixty known superconductors with Tc > 10 K successfully into three small volumes which occupy about 1% of elemental configuration space and suggested that compounds in the third volumes with formulae near YBa_2CuO_7 are promising candidates for new high Tc superconductors. Undoubtedly these semi-empirical criterions are helpful for the searching of new superconductors.

In this short communication we report the correlation between mean ionization potential and superconductivity of oxides. The mean ionization potential of a crystal is defined as

$$<Uo> \ = \ \frac{\Sigma X_i Uoi}{\Sigma X_i}$$

where Uoi is the first ionization potential of a given element i and Xi denotes the number of atoms of a given element i in the molecular formula of a crystal. It is considered by the authors that $<Uo>$ represents the excitation energy of a single electron from its valency state in the crystal lattice to the free state. Thus, similar to electronegativity, $<Uo>$ may indicate the electrostatic action of the electron in the crystal lattice, and may be used to characterize the behaviour of electron motion or transformation in crystal lattice. We deduce that $<Uo>$ is correlated with the superconductivity of materials.

The calculated $<Uo>$ values of 122 oxides (including 77 superconducting oxides and 45 non-superconducting oxides) are given in Table 1. The first ionization potential of free atoms is taken from Fyfe[8], and the Tcs from published reviews[9,10]. The result shows that the $<Uo>$ values of almost all superconducting oxides concentrate in a narrow range from 10.20 to 10.50 eV, while that of the non-superconducting oxides scatter outside this region (Fig. 1). This characteristic may be taken as a new criterion for superconductivity of oxides.

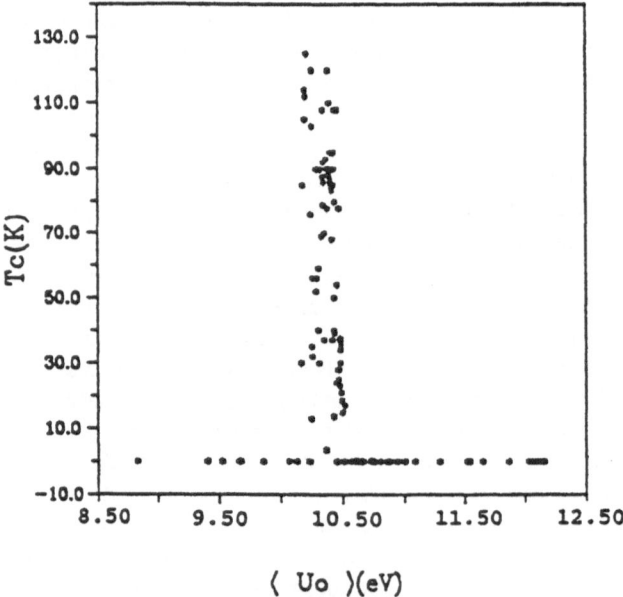

Fig. 1 : A distribution of mean ionization potential <Uo> vs superconducting critical temperature (Tc) among 122 oxides

222

Table 1 : The mean ionization potential $<Uo>$ of oxides of which 77 kinds are superconducting and 45 kinds are non-superconducting.

Oxides	$<Uo>$ (eV)	Tc(K)
$La_2CuO_{4.03}$	10.427	39.0
$La_2CuO_{4.04}$	10.432	39.0
$La_{1.35}Sr_{.15}CuO_4$	10.479	37.5
$La_{1.325}Sr_{.175}CuO_4$	10.480	36.0
$La_{2.825}Sr_{.175}CuO_4$	10.480	35.0
$La_{1.83}Sr_{.17}CuO_4$	10.417	37.0
$La_{1.85}Ba_{.15}CuO_4$	10.469	28.0
$La_{1.85}Ca_{.15}CuO_4$	10.514	17.0
$La_{1.8}Ba_{.2}CuO_4$	10.466	25.6
$Nd_{1.85}Ce_{.15}CuO_4$	10.451	24.0
$Nd_{1.4}Ca_{.2}Sr_{.4}CuO_4$	10.463	28.0
$Nd_{1.35}Tho_{.15}CuO_{4.048}$	10.500	15.0
$(La,Nd,Ce)_2CuO_4$	10.477	23.0
$(La,Ce,Y)CuO_4$	11.290	0.0
$YBa_2Cu_3O_7$	10.38	95.0
$YBa_2Cu_3O_{8.95}$	10.385	90.
$YBa_2Cu_3O_{8.84}$	10.358	88.0
$YBa_2Cu_3O_{8.81}$	10.350	86.0
$YBa_2Cu_3O_{6.78}$	10.342	79.0
$YBa_2Cu_3O_{6.73}$	10.300	69.0
$YBa_2Cu_3O_{6.84}$	10.306	59.0
$YBa_2Cu_3O_{6.53}$	10.290	56.0
$YBa_2Cu_3O_{.45}$	10.256	56.0
$YB_2Cu_3O_{6.35}$	10.229	0.0
$YBa_2Cu_3O_8$	10.130	0.0
$YBa_2Cu_4O_8$	10.432	80.0
$Y_2Ba_4Cu_7O_{14.8}$	10.426	40.0
$Y_{0.35}Ba_{0.85}CuO_{2.3}$	10.319	90.0
$YBa_2(Cu_{.96}Fe_{.04})_3O_{6.8}$	10.240	76.1
$YbBa_2Cu_3O_7$	10.323	90.0
$Y_2Ba_{1.5}Ca_{0.5}Cu_3O_8$	10.375	78.0
$SmBa_2Cu_3O_7$	10.336	87.7
$GdBa_2Cu_3O_7$	10.379	90.0
$DyBa_2Cu_3O_7$	10.428	90.0
$EuBa_2Cu_3O_7$	10.344	92.0
$HoBa_2Cu_3O_7$	10.367	93.0
$YBa_2Cu_3O_{8.948}$	10.385	90.4
$YBa_{1.8}Sr_{.2}Cu_3O_{6.948}$	10.393	88.4
$YBa1._7Sr_{.3}Cu_3O_{8.948}$	10.397	87.4
$YBa_{1.8}Sr_{.4}Cu_3O_{6.948}$	10.400	86.4
$YBa_{1.4}Sr_{.5}Cu_3O_{6.948}$	10.404	85.4
$YBa_{1.4}Sr_{.6}Cu_3O_{6.948}$	10.408	84.4

Table 1 (cont'd)

$Yba_{1.3}Sr_{.7}Cu_3O_{6.948}$	10.412	83.4
$Sm0.5Y_{0.5}Ba_2Cu_3O_7$	10.308	90.0
$Sm0.5HO_{0.5}Ba_2Cu_3O_7$	10.292	90.0
$Bi_2Sr_2CuO_8$	10.487	21.0
$Bi_2Sr_2CaCu_2O_8$	10.425	85.0
$Bi_2Sr_2Ca_2Cu_3O_{10}$	10.389	100.0
$Bi_{1.7}Pb_{.2}In.1Sr_2Ca_2Cu_3O_{10}$	10.425	95.0
$Bi_{1.8}Pb_{.4}Sr_{1.75}Ca_2Cu_3O_{10.05}$	10.450	108.0
$Bi_{1.8}Pb_{.4}Sr_2Ca_2Cu_3O_{10.3}$	10.430	108.0
$Bi_{1.4}Pb_{.8}Sr_2CaCu_2O_8$	10.469	78.0
$Bi_{.5}Pb_{.5}Sr_{1.5}Ca_3Cu_4O_{10}$	10.338	108.0
$TlBa_2CaCu_2O_7$	10.252	103.0
$TlBa_2Ca_2Cu_3O_9$	10.252	120.0
$Tl_2Ba_2CuO_8$	10.178	85.0
$Tl_2Ba_2CaCu_2O_8$	10.199	112.0
$Tl_2Ba_2Ca_2Cu_3O_{10}$	10.211	125.0
$TlBa_2Ca_4Cu_5O_{13}$	10.196	114.0
$TlBa_2Ca_5Cu_6O_{15}$	10.197	105.0
$(Tl,Bi)Sr_2CuO_5$	10.427	50.0
$TlLaSrCuO_5$	10.350	37.0
$TlBa_{1.2}La_{.8}CuO_5$	10.286	52.0
$(Tl_{.5}Pb_{.5})Sr_2Ca_2Cu_3O_9$	10.379	120.0
$(Tl,Bi)Sr_2CaCu_2O_7$	10.375	90.0
$TlBa_2Ca_3Cu_4O_{11}$	10.253	120.0
$TlGdSrCaCuO_6$	10.311	30.0
$TlSmSrCaCuO_6$	10.260	32.0
$TlErSrCaCuO_6$	10.305	40.0
$TlNdSrCaCuO_8$	10.251	35.0
$Pb_2Sr_2YCu_3O_8$	10.411	68.0
$Pb_2Sr_2NdCu_3O_8$	10.355	70.0
$Pb_{.7}In_{.3}Sr_2Y_{.2}Ca_{.8}Cu_2O_7$	10.450	54.0
$Ba_{.7}K_{.3}BiO_{2.35}$	10.167	30.0
$BaPb_{.7}Bi.3O_{2.5}$	10.248	13.0
$BaPb_{.75}Sb_{.25}O_{2.5}$	10.362	3.5
$LiTi_3$	10.427	13.7
$SrTiO_3$	10.600	0.3
$NaWO_3$	10.728	0.3
$(Eu, La)_{1.85}Ce_{.15}CuO_4$	10.490	18.5
$PbBaYSrCu_3O_8$	10.512	
$BaPbO_3$	10.866	
$Ba_4CaCu_3O_{8.81}$	10.065	
$MgCu_2O_3$	10.845	
$BiCuO_5$	11.864	

Table 1 (cont'd)

$BiPbSr_2BiFe_2O_{9.5}$	10.878
$Bi_{1.5}Pb_{0.5}Sr_2BiFe_2O_{9.5}$	10.845
$BiPbSr_2Bi_2Fe_3O_{12.5}$	10.955
$Bi_{1.5}Pb_{0.5}Sr_2Bi_2Fe_2O_{12.5}$	10.929
$Bi_{10}Sr_{15}Fe_{10}O_{48}$	10.661
$BaBi.0T_{1.103}$	10.642
$BaBi._7t_{1.103}$	10.599
$Bi_2Sr_4Fe_3O_{12}$	10.755
$CaNbBio_9$	11.515
$Bi_2Ca_3CoO_4$	9.520
$Bi_2Sr_3CoO_4$	9.400
$Bi_2Ba_3CoO_4$	9.250
Gd_2NiO	10.623
La_2NiO_4	10.463
$DyNiO_4$	10.806
$(Y,Ca)(Bi,Pb)O_7$	12.152
$(Y,Sr)(Bi,Pb)O_7$	12.130
$(Y,Ba)(Bi,Pb)O_7$	12.102
$(Pr,Ca)(Bi,Pb)O_7$	12.078
$(Pr,Sr)(Bi,Pb)O_7$	12.056
$(Pr,Ba)(Bi,Pb)O_7$	12.028
MgO	10.625
Cu_2O	9.670
CuO	10.655
SnO_2	11.520
PbO	11.005
Bi_2O_3	11.086
Eu_2O_3	10.446
Y_2O_3	10.726
CaO	9.855
SrO	9.655
BaO	9.405
ZnO	10.505
V_2O_5	11.647
MnO_2	11.540
Al_2O_3	10.566
P_2O_5	8.833

It is also found from further study that in $YBa_2Cu_3O_y$, the $<Uo>$ increases linearly with the increase in oxygen content y, while Tc-y and Tc-$<Uo>$ curves variate synchronously. The above indicate that the mean ionization potential $<Uo>$ of crystal is closely correlated with the superconductivity of oxides. Details on these will be discussed in a forthcoming paper.

It must, however, be pointed out that there is also a limit on the use of mean ionization potential as a new criterion for superconductivity of oxides. This is because it does not take account of crystal structure, which is a very important factor in determining the superconductivity of oxides.

References

1. J.G. Bednorz and K.A. Muller, *Z. Phys.*, **B64** (1986) 189.
2. V.L. Ginzburg, *Contemporary Physics*, **33** (1992) 15.
3. G. Groetzinger, D. Kahn and P. Schwed, *Phys. Rev.*, **96** (1954) 887.
4. D. Pines, *Phys. Rev.*, **109** (1958) 280.
5. Q.G. Luo and R.Y. Wang, *J. Phys. Chem. Solids*, **48** (1987) 415.
6. R. Jayaprakash and J. Shankar, *J. Phys. Chem. Solids*, **54** (1993) 365.
7. P. Villars and J.C. Pillips, *Phys. Rev.*, **B37** (1988) 2345.
8. W.S. Fyfe, Geochemistry of Solids, 1964, P78, McGraw Hill, New York - San Francisco - Toronto - London.
9. T. Hirata and Y. Asada, *J. Superconductivity*, **4** (1991) 171.
10. T.A. Vanderah, Chemistry of Superconductor Materials, Chapter 1, 1992, Noyes Publications, New Jersey.

Main Group Elements and Their Compounds
V.G. Kumar Das (Ed)
Copyright © 1996 Narosa Publishing House, New Delhi, India

Surface Segregation and Concentration fluctuations of Compound-forming Binary Alloys

L.C. Prasad[a] and R.N. Singh[b]

[a]Department of Chemistry, T.N.B. College, Bhagalpur University
Bhagalpur - 812007, India
[b]Department of Physics, College of Science, Sultan Qaboos University,
MUSCAT, Sultanate of Oman

It is well established that the equilibrium composition of an alloy surface differs from bulk composition. Because of the segregation of one component of the binary alloy to its surface, the composition of the alloy surface is not same as that of the bulk composition. This phenomenon, known as surface segregation has considerable importance in metallurgy[1] and catalysis[2-4]. This has inspired theoretical and experimental studies aimed at understanding surface segregation in terms of bulk properties.

Crudely, the excess surface concentration[5] of one component of an alloy over its bulk concentration[5], Γ is related to composition dependence of surface tension $(d\sigma/dc)$. As very little is known about the composition dependence of surface tension, not much success has been achieved. Recent developments in experimental techniques such as Auger-electron spectroscopy[6-7], X-ray and ultraviolet photoemission spectroscopy[8,9], titrimetry[3,4], low energy ion scattering[10] and time of flight atom probe[11] have aroused the interest of theoreticians to provide possible explanations of this problem by using various theoretical models.[12-15] Of all the models, statistical mechanical theory[16], which is based on the concept of layered structure near the interface , has been widely used.[17-20] Likewise, bulk grand partition function for the surface has been constructed and solved under the framework of a quasi-lattice model.[21,22] An expression for the surface tension of binary molten alloys is obtained through the solution of grand partition function in terms of surface composition, surface tension of pure components, atomic surface area, surface coordination fractions and order energy. The above expression has been improved upon to incorporate the effect of strong interaction due to which intermetallic compounds exist at different stoichiometric compositions.

In this paper, an attempt has been made to understand the phenomenon of surface segregation in terms of surface concentration fluctuations [$Scc^s(O)$]. In recent years, bulk theory[22-24] based on concentration fluctuations, [$Scc^b(O)$], has emerged as a powerful tool to understand the alloying behaviour of bulk molten alloys. It not only tells us about the stability of the system but also demonstrates the nature of atomic interaction in the bulk of an alloy. Usually the characteristic behaviour of the bulk is gauged through the deviation of $Scc(O)$ from the ideal values; $Scc(O, ideal) = C_A.C_B$. A composition for which $Scc^b(O) > Scc^b(O, ideal)$

is indicative of self coordination (segregation), while for the case $Scc^b(O) < Scc^b(O, id)$ this is indicative of atomic ordering in the alloy. Keeping in mind the importance of $Scc^b(O)$, we have derived an expression for the surface concentration fluctuations $[Scc^s(O)]$ in terms of surface free energy of mixing.

The above theory has been utilised to understand the surface properties such as surface segregation and surface tension of Au-Zn molten alloys. The excess free energy of mixing (G_M^{XS}) for the Au-Zn system has a large negative value. Its phase diagram (liquidus), which shows a distinctive maximum at $C = 0.5$ and large negative value of G_M^{XS} suggests[25] that a 1:1 compound (AuZn) exists at equiatomic composition.

Theory

A quasi-lattice model for the surfaces of binary molten alloys

Viewed as a distinct phase, a surface may be defined as all that volume of materials surrounding an interface between two bulk phases, namely the liquid phase and the vapour phase. The vapour phase contribution in liquid vapour interface is negligible and the liquid (bulk) just below the surface is a major contributing phase. The derivation of our expressions for surface segregation is based on the assumptions that a surface of area, A, is created after slicing the bulk liquid having a cylindrical structure of cross-sectional area ½A into two parts. The distribution of the atoms at the surface of uniatomic dimension is presumed to be same as that of the bulk counterpart. Further, if the concept of layered structure of liquids is accepted, it may be assumed that any particular atom is influenced not only by the layer in which the atom lies but also by the two layers adjacent to it. In other words, the coordination number of the atom will be affected.

If p and q, termed as coordination fractions, be the contributions of the layer in which the atom lies and of the layer adjacent to it, then for the bulk atom, there will be two adjacent layers (one p and two q's), while for the surface atom there will be only one adjacent layer (one p and one q) because of negligible contribution of one of the adjacent vapour phases.

On the basis of the above assumptions, it is possible to express bulk coordination number Z^b and surface coordination number Z^s as follows:

$$Z^b = (p + 2q)\, Z \tag{1}$$

$$Z^s = (p + q)\, Z \tag{2}$$

For a closed-packed structure,

$$Z = 12, \ p = \tfrac{1}{2} \ \text{and} \ q = \tfrac{1}{4}, \ \text{so that } Z^b = Z = 12, \ \text{and } Z^s = 9$$

If the components in the surface and the bulk phase are in thermodynamic equilibrium, then $\mu_i^b = \mu_i^s$ where μ_i is the chemical potential of the component in the alloy and superscripts 'b' and 's' denote bulk and surface respectively. From thermodynamic considerations, it can be said that the activity coefficients will be

228

different for bulk and surface. We have here assumed that the bulk activity coefficient (Y_i) of a component and its activity coefficient at the surface (Y_i^s) are related to the coordination fractions p and q , as follows:

$$\ln Y_i = p[\ln Y_i \text{ containing surface concentration in place of bulk concentration}) + q \ln Y_i \qquad (3)$$

Construction of grand partition function for the surface

The grand partition function of the bulk (Ξ^b)[21,22] of an alloy A-B, which consists of $N_A[=NC_A]$ atoms of A and $N_B[=NC_B]$ atoms of B respectively, [N being the total number of atoms equal to $N_A + N_B$, and C_A and C_B being the mole fractions respectively of A and B] can be constructed as shown in equation (4):

$$\Xi^b = [\lambda_A q_A(T)]^{N_A} [\lambda_B q_B(T)]^{N_B} \exp(-E/K_B T) \qquad (4)$$

where E is bulk configurational energy and $q_i(T)$ are atomic partition function of i (A or B) atoms of the bulk alloy. λ_A and λ_B are related to chemical potentials through the relation

$$\lambda_A = \exp(\mu_A/K_B T)$$

$$\lambda_B = \exp(\mu_B/K_B T) \qquad (5)$$

As the distributions of atoms in the bulk and at the surface is the same, it is safe to assume that order energy W $[= Z[\varepsilon_{AB} - (t_{AA} + \varepsilon_{BB})/2]$; ε_{AA}, ε_{BB}, ε_{AB} are bond energies of A-A, B-B and A-B bonds respectively] will be the same for bulk and surface, that is

$$W^b = W^s = W \qquad (6)$$

Using equation (2), the grand partition function for the surface (Ξ^s) can be constructed in the same way as in equation (4), viz:

$$\Xi^s = [\lambda_A q_A^s(T)]^{N_A^s} [\lambda_B q_B^s(T)]^{N_B^s} \exp(-E^s/K_B T) \qquad (7)$$

where $q_i^s(T)$ refers to the surface and total number of atoms at the surface, $N^s = N_B^s + N_B^s [N_A^s = N^s C_A^s$ and $N_B^s = N^s C_B^s$; C_B^s and C_B^s are surface concentrations of A and B components of the alloy].

The main problem in the solution of equation (7) is the definition and assignment of the value of surface configurational energy E^s. Prasad and Singh[20] have expressed E^s as the product of order energy (W/Z^s) and effective number of A-B contacts at the surface (N_{AB}), as shown in equation (8):

$$E' = N'W[pC_A^sC_B^s + q\{C_AC_B^s + C_B(C_A^s - C_A)\}] \tag{8}$$

Equation (7) is solved by substituting for the value of E' from equation (8), replacing the sum by the greatest term and equating the first differential of Ξ' with respect to C_A^s to zero ($\delta\Xi'/\delta C_A^s = 0$), i.e.,

$$\Xi' - [\lambda_A q_A^S(T) \exp\{-[p(C_B^S)^2 + q(C_B)^2]\frac{W}{\kappa_B T}\}/C_A^s]^{N\,s} \tag{9a}$$

$$- [\lambda_B q_B^S(T) \exp\{-[p(C_A^S)^2 + q(C_A)^2]\frac{W}{\kappa_B T}\}/C_A^s]^{N\,s} \tag{9b}$$

Standard thermodynamic relations and expressions for surface tension of binary alloys

Ξ' is related to the surface tension of the alloy (σ) and surface area (A) through the relation,

$$\Xi' = \exp(-\delta A/K_B T) = \exp(-N'\sigma\alpha/K_B T) \tag{10}$$

where α ($= A/N'$) is the mean atomic surface area.

Further, it is very simple to express surface tension of the i-component (σ_i) in terms of the atomic partition function of the bulk and the surface as

$$\sigma_i = (K_B T/\alpha) \cdot \ln[q_i(T)/q_i'(T)] \tag{11}$$

Using equations (3), (9), (10) and (11), we obtain a pair of equations for surface tension of the alloy, viz:

$$\sigma = \sigma_A + [K_B T/\alpha]\ln[C_A/C_A^s] + [K_B T/\alpha]\ln[Y_A^s/Y_A] \tag{12a}$$

$$= \sigma_B + [K_B T/\alpha]\ln[C_B/C_B^s] + [K_B T/\alpha]\ln[Y_B^s/Y_B] \tag{12a}$$

Expression for surface tension of compound forming alloys

In compound forming alloys, the constituent atoms A and B of a binary molten alloy AB preferentially arrange to form complex $A_\mu B_\nu$, i.e.

$$\mu_A + \nu_B \rightleftharpoons A_\mu B_\nu \tag{13}$$

where μ and ν are small integers. Thus a compound forming alloy can be understood to consist of mixtures of A atoms, B atoms and number of chemical complexes $A_\mu B_\nu$, all in equilibrium. Bhatia and Singh[26] solved for the bulk grand partition function (equation 4) by assuming that if ε_{ij} ($i = A, j = B$) is the energy of the i-j bond when it is free, then $\varepsilon_{ij} + \Delta\varepsilon_{ij}$ stands for the energy when i-j bond is one of the bond in the complex $A_\mu B_\nu$. On this basis, the bulk configurational energy, E, becomes:

$$E = \sum \varepsilon_{ij} + P_{ij}\,\Delta\varepsilon_{ij} \tag{14}$$

where P_{ij} denotes the probability that i-j bond in the bulk is part of the complex and can be easily determined for different sets of μ and ν by using simple arguments.

On the basis of above model, they obtained an expression for the ratio of bulk activity coefficients $Y(Y_A/Y_B)$ which is related to excess free energy of mixing, G_M^{XS}, as follows:

$$\frac{G_M^{XS}}{NK_BT} \; \cdot \; \int_0^c \ln Y \, dc \tag{15a}$$

$$= Z\int[\ln h + (ZK_BT)^{-1}\,(P_{AA}\Delta\varepsilon_{AA} - P_{BB}\Delta\varepsilon_{BB})]d_c + \phi c \tag{15b}$$

where ϕ is a constant independent of concentration but may depend upon the temperature and pressure. It is determined from the requirement that $G_M^{XS} = 0$ at $C = 0$ and $C = 1$, and h is given by

$$h = \frac{2C_A - 1 + \beta}{2C_A\eta} \tag{16}$$

with

$$\beta = (1 + 4C_AC_B\,(\eta^2 - 1)^{\frac{1}{2}} \tag{17}$$

and

$$\eta^2 \; \cdot \; \exp(\frac{2W}{ZK_BT})\exp[\frac{2P_{AB}\Delta e_{AB} - P_{AA}\Delta e_{AA} - P_{BB}\Delta e_{BB}}{K_BT}] \tag{18}$$

Further, G_M^{XS} is related to the activity coefficient Y_A and Y_B through the relations

$$\ln Y_A = f(c) + Cbf'(c)$$

$$\ln Y_B = f(c) - C_Af'(c) \tag{19}$$

where

$$f(c) = G_M^{XS}/NK_BT$$

and

$$f'(c) = df/dc \tag{20}$$

Various expressions for bulk activity coefficients (Y_A and Y_B) are obtained for

231

different sets of μ and ν by using equations (15) to (19). For the 1:1 complex AB,

$$\mu = 1 \quad \text{and} \quad \nu = 1$$

and $$P_{AB} = 1 \quad \text{and} \quad P_{AA} = P_{BB} = 0$$

Hence

$$\frac{G_M^{xs}}{NK_BT} = \frac{Z}{2}(C_A \ln(\frac{\beta-1+2C_A}{C_A(1+\beta)}) + C_B \ln(\frac{\beta+1-2C_A}{C_B(1+\beta)})) \tag{21}$$

with

$$\eta^2 = \exp\left[\frac{2(W + Z\Delta\varepsilon_{AB})}{ZK_BT}\right] \tag{22}$$

and

$$\ln Y_A = Z/2 \ln\left[\frac{\beta-1+2C_A}{C_A(1+\beta)}\right] \tag{23}$$

$$\ln Y_B = Z/2 \ln\left[\frac{\beta+1-2C_A}{C_B(1+\beta)}\right] \tag{24}$$

where β is given by equation (17) with η defined by equation (22). For conformal solution (Zeroth approximation), equations (23) and (24) reduce to

$$\ln Y_A = C_B^2 \cdot \frac{(W + Z\Delta\varepsilon_{AB})}{K_BT} \tag{25}$$

$$\ln Y_B = C_A^2 \cdot \frac{(W + Z\Delta\varepsilon_{AB})}{K_BT} \tag{26}$$

Equations (25) and (26) can be used in equation (3) to obtain expressions for activity coefficient at the surface (Y_A^s and Y_B^s). The above expressions obtained are used in equation (12) to obtain expressions for surface tension of the binary alloy when a 1:1 compound ($\mu = 1$, $\nu = 1$) exists.

$$\sigma = \sigma_A + \frac{K_BT}{\alpha} \ln \frac{C_A}{C_A'} + [p(C_B')^2 - (p+q)(C_B)^2] \frac{W+Z\Delta\varepsilon_{AB}}{\alpha} \tag{27a}$$

$$= \sigma_B + \frac{K_BT}{\alpha} \ln \frac{C_B}{C_B'} + [p(C_A')^2 - (p+q)(C_A)^2] \frac{W+Z\Delta\varepsilon_{AB}}{\alpha} \tag{27b}$$

The above pair of equations can be solved numerically to obtain σ and surface composition (C_i') as a function of bulk composition C_i.

Concentration fluctuations in the long wavelength limit at surfaces of binary molten alloys

The bulk theory[22] for molten alloys relates concentration fluctuations $(q \to 0)$ $Scc^b(0)$, free energy of mixing, G_M and activity a_i of i (A or B) component of the alloy through the relation

$$Scc^b(0) = \frac{NK_BT}{(\partial^2 G_M/\partial C^2)_{T,P,N}} \tag{28a}$$

$$= \frac{C_B a_A}{(\partial a_A/\partial C_A)_{T,P,N}} = \frac{C_A a_B}{(\partial a_B/\partial C_B)_{T,P,N}} \tag{28b}$$

The activity a_i is related to the activity coefficient through the relation $a_i = C_i Y_i$. In order to obtain bulk $Scc^b(0)$, we use equation (21) for the expression of G_M in equation (29). This yields

$$Scc^b(0) = \frac{C_A \cdot C_B}{1 + Z/2[(1 - \beta)/\beta]} \tag{29}$$

where β is defined by equations (17) and (22). For an ideal mixture $(W = 0)$, the above equation reduces to

$$Scc^b(0, ideal) = C_A C_B \tag{30}$$

Following bulk theory[21-25], the grand partition function for the surface Ξ', (equation 7) has been solved to obtain an expression for the surface free energy of mixing. This upon using equation (28) yields the following expression for the surface concentration fluctuation, $Scc'(0)$:

$$Scc'(0) = \frac{C_A' C_B'}{1 + Z'/2[(1 - \beta')/\beta']} \tag{31}$$

where $\quad \beta' = \{1 + 4 C_A' C_b' [\exp 2 (\frac{W + Z\Delta \epsilon C_{AB}}{K_B T}) - 1]\}^{\frac{1}{2}}$ \hfill (32)

The surface concentration fluctuation for an ideal mixture is thus simply

$$Scc'(0, ideal) = C_A' C_B' \tag{33}$$

Results and Discussion

Equation (27) been utilised to evaluate surface tension (σ) and surface composition (C_i^s) of Au-Zn molten alloys where 1:1 complex (AuZn) formation has been found to exist at equiatomic composition as a function of bulk concentration (C_i). The basic inputs for the above computation are the energy parameter $W = Z\Delta\varepsilon_{AB}$, surface coordination fractions p and q, atomic surface area, α, and surface tension of pure components (Au and Zn). In principle, ($W + Z\Delta e_{AB}$) can be determined by using pairwise interactions of the pseudopotential method [27,28], but very little work has been done in this direction. Instead, a simpler method[21] has been used to evaluate energy parameters from the observed free energy of mixing or activity data at $C = 0.5$. For equiatomic composition, G_i^{XS} (equation 21) reduces to

$$\frac{G_M^{XS}}{NK_BT} = \ln 2^{Z/2}[1 + \exp(\frac{-(W + Z\Delta\epsilon_{AB})}{ZK_BT})]^{-Z/2} \tag{34}$$

where Z is usually determined from the neutron or X-ray diffraction experiment. Since atoms in the liquid and amorphous states are randomly distributed in closed-packed structures and the mean free path is short and comparable to atomic size, the value of Z has been suggested to be 10. The observed value of G_M^{XS} (equation 28) has been used in equation 34 to evaluate $W + Z\Delta\varepsilon_{AB}$ at 1080 K.

$$\frac{W + Z\Delta\varepsilon_{AB}}{K_B T} = -7.6936 \text{ (at 1080 K)}$$

Surface coordination fractions p and q are treated as parameters and taken at $p = \frac{1}{2}$ and $q = \frac{1}{4}$. For the evaluation of mean atomic surface area, α, the following equation[30] is used:

$$\alpha = 1.102 \times N^{-2/3}\Omega \tag{35}$$

where Ω is volume of the alloy.

The surface tension of pure components are taken from Smithel's reference book, viz

$$\sigma_{Au} = 1.140 \text{ Nm}^{-1}$$

$$\sigma_{Zn} = 0.716 \text{ Nm}^{-1}$$

The computed values of surface composition (C_i^s) and surface tension (σ) have been graphically presented in Figures 1 and 2, respectively, as a function of bulk concentration (C_i).

For the evaluation of concentration fluctuations ($Scc^b(0)$ and $Scc^s(0)$) we have used equations (29) and (31). The energy parameter, $W + Z\Delta\varepsilon_{AB}$, required for this is the same as that used for the computation of σ and C_i^s. Z^s has been determined for surface coordination fractions (equation 2). C_i^s has been taken from the solution of equation (27). The computed values of $Scc^b(0)$ and $Scc^s(0)$ of Au-Zn molten alloys are plotted as a function of bulk concentration (C_i) in Figure 3. In principle, $Scc^b(0)$ may be determined experimentally from the knowledge of partial structure factors at the long wavelength limit, but evaluation of $Scc(0)$ experimentally from diffraction experiments continue to elude us. Instead, $Scc(0)$ determined from the numerical differentiation [equation (28b)] of measured activity data is accepted as the experimental value. We have compared our theoretical results with the experimental values of $Scc(0)$ obtained from above method.

Our results suggest that Zn atoms segregate at the surface for all bulk compositions of Au-Zn molten alloys (Fig. 1). The segregation is more visible towards Zn-rich ends. The surface tension of the alloys (σ) (Fig. 2) has been found to be greater than the ideal value σ_{id} (alloy) [$= C_{Au}\sigma_{Au} + C_{Zn}\sigma_{Zn}$] in the concentration range of $0 < C_{Zn} < 0.77$ but less than the ideal value in the range $0.77 < C_{Zn} < 1.0$.

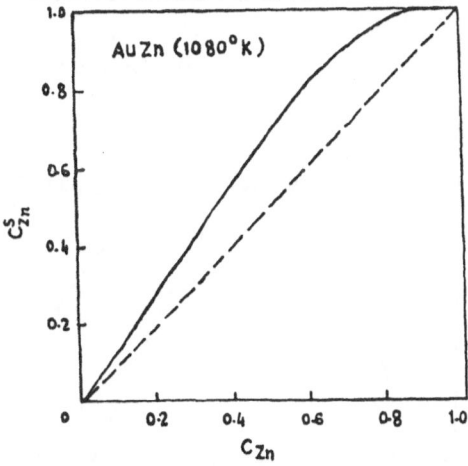

Fig. 1 : Surface concentration (C_{Zn}^s) vs bulk concentration (C_{Zn}) for Au-Zn system. (—) refers to present theory

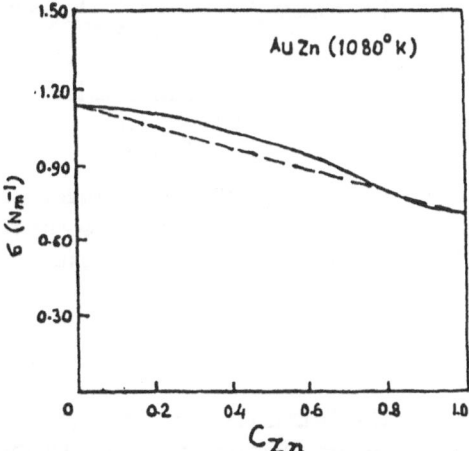

Fig. 2 : Surface tension (σ) vs bulk concentration (C_{Zn}) for Au-Zn system. (—) and (— —) refer to present theory and ideal values, respectively.

Our results for Scc(0) indicate that the concentration fluctuations at the surface (Fig. 3) for this system differ in a major way from the bulk values towards Zn rich end. The maximum of $Scc^s(0, ideal)$ shifts towards Au rich end. It is also interesting to note that $Scc^s(0)$ do not differ appreciably from the bulk values. It may be mentioned that $Scc^s(0) < Scc^s(0, ideal)$ indicates heterocoordination (preference for unlike atoms pairing) in contrast to the earlier observation of surface composition, where there is evidence of surface segregation. This may be explained by the fact that surface concentration (C_i^s) is obtained using equation (27) where the energy parameter and surface tension control surface segregation. By way of contrast, our expression for $Scc^s(0)$ involves only the energy parameter. This suggests that surface segregation in the Au-Zn system mainly occurs due to the surface tension effect. As the energy parameter has a negative value, $Scc^s(0)$ is not greater than $Scc^s(0, ideal)$.

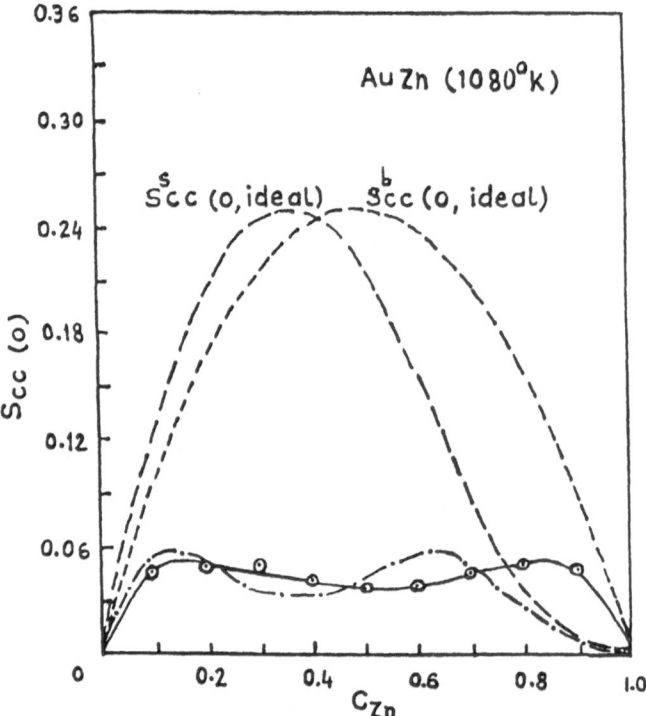

Figure 3 : Concentration fluctuations [Scc(0)] vs bulk concentration (C_{Zn}) for Au-Zn system: —— $Scc^b(0)$ theory; ⊙⊙⊙ experimental points[29] for $Scc^b(0)$; -•-•-•- $Scc^s(0)$ theory; ---- refers to ideal values

Acknowlegement

One of the authors (LCP) is thankful to CSIR (Council of Scientific and Industrial Research) India for providing financial support to the research work vide letter no. 03(0728)93/EMRII.

References

1. M.P. Seah, *Surf. Sci.* **53** (1975) 168.
2. J.J. Burton and E. Hyman, *J. Catal.*, **3** (1975) 114.
3. W.M.H. Sachter and P. Vander Plank, *Surf. Sci.*, **18** (1969) 62.
4. J.H. Sinfelt, J.L. Carter and D.J.C. Yates, *J. Catal.*, **24** (1972) 280.
5. J.W. Gibbs, *Trans. Conn. Acad. Arts Sci.*, **3** (1875/76) 108, **3** (1877/78) 343.
6. C.R. Helms, *J. Catal.*, **36** (1975) 114.
7. K. Watanabe, M. Hashiba and T. Yemashina, *Surf. Sci.*, **61** (1976) 483.
8. P.J. Durhan, R.G. Jordan, G.S. Sohal and L.T. Wille, *Phys. Rev. Lett.*, **53** (1984) 2038.
9. K. Wandelt and C.R. Brundle, *Phys. Rev. Lett.*, **46** (1981) 1529.
10. H.H. Brongersma, M.J. Sparnaay and T.M. Bunch, *Surf. Sci.*, **71** (1978) 657.
11. Y.S. Ng, T.S. Tsong and S.B. McLane, Jr., *Phys. Rev. Lett.*, **42** (1979) 588.
12. F.L. Williams and D. Nason, *Surf. Sci.*, **45** (1974) 377.
13. W.F. Egelhoff, Jr., *Phys. Rev. Lett.*, **50** (1983) 587.
14. S.H. Overbury, P.A. Bertrand and G.A. Somorjai, *Chem. Rev.*, **75** (1975) 547.
15. G. Kerker, J.L. Moran Lopez and K.H. Bennemann, *Phys. Rev.*, **B15** (1977) 638.
16. E.A. Guggenheim, Mixtures, Oxford University Press, Oxford, England, 1952.
17. K.S. Yeum, R. Speiser and D.R. Poivier, *Metall. Trans.*, **20B** (1989) 693.
18. D.R. Poivier and R. Speiser, *Metall. Trans.*, **18A** (1987) 1156.
19. Yi-Chen Cheng, *Phys. Rev.*, **B28** (1983) 2937.
20. L.C. Prasad and R.N. Singh, *Phys. Rev.*, **B44** (1991) 24, 13768.
21. R.N. Singh, I.K. Mishra and V.N. Singh, *Phys. Rev. Condens. Matter*, **2** (1990) 8457.
22. R.N. Singh, *Can. J. Phys.*, **65** (1976) 309.
23. A.B. Bhatia and D.E. Thornton, *Phys. Rev.*, **B2** (1970) 3004.
24. R.N. Singh and I.K. Mishra, *Phys. Chem. Liq.*, **18** (1988) 303.
25. L.C. Prasad and R.N. Singh, *Phys. Chem. Liq.*, **22** (1990) 1.
26. A.B. Bhatia and R.N. Singh, *Phys. Chem. Liq.*, **11** (1982) 285, 343.
27. H.J. Hafner and V. Heine, *J. Phys. F. Met. Phys.* **13** (1983) 2479.
28. F.A. Khvaya, A.A. Katnelson, V.M. Silonov and M.M. Khrushchov, *Phys. Stad. Sol.*, **82b** (1977) 701.
29. R. Hultgren, P.D. Desai, D.T. Hawkins, M. Gleiser and K.K. Kelley, Selected values of the Thermodynamic Properties of Binary Alloys (American Society for Metals, Metal Park OH) 1973.
30. P. Laty, J.C. Joud and P. Desre, *Surf. Sci.*, **60** (1976) 109.

Main Group Elements and Their Compounds
V.G. Kumar Das (Ed)

Photoelectrochemistry of H₂- and Zn-based sulfophthalocyanines in an Ion-exchange Polymer Blend

Ishmael D. Ordoñez[a] and Marcus F. Lawrence[b]

[a]Institute of Chemistry, University of the Philippines, Diliman, Quezon City 1101, Philippines, [b]Department of Chemistry and Biochemistry, Concordia University, 1455 de Maisonneuve Blvd. W., Montréal, Québec, Canada H3G 1M8

Dye sensitization of semiconductors to visible light has been of interest to exploit the use and application of the durable and inexpensive, albeit large bandgap, metal oxide semiconductor materials (e.g. SnO_2, TiO_2, ZnO) for solar energy conversion devices and/or photocatalysis.[1-5] Phthalocyanines show potential as dye sensitizers since they absorb strongly in the 600-700 nm region (Q band) of the visible spectrum and are relatively inexpensive and chemically stable. However, phthalocyanines are insoluble in many common laboratory solvents which necessitated the use of the water soluble anionic tetrasulfonated analogs (MPcTS). In this context, polymers have also gained attention for the chemical modification of electrodes.[6-10] Our interest is to utilize conducting polymers as a support material for immobilizing the dyes near the semiconductor electrode surface. The ion-exchange polymer blend used in this study possesses positively charged sites which provide electrodes with surfaces having a high affinity for anionic redox reactants. Under illumination, the dye/polymer system should ideally be capable of performing the spatial separation of photogenerated charges using both ionic and electronic conduction mechanisms.

Previous studies from our laboratory proposed a model based on the energetic relation between the dye, polymer and semiconductor to account for observed photoinduced charge migration.[1,2,11] The results presented in this article reflect a systematic dependence of the photoelectrochemical behaviour of the polymer/MPcTS (M = Zn, H_2) films on SnO_2 electrodes on the photophysical properties of the dyes.

Experimental

Materials

Synthesis of sulfonated metallophthalocyanines [MPcTS⁴⁻]

The ammonium salts of MPcTS⁴⁻ were synthesized according to the method developed by Weber and Busch.[12] The products were repeatedly recrystallized and washed in 95% ethanol until a uniform blue colour was observed. The MPcTS^{4-}'s were determined by UV-vis spectroscopy and compared to literature.[5] The chemical formula of MPc[SO$_3^-$NH$_4^+$]$_4$ is shown in Fig. 1A.

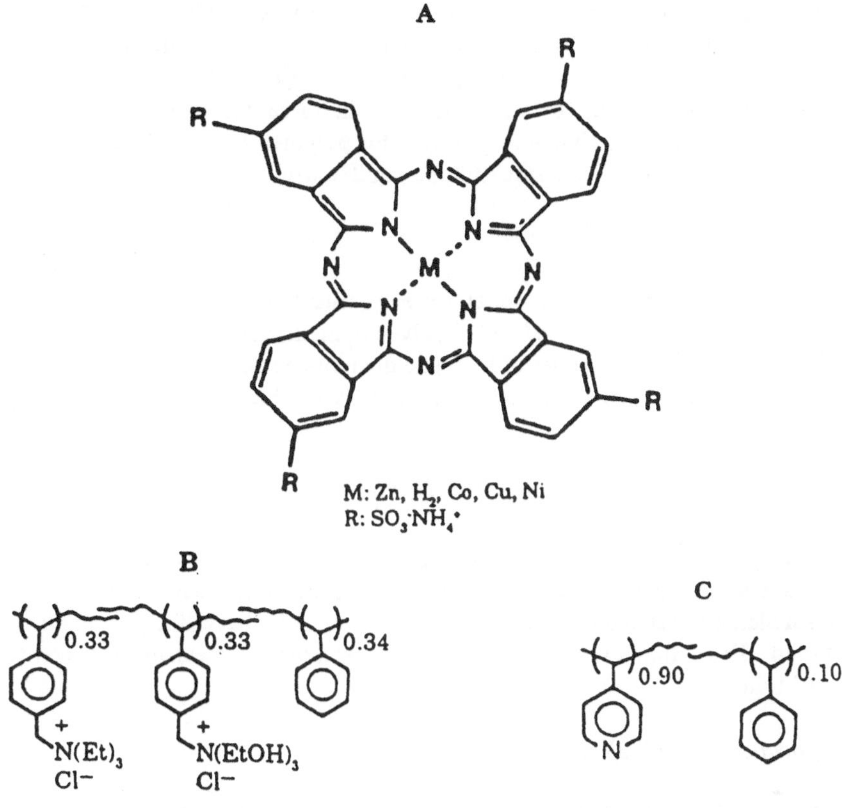

A

M: Zn, H$_2$, Co, Cu, Ni
R: SO$_3^-$NH$_4^+$

B

C

Fig. 1 : **Molecular structures of the metallophthalocyanine tetrasulfonates and polymers used in this study: (A) MPcTS^{4-}; (B) random quaternary copolymer; (C) poly[(4-vinylpyridine)-co-styrene] (PVP)**

Synthesis of H$_2$PcS

H$_2$PcS was prepared according to the procedure of Linstead and Weiss.[13] H$_2$Pc (2 g, Aldrich, 98%) was refluxed in 40 mL H$_2$SO$_4$ (20% free SO$_3$, Aldrich) at 80-85 °C with stirring for 14 h. The brown fuming solution was poured slowly over crushed ice (100 g) while maintaining the solution temperature close to 0 °C during neutralization. The solution was filtered to remove any water insoluble residue and boiled down to preconcentrate the product. Excess salts formed during the neutralization and lower molecular weight impurities were removed by gel chromatography (Sephadex G-10, Pharmacia). The product was then recrystallized and purified using the same procedure as for the MPcTS's. According to this synthetic procedure, the presence of mono- up to tetra-sulfonated H$_2$Pc's cannot be excluded.

Preparation of random copolymer

The preparation of the random ternary copolymer has been described in detail previously.[1] Elemental analysis performed on the final product used in this study gave the following weight percentages: C (69.37%), H (8.82%), N (4.38%), Cl (10.37%) and O (7.06% by difference), which agree reasonably well with the theoretical percentages of C (69.20%), H (8.57%), N (4.24%), Cl (10.76%) and O (7.23%) that one expects for a 1:1:1 ratio of the three different pendant groups present in the ternary copolymer (Fig. 1B). This polymer was dissolved in spectral grade methanol to make up a stock solution of 2% w/v.

Preparation of polymer blend

Poly[(4-vinylpyridine)-co-styrene], styrene content 10% (PVP, Aldrich) shown in Fig. 1C, was used as received and dissolved in methanol to give a stock solution of 2% PVP w/v. Equal volumes of PVP and quaternized copolymer solutions were thoroughly mixed together to constitute the polymer blend referred to below as *polyXIO* for simplicity.

Preparation of dye-polymer blend (MPcTS^{4-}/polyXIO)

The dye was first dissolved in water to give a concentration of *ca.* 20 mM. An equal amount of methanol was added to further dilute by one-half. The volume of this solution that was added to the polymer blend was 10% that of the polymer blend solution. The solution was then vigorously shaken using a mechanical shaker for *ca.* 15 min.

Preparation of SnO$_2$ electrodes

Rectangular SnO$_2$ optically transparent electrodes (1.5 x 5.0 x 0.32 cm) were cut from SnO$_2$ coated IR reflective Pyrex plates (Swift Glass Co.). The SnO$_2$ had typical resistances of *ca.* 30 ohms/cm. Prior to use they were soaked in sulfochromic acid for 20 min (Chromerge, Fisher). After throughly rinsing with distilled water the electrodes were dried in a stream of prepurified nitrogen gas.

Preparation of MPcTS^{4-}/polyXIO films (electrode modification)

Depending on the desired film thickness, films were cast by depositing 0.1 to 1.0 mL of the prepared dye-polymer blend solution onto the SnO$_2$ OTE surface. To prevent excessive use of the casting solutions, barriers made from epoxy were created in the middle of the SnO$_2$ so that only one-half of the electrode was coated with dye-polymer blend. All the modified electrodes were prepared under an inverted crystallization dish in the presence of CaCl$_2$ and the resulting films had uniform thicknesses in the range of 1-20 μm. Thicknesses were measured with a Dek-Tak surface profilometer (Sloan). Prior to use, the outer edges of the modified electrodes were coated with clear nail polish leaving an area of *ca.* 1 cm^2 which was exposed to solution.

Apparatus and Procedures

Photoelectrochemistry (PEC)

The photoelectrochemical cell was made of Pyrex with a quartz glass window for electrode illumination and a side chamber for the Ag/AgCl, saturated KCl reference electrode. The counter electrode was Pt foil. The aqueous solutions were 1.0 M KCl and 2 mM $K_3Fe(CN)_6$/2 mM $K_4Fe(CN)_6$ in 1.0 M KCl (pH 6.7). The PEC cell was completely encased in an aluminum Faraday cage.

Potential were controlled with an EG&G Princeton Applied Research Scanning Potentiostat (Model 362). Cyclic voltammograms were recorded on an X-Y recorder (HP 7004 B). Photocurrent measurements were monitored by a Keithley Programmable Electrometer (Model 617) coupled to a strip chart recorder (Kipp and Zonen BD 40).

The light source used was a Kodak (Model 4200) slide projector. A 12 cm long water cell was employed as an IR filter in between the light source and focusing lens. Most irradiations used a 495 nm cutoff filter. The position of the light source and lens was adjusted so that the light intensity was 80 mW cm^{-2}. Light intensities were measured using a Photodyne (Model 88XLC) photometer/radiometer or with a Coherent Power Meter (Model 210). Absorption spectra were obtained using a HP 8452 UV-vis diode array spectrophotometer. In all cases, illumination was initiated from the dye/polymer film side of the modified electrodes. All measurements were carried out at room temperature.

Fluorescence measurements

MPcTS^{4-} solutions in methanol were prepared to give a dye concentration of 4 x 10^{-5} M. The corresponding MPcTS^{4-}/polyXIO solutions described previously were also diluted with methanol to give a concentration of 4 x 10^{-5} M of dye solution and MPcTS^{4-}/polyXIO films were cast on 1.0 x 5.0 x 0.32 cm SnO$_2$ OTE's in the usual manner. The same experiments were performed using quartz plates as substrates.

Solution and solid state fluorescence measurements were obtained using a Perkin Elmer MPF 44B spectrofluorimeter. Solid state measurements required a special sample holder allowing the incident excitation wavelength to impinge at a 30° angle to the surface of the sample.

Fluorescence decay for both solutions and solids were measured at the Canadian Centre for Picosecond Flash Photolysis located at Concordia University. The excitation beam was a 30 ps pulse at 355 nm obtained from a neodymium-doped YAG laser. Fluorescence traces were monitored with a streak camera (Hammamatsu) having a resolution of 10 ps. Glass filters were placed before the entrance slit of the streak camera to reject scattered radiation. All fluorescence measurements were obtained at room temperature.

Electrochemistry of MPcTS^{4-}

Voltammetric studies were carried out using a SnO$_2$ OTE as the working electrode. Cleaning of the SnO$_2$ OTE and the electrochemical apparatus used are

241

as described in the foregoing sections. Cyclic voltammograms were measured in *ca*. 1.5 x 10^{-4} M solutions of MPcTS^{4-} (1 M KCl) at pH 6.7. All solutions were deoxygenated by bubbling with prepurified nitrogen gas for at least 30 min.

Results and Discussion

Photoelectrochemistry

The absorption spectra of typical MPcTS^{4-}/polyXIO films on SnO$_2$ OTE's are shown in Fig. 2 for ZnPcTS^{4-} and H$_2$PcS. The appearance of a sharp peak at around 684 nm in the absorption spectrum of ZnPcTS^{4-}/polyXIO films, for example, is attributed to the monomeric form of the dye,[5] which suggests that the pyridine groups in PVP may be axially ligated to the metal centres of the dyes. This behaviour was also noted when a small amount of pyridine (5% v/v) was added to aqueous solutions of the MPcTS^{4-}'s leading to a similar sharp peak in the 670 to 680 nm range.

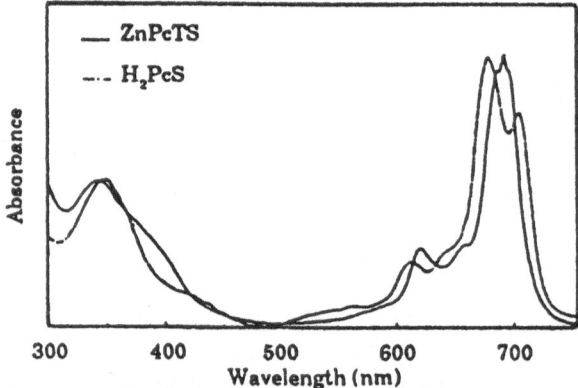

Fig. 2 : **Absorption spectra of ZnPcTS^{4-} and H$_2$PcS incorporated in the polymer blend as a film on SnO$_2$ OTE. The absorbance scale is arbitrary.**

Films prepared by slow solvent evaporation of the dye/polymer blend solutions deposited onto the SnO$_2$ electrode surface produced visually smooth solid films. These films are translucent and retain remarkable solution-like visible absorption properties as seen in Fig. 2. The ion-exchange polymer blend itself does not absorb in the visible and provides a robust support for these anionic dyes giving the modified electrodes durability.

The dark electrochemical behaviour of these MPcTS^{4-}/polymer blend films is similar to that described in previous articles dealing with different dyes[1,2] where the systems were shown to exhibit quasi-reversible voltammograms for hexacyanoferrate (II, III). The transport of Fe(CN)$_6^{3-/4-}$ ions in the film is diffusion controlled and exhibits nearly solution like behaviour.

The intense absorption band in the red region of the visible spectrum (Q band) of the MPcTS's makes them good candidates for photosensitization. To this extent, light of wavelengths greater than 495 nm was employed through the use of a filter to ensure that only the Q band of the MPcTS's was excited and also to prevent any

242

excitation of the SnO_2.

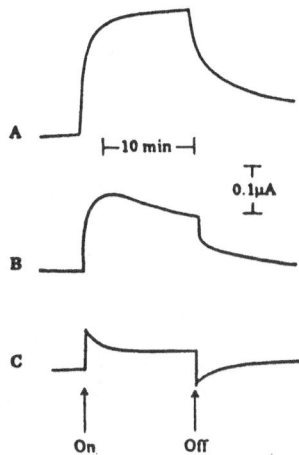

Fig. 3 : Different types of short circuit photocurrent responses obtained as a function of illumination time for the various MpcTS^{4-}/polymer blend modified SnO$_2$ OTE's in contact with 2 mM Fe(CN)$_6^{3-/4-}$ (1 M KCl) under illumination with light >495 nm: (A) ZnPcTS^{4-} and H$_2$PcS; (B) CoPcTS^{4-}; (C) CuPcTS^{4-} and NiPcTS^{4-}

Fig. 3 shows the types of short-circuit photocurrent-time responses observed with the different MPcTS^{4-}/polyXIO films on SnO$_2$ OTE when illuminated (>495 nm) in contact with 2 mM Fe(CN)$_6^{3-/4-}$ (1 M KCl). In all cases the photocurrents were anodic corresponding to electron injection into the SnO$_2$ conduction band. Fig. 3A is typical of the short-circuit photocurrent response observed for ZnPcTS and H$_2$PcS systems which exhibit a delay in reaching maximum photocurrent during the on-off light sequence. This may indicate that a key event in the production of photocurrent of the dye/polyXIO structures of Fig. 3A could be the formation of a charge-transfer state between the dye and a coordinated pyridine group.[1,14] Further separation of charge following the formation of (dye$^+$-py$^-$) ion pairs would then lead to migration of electrons to the SnO$_2$ electrode through hopping between fixed redox sites in the pendant group polymer[15,16] and migration of holes that end up oxidizing the Fe(CN)$_6^{4-}$ species present in the electrolyte. The reduction of Fe(CN)$_6^{3-}$ at the Pt counter electrode thus completes the light driven cycle. This cycle is presented schematically in Fig. 4. Recently, a submitted study concerning photocurrent generation and charge transport in dry SnO$_2$/polyXIO-ZnTPP/Au cells[11] determined that the photocurrent-time responses in this case paralleled those obtained with the wet cells as shown in Fig. 3A. The time constants for dry and wet cells using dyes that produce stable photocurrents are not experimentally distinguishable. This lends further support for the description of charge transport in these systems as being controlled by hopping between fixed sites associated to the polymer (pendant groups) rather than by diffusion of redox species (Fe(CN)$_6^{3-/4-}$) through the hydrophilic domains of the polymer film. This also tends to rule out the possibility of direct electron transfer to Fe(CN)$_6^{3-}$ as being responsible for the photocurrent generation.

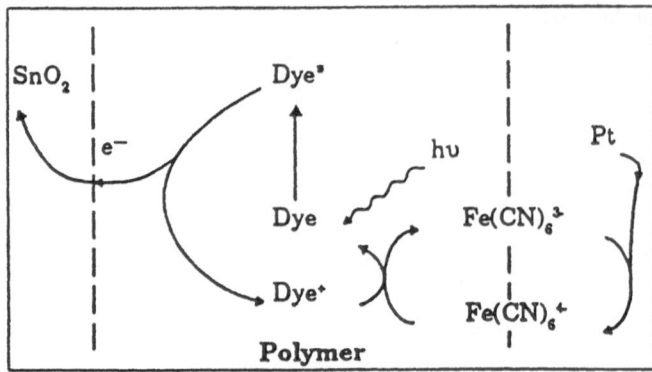

Fig. 4 : Schematic representation of the photoinduced electron flow originating at the irradiated MPcTS^{4-}/polymer blend modified SnO$_2$ OTE in contact with Fe(CN)$_6^{3-/4-}$ redox couple

Fig. 3C, on the other hand, shows the type of short-circuit photocurrent response obtained for CuPcTS^{4-} and NiPcTS^{4-} systems. A charging and discharging type of transient behaviour is observed during the on-off light sequence. The response shown in Fig. 3C would also indicate that the CuPcTS^{4-} and NiPcTS^{4-} molecules throughout the bulk of the polymer film are unable to contribute to a steady-state photocurrent when excited in the red.

Fig. 3B shows the response obtained for the CoPcTS^{4-}/polyXIO system. The appearance is intermediate between that obtained for ZnPcTS^{4-} and H$_2$PcS systems (Fig. 3A) and the response obtained for the CuPcTS^{4-} and NiPcTS^{4-} systems (Fig. 3C). This response may be the result of oxidation of the metal centre to Co(III) which is a rather stable species that can accumulate in the system. The photocurrent response in this case would be expected to show an initial rise followed by a slow decrease. When the light is switched off the current decreases and continues to flow in the same direction until the cobalt centres have been reduced again. For these reasons, only the Zn- and H$_2$-based sulfophthalocyanines were studied further.

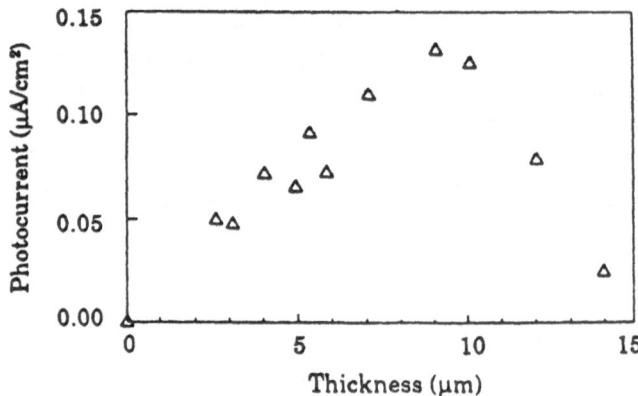

Fig. 5 : Maximum short-circuit photocurrent as a function of film thickness for ZnPcTS^{4-}/polyXIO films in contact with 2 mM Fe(CN)$_6^{3-/4-}$ redox couple

In Fig. 5 the maximum short-circuit photocurrents are plotted as a function of film thickness for ZnPcTS^{4-}/polyXIO films. The photocurrents increase linearly up to a thickness of about 9 μm which indicates that photocurrent generation is originated within the bulk of the dye/polymer films as well as the film-electrolyte interface. Absorbance measurements taken at 684 nm for these films show that practically 100% of the light is absorbed when the film thickness reaches 9 μm.

Energetic Considerations

Photophysical properties of the MPcTS^{4-} dyes

In Table 1, experimental values of the relevant photophysical properties for the two dyes showing good photocurrent-time responses (ZnPcTS^{4-} and H$_2$PcS, Fig. 3A) are compared to those of a dye presenting a poor response (CuPcTS^{4-}, Fig. 3C). Other values in Table 1, such as fluorescence and triplet quantum yields, were taken from available data in the literature.[5,17] Fluorescence measurements show that only ZnPcTS^{4-} and H$_2$PcS show strong singlet emission at around 700 nm, whereas CuPcTS^{4-}, NiPcTS^{4-} and CoPcTS^{4-} showed very little, if any, singlet emission in this region. Looking at the quantum yield for triplet formation, CuPcTS^{4-} has a value of 0.92. Since NiPcTS^{4-} and H$_2$PcS show strong singlet emission at around 700 nm, whereas CuPcTS^{4-}, NiPcTS^{4-} and CoPcTS^{4-} do not fluoresce at 700 nm it is assumed that their yields for triplet formation would also be quite high; ZnPcTS^{4-} and H$_2$PcS, on the other hand, show appreciable quantum yields for fluorescence (0.32 and 0.62, respectively).

The lifetimes of the singlet states were measured from picosecond emission decay measurements which reveal that ZnPcTS^{4-} and H$_2$PcS have singlet lifetimes of 2.7 and 3.4 ns respectively in methanol alone (Table 1). In methanol solution with polyXIO present, the lifetimes decrease to about 2 ns and the fluorescence maxima are red-shifted. These lifetimes are further shortened to 40-50 ps when these dyes are immobilized in the polymer blend as a solid film on SnO$_2$OTE. Similar experiments performed with the dye/polymer films cast onto quartz substrates indicated that the SnO$_2$ substrates had no influence on observed lifetimes.

From these results, it can be seen that the ability of the dyes such as ZnPcTS^{4-} and H$_2$PcS to fluoresce provides some insight into the results obtained for photocurrent generation. The energy available from the first excited singlet state is higher than the energy available from the first excited triplet state by about 0.6 eV. Piechowski and co-workers[18] correlated solution fluorescence yield with photovoltaic cell performance in the course of their evaluation of cyanine and merocyanine dyes. They determined that it was advantageous to use dyes which do not have a facile route for direct internal conversion from the first excited singlet-state to the ground state and that, in general, good materials will exhibit a high quantum yield of fluorescence in fluid solutions at room temperature. This would seem to be the case for ZnPcTS^{4-} and H$_2$PcS in explaining their ability to produce photocurrents as compared to the MPcTS^{4-}'s that do not fluoresce (M = Cu, Ni and Co). Darwent et al.[5] assert that the utility of the excited singlet-state of the MPcTS^{4-} for photoredox processes necessitates very high concentrations of quencher before intermolecular quenching can compete with the intrinsic deactivation of the excited singlet-state.

Table 1 : Photophysical properties of the MPcTS^{4-} dyes under various conditions

Compound	$\gamma_{emission}$ (nm)	Γ_s (ns)	ϕ_F^a	Γ_T^a (μs)	ϕ_T^a	E_T^a (eV)	E_S^b (eV)
ZnPcTS^{4-} in MeOH	690	2.7	0.32	245	0.56	1.12	1.80
ZnPcTS^{4-}/polyXIO	699	2.0	-	-	-	-	1.77
ZnPcTS^{4-}/polyXIO/SnO$_2$	700	0.042	-	-	-	-	1.77
H$_2$PcS in MeOH	688	3.4	0.62	170	0.22	1.24^d	1.80
H$_2$PcS/polyXIO in MeOH	705	1.8	-	-	-	-	1.76
H$_2$PcS/polyXIO/SnO$_2$	709	0.05	-	-	-	-	1.75
CuPcTS^{4-} in MeOH	-	-	<10^{-4}	0.065	0.92	1.16	1.85
CuPcTS^{4-}/polyXIO in MeOH	-	-	-	-	-	-	-
CuPcTS^{4-}/polyXIO/SnO$_2$	-	-	-	-	-	-	1.84^c

^{a}Ref. 5.
^{b}calculated from fluorescence wavelength maxima
^{c}estimated from absorption wavelength maxima
^{d}value taken from unsubstitued H$_2$Pc data from Ref. 5

A modest decrease in fluorescence intensity and, coincidentally, fluorescence lifetime (Table 1) was observed for ZnPcTS^{4-} and H$_2$PcS in the presence of polyXIO in solution. In the solid state, the lifetimes are more profoundly shortened for these dyes. Since the MPcTS^{4-}/polyXIO solutions used for the fluorescence measurements were diluted working solutions for electrode modification, the number of MPcTS^{4-} molecules per unit polyXIO should be the same in the solid state as in solution. In the solid state, the MPcTS^{4-} molecules are in a position where spatial separation from the polymer is very much smaller as compared to the situation in solution. The solid state is akin to having a high concentration of polymer surrounding the MPcTS^{4-}. One possible explanation for the quenching of the MPcTS^{4-} (M = Zn, H$_2$) singlet is that the rapid decay of fluorescence intensity may be associated with the rate of photoinduced transfer of an electron from the MPcTS^{4-} excited singlet-state to an axially coordinated pyridine group of the polymer matrix creating a charge-transfer state. Although the influence of impurities associated to the polymer cannot be ruled out, aggregation of the MPcTS^{4-} molecules, known to lead to a significant decrease in fluorescence lifetime,[4,5] is not important since the absorption spectra of the MPcTS^{4-}/polyXIO/SnO$_2$ electrodes show a strong and sharp absorption band at around 684 nm which is attributed to the monomeric form of the dye (See Fig. 2).

The rate constants for the non-radiative intersystem crossing process (k_{ISC}) have been calculated by Harriman and Richoux[17] to be 3.9×10^7 s^{-1}, 2.3×10^8 s^{-1} and $> 10^{12}$ s^{-1} for H$_2$PcS, ZnPcTS^{4-} and CuPcTS^{4-} in solution, respectively. A rate constant of $> 10^{12}$ s^{-1} would most likely be the value for NiPcTS^{4-} and CoPcTS^{4-} since they, like CuPcTS^{4-}, exhibit very little fluorescence ($\phi_F < 10^{-4}$) and minimal internal conversion back to the ground state.[4,5] Since the process of direct internal conversion, S$_1 \rightarrow$ S$_0$, is not a significant factor, the main competing process in the formation of a MPcTS^{4-}-polymer charge transfer state would be the rate of intersystem crossing to the triplet state. The rate constants, k$_{ISC}$, reported above for H$_2$PcS and ZnPcTS^{4-} indicate that their excited singlet-state may be reasonably accessible for the formation of a charge-transfer state and, consequently, photocurrent generation in the polymer blend film. Additional support for this interpretation is obtained by determining the positions of the ground and excited state energy levels of the dyes on the electrochemical scale.

Redox potentials of ground and excited state energy levels of the MPcTS^{4-} dyes

The excited state redox potentials for the MPcTS^{4-} dyes used were estimated from equation (1):

$$E(Pc^+Pc_s^*) = E(Pc^+/Pc) - E_s \qquad (1)$$

where $E(Pc^+/Pc_s^*)$ is the oxidation potential of the first singlet-state, $E(Pc^+/Pc)$ is the oxidation potential in the ground state and Es is the singlet energy. Table 2 lists the ground and excited state oxidation potentials of the MPcTS^{4-} dyes used in this study.

Table 2 : Ground and excited state redox potentials (V vs. NHE) for the MPcTS^{4-} dyes used in the ion-exchange polymer blend films where Pc$_s$* refers to the singlet excited state and Pc$_T$* refers to the excited triplet state

Compounds	$E(Pc^+/Pc)$	$E(Pc^+/Pc_S^*)$	$E(Pc^+/Pc_T^*)$
ZnPcTS^{4-}	+0.90[a]	-0.90	0.22
H$_2$PcS	+0.87[a] +1.14[b]	-0.93	-0.37
CoPcTS^{4-}	+0.69[b] +1.16[c]	-1.15	-0.46
CuPcTS^{4-}	+1.19[a] +1.37[c] +1.11[b]	-0.65	-0.03
NiPcTS^{4-}	+0.94[a] +1.22[b]	-0.90	-0.22

[a]measured in 1 M KCl (pH 6.7) using SnO$_2$ as the working electrode
[b]Ref. 19
[c]Ref. 20

The ground state first oxidation potentials of the MPcTS^{4-} dyes were measured in aqueous solution using bare SnO$_2$ OTE's as the working electrode. Other values obtained from available literature are included for comparison which show the differences observed when various electrochemical conditions are used. Rollman and Iwamoto[19] obtained their values using a rotating platinum electrode (rpe) and DMSO as the solvent. Shepard and Armstrong[23] did their measurements on SnO$_2$ in aqueous solutions but primarily modified the electrode surface with organooxysilane derivatives to bind the MPcTS^{4-} to the surface. For purposes of this investigation, the ground state oxidation potentials used in equation (1) were obtained by measuring the half-wave potentials in aqueous solution (1 M KCl) and bare SnO$_2$ since this best matches the conditions prevalent in the photoelectrochemical studies.

As expected, the triplet excited state redox potentials (E(Pc$^+$/Pc$_T$*)) are much lower in energy than those for the corresponding singlet excited-state redox potentials (E(Pc$^+$/Pc$_s$*)). However, both ZnPcTS^{4-} and H$_2$PcS may benefit from a lower intersystem crossing rate to the triplet state enabling the singlet excited state energy to dominate in photocurrent generation. It is worth noting also that according to these measurements, the first excited state redox potential for CoPcTS^{4-} is about 0.25 V more negative than the ones determined for ZnPcTS^{4-} and H$_2$PcS, which may explain its partial success at photocurrent generation even though it is characterized by a high rate of intersystem crossing.

In summary, the photophysical properties of the MPcTS^{4-} dyes play an important role in the photoelectrochemical behaviour of the MPcTS^{4-} loaded ion-exchange polymer blend films used in this study. ZnPcTS^{4-} and H$_2$PcS appear to benefit from lower intersystem crossing rates to the lower energy triplet-state, enabling their more energetic singlet-state to participate in photocurrent generation. Picosecond resolved emission decay measurements indicate that the singlet lifetimes of ZnPcTS^{4-} and H$_2$PcS are profoundly shortened when these dyes are incorporated into the ion-exchange polymer blend as a solid film on SnO$_2$ OTE's. One possible explanation for this quenching is that the rapid fluorescence decay may be related to the rate of electron transfer from the dye's excited singlet-state to the polymer matrix via the pyridine moiety. The singlet excited state energy levels of CuPcTS^{4-}, NiPcTS^{4-} and CoPcTS^{4-} are also located at energies which favour electron transfer to the polymer, however, their high intersystem crossing rates appear to prohibit the more energetic singlet excited state from participating in photocurrent generation.

Acknowledgement

This work was supported under operating and strategic grants from the Natural Sciences and Engineering Research Council of Canada. We also acknowledge the support of the MESS of the province of Quebec.

References

1. A.M. Crouch, I.D. Ordoñez, C.H. Langord and M.F. Lawrence, *J. Phys. Chem.*, **92** (1988) 6058.

2. D.A. Biro and C.H. Langford, *Inorg. Chem.*, **27** (1988) 3601.

3. M. Gratzel, ed., *Engery Resources through Photochemistry and Catalysis*, ·Academic Press, New York, 1983.

4. J. Simon and J.-J. Andre, *Molecular Semiconductors*, Springer-Verlag, Berlin, 1985.

5. J.R. Darwent, P. Douglas, A. Harriman, G. Porter and M.-C. Richoux, *Coord. Chem. Rev.*, **44** (1982) 83.

6. M.S. Wrighton, *Science*, **231** (1986) 32.

7. H.D. Abruña, *Coord. Chem. Rev.*, **86** (1988) 135.

8. T. Inoue and F.C. Anson, *J. Phys.Chem.*, **91** (1987) 1519.

9. K. Sumi and F.C. Anson, *J. Phys. Chem.*, **90** (1986) 3845.

10. D.D. Montgomery and F.C. Anson, *J. Amer. Chem.Soc.*, **107** (1985) 3431.

11. M.F. Lawrence, Z. Huang, C.H. Langford and I.D. Ordoñez, *J. Phys. Chem.*, **97** (1993) 944.

12. J.H. Weber and D.H. Busch, *Inorg. Chem.*, **4** (1965) 469.

13. R.P. Linstead and F.T. Weiss, *J. Chem. Soc.*, (1950) 2975.

14. A.M. Crouch, D.K. Sharma and C.H. Langford, *J. Chem. Soc., Chem. Commun.*, (1988) 307.

15. D.A. Buttry and F.C. Anson, *J. Amer. Chem.Soc.*, **105** (1983) 685.

16. R.W. Murray, In: *Electroanalytical Chemistry*, ed., A.J. Bard, Dekker, New York, 1984, p. 191.

17. A. Harriman and M.-P. Richoux, *J. Chem. Soc. Faraday Trans. II*, **76** (1980) 1618.

18. A.P. Piechowski, G.R. Bird, D.L. Morel and E.L. Stogryn, *J. Phys. Chem.*, **88** (1984) 934.

19. L.D. Rollman and R.T. Iwamoto, *J. Amer. Chem. Soc.*, **90** (1968) 1455.

20. V.G. Shepard, Jr. and N.R. Armstrong, *J. Phys. Chem.*, **83** (1979) 1268.

Main Group Elements and Their Compounds
V.G. Kumar Das (Ed)
Copyright © 1996 Narosa Publishing House, New Delhi, India

Fiber Optic Reflectometric Investigation
of Oxidation-Reduction Equilibria
# of Immobilized Reagents

Edna C. Quinto and Fortunato Sevilla III

Research Center for the Natural Sciences,
University of Santo Tomas Espana, Manila, Philippines

The immobilization of reagents is an important operation in the preparation of chemical sensors. These reagents which are often localized proximate to the sensing element react with the analyte and generate species which can be detected by the transducer system of the sensor.

Immobilization can be carried out through physical and chemical methods. The physical methods employed are adsorption, electrostatic attraction and encapsulation. Chemical immobiliation involves the formation of covalent bonds between the reagent and the solid support.

Immobilization has been observed to cause changes in some of the properties of the reagent. The acidity of some pH indicators was changed upon immobilization, decreasing when the reagent was adsorbed on a non-ionic polymer[1-4]and increasing when it was immobilized on an ion-exchange resin[5]. The reduction potential of several oxidation-reduction indicators was lowered when these reagents were adsorbed on a non-ionic polymer.[6] The selectivity of chelating agents was enchanced upon immobilization on an ion-exchange resin.[7]

In this work, the effect of the nature of the immobilizing agent on the modification of the equilibrium characteristics of oxidation-reduction indicators was investigated using a reflectometric method involving optical fibers.

Experimental

Instrumentation

A schematic diagram of the instrumentation system employed is presented in Figure 1. The instrumentation was based on a UV-VIS spectrophotometer (Hitachi Model 139). A bifurcated optical fiber system (Matec LG 3203-048) was interposed between the lamp housing and the monochromator-detector system of the spectrophotometer. Light from the tungsten lamp was focused on the optical fiber and directed to the immobilized reagent system which was localized on the distal end of the optical fiber. The radiation was reflected by the reagent phase to the other branch of the fiber system and led to the PMT detector system which measured the intensity of the radiation. The output of the detector system was amplified by an op-amp circuit and displayed on a digital multimeter (Wheeler WD 7745) and on a chart recorder (Lloyd Graphic).

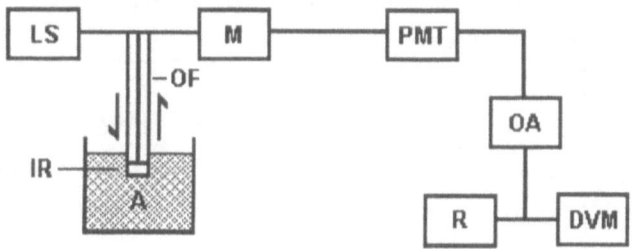

FIGURE 1. Instrumentation system: LS, light source; OF, optical fiber; IR, immobilized reagent; A, analyte; M, monochromator; PMT, photo-multiplier tube; OA, op amp circuit; R, recorder; DVM, digital voltmeter

Reagents

All reagents were of analytical grade. The immobilizing agents used were Amberlite XAD-2, a non-ionic styrene/divinylbenzene copolymer, and Amberlite IRC-50, a weakly acidic anion-exchange resin. The polymeric microspheres were used without further sieving.

The indicators studied were dyes containing the diphenylamine grouping: 2,6-dichlorophenol-indophenol (DCP), diphenylamine (DA) and diphenylamine sulfonate (DAS). These indicators were immobilized from 0.10% aqueous solutions.

The potential-buffer or potentiopoised solutions were prepared by mixing 0.02 M solutions of ferric ammonium sulfate and of ferrous ammonium sulfate in 0.5 M sulfuric acid. These solutions provide potentials ranging from 590 to 890 mV. A pH meter (Metrohm 691) equipped with a platinum electrode (Corning) and a saturated calomel electrode (Corning) was used to ascertain the potential of the mixture.

Measurement

The polymer microbeads containing the immobilized reagents were packed tightly on the distal end of the bifurcated optical fiber by means of a nylon mesh secured by O-rings. The resulting optical fiber probe was immersed in the potentiopoised solution, and the reflectance of the reagent phase at a wavelength of 520 nm was monitored.

Results and Discussion

Reflectance spectra

For all the three indicators studied, the oxidized form exhibited a lower reflectance than the reduced form at all wavelengths. This behavior can be attributed to the greater absorption of radiation by the darkly colored reduced form, compared to the almost colorless reduced form. Figure 2 shows the reflectance spectra obtained for DA immobilized on XAD-2 and on IRC-50.

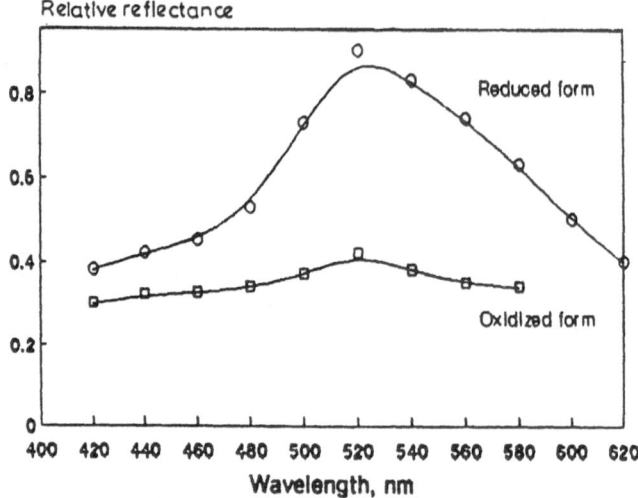

Fig. 2A : Reflectance spectra of DA on XAD-2 at pH 0

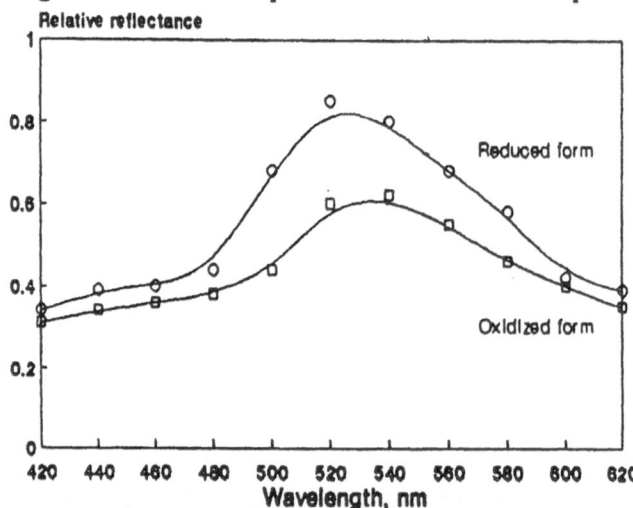

Fig. 2B : Reflectance spectra of DA on IRC-50 at pH 0

Inspection of the spectra reveals that the maximum change in the reflectance of the reagent occurs at 520 nm. The reflectance at this wavelength is highly sensitive to changes in the potential of the medium to which the immobilized reagent is subjected. As a result, the subsequent measurements were carried out at this wavelength.

Effect of potential

The potential of the imobilized indicator was varied by allowing it to equilibrate with different potentiopoised solutions based on the Fe(II)-Fe(III) redox couple.

For the three reagents studied, the measured reflectance was found to decrease as the potential increased.

Figure 3 depicts the variation of the Kubelka-Munk function corresponding to each reflectance data of the immobilized indicators with potential. The Kubelka-Munk function, which is defined by the following equation

$$F(R) = (1 - R)^2 / 2R = kC$$

was used to describe the immobilized system, since it is linearly related to the concentration of the absorbing species.[8] The sigmoid trend resembles that of a potentiometric titration curve which follows the Nernst equation.

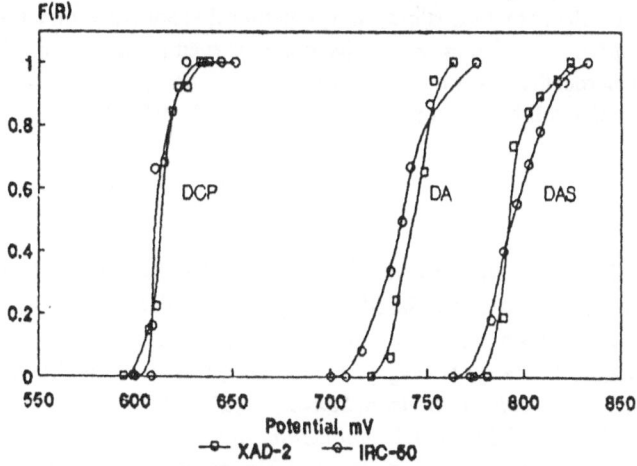

Fig. 3 : Variation of the Kubelka-Munk function with potential at pH 0

Comparison of formal potentials

The formal potential of each indicator was obtained by plotting the degree of reduction, $\alpha = C_{red}/C_{total}$, against the potential and determining the potential when $\alpha = 0.50$. Figure 4 presents the plot obtained for DAP immobilized on IRC-50. The values obtained for each indicator immobilized in the nonionic polymer and ion-exchange polymer is given in Table 1.

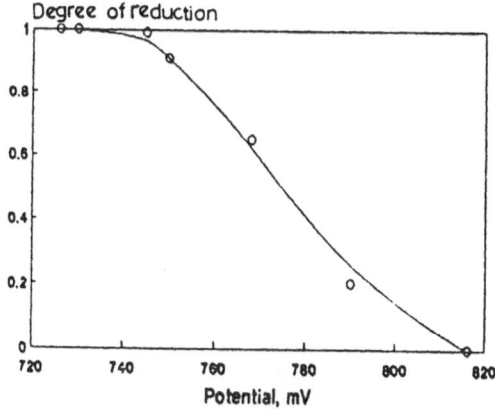

Fig. 4 : Variation of the degree of reduction for DA on IRC-50 at pH 0

Statistical tests show that there is no significant difference between the formal potentials of the indicator immobilized on a nonionic polymer and on an ion-exchange resin. Furthermore, the potential of the reagent in the immobilized state is lower that that of the reagent dissolved in water. This behavior is consistent with the solid solvent model proposed previously.[6]

Conclusion

This study shows that the nature of the immobilizing agent does not affect the equilibrium characteristics of a redox indicators. The measured formal potential was similar for the immobilized indicator, whether it is adsorbed on a non-polar or on a polar surface. This behavior is unlike that observed for immobilized acid-base indicators wherein there was a marked difference in the acidity constants obtained from a nonpolar solid support and from a polar solid surface.

References

1. G.F.Kirkbright, R.Narayanaswamy and N.A.Welti. *Analyst.*, **108** (1984) 15.
2. R.Narayanaswamy and F.Sevilla III. *Anal.Chim.Acta.*, **189** (1986) 365.
3. M.Bacci, F.Baldini and A.M.Scheggi. *Anal.Chim.Acta.*, **207** (1988) 343.
4. G.Boisdem B.Biatry, B.Magay, B.Dureault, F.Blanc and B.Sebille. *Proc. SPIE (Chem Biochem. Environ. Sensors)*, **1172** (1989) 239.
5. R.T.Andres and F.Sevilla III. *Anal.Chim.Acta.*, **251** (1991) 165.
6. R.Narayanaswamy and F.Sevilla III. *Mikrochim.Acta* I, (1989) 239.
7. M.A.Ditzler, H.Pierre-Jacques, S.A.Harrington. *Anal.Chem.*, **58** (1986) 195.
8. A.Guthrie, R.Narayanaswamy and D.Russell. *Analyst*, **113** (1988) 457.

Main Group Elements and Their Compounds
V.G. Kumar Das (Ed)
Copyright © 1996 Narosa Publishing House, New Delhi, India

Functionalization of Initiation Chain Ends of Isotactic Polypropylene using Modified Cocatalysts

Takeshi Shiono, Hiroki Kurosawa and Kazuo Soga

*Research Laboratory of Resources Utilization, Tokyo Institute of Technology,
Nagatuta 4259, Midori-ku, Yokohama 227, Japan*

Terminally functionalized polymers are useful not only for the synthesis of block copolymers but also for the improvement of polymer properties. Functionalization of isotactic polypropylene (PP), which is one of the most important plastics, offers the possibility of improving its compatibility with other materials. Conceivably, terminally functionalized PP may be obtained using 'living polymerization' methods[1] or chain transfer reactions[2]. In carrying out propene polymerization with the $TiCl_3$-$AlEt_2Cl$ catalyst system containing a large amount of $ZnEt_2$, we observed that almost all polymer chain ends were terminated by metal atoms (mainly Zn). These metal-polymer bonds were then subjected to reactions with oxygen, carbon dioxide, chlorine or allyl bromide to obtain respectively the hydroxyl-, carboxyl-, chlorine- or vinyl-terminated isotactic PP.[3-5] However, the above method does not yield a polymer with a functional group at both ends of the polymer chain. In this paper, several dialkylzinc compounds having functionalized organic groups were synthesized and used as co-catalysts with $TiCl_3$ to obtain functionalized initiation chain ends.

Experimental

Materials

Propene (Mitsubishi Petrochemical Co.) was purified by passing through columns of $CaCl_2$, P_2O_5 and molecular sieve 3A. $TiCl_3$ (AA type, Toho Titanium Co.), $AlEt_2Cl$ and $Al(i\text{-}Bu)_2H$ (Tosoh Akuzo Chemical Co.) were used without further purification. Research grade pentane and heptane were commercially obtained and dried over CaH_2 under reflux for 24 h, and distilled before use. Diethyl ether was successively dried over CaH_2 and sodium metal under reflux for 12 h and distilled before use. Magnesium (turnings for Grignard reagent, from Wako Pure Chemical Ind., Ltd.) was dried under vacuum. $ZnCl_2$ (10 g) was treated with 25 ml of thionyl chloride under reflux for 2 h. After decantation, the remaining thionyl chloride was removed under vacuum. Commercially obtained 1,7-octadiene and N-methylimidazole were distilled and dried over molecular sieve 4A. 4-Bromo-1-butene and 5-bromo-2-methyl-2-pentene (Aldrich Chemical Co., Inc.) were dried over molecular sieve 4A. Argon (99.9995%) and nitrogen (99.9995%) were used without further purification.

Synthesis of bis(3-butenyl)zinc (BBZ)

It was reported by Denis *et al.*[6] that BBZ could be synthesized by metal exchange in the reaction of zinc with of bis(3-butenyl)mercury. In this paper, however, BBZ was prepared from 3-butenylmagnesium bromide and $ZnCl_2$ as follows:

In a 500 ml three-necked flask equipped with a mechanical stirrer were placed 9 g of magnesium and 170 mL of ether under an argon atmosphere. 50 g of 4-bromo-1-butene in 50 mL of diethyl ether was added dropwise to the mixture at room temperature, after which the temperature of the reaction mixture was raised to reflux. After stirring for 2 h, a small portion of the reaction mixture was hydrolysed and titrated with hydrochloric acid to estimate the concentration of 3-butenylmagnesium bromide. 21 g of $ZnCl_2$ in 200 mL of ether was then added dropwise to the Grignard reagent and the mixture was stirred for 12 h at 30 °C. The reaction mixture was then filtered, and the filtrate concentrated by stripping off the diethyl ether under vacuum. The resulting product was treated with pentane to precipitate magnesium halides. Following filtration, the pentane was evaporated and the resulting residue dried under vacuum to obtain BBZ in 90% yield as a reddish brown liquid. Analytical data: ^1H NMR (C_6D_6): 0.2-0.4 (2H, t, CH_2Zn), 2.0-2.4 (2H, q, CH_2), 4.8-5.2 (2H, t, $=CH_2$), 5.8-6.2 (1H, m, $-CH=$); ^{13}C NMR (C_6D_6): 14.7 (CH_2Zn), 30.7 (CH_2), 112.1 ($=CH_2$) and 114.8 ($-CH=$). These values were in good agreement with those reported previously.[7]

Synthesis of bis(4-methyl-3-pentyl)zinc (BMPZ)

BMPZ was prepared by the reaction of 4-methyl-3-pentylmagnesium bromide and $ZnCl_2$, duplicating the procedure described above. BMPZ was obtained in 50% yield as a reddish brown liquid. Analytical data: ^1H NMR (C_6D_6): 0.4-0.6 (2H, t, CH_2Zn), 1.5-1.7 (6H, d, $=C(CH_3)_2$), 2.0-2.4 (2H, q, CH_2), 5.1-5.4 (1H, t, $-CH=$); ^{13}C NMR (C_6D_6) 16.3 (CH_2Zn), 17.7 ($=CCH_3$, *cis*), 25.1 (CH_2), 25.9 ($=CCH_3$, *trans*), 129.0 ($-CH=$), 130.9 ($=C<$).

Synthesis of tris(7-octenyl)aluminum (TOA)

In a 50 mL Schlenk tube under argon atmosphere was placed 15 mL of 1,7-octadiene and 2 mL of $Al(i-Bu)_2H$. The mixture was heated under reflux for 5 h. The excess amount of 1,7-octadiene was removed under reduced pressure to obtain 5.3 g of a viscous liquid. The ^1H and ^{13}C NMR spectra of the product displayed resonances assignable not only to a vinyl group but also to a methylene group bonded to Al (broad resonances at 0.5 ppm in ^1H NMR and at 11 ppm in ^{13}C NMR, respectively). It proved difficult to purify TOA and, hence it was used as obtained.

Disporportionation of TOA with $ZnCl_2$

In a 100 mL Schlenk tube was placed 30 mL of 1,7-octadiene and 4 mL of $Al(i-Bu)_2H$ under an argon atmosphere. The mixture was heated under reflux for 5 h. The excess amount of 1,7-octadiene was removed under reduced pressure, and 30 mL of heptane and 6.5 g of $ZnCl_2$ were added to the residue. The mixture was stirred for 4 h at the reflux temperature, and then filtered. Upon stripping off the solvent from the filtrate under vacuum, there was obtained 14.4 g of a viscous liquid. The product was diluted with 30 mL of heptane and used for the polymerization experiments.

Polymerization procedure

Propene polymerization was conducted using a 200 mL glass reactor or a 50 mL stainless steel autoclave equipped with a magnetic stirrer. For the case of the glass reactor, after the measured amounts of $TiCl_3$ and heptane were added to the reactor under a nitrogen atmosphere, propene was introduced at the polymerization temperature until the solvent became saturated with propene. Polymerization was started by adding the cocatalyst (Al or Zn compound). On the other hand, with the autoclave, measured amounts of $TiCl_3$, heptane and cocatalyst were added in turn under nitrogen, and then 3 L(STP) of propene were condensed into the autoclave at liquid nitrogen temperature. Polymerization was started by setting the autoclave at the polymerization temperature. The polymerization was terminated by adding a hydrochloric acid solution in ethanol. The precipitated polymer was filtered and dried under vacuum at 60 °C for 8 h.

Coupling reaction with allyl bromide

After the polymerization, the remaining propene monomer was evacuated and nitrogen was introduced. N-methylimidazole (25 mmol) and allyl bromide (25 mmol) were then added into the autoclave and the mixture was vigorously stirred at around 130 °C for 3 h. The reaction was terminated by pouring the mixture into a dilute solution of hydrochloric acid in ethanol.

Analytical procedures

^1H NMR spectra of samples were recorded on a JEOL EX-90 spectrometer operated at 89.45 MHz in the pulse Fourier transfer (FT) mode. ^{13}C NMR spectra were recorded on a JEOL EX-90 spectrometer operated at 22.4 MHz or on a JEOL GX-500 spectrometer operated at 125.65 MHz in the pulse FT mode. In the ^1H NMR measurements, the pulse angle was 45°, and 100-500 scans were accumulated in 9s of pulse repetition. In the ^{13}C NMR measurements, broad band coupling was used to remove the ^{13}C-^1H coupling. The pulse angle was 45°, and 6000-8000 scans were accumulated in 8s pulse repetition. The spectra were obtained at room temperature or at 130 °C in benzene-d_6 or 1,1,2,2-tetrachlorothane-d_2 solution (2 wt% for ^1H NMR and 15 wt% for ^{13}C NMR in a 5 mm o.d. tube). Hexamethyldisiloxane was used as internal reference (0 ppm for ^1H NMR; 2.03 ppm for ^{13}C NMR, respectively).

Differential scanning calorimetry (DSC) measurements were made on a Seiko DSC-220. Polymer samples (ca. 4 mg) were encapsulated in aluminum pans and scanned at 10 °C/min. IR spectra of the polymers were recorded on a JASXO FT/IR-3 spectrometer. Sample films were made with a hotpress at 100 °C under 100 kg/cm^2 of pressure.

The vinyl content in the polymer was estimated from IR spectra by noting the ratio of the absorbance of C=C stretching band (1643 cm^{-1}) of the vinyl group to that of the C-H deformation bands (1303, 1256, 1220, 1167 and 973 cm^{-1}). Calibration curves were obtained with vinyl-terminated isotactic PPs containing 0.076, 0.12 and 0.18 mmol of vinyl groups per gm.-PP, which were prepared by the coupling reaction between Zn-polymer bonds and allyl bromide and characterized by ^{1}H and ^{13}C NMR.[5] Gel-permeation chromatography of the polymers was performed on a Waters 150C instrument at 140 °C in o-dichlorobenzene as solvent.

Results and Discussion

Bis(3-butenyl)zinc (BBZ) was synthesized according to the procedure described in the experimental section. Propene polymerization was conducted with the TiCl$_3$-BBZ catalyst system at 40 °C for 1 h, the results of which are summarized in Table 1. The polymer yield increased with increase in the concentration of BBZ. Propene polymerization was also conducted with the conventional TiCl$_3$-AlEt$_2$Cl catalyst system for purposes of comparison. The TiCl$_3$-BBZ catalyst system showed lower activity than TiCl$_3$-AlEt$_2$Cl.

The thermal property of PP obtained with these catalysts was investigated by DSC. The melting points (T$_m$) and heats of fusion (ΔH) of the polymers produced are shown in Table 1. Both T$_m$ and ΔH do not differ very much between the two catalyst systems, suggesting that the isotacticities of the PP's derived using the modified and conventional catalysts are comparable. Addition of AlEt$_2$Cl to the TiCl$_3$-BBZ catalyst caused a slight decrease in the ΔH value without changing the polymerization activity.

Inasmuch as BBZ initiates the isospecific polymerization of propene, it may *a priori* be anticipated that the initiation ends of the produced polymers be at least partly functionalized by a vinyl group. IR and ^{1}H NMR spectra of typical polymers were procured in order to characterize the microstructure of the polymers obtained with the TiCl$_3$-BBZ and TiCl$_3$-BBZ-AlEt$_2$Cl catalysts. However, no vinyl groups could be detected both in the IR (Fig. 1a) and ^{1}H NMR (Fig. 2a) spectra. We have previously reported that the metal-polymer bonds of PP obtained with the TiCl$_3$-AlEt$_2$Cl-ZnEt$_2$ catalyst could be converted to vinyl groups in fairly good yield by a coupling reaction with allyl bromide.[5] Therefore, this method was also applied to the present system.

Figs. 1b and 2b show the IR and ^{1}H NMR spectra of the polymer after being subject to the coupling reaction (Run no. BBZ6); these display the peaks readily attributable to the vinyl group. The content of vinyl group in PP as determined from IR spectral analysis increased with an increase in the concentration of BBZ, indicating that BBZ acts as an initiator and as a chain transfer agent. It is not clear, however, why the vinyl group is not introduced at the initiation chain end.

To investigate the structure of the chain ends in more detail, ^{13}C NMR spectra of the polymers obtained with the TiCl$_3$-BBZ-AlEt$_2$Cl catalyst were procured before and after the coupling reaction (Runs no. BBZ5 and BBZ6). Fig. 3a shows the ^{13}C NMR spectrum of the ethanol-quenched PP. Besides the three strong resonances of the main chain carbons, weak resonances attributable to the iso-butyl end group are observed. However, the spectra do not display any resonances assignable to ethyl, vinyl, propyl and vinylidene end groups, which should be formed by the

Table 1 : Results of propene polymerization with the $TiCl_3/Al(C_2H_5)_2Cl/Zn(CH_2CH_2CH_2CH_2CH=CH_2)_2$

Run no.	$TiCl_3$ [mmol]	$AlEt_2Cl$ [mmol/L]	$Zn(CCC=C)_2$ [mmol/L]	Terminator	Yield [g/mmol-Ti]	$M_n^a/10^3$	Tm [°C]	ΔH [J/g]	Content of[b] -C=C [mmol/g-PP]
BBZ1	0.76	0	40	C_2H_5OH	0.38	8.4	162	90.9	n.d.
BBZ2	0.61	0	120	C_2H_5OH	0.46	-	162	92.5	n.d.
BBZ3	0.82	0	120	Br-CC=C	0.61	-	-	-	0.12 - 0.13
BBZ4	0.78	0	240	Br-CC=C	0.69	5.1	159	81.6	0.18 - 0.41
BBZ5	0.70	20	240	C_2H_5OH	0.62	6.1	158	69.4	n.d.
BBZ6	0.80	20	240	Br-CC=C	0.85	3.9	-	-	0.25 - 0.45
DEAC	0.34	20	0	C_2H_5OH	5.7	210	164	85.5	-

[a]measured by GPC
[b]determined by IR; n.d. = not done

259

initiation and/or chain transfer with AlEt$_2$Cl and BBZ, and chain transfer via β-hydrogen abstraction or by monomer, respectively (Scheme 1).

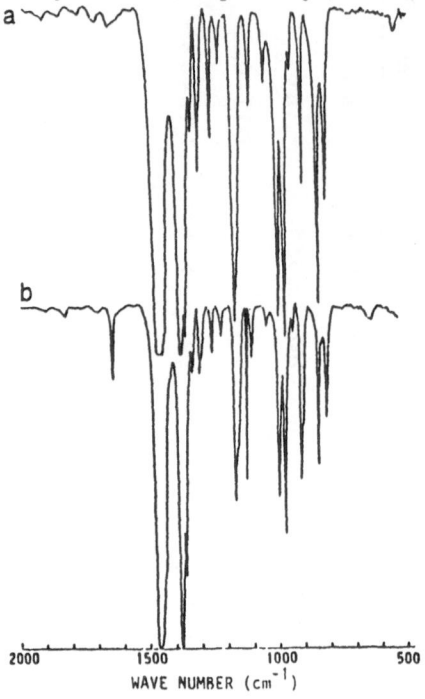

Fig. 1 : IR spectra of PP obtained with the TiCl$_3$-BBZ catalyst: (a) ethanol-quenched, (b) allyl bromide-quenched

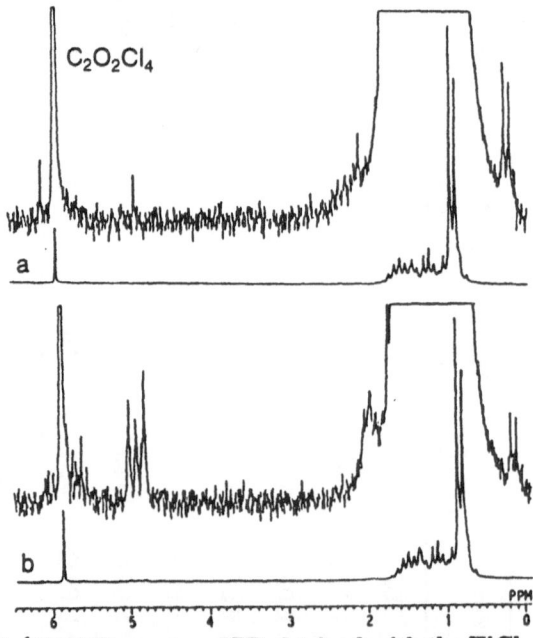

Fig. 2 : 90-MHz ^{1}H NMR spectra of PP obtained with the TiCl$_3$-BBZ-AlEt$_2$Cl catalyst : (a) ethanol-quenched, (b) allyl bromide-quenched

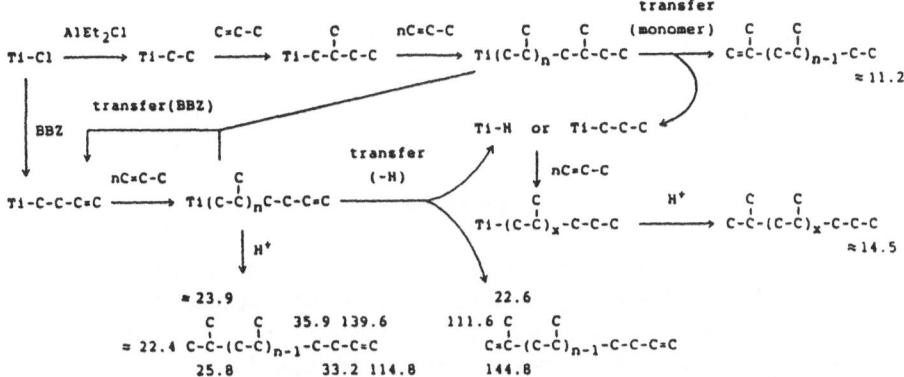

Scheme 1 : Plausible structures of chain end groups and their ^{13}C chemical shifts (ppm from SiMe$_4$)

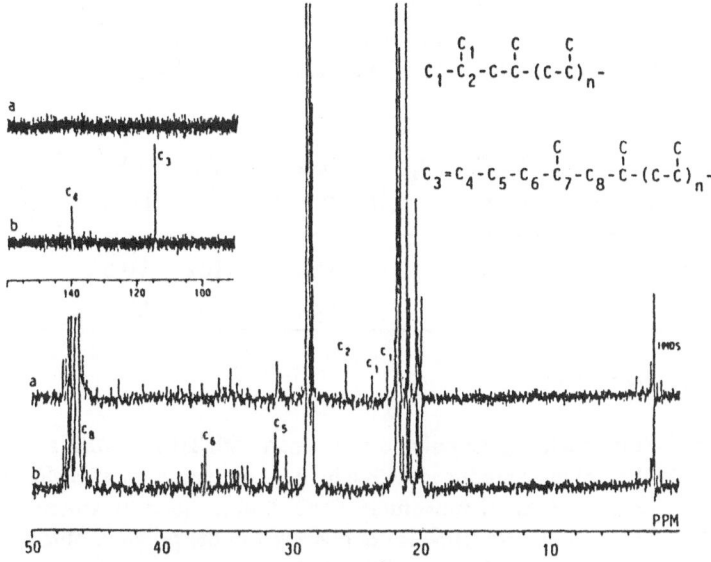

Fig. 3 : ^{13}C NMR spectra of PP obtained using the catalyst system TiCl$_3$-BBZ-AlEt$_2$Cl: (a) ethanol-quenched, (b) allyl bromide-quenched

After the coupling reaction with allyl bromide, the resonances due to the iso-butyl end group disappeared completely and those assignable to the vinyl end group newly appeared, suggesting that the iso-butyl group derived from the metal-polymer bonds is quantitatively converted to the vinyl group. The absence of vinyl group at the initiation chain end may be explained in terms of the model shown in Scheme 2. BBZ acts as an initiator as well as a chain transfer reagent. When the first propene monomer inserts into the Ti-(3-butenyl) bond, however, a pentacyclo end group is formed as shown in Scheme 2. The weak resonances between 30 and 40 ppm may be assigned to this cyclic structure. The formation of a similar five membered ring has been confirmed in 1,5-hexadiene polymerization with V(acac)$_3$-AlEt$_2$Cl, TiCl$_3$-AlEt$_2$Cl or Cp$_2$ZrCl$_2$-methylaminoxane catalyst systems.[8-10]

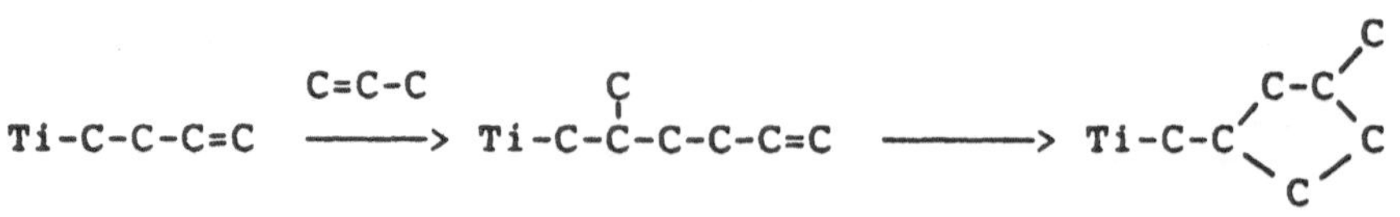

$$\text{Ti-C-C-C≡C} \xrightarrow{\;\text{C=C-C}\;} \text{Ti-C-C-C-C-C≡C} \xrightarrow{} \text{Ti-C-C}\overset{\displaystyle C-C\overset{\displaystyle C}{\diagup}}{\underset{\displaystyle C}{\diagdown}}$$

Scheme 2

Judging from the fact that di-substituted olefins are not polymerized with conventional Ziegler-Natta catalysts, such a cyclization pathway might be prevented by changing the vinyl group to a sterically more hindered one. To verify this, bis(4-methyl-3-pentyl)zinc (BMPZ) was synthesized and applied as a cocatalyst. The results of propene polymerization with the $TiCl_3$-BMPZ catalyst are summarized in Table 2. Addition of more BMPZ caused an increase in the polymer yield, with the isospecificity remaining unchanged.

Table 2 : Results of propene polymerization with the $TiCl_3/Zn(CH_2CH_2CH=C(CH_3)_2)_2$ catalyst system

Run no.	$TiCl_3$ [mmol]	$Zn(CCC=C<)_2$ [mol/L]	Yield [g/mmol-Ti]	Tm [°C]	ΔH [J/g]	Content of[a] -C=C< [g/mmol-Ti]
BMPZ1	0.80	40	0.83	162	90.6	-
BMPZ2	0.83	240	1.44	161	88.6	0.092

[a]determined by ¹H NMR

The ¹H NMR spectra of the polymers obtained with $TiCl_3$-BMPZ are shown in Fig. 4. Only the polymer produced with a higher concentration of BMPZ displays the proton connected to tri-substituted C=C double bond at around 5.3 ppm (BMPZ2). This proton resonance is not observed in the polymer obtained with a lower concentration of BMPZ (BMPZ1), which may be due to an increase in molecular weight. The ¹³C NMR spectrum of the polymer obtained in run no. BMPZ2 is shown in Fig. 5, which displays the resonances attributable to the carbons of iso-butyl and 2-methyl-2-pentenyl ($-CH_2CH_2CH_2=C(CH_3)_2$) chain end groups in approximately equal intensity. No other resonances assignable to the chain end structure could be discerned in the spectrum. These results strongly indicate that BMPZ acts both as an initiator and a chain transfer agent.

The alkylzinc compounds used in the above experiments were synthesized via alkylation of $ZnCl_2$ with the Grignard reagent derived from the corresponding bromide. In seeking to simplify the preparation of functionalized cocatalysts, we have explored the synthesis of (7-octenyl)₃Al (TOA) by hydroalumination of 1,7-octadiene (see experimental section) and used the resulting product as cocatalyst. The disproportionation product between TOA and $ZnCl_2$ (TOA-ZC) was also used as cocatalyst.

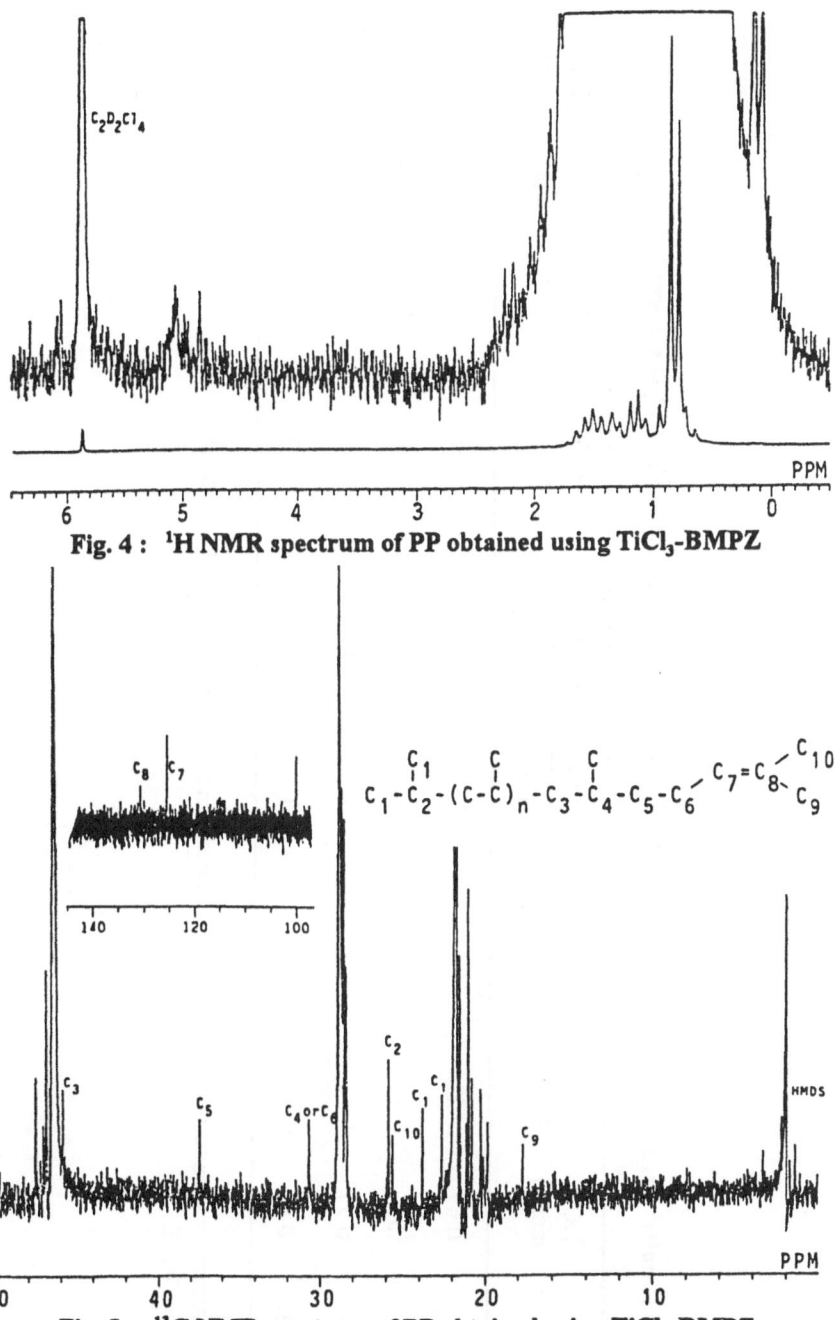

Fig. 4 : ^{1}H NMR spectrum of PP obtained using TiCl$_3$-BMPZ

Fig. 5 : ^{13}C NMR spectrum of PP obtained using TiCl$_3$-BMPZ

The results of propene polymerization using these cocatalysts are summarized in Table 3. The TiCl$_3$-TOA catalyst showed a higher activity than the TiCl$_3$-TOA-ZC catalyst. The increase in the concentration of TOA caused a marked increase in

Table 3 : Results of propene polymerization with the TiCl₃/TOA and TiCl₃/TOA-ZC catalyst systems

| Run no. | TiCl₃ [mmol] | TOA [mol/l] | TOA-ZN [ml] | Polymerization conditions | | Yield [g/mol-Ti·h] | Tm [°C] | DH [J/g] | Content of[c] -C=C [mmol/g-PP] |
				Temp [°C]	Time [h]				
TOA1[a]	0.65	220	-	40	1	0.98	160	68.4	-
TOA2[a]	3.1	100	-	40	0.5	2.4	160	63.4	0.05 - 0.08
TOAZC1[a]	0.77	-	1.1	40	1	0.039	159	57.7	0.11 - 0.16
TOAZC2[b]	0.83	-	1.1	40	1	0.23	161	64.7	0.13 - 0.20

Polymerization was conducted in a 200 ml glass reactor,[a] in a 50 ml stainless-steel autoclave[b]

[c]determined by IR

the polymer yield without changing the T_m and ΔH values.

The IR spectra of the polymers obtained are illustrated in Fig. 6, which displays the C=C stretching band of the vinyl group at around 1643 cm^{-1}. The vinyl group contents were determined from the IR (see Table 3). It was thus demonstrated that TOA is also useful for the functionalization of initiation chain ends, although pure TOA could not presently be isolated.

In conclusion, it may be stated that the functionalization of the initiation end of isotactic PP may be achieved by modifying the cocatalyst in the conventional Ziegler-Natta catalysts. A more detailed study including the development of other functionalization methods is now in progress.

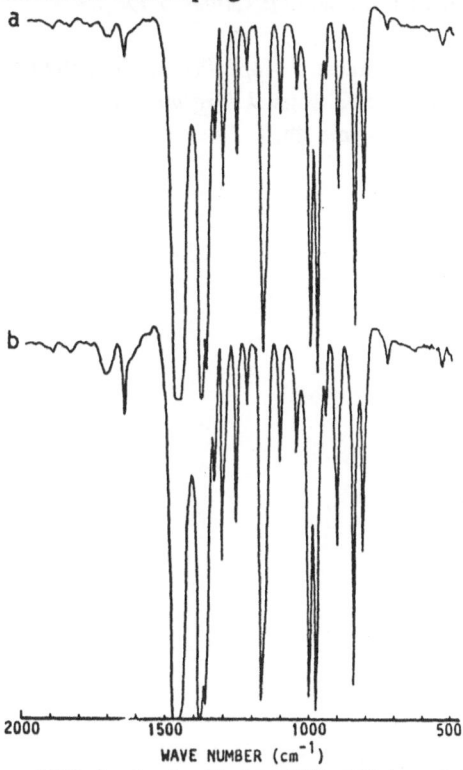

Fig. 6 : **IR spectra of PP obtained with (a) TiCl₃-TOA and (b) TiCl₃-TOA-ZC catalysts**

References

1. Y. Doi and T. Keii, *Adv. Polym. Sci.*, **73/74** (1986) 201.
2. (a) E. Agouri, C. Parlant, P. Monet, J. Rideau and J.F. Teitgen, *Makromol. Chem.*, **137** (1970) 229; (b) Y. Doi, K. Soga, M. Murata and Y. Ono, *Makromol. Chem., Rapid Commun.*, 4 (1983) 789; (c) D.R. Burfield, Polymer, **25** (1984) 1817; (d) G. Redina and E. Albizzati, *Eur. Pat. Appl.*, **350059** (1989); (e) R. Mulhaupt, T. Duschek and B. Rieger, *Makromol. Chem. Symp.*, **48/49** (1991) 317; (f) T. Shiono and K. Soga, *Macromol. Symp.*, **48/49** (1991) 317; (f) T. Shiono and K. Soga, *Macromolecules*, **25** (1992) 3356;

(g) T. Shiono and K. Soga, *Makromol. Chem., Rapid Commun.*, **13** (1992) 371; (h) T. Shiono, H. Kurosawa, O. Ishida and K. Soga, *Macromolecules*, **226** (1993) 2085.

3. T. Shiono, K.Yoshida and K. Soga, *Makromol. Chem., Rapid Commun.*, **11** (1990) 169.
4. T. Shiono, H.Kurosawa and K. Soga, *Makromol. Chem.*, **192** (1992) 2751.
5. H. Kurosawa, T. Shiono and K. Soga, *ibid.*, to be submitted.
6. J. St. Denis, J.P. Oliver, T.W. Dolzine and J.B. Smart, *J. Organomet. Chem.*, **71** (1974) 71.
7. M.J. Albright, J.N. St. Denis and J.P. Oliver, *J. Organomet. Chem.*, **125** (1977) 1.
8. H.N. Cheng and N.P. Khasat, *J. Appl. Polym. Sci.*, **35** (1988) 825.
9. Y. Doi, N. Tokuhiro and K. Soga, *J. Organomet. Chem.*, **190** (1989) 643.
10. L. Resconi and R.M. Waymouth, *J. Amer. Chem. Soc.*, **112** (1990) 4953.

Main Group Elements and Their Compounds
V.G. Kumar Das (Ed)
Copyright © 1996 Narosa Publishing House, New Delhi, India

Copolymerisation of Substituted Tributyltin Cinnamates with Styrene. Syntheses, Reactivity Ratios, Relative Reactivities, NMR and Mössbauer Assignments and Thermal Analysis

Lay-Foong Siah[a], V.G. Kumar Das[b], and Seng-Neon Gan[b]

*[a]Institute of Advanced Studies and [b]Department of Chemistry,
University of Malaya, 59100 Kuala Lumpur, Malaysia*

Copolymer systems with pendant triorganostannyl groups have received much attention in the chemical literature on account of their appeal on environmental grounds as controlled-release marine antifoulants.[1-4]

A number of chemical studies have focussed on tributyltin- and triphenyltin methacrylates, -acrylates and -maleates in their binary and ternary polymerisations with a range of vinyl monomers.[5-7] In general, the organotin monomer activity was found to be in the order, tributyltin methacrylate (TBTM) >> tributyltin acrylate (TBTA) > bis(tributyltin) maleate (BTM). In those cases where the organotin monomer does not undergo homopolymerisation, as with BTM, block polymers are invariably produced in which large blocks of the comonomer units are interrupted by single molecules of BTM.[8] On the other hand, the copolymerisation of TBTM with methylmethacrylate was shown to yield a random distribution of the monomers typical of an ideal copolymerisation system in which both comonomers are of comparable reactivity.[9] In copolymerisation studies of TBTM and TBTA with alkylmethacrylates and acrylates, $CH_2=C(R)CO_2R'$ (R = Me, H respectively), it was observed that the tendency towards formation of an alternating copolymer increased with increasing alkyl (R') chain length of the comonomer. Thus, TBTM yielded an azeotropic copolymer with propylmethacrylate with a molar ratio of 19.2:80.8, while the azeotropic composition of TBTM-*co*-butylmethacrylate[5] was 47:53.

Alternating copolymerisation has also been reported for poly(TBTM-*co*-styrene), with observations of a near-equal reactivity ratio of 0.5 for either monomer and of *Q-e* values of 0.78 and 0.38, respectively, for TBTM.[10]

The present study is aimed at extending our understanding of the polymerisation behaviour of unsaturated triorganotin monomers by a consideration of the tributyltin cinnamate-*co*-styrene system which has hitherto not been described.

Experimental

Preparation and purification of materials

Styrene (monomer 1, M_1) was purified by distillation under reduced pressure and the centre fraction was retained for copolymerisation studies (bp$_{18}$: 42 - 43°C).

All the five tributyltin cinnamates (monomer 2, M$_2$ = p-Z-C$_6$H$_4$CH=CHC(O)OSn(n-C$_4$H$_9$)$_3$, where Z = OCH$_3$, CH$_3$, H , Cl and NO$_2$) were prepared by the esterification of bis(tributyltin) oxide with the respective cinnamic acids and analytically pure grades were obtained upon recrystallisation from hot hexane.

Azobisisobutyronitrile (AIBN), recrystallised thrice from ethanol, was used as initiator. Toluene, used as solvent medium for the copolymerisation reactions, was purified and dried according to a standard procedure.[11]

Copolymerisations

The solution copolymerisations of tributyltin cinnamates (TBTC(Z)) with styrene (STY) were carried out in toluene as described below:[5,7]

Pre-determined amounts of the tin monomers over a wide range of mole ratios were weighed into pyrex tubes. The requisite amount of styrene was added into each tube via a syringe followed by 1 mol% of AIBN and toluene was then syringed in to obtain a total monomer solution with a concentration of 2 mol L^{-1}.

The contents were degassed by three freeze-thaw cycles by repeatedly dipping the prepared tubes into liquid nitrogen and then thawing to room temperature. The tubes were next flushed with nitrogen for about 10 minutes and then capped tightly with rubber septa before being placed in a water-bath maintained at a temperature of 60±0.5°C. Initial expanded vapour was allowed to escape via a syringe needle inserted through the stopper. The needle was removed after allowing for an equilibration time of 5 minutes.

The polymerisation was allowed to proceed and after some known time interval when a certain amount of contraction of the reaction mixture was observed, each tube was separately taken out of the water-bath and its contents completely emptied into a flask containing a ten-fold excess of methanol:water (9:1 v/v). The copolymer formed precipitated out immediately while the unreacted monomer remained in solution. The precipitate was isolated quantitatively by centrifugation. It was dissolved in a little chloroform, reprecipitated with 90% methanol and isolated again by means of centrifugation. The whole process was repeated a second time. The purified copolymer, free of unreacted monomers, was dried under vacuum to constant weight.

The copolymerisations were carried out to low (10-15%) conversions thereby enabling the use of linear methods for the determination of the reactivity ratios as described by the classical copolymerisation equation,

$$y = x. \frac{r_1x + 1}{r_2 + x}$$

where y = m$_1$/m$_2$, x = M$_1$/M$_2$, r$_1$ = k$_{11}$/k$_{12}$ and r$_2$ = k$_{22}$/k$_{21}$. The linear methods used herein include both graphical (Fineman-Ross[12] and Kelen-Tudos[13-14]) and numerical (Yezrielev-Brokhina-Roskin, YBR[15]) methods (vide infra). In experiments where copolymer yields far exceeded the 15% conversion limit, the copolymerisation reaction was repeated over a shorter time period.

Copolymer compositions

The copolymer compositions were determined by ^{1}H-NMR technique using a JEOL JNX270 instrument operating at 270.1 MHz with the solvent, CDCl$_3$, also serving as an internal lock. The ratio of the integrated peak intensities of the aliphatic to aromatic protons present in the copolymer (see Fig. 1) was employed to calculate the mole fraction of tributyltin monomers (m_2) in the copolymer according to the following expressions:

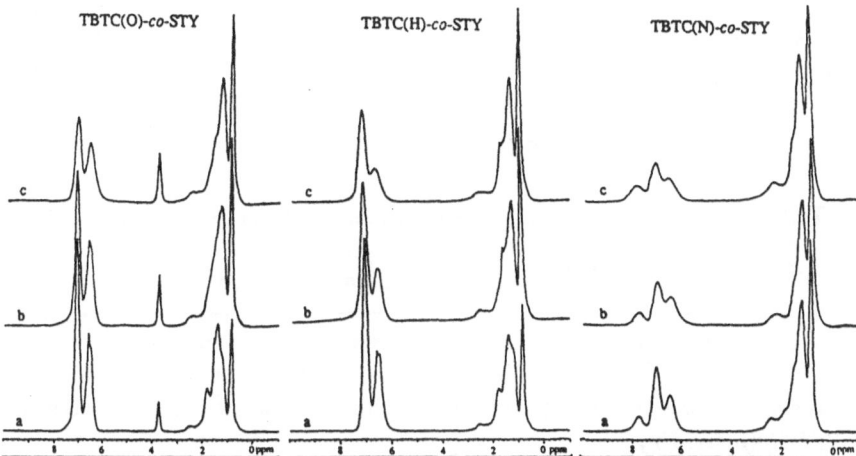

Fig. 1 : ^{1}H NMR spectra of tributyltin cinnamate-*co*-styrene copolymer samples (in CDCl$_3$) showing various composition with different M_2:M_1 feed ratios: a) 3:7 ; b) 5:5 ; c) 7:3 .

Tributyltin *p*-methoxycinnamate-*co*-styrene, (TBTC(O)-*co*-STY):

$$\frac{\text{Aliphatic }^1\text{H}}{\text{Aromatic }^1\text{H}} = \frac{3\, m_1 + 32\, m_2}{5\, m_1 + 4\, m_2} = \frac{3 + 29\, m_2}{5 - m_2}$$

A similar expression also holds for **tributyltin *p*-methylcinnamate-*co*-styrene, (TBTC(M)-*co*-STY)**.

Tributyltin cinnamate-*co*-styrene, (TBTC(H)-*co*-STY):

$$\frac{\text{Aliphatic }^1\text{H}}{\text{Aromatic }^1\text{H}} = \frac{3\, m_1 + 29\, m_2}{5\, m_1 + 5\, m_2} = \frac{3 + 29\, m_2}{5}$$

269

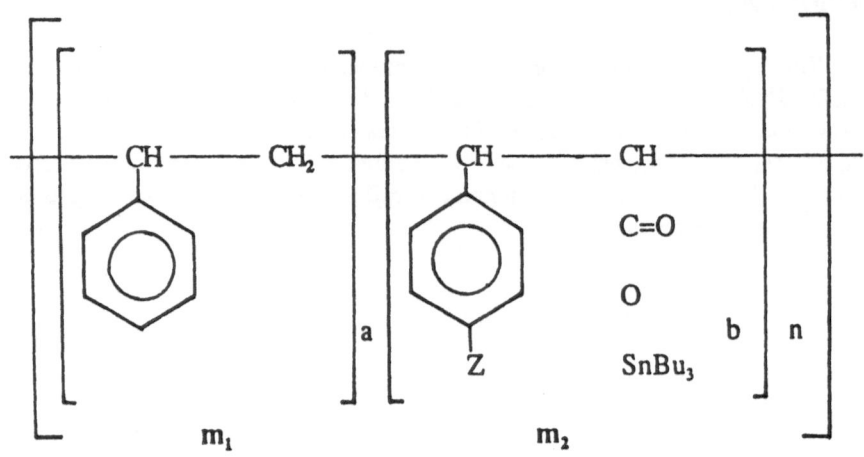

where $(m_1 = 1 - m_2)$

Tributyltin *p*-chlorocinnamate-*co*-styrene, (TBTC(C)-*co*-STY):

$$\frac{\text{Aliphatic } {}^1\text{H}}{\text{Aromatic } {}^1\text{H}} = \frac{3\,m_1 + 29\,m_2}{5\,m_1 + 4\,m_2} = \frac{3 + 26\,m_2}{5 - m_2}$$

A similar expression also holds for **tributyltin *p*-nitrocinnamate-*co*- styrene, (TBTC(N)-*co*-STY).**

The ^{119}Sn NMR spectra were recorded in CDCl$_3$ solutions with the instrument operating at 100.55 MHz under nuclear Overhauser supressed conditions. $\delta({}^{119}\text{Sn})$ are quoted relative to Me$_4$Sn and are accurate to $\pm\,0.1$ ppm. The data are given in Table 1.

Table 1 : Tin-119 NMR data for TBTC(Z)-*co*-STY[a] (in CDCl$_3$)

Copolymer system	Pure TBTC(Z) monomer δ (ppm)	Copolymer	
		$\delta^{[b]}$ (ppm)	width (Hz)
TBTC(O)-*co*-STY	+ 106.0	+ 98.4	1007
TBTC(M)-*co*-STY	+ 107.2	+ 98.2	1209
TBTC(H)-*co*-STY	+ 109.4	+ 96.3	1309
TBTC(C)-*co*-STY	+111.5	+ 101.2	1209
TBTC(N)-*co*-STY	+ 118.0	+ 107.9	1410

[a]Concentration of solutions in CDCl$_3$: 0.01 g ml^{-1} [b]Position of peak maximum

The ^{119m}Sn Mössbauer spectra were measured at 80K using a Cryophysics constant-acceleration microprocessor Mössbauer spectrometer with a 512-channel data store. The velocity range was calibrated with β-tin and calcium stannate against the Ca^{119m}SnO$_3$ source and the spectra were fitted with a Lorentzian curve-fitting programme supplied by the manufacturer. The Mössbauer data are listed in Table 2.

Table 2 : Tin-119^m Mössbauer data (mm s^{-1}, 80 K) of TBTC(Z)-co-STY

	Pure monomer TBTC(Z), Z =			TBTC(Z)-co-styrene, Z =		
	O	H	N	O	H	N
IS	1.44	1.45	1.47	1.31	1.31	1.33
QS	3.62	3.65	3.73	2.61	2.64	2.74
Γ_1	0.83	0.96	0.92	0.85	0.86	0.82
Γ_2	0.89	0.97	0.93	0.88	0.87	0.87
ρ	2.51	2.52	2.54	1.99	2.01	2.06
CN*	5	5	5	4	4	4

*CN : Coordination number of tin

Differential scanning calorimetric analysis (DSC) of selected copolymer samples (see Fig. 5) were recorded on a Perkin-Elmer DSC-2C calorimeter in nitrogen atmosphere at a heating rate of 20 K min^{-1}.

Results and Discussion

Reactivity ratios

The tributyltin cinnamates (M$_2$) described herein do not homopolymerise but yield copolymers with styrene (M$_1$). With the exception of TBTC(N)-co-STY which is a yellow solid, the rest of the copolymers are white solids. The monomer feed and copolymer composition data are given in Table 3; the composition curves derived from these data are depicted in Fig. 2 (curves a-e).

Curves 2a-2c are typical of a classical terminal copolymerisation model with r$_2$=0 and r$_1$>2;[16] the copolymers are essentially non-alternating being composed of blocks of polystyrene units randomly interrupted by M$_2$. The composition curve 2d for TBTC(C)-co-STY reveals an almost linear system with slope 0.5. For this system, r$_1$ is expected to be about 2.[16] In contrast, for TBTC(N)-co-STY, the composition curve 2e is S-shaped and typifies the situation where a penultimate unit effect operates.[16] For this latter case, the composition diagram can be analysed by the following equation :

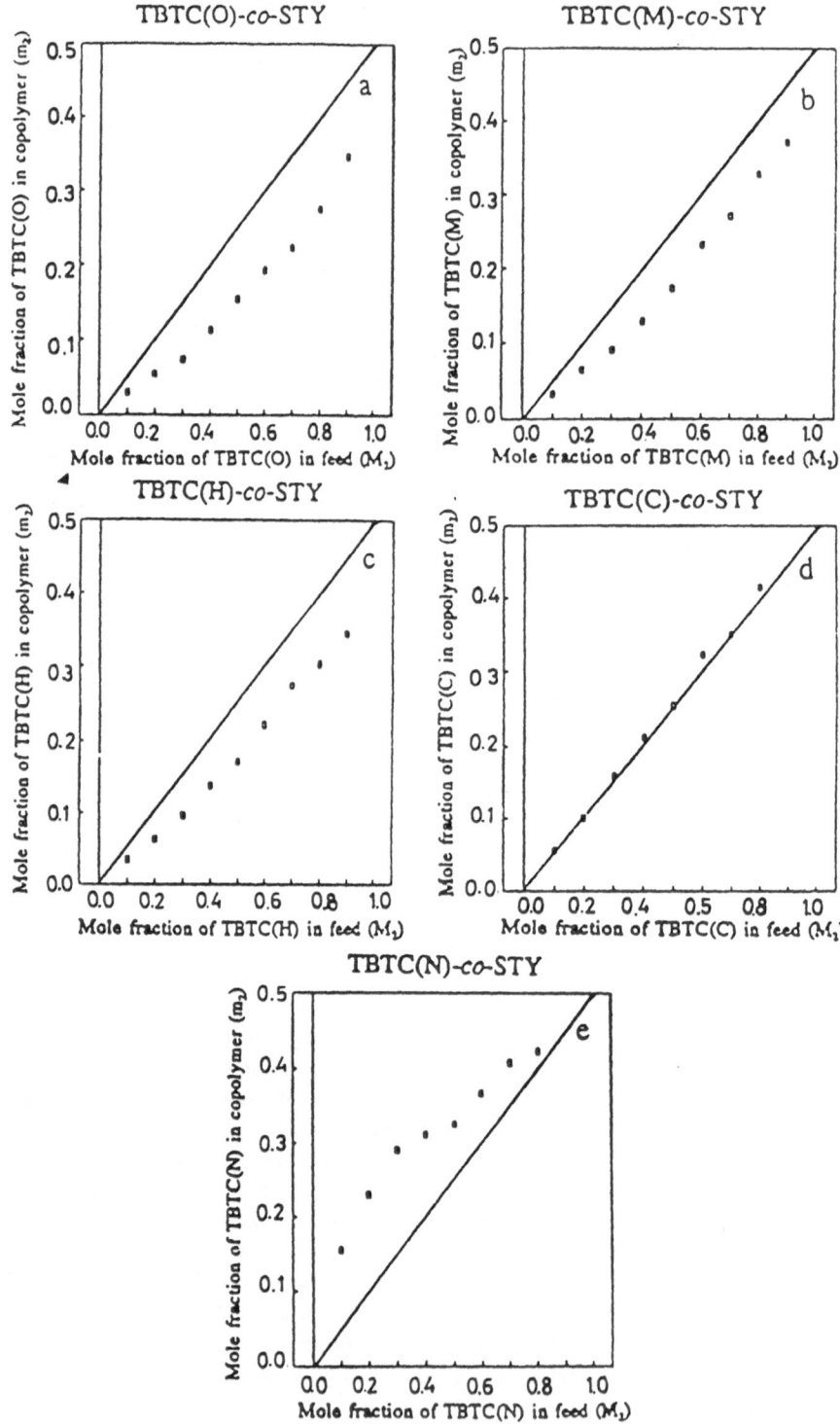

Fig. 2 : Composition diagrams of TBTC(Z)-co-STY systems

Table 3 : Copolymerisation data for tributyltin *p*-methoxycinnamate-*co*-styrene, tributyltin *p*-methylcinnamate-*co*-styrene, tributyltin cinnamate-*co*-styrene, tributyltin *p*-chlorocinnamate -*co*-styrene and tributyltin *p*-nitrocinnamate-*co*-styrene

$M_2:M_1$ ratio	Feed M_1	M_2	Copolymer m_1	m_2	Conv. % wt.	Time (h)
tributyltin *p*-methoxycinnamate-*co*-styrene						
1:9	0.900	0.100	0.971	0.029	9.9	6.0
2:8	0.800	0.200	0.946	0.054	8.3	7.0
3:7	0.700	0.300	0.913	0.087	7.0	8.0
4:6	0.600	0.400	0.884	0.116	6.3	9.5
5:5	0.501	0.499	0.847	0.153	9.8	24.0
6:4	0.401	0.599	0.798	0.202	7.7	26.0
7:3	0.303	0.697	0.778	0.222	5.4	28.0
8:2	0.203	0.797	0.726	0.274	3.6	32.5
9:1	0.100	0.900	0.654	0.346	1.7	48.0
tributyltin *p*-methylcinnamate-*co*-styrene						
1:9	0.900	0.100	0.967	0.033	8.3	6.0
2:8	0.800	0.200	0.935	0.065	8.0	7.0
3:7	0.700	0.300	0.909	0.091	6.7	8.0
4:6	0.600	0.400	0.871	0.129	5.5	9.0
5:5	0.501	0.499	0.827	0.173	10.0	24.0
6:4	0.399	0.601	0.769	0.231	8.0	27.0
7:3	0.302	0.698	0.730	0.270	6.2	30.0
8:2	0.202	0.798	0.673	0.327	4.6	39.0
9:1	0.100	0.900	0.629	0.371	1.6	44.0
tributyltin cinnamate-*co*-styrene						
1:9	0.900	0.100	0.965	0.035	10.7	7.0
2:8	0.801	0.199	0.937	0.063	14.4	14.0
3:7	0.700	0.300	0.904	0.096	11.6	14.0
4:6	0.601	0.399	0.862	0.138	10.9	21.8
5:5	0.500	0.500	0.829	0.171	8.9	24.5
6:4	0.403	0.597	0.779	0.221	7.0	28.3
7:3	0.302	0.698	0.726	0.274	4.5	29.0
8:2	0.203	0.797	0.698	0.302	2.1	30.0
9:1	0.101	0.899	0.656	0.344	0.3	44.0

Table 3. (cont.)

---

tributyltin p-chlorocinnamate-co-styrene

1:9	0.900	0.100	0.945	0.055	10.1	6.0
2:8	0.802	0.198	0.899	0.101	9.2	7.0
3:7	0.699	0.301	0.840	0.160	8.0	8.0
4:6	0.600	0.400	0.788	0.212	7.3	9.0
5:5	0.501	0.499	0.746	0.254	13.3	24.0
6:4	0.400	0.600	0.677	0.323	11.7	27.0
7:3	0.303	0.697	0.649	0.351	9.9	30.0
8:2	0.201	0.799	0.585	0.415	7.8	39.0
9:1	0.105	0.895	0.524	0.476	4.5	44.0

butyltin p-nitrocinnamate-co-styrene

1:9	0.900	0.100	0.845	0.155	3.8	5.5
2:8	0.800	0.200	0.770	0.230	2.4	9.0
3:7	0.700	0.300	0.709	0.291	3.5	10.0
4:6	0.599	0.401	0.688	0.312	4.8	20.2
5:5	0.498	0.502	0.674	0.326	6.1	24.0
6:4	0.403	0.597	0.633	0.367	4.4	28.0
7:3	0.300	0.700	0.592	0.408	3.0	44.0
8:2	0.200	0.800	0.577	0.423	2.9	52.0
9:1	0.107	0.893	0.573	0.427	1.5	65.0

$$y = 1 + r_1' \cdot \frac{r_1 x + 1}{r_1' x + 1}$$

where $y = m_1/m_2$, $x = M_1/M_2$.

The copolymerisation reactivity ratios are:

$$r_1 = \frac{k_{111}}{k_{112}} \quad \text{and} \quad r_1' = \frac{k_{211}}{k_{212}} \; ; \quad r_2 = r_2' = 0.$$

Linearisation of this equation may be achieved by the Kelen-Tudos transformation:[16]

$$\eta = (r_1 + \frac{1}{\alpha r_1'}) \xi - \frac{1}{\alpha r_1'}$$

274

with $\qquad \eta = \dfrac{x.\dfrac{y-2}{y-1}}{\alpha + \dfrac{x^2}{y-1}} \qquad\qquad \xi = \dfrac{\dfrac{x^2}{y-1}}{\alpha + \dfrac{x^2}{y-1}}$

where α is an arbitrary parameter.

     Comparison of the graphical plots (Fig. 3 a and b) linearised to the Kelen-Tudos method indicates that a better fit is obtained if cognizance is indeed taken of penultimate unit effects for TBTC(N)-*co*-STY. The deviation from classical behaviour is, however, less discernible using the more simplistic Fineman-Ross plot for the system (Fig. 4). For the other copolymers, good correlation between η and ξ was achieved with KT-plots based on the terminal model.

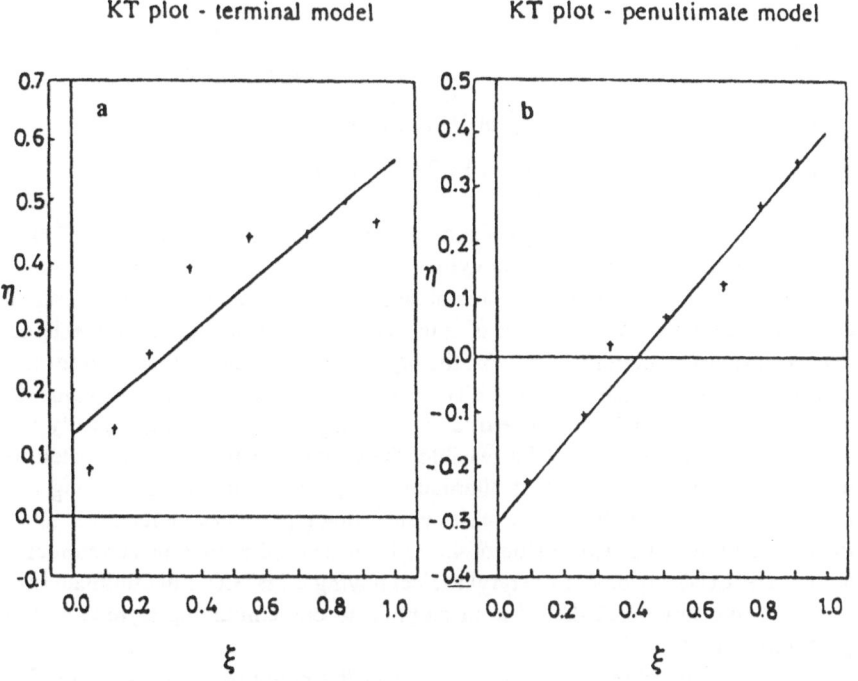

Fig. 3 : Kelen-Tudos plots of the TBTC(N)-*co*-STY system

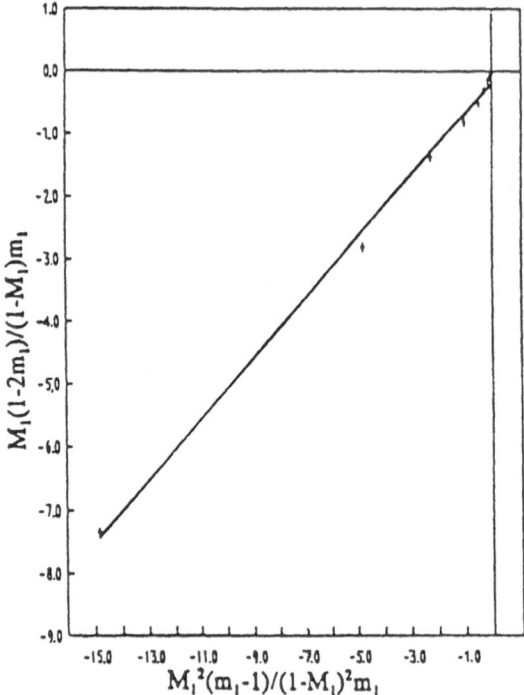

Fig. 4 : Fineman-Ross plot of the TBTC(N)-*co*-STY system

      The reactivity ratios estimated by the Fineman-Ross, Kelen-Tudos and YBR methods for the five copolymer systems are compared in Tables 4 and 5. For TBTC(C)-*co*-STY, the r_1 value obtained by the various methods averages to 1.845, close to the predicted value of 2; this copolymer system is thus considered to have the most random placement of the organotin monomer in the chain for the series under investigation. For TBTC(N)-*co*-STY, the estimated r_1 value is significantly less than 2, indicative of an alternating system. Although a truly 1:1 alternating polymer may not be feasible, the copolymer formed is expected to contain a significant amount of alternating monomer unit sequence along the polymer chain especially in those cases containing higher feed concentrations of the tin monomer. The r_1/r_1' value of 4.69 (Table 5) implies that the addition of a second TBTC(N) monomer to a styryl radical having a TBTC(N) penultimate unit is about five times less than that of addition to one containing styrene as the penultimate unit.

      The observation of a plateau in the composition curve (Fig. 2e) further suggests for the TBTC(N)-*co*-STY system the possibility of a charge transfer complex formation between the two monomers which could give rise to an azeotropic composition given by[16]

$$\frac{M_1}{M_2} = \frac{1 - 2/r_1'}{2 - r_1}$$

276

Table 4 : Reactivity ratios of TBTC(Z)-*co*-STY

Copolymer system	Kelen-Tudos terminal model	Fineman-Ross method	YBR method
TBTC(O)-*co*-STY			
r_1	4.14 ± 0.44	3.60 ± 0.07	3.92 ± 0.17
r_2	-0.06 ± 0.50	-0.16 ± 0.16	-0.05 ± 0.02
α	0.126	-	-
corr. coeff	0.943	0.997	-
TBTC(M)-*co*-STY			
r_1	3.45 ± 0.27	3.13 ± 0.05	3.17 ± 0.10
r_2	-0.03 ± 0.31	-0.13 ± 0.14	-0.04 ± 0.02
a	0.142	-	-
corr. coeff	0.971	0.998	-
TBTC(H)-*co*-STY			
r_1	3.31 ± 0.25	2.95 ± 0.07	3.17 ± 0.10
r_2	-0.05 ± 0.29	-0.17 ± 0.19	-0.04 ± 0.02
α	0.139	-	-
corr. coeff	0.971	0.996	-
TBTC(C)-*co*-STY			
r_1	1.86 ± 0.09	1.81 ± 0.03	1.84 ± 0.03
r_2	0.01 ± 0.09	-0.03 ± 0.11	-0.00 ± 0.02
α	0.460	-	-
corr. coeff	0.993	0.99	-

Table 5 : Reactivity ratios of TBTC(N)-*co*-STY

	Kelen-Tudos penultimate model	Kelen-Tudos terminal model	Fineman-Ross method	YBR method
r_1	0.40 ± 0.09	0.57 ± 0.12	0.49 ± 0.01	0.54 ± 0.04
r_2	0	-0.11 ± 0.11	0.22 ± 0.16	-0.08 ± 0.03
r_1	1.90 ± 0.54	-	-	-
r_2	0	-	-	-
α	1.767	0.825	-	-
corr. coeff	0.954	0.833	0.996	-

Whereas the estimated value of r_1 is unequivocally less than 2, the value of 1.90 together with a relatively large error for r_1' does not, however, permit a rigorous conclusion to be drawn concerning the presence of an azeotropic composition.

The study of tributyltin cinnamates in forming copolymers with styrene reported herein has a precedence in somewhat similar studies using other organotin monomers such as trimethyltin methacrylate (TMTM),[17] tributyltin acrylate (TBTA),[5] tributyltin methacrylate (TBTM),[5,18] triphenyltin methacrylate (TPTM)[6] and bis(tributyltin)maleate (BTM).[8]

The reactivity ratios for the TMTM-, TBTA- and TBTM-co-STY systems indicate a random distribution of monomer units with a tendency toward alternation in the resultant copolymer. In the TPTM-co-STY system, however, the copolymer has an azeotropic composition containing 0.69 mol fraction of TPTM.[6] The BTM monomer which does not undergo homopolymerisation ($r_2=0$) yields a block copolymer with styrene[8] somewhat similar to the TBTC(Z)-co-STY systems (Z = H, O, M) studied here.

Relative reactivities

The reciprocal of the reactivity ratio ($1/r_1$) is, by definition, a measure of the relative reactivities of a series of monomers (M_2) towards a reference comonomer (M_1).[19-21]

Generally, the reactivity of a monomer in a copolymerisation reaction is influenced by such factors as (i) steric hindrance, (ii) the polarity of the C=C bond and (iii) the stability of the free radical adduct formed by addition of the monomer to the growing polymer chain (resonance stabilization).[19-21]

Thus, while the bulkiness of the tributyltin cinnamates may be expected to depress somewhat their reactivity towards the styrene comonomer, the other two factors may be anticipated to exert their influence more strongly on the composition of the resultant copolymer.

Table 6 shows the values of $1/r_1$, calculated from r_1 determined using the YBR method which has been reported[7,13,15] to give a very balanced average reactivity ratio in spite of any stray experimental error.

Table 6 : Relative reactivities of the styryl radical towards substituted tributyltin cinnamates and substituted styrenes

Substituent on aromatic ring	($1/r_1$)	
	substituted tributyltin cinnamates	substituted styrene[19]
p-OCH$_3$	0.255	0.862
p-CH$_3$	0.302	1.205
H	0.315	1.000
p-Cl	0.543	1.348
p-NO$_2$	1.852	5.263

For the tributyltin cinnamates, it is seen that the presence of electron-withdrawing (chloro- and nitro-) ring substituents enhances the reactivity of the tin monomers in the copolymerisation reaction with styrene in comparison with the case where the substituents (methyl- and methoxy-) are electron-donating. The parent cinnamate has a $1/r_1$ value intermediate between these two groups.

The highest reactivity is shown by TBTC(N). This is attributable to the strong electron-withdrawing character of the NO_2 group which makes the conjugated exo-cyclic C=C bond of the monomer more electrophilic, thereby favouring a strongly polar interaction with styrene which has a high electron-availability at its olefinic double bond (with a high e-value[22] of -0.80).[19] This polarity is expected to -manifest itself in a strong tendency on the part of TBTC(N) to form an alternating copolymer with styrene especially at higher TBTC(N) feed concentrations. In addition, a penultimate unit effect may also be expected for the system on account of repulsive forces between the highly polar NO_2 groups which prevent the addition of a TBTC(N) monomer to a polymer chain containing a TBTC(N) unit near the growing end.

In TBTC(C), the electron-withdrawing halogen group is expected to dictate the course of the copolymerisation reaction, though perhaps not as markedly as in the TBTC(N) case since the TBTC(C) monomer would be expected to have lower tendency to alternate compared to the more polar TBTC(N).[19,23]

The reduced reactivity of TBTC(O) (and TBTC(M)) with styrene, on the other hand, is explicable in terms of the reduced electrophilic character of the cinnamyl double bond as a result of the electron-donating methoxy- (and methyl-) group. Indeed, it has been observed that pairs of monomers which most readily copolymerise are those in which an electron-rich double bond is present in one monomer and an electron-poor double bond in the other.[24] The copolymers of TBTC(O) and TBTC(M) with styrene, as with TBTC(H), consist therefore[19] of blocks of polystyrene with random placement of single molecules of the tin monomers along the polymeric chain.

The reactivity trends of the TBTC(Z) monomers in copolymerisation with styrene are seen to parallel those previously reported[19] for the copolymerisation of p-substituted styrene monomers with styrene (Table 6), although the reactivities of the bulkier tributyltin cinnamate monomers towards styrene are about 3 to 4 times lower than those of the similarly substituted styrene monomers.

Characterisation Studies

^{119}Sn NMR analysis

Tin-119 NMR spectra were obtained in $CDCl_3$ for all the five tributyltin cinnamate-co-styrene polymers prepared using 1:1 molar feed ratios, and the data are tabulated in Table 1 along with the chemical shifts of the pure TBTC(Z) monomers. Relative to the tin monomers which yielded sharp resonances, the copolymers yielded broad signals with an average width of 1229 Hz, shifted by 8 - 13 ppm to higher frequencies. The tin chemical shifts of the nitro- and chloro-substituted tin monomers occur at lower fields compared to the other cases; this trend is also seen for the pure tributyltin monomers.

The broad unresolved tin peaks for poly(TBTC(Z)-co-STY) contrast with those reported for the 1:1 and 1:2 poly(methyl methacrylate-*co*-tributyltin methacrylate) which exhibit triplet [119]Sn resonances that have been ascribed to compositional sequencing of the following TBTM-centred triads: MMA-TBTM-MMA, MMA-TBTM-TBTM and TBTM-TBTM-TBTM.[25] It is suggested that the unresolved tin signals in the present study, in the absence of homopolymerisation of TBTC(Z), is probably the consequence of very randomly assembled monomer sequences, $(STY)_m$-TBTC(Z)-$(STY)_n$ (m,n >1), although other factors may also be contributory.

[119m]Sn Mössbauer Analysis

[119m]Sn Mössbauer spectral data for three representative TBTC(Z)-*co*-STY systems (Z = O, H, M) are tabulated in Table 2, together with data for the corresponding monomers. Asymmetrical doublets characteristic of polymeric organotin species were obtained for all the three copolymers.

Perusal of the data shows that when the respective tin monomers are copolymerised with styrene, there is a marginal decrease of between 0.13-0.14 mm s^{-1} in the IS values. This suggests that the *s*-electron density at the tin atom is little altered as a result of the polymerisation.

In contrast, the QS values for the copolymers show a marked decrease of about 1.0 mm s^{-1} compared to the respective tin monomers. The QS values for the monomers in the range 3.60 - 3.75 mm s^{-1} are diagnostic of pentacoordinated tin structures with *trans*-$[R_3SnO_2]$ trigonal bipyramidal geometries.[26] The result is typical of trialkyltin carboxylates which yield carboxylate-bridged polymers in the solid state.[27] These **coordinate** polymers often, however, dissociate into monomeric four-coordinated species in $CDCl_3$ solution.

For the **covalent** TBTC(Z)-*co*-STY polymers, the lower QS values of 2.61-2.74 mm s^{-1} are suggestive of four-fold coordination at tin and this is also reflected in the magnitude of the QS/IS ratio which is < 2.1 in each case.[28]

In view of the polymerisation behaviour of the TBTC(Z) monomers with styrene which leads to monomer sequences of the form STY_m-TBTC(Z)-STY_n (where m, n > 1), a tetrahedral configuration at the tin atoms is not unreasonable. Since the TBTC(Z) units would be randomly assembled in the copolymer and separated by blocks of styryl units, the possibility of intermolecular Sn←:O coordinative linkages is significantly reduced, including for the case of TBTC(N)-*co*-STY which has the highest degree of alternating units in the series.

Thermal Analysis

The Differential Scanning Calorimetry (DSC) curves of the 1:1 poly(TBTC(Z)-*co*-STY) are depicted in Fig. 5. Observed T_m and ΔH_m values together with the associated copolymer compositions are assembled in Table 7.

All five thermograms exhibit neither a glass transition temperature around 375K nor a melting peak around 500 K. This suggests that the copolymers are not contaminated to any significant extent by the polystyrene homopolymer. Their melting points, however, are lowered by 150-160 K from the T_m value of 500 K, reflecting a decrease in crystallinity[21] attending the incorporation of the tributyltin groups into the polystyrene chain.

Table 7 : T_m and ΔH_m of copolymers of tributyltin cinnamates with styrene

Copolymer system	Composition of styrene in copolymer (m_1)		T_m (K)	ΔH_m (cal g^{-1})
	Mole* fraction	Average %		
TBTC(O)-*co*-STY	0.847	55.0	346	0.873
TBTC(M)-*co*-STY	0829	54.0	340	0.825
TBTC(H)-*co*-STY	0.827	52.0	345	0.771
TBTC(C)-*co*-STY	0.746	39.0	342	0.507
TBTC(N)-*co*-STY	0.674	31.0	347	0.245

*estimated from ^{1}H NMR analysis (see Table 3)

Since T_m is the melting point of the polystyrenyl crystalline regions, it follows that the higher the mole fraction of the tributyltin cinnamate monomer in the copolymer, the more diffuse is the T_m peak that is observed. Indeed, for poly(TBTC(N)-*co*-STY), which has the highest mole fraction (0.326) of tributyltin monomer incorporated into the copolymer, T_m is barely located at 347 K (Fig. 5, curve e). By way of contrast, sharper T_m peaks are located in the DSC curves of TBTC(O)-, TBTC(M)-, TBTC(H)- and TBTC(C)-*co*-STY systems, which have lower tributyltin monomer contents (Fig. 5, curves a-d).

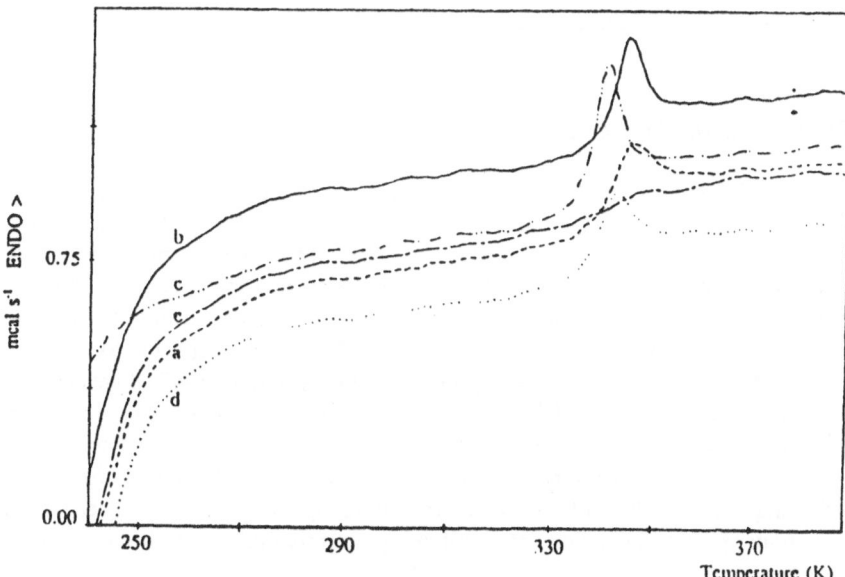

Fig. 5 : DSC curves for poly(TBTC(Z)-co-STY)
 a : Z = OCH$_3$ (-··-, 10.12 mg) d : Z = Cl (....., 10.23 mg)
 b : Z = CH$_3$ (____ , 10.25 mg) e : Z = NO$_2$ (--·-, 10.23 mg)
 c : Z = H (-···-,10.20 mg)

281

Similarly, the heat of fusion, ΔH_m, is also related to the extent of crystallinity in the copolymer,[21,29] i.e. to the amount of polystyrene blocks present. Thus, higher amounts of styrenyl units in the copolymer will yield higher values for ΔH_m, as exemplified by the higher value of 0.873 cal g^{-1} obtained for the TBTC(O)-*co*-STY than for TBTC(N)-*co*-STY (0.245 cal g^{-1}).

A plot of the average percentage of styrene present in the five TBTC(Z)-STY copolymers versus ΔH_m revealed an almost linear relationship, as shown in Fig. 6.

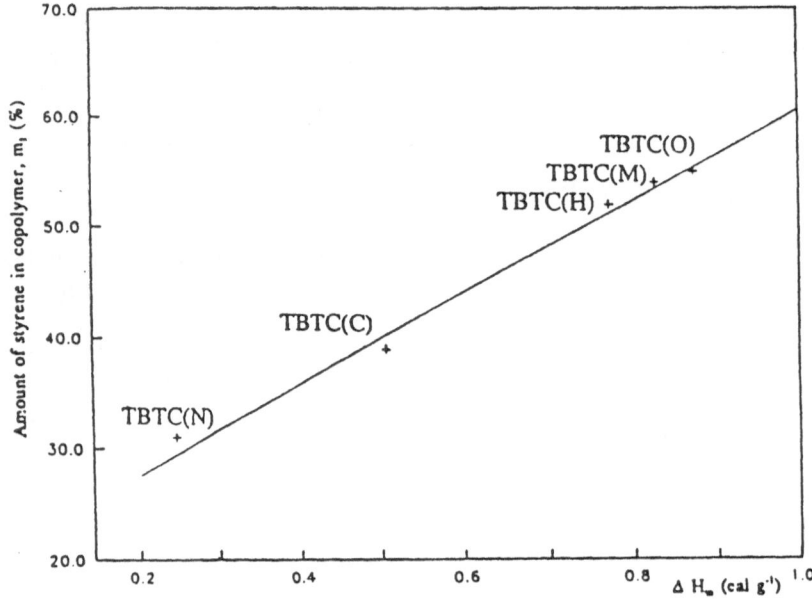

Fig. 6 : Plot of ΔH_m versus % of styrene in TBTC(Z)-*co*-STY

The magnitudes of ΔH_m for the copolymers studied were found to increase in the order TBTC(O) > TBTC(M) ~ TBTC(H) > TBTC(C) >> TBTC(N). This agrees well with the observed tributyltin monomer content in the copolymers determined from ^{1}H-NMR analysis. The above is also in agreement with the inverse trend in the tributyltincinnamate monomer reactivities towards styrene, *viz.* TBTC(N) >> TBTC(C) > TBTC(M) ≈ TBTC(H) > TBTC(O) (Table 6).

Recently,[30] in studies on ethylene-propylene copolymers by DSC, it was shown that the enthalpies of fusion of the ethylene and propylene segments were functions of their respective sequence lengths. Consequently, it was of some interest to see whether a similar dependency existed for the TBTC(Z)-STY copolymers. By assuming that the melting peaks are due to uninterrupted styrene sequences which can undergo crystallisation, the following correlation[30] may be tentatively proposed:

$$\Delta H_m = k \, m_i^n$$

where m_i represents the mole fraction of styrene in the copolymer, n is the minimum sequence length for crystallisation to occur and k is the proportionality constant. A linear plot is indeed realised as shown in Figure 7 when log ΔH_m is plotted against the styrene content. This yields a value of n = 3.2 for the slope, indicating that the styrene blocks in the copolymer require a styrene sequence length of at least 3 units for crystallisation to occur.

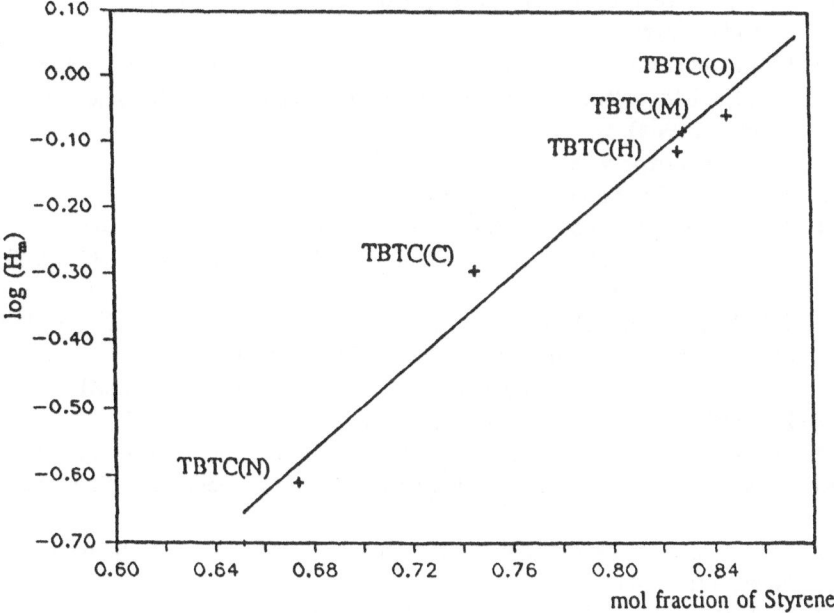

Fig 7 :   **Plot of mole fraction of styrene in TBTC(Z)-*co*-STY versus log (ΔH_m)**

Acknowledgements

We thank the the National Science Council for Research and Development, Malaysia (Grant No. 2-07-04-06) for supporting this work. One of us (SLF) is grateful to the University of Malaya for a Postgraduate Fellowship Award.

References

1.   J. Kowalski, W. Stancyk and J. Chojnowski, *Rev. Si, Ge, Sn and Pb Compds.*, **6** (1982) 225.

2.   R.V. Subramanian and K.N. Somasekharan, *J. Macromol. Sci. -Chem.*, **A16** (1981) 73.

3.   D. Atherton, J. Verborgt and M.A.M. Winkler, *J. Coat. Technol.*, **51** (1979) 823.

4.   C.J. Evans and R. Hill, *Rev. Si, Ge, Sn and Pb Compds.*, **7** (1983) 57.

5. N.A. Ghanem, N.N. Messiha, N.E. Ikladious and A.F. Shaaban, *Eur. Polym. J.*, **16** (1980) 339, 1047; *idem.*, *J. Appl. Polym. Sci.*, **26** (1981) 97.

6. N.E. Ikladious, N.N. Messiha, and A.F. Shaaban, A F *Eur. Polym. J.*, **20** (1984) 625

7. B.K. Garg, J. Corredor and R.V. Subramanian, *J. Makromol. Sci. Chem.*, **A11** (1977) 1567.

8. N.E. Ikladious and A.F. Shaaban, *Polymer*, **24** (1983) 1635.

9. N.A. Ghanem, N.N. Messiha, N.E. Ikladious and A.F. Shaaban, *Eur. Polym. J.*, **15** (1979) 823.

10. M. Zeldin and J.J. Lin, *J. Polym. Sci., Polym. Chem. Ed.*, **23** (1985) 2333.

11. Vogel, A I "*A textbook of Practical Organic Chemistry*", 3rd ed., Longmans, London, (1964), pp. 173-174.

12. M. Fineman and S.D. Ross, *J. Polym. Sci.*, **5** (1950) 259.

13. T. Kelen and F. Tüdős, *J. Macromol. Sci. Chem.*, **A9** (1975) 1.

14. T. Kelen and F. Tüdős, *J. Macromol. Sci. Chem.*, **A16** (1981) 1283.

15. A.I. Yezrielev, E.L. Brokhina and Y.S. Roskin, *Vysokomol. Soed.*, **A11** (1969) 1670; see also ref. 7.

16. T. Kelen, F. Tüdős, D. Braun and W.K. Czerwinski, *Makromol. Chem.*, **191** (1990) 1853; *idem.*, *Makromol. Chem.*, **191** (1990) 1863.

17. B. Yamada, H. Yoneno and T. Otsu, *J. Polym. Sci.*, **A1** (1970) 2021.

18. P.C. Deb and A.B. Samui, *Angew. Makromol. Chem.*, **112** (1983) 15.

19. G.E. Hams, E "*Copolymerization*", Interscience Publishers, New York, 1964.

20. G.M. Burnett, "*Mechanism of Polymer reactions*", Interscience Publishers, New York, 1954.

21. F.W. Billmeyer, "*Textbook of Polymer Science*", 3rd ed., Wiley-Interscience, New York, 1984.

22. T. Alfrey and C.C. Price, *J. Polym. Sci.*, **2** (1947) 101.

23. F.R. Mayo and C. Walling, *Chem. Rev.*, **46** (1950) 191.

24. C.C. Price, *J. Polym. Sci.*, **3** (1948) 772.

25. W.F. Manders, J.M. Bellama, R.B. Johannesen, E.J. Parks and F.E. Brinckman, *J. Polym. Sci., Polym. Chem.*, **A25** (1987) 3469.

26. G.M. Bancroft, V.G. Kumar Das, T.K. Sham and M.G. Clark, *J Chem. Soc., Dalton Trans.*, (1976) 643.

27. E.R.T. Tiekink, *Appl. Organomet. Chem.*, **5** (1991) 1.

28. R.H. Herber, H.A. Stöckler and W.T. Reichle, *J. Chem. Phys.*, **42** (1965) 2447.

29. P.J. Flory, *J. Chem. Phys.*, **17** (1949) 223.

30. S.N. Gan, D.R. Burfield and K. Soga, *Macromolecules*, **18** (1985) 2684.

Main Group Elements and Their Compounds
V.G. Kumar Das (Ed)

CO₂ - Laser-induced Surface Grafting of 2-Hydroxyethyl methacrylate and N-vinylpyrrolidone onto Ethylene-Propylene Rubber for Use as Biomaterial

H. Mirzadeh[a], A.A. Katbab[a] and R.P. Burford[b]
*[a]Department of Polymer Engineering, Amirkabir University, Tehran, Iraq,
and [b]Polymer Science Department, University of South Wales, Sydney,
Australia*

Several reports have been published dealing with laser initiated polymerization of vinyl monomers.[1-4] The advantage here is that the laser can be tuned to a specific wavelength thereby exciting a particular absorption band in the monomer without dispersing wasteful energy throughout the medium. Consequent scission of the relevant bond initiates the polymerization. An infrared laser beam, for example, has been successfully used for the imidization of polyamide acids.[5] We have recently described the use of CO_2 pulsed lasers for polymerization and grafting of vinyl monomers onto polymer substrates.[6] Poly(2-hydroxyethyl methacrylate) (PolyHEMA) and polyN-vinylpyrrolidine (polyNVP) are hydrogels with high hydrophilicity and good biocompatibility, but show weak mechanical properties when swelled in water.[7] A balance between hydrophilic and hydrophobic properties is necessary to optimize biocompatibility.[8] In the present work, we report on the grafting of HEMA and NVP onto the surface of hydrophobic ethylene-propylene rubber EPR (Vistalon 808) using a line tunable CO_2 pulsed laser (TEA CO_2 laser Lumonics-103-2), and the surface characterization, water compatibility and tissue compatibility of the resulting grafted samples.

Experimental

The grafting experiment was performed applying the simultaneous technique. A special cylindrical pyrex reactor equipped with NaCl window was charged with an aqueous solution of HEMA or NVP, photosensitizing system (typically benzophenone/thf) and extracted dried films of vulcanized EPR (0.3 mm thick). Oxygen was removed by purging the solution with nitrogen for 10 minutes. The system was then exposed to the CO_2 laser pulses of selected fluence (output power, J/cm²) and repetition rate for predetermined times at room temperature. The laser was tuned to the appropriate wavelength corresponding to the absorption maxima of C-H bond of HEMA, BZP (10.6 μm) or NVP (0.5 μm). After each treatment, samples were removed, cleaned free of the external homopolymer, and washed with acetone/water mixture. The samples were extracted and then dried. Surface characterization was performed using ATR-IR (Brucker 90, KRS-5), EDXA (AN-10,000) coupled with a scanning electron microscope SEM (Cambridge-S-360). Water compatibility of the grafted samples was evaluated by measuring the extent of attachment of macrophage cells (Alveolar from mice) and their degree of spreading.[9] Adhesion experiments were carried out by seeding the samples at the bottom of each well of a 96 well microtitre plate (NUNC) for a predetermined time at 37 °C. Adherent cells were examined by scanning electron microscopy at 10 kV.

Results and Discussion

Fig. 1 depicts the ATR-IR spectra of EPR samples grafted with HEMA in the presence of benzophenone (BZP) using laser pulses of 10.6 μm wavelength (corresponding to the maximum absorption of both HEMA and BZP). Variation in the intensity of the carbonyl stretching frequency (1722 cm^{-1}) of the grafted Poly(HEMA) with the laser repetition rate is also demonstrated in Fig. 1. It can be seen that the pulsing rate has a profound effect upon the degree of grafting, and also that the trend is not linear. The effect of laser fluence upon the degree of grafting of HEMA and NVP is shown in Figs. 2 and 3, respectively.

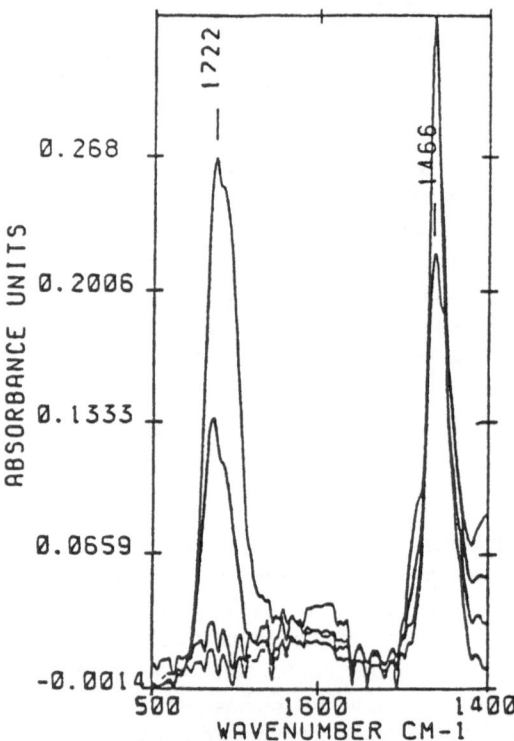

Fig. 1 : Comparison of ATR-FT-IR spectra of EPR grafted with HEMA in the presence of the BZP using CO$_2$ laser pulses of λ 10.6 μm at a repetition rate of (a) 1, (b) 0.75, (c) 0.5 and (d) 0.3 Hz

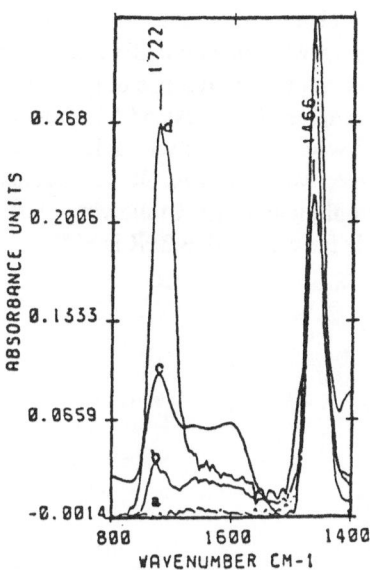

Fig. 2 : ATR-FT-IR spectra of unmodified (a) and HEMA modified (b-d) EPR samples. Laser (λ 10.6 μm) pulse fluence: b) 1 J cm^{-2}, c) 3 J cm^{-2}, d) 5 J cm^{-2}

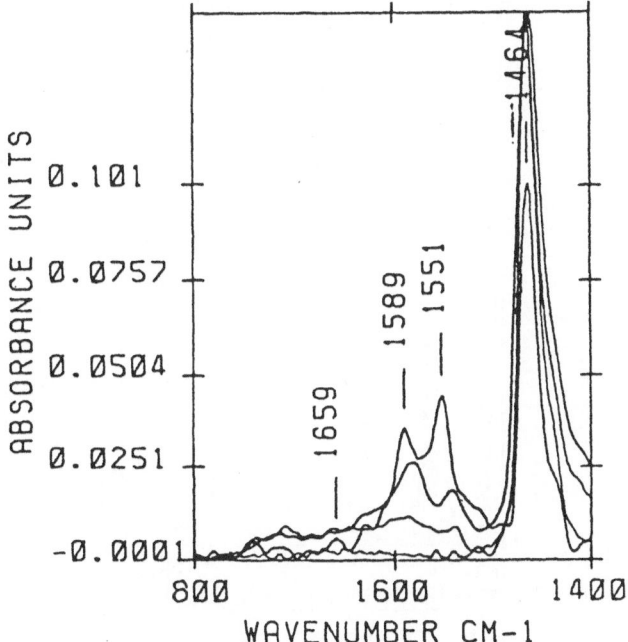

Fig. 3 : ATR-FT-IR spectra of unmodified (a) and NVP modified (b-d) EPR samples. Laser (λ 9.58 μm) pulse fluence: b) 1 J cm^{-2}, c) 3 J cm^{-2}, c) 5 J cm^{-2}

Fig. 4 shows the EDXA of the HEMA modified EPR samples. The higher percentage of oxygen observed on the surface of the grafted samples (hydrophilic sites) compared with the unmodified sample is consistent with the ATR-IR results, and attests to the presence of grafted HEMA groups on the surface of the treated samples. Figs. 5A and 5B show the variation with the level of grafting of the water drop contact angle for samples modified with HEMA and NVP, respectively. The results are explicable in terms of the increased surface hydrophilicity of EPR following grafting with HEMA and NVP. The contact angle for unmodified EPR is 89°).

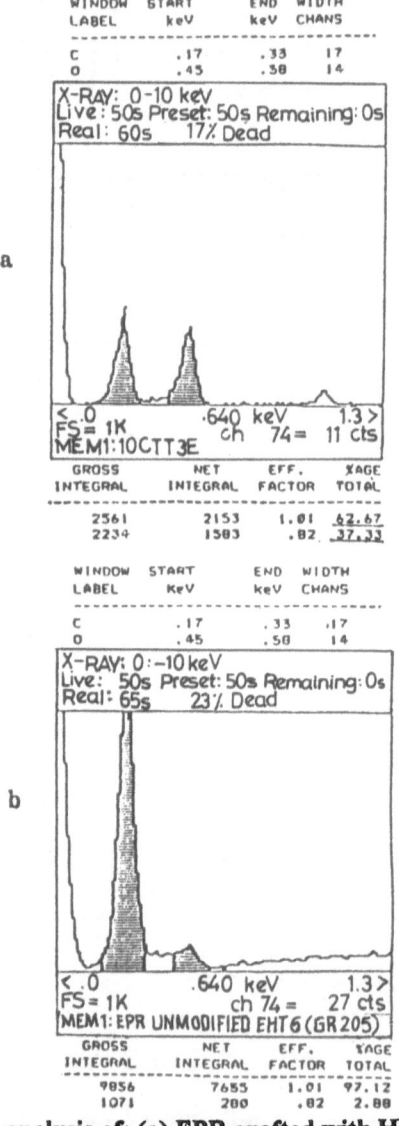

Fig. 4 : EDXA analysis of: (a) EPR grafted with HEMA in the presence BZP using CO$_2$ laser pulses of λ 10.6 μm and repetition rate of 0.5 Hz, (b) unmodified EPR

288

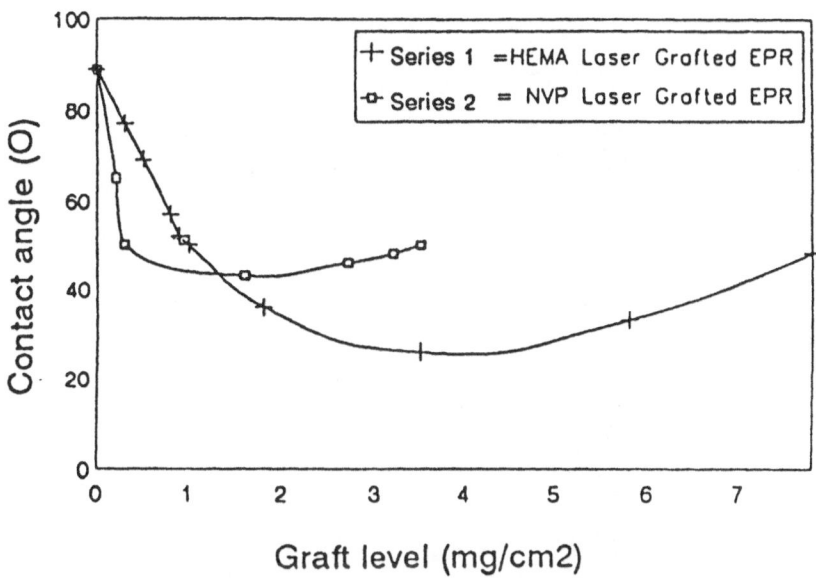

Fig. 5 : **Variation of water drop contact angle for samples modified with HEMA or NVP with the extent of grafting**

Fig. 6 plots the number of macrophages adhering to the surface of HEMA grafted EPR as a function of graft level. It can be seen that although the number of adherent cells is higher for the modified samples, the surface with high wettability does not necessarily lead to better tissue compatibility. In other words, surface wettability as well as the ratio of hydrophobic to hydrophilic sites needs to be optimized for desired tissue compatibility. Typical SEM micrographs of adhered macrophages on HEMA grafted EPR sample surface have also been taken and compared with those involving the unmodified surface (Fig. 7). Cell adhesion with better flattening on the HEMA grafted EPR is indicative of better tissue compatibility.

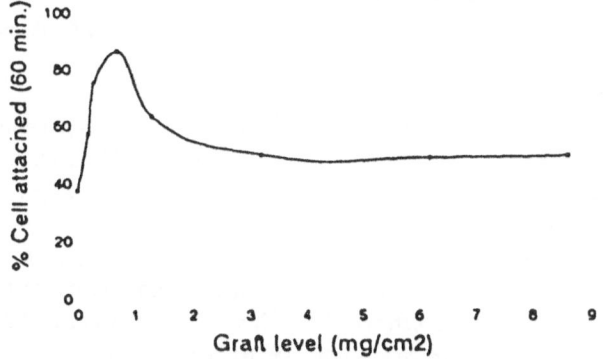

Fig. 6 : **Influence of HEMA grafting on the surface attachment of macrophage cells onto EPR**

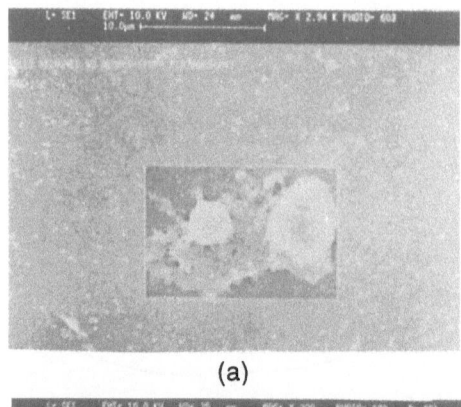

(a)

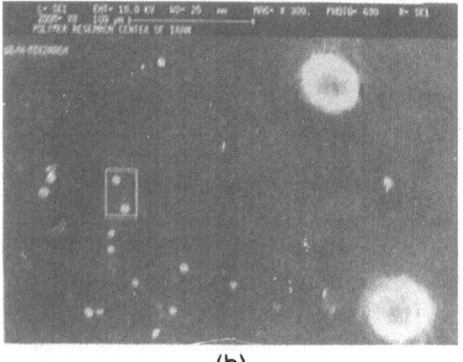

(b)

Fig. 7 : SEM photomicrograph of cells attached to the surface of (a)
unmodified EPR, and (b) laser HEMA grafted EPR

References

1. R.K. Sadhir, G.D.B. Smith and P.M. Castle, *J. Polym. Sci. Polym.Chem.*, **23** (1985) 411.
2. C. Decker and K. Moussa, *Radiation curing of polymeric materials*, John Wiley, N.Y., 1990, p. 439.
3. D.J. Lougnot and J.P. Fouassier, *Makrom. Chem. Rapid Commun.*, **4** (1983) 11.
4. C. Decker, *J. Coat. Tech.*, **56** (1984) 29.
5. S.G. Il'yasov, N. Kalvina, G.A. Kyulyan, V.F. Moskalenko and E.P. Ostapchenko, *Sov. J. Quant. Electron*, **4**(10) (1975) 1287.
6. H. Mirzadeh, A.A. Katbab and R.P. Burford, *Radiat. Phys. Chem.*, **41**(3) (1993) 507.
7. B.D. Ratner and A.S. Hoffman, *Synthetic Hydrogels for Biomedical Application*, ed., J.D. Andrade, ACS Symposium Series, **31** (1976) 1.
8. B.D. Ratner, A.S. Hoffman, S.R. Hanso, L.A. Harker and J.D. Whiffen, *J. Polym. Sci. Polym. Symp.*, **66** (1976) 363.
9. G.K. Brand, *Foreign body induced sarcomas in cancer: A comprehensive Treatise*, Vol. 1, ed., F.F. Becker, Plenum Press, New York, 1975.

Main Group Elements and Their Compóunds
V.G. Kumar Das (Ed)

Improved Polycaprolactone-polyether Block Copolymer and its Drug Release Property

Shenguo Wang, Jianzhoug Bei and Zhifeng Wang

Institute of Chemistry, Academia Sinica, Beijing 100080, P.R. China

Biodegradable polymers are of special interest in the development of efficacious controlled release drug carriers. This is because unlike in a general polymer where the drug release rate decreases with depletion of drug content, a constant release rate is, in principle, achievable with biodegradable polymers on account of their degradation *in vivo*.[1,2] Among the several class of biodegradable polymer materials receiving study are the aliphatic polyesters which can be readily degraded through hydrolysis of ester bonds.

Polycaprolactone (PCL) is one of the aliphatic polyesters with good drug transportability, but the biodegradation rate of PCL is not fast because of its high crystallinity.[3] In an effort to improve and control the degradability of the PCL, a number of copolymers such as random or block copolymers of caprolactone-*co*-lactide[4], poly(caprolactone)-β-polylactide,[5] polylactide-β-poly(ethylene glycol)[6] etc., have been synthesized and studied.

In this paper we describe the synthesis of a new biodegradable polymer - poly(caprolactone)-poly(ethylene glycol) block copolymer (PCE) - and the relationships between its drug release rate, using 5-Fluoro-uracil as a model drug, and other parameters such as composition, hydrophilicity, crystallinity and biodegradability which have been investigated using by ^{1}H-NMR, X-ray diffraction, DSC, solution viscosity, and other techniques.

Experimental

Materials

ε-Caprolactone, from Aldrich Chemical Company, Inc., was purified by drying over calcium hydride for 24 hr followed by distillation (70 °C/1 mm Hg).

Poly(ethylene glycol), Mr 6000 was obtained from Tianjin Eastern Health Materials Factory.

Synthesis and purification: A mixture of ε-caprolactone and polyethylene glycol (in different compositional amounts for each experiment so as to form a series) was heated at 160 °C for 10 h under an argon atmosphere in the presence of Ti(OBu)$_4$ as catalyst. The crude PCE product formed was taken up in CHCl$_3$, reprecipitated with petroleum ether, and following extraction with cold ethanol was dried in vacuum at room temperature.[7,8]

Water Sorption: The water sorption of the copolymer was determined by the method reported previously:[9]

$$\text{Water sorption (\%)} = [W_1 - W_2]/W_2 \times 100\%$$

where W_1 is the weight of the copolymer after water sorption, and W_2 is the weight of the dried copolymer.

Copolymer composition : The composition of the PCE copolymer was determined by ^{1}H-NMR spectrometry using a JEOL JNM-FX100 spectrometer with deuterated-chloroform as solvent at room temperature.

Crystallinity of the copolymer : This was studied by means of X-ray diffraction (Rigaku Dmax-3B X-ray diffractometer) and differential scanning calorimetry (Perkin-Elmer DSC-7 at a heating rate of 5 °C/min[10]).

In vitro degradation of PCE copolymer : 200 mg of PCE copolymer films (thickness about 0.15 mm) were placed into flasks containing 150 mL of a phosphate buffer of pH 7.2 at 37 ± 1 °C, with and without 15 mg of lipase. At predetermined time intervals, the film samples were removed from the flask and were rinsed with distilled water. The samples were dried over P_2O_5 in vacuum to constant weight. The extent of *in vitro* (percent) degradation was evaluated from viscosity measurements:

$$[\Delta\eta]\,(\%) = ([\eta]_o - [\eta]_t)/[\eta]_o \times 100\%$$

where $[\eta]_o$ and $[\eta]_t$ are inherent viscosities of the copolymer before and after degradation, respectively. The inherent viscosity measurements were made at 30 °C, and using the "One Point Method"[11].

Drug release rates: 5-Fluoro-uracil (5-Fu) was employed as the model drug in this experiment. A mixture of 20 mg of 5-Fu and 200 mg of PCE copolymer was charged into a poly(tetrafluoroethylene) mould and heated to 70 °C to obtain tablets of size 10 mm diameter and 3 mm thickness. The tablet samples were placed into flasks filled with 120 mL buffer medium of pH 7.4 and the drug release test was carried out at 37 ± 1 °C. At specific time intervals, the medium was sampled and the concentration of 5-Fu was detected by UV spectral analysis (λ 268nm). The drug release rate was calculated using the equation:

$$D(\%) = [C_t \times 120]/W_o \times 100\%$$

where W_o is the original weight of 5-Fu and C_t its weight determined at time t..

Results and Discussion

A new polymer - polycaprolactone-poly(ethylene glycol) block copolymer was synthesized from ε-caprolactone (CL) and poly(ethylene glycol) (PEG) by direct copolycondensation in the presence of Ti(OBu)$_4$ as catalyst at 160 °C. A series of PCE copolymers with different compositional PCL/PEG ratio was thus prepared.

$$H[O(CH_2)_5CO]_n- \quad -[O(CH_2CH_2O)_m]- \quad -[OC(CH_2)_5O]_nH \qquad (1)$$

PCL   PEG   PCL

Polycaprolactone-poly(ethylene glycol) block copolymer (PCE)

The crystallinity of the copolymer was determined from X-ray diffraction. The crystallinity of the copolymer was found to decrease with increasing PEG content (Fig. 1). This revealed that the crystallinity of the copolymer could be easily adjusted by changing the content of the PEG component. It is also possible to adjust the crystallinity of the PCE copolymer by using different molecular weight PEG segments, as shown in Table 1.

Table 1 : Dependence of crystallinity of PCE copolymer on the molecular weight of PEG segment

No.	Segment molecular weight		ΔH (J/g)*
	PCL	PEG	
PCE-12	20,000	1,000	64.40
PCE-3	20,000	6,000	56.34
PCE-7	20,000	10,000	53.34

*determined from DSC measurements

Hydrophilicity of a polymer is not only an important factor in influencing its biodegradability via hydrolysis, but also in influencing its drug release property via drug dissolution and diffusion. The hydrophilicity of the PCE copolymer was determined by water sorption measurements. The relationship between crystallinity and hydrophilicity was also established in the PCE copolymer. Thus it was found that both crystallinity and hydrophilicity are closely dependent on the composition of the PCE copolymer; the greater the PEG content, the less crystalline and more hydrophilic is the copolymer (Fig. 1).

The degradability of the copolymer *in vitro* was characterized and compared by measuring the inherent viscosity of the copolymer. According to the dynamic equation for hydrolysis of common esters[3]:

$$[\eta] = [\eta]_o \exp(-\alpha K t)$$

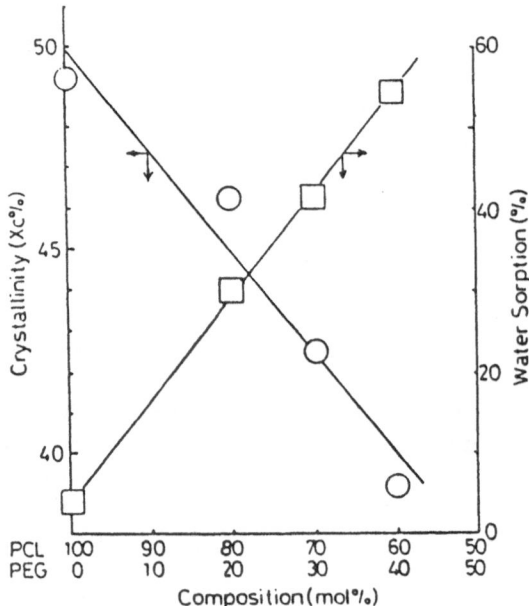

Fig. 1 : Dependence of crystallinity and water sorption on the composition of PCE copolymer

where α is the exponent in the Mark-Honwink equation, $[\eta] = Km^{\alpha}$; K is a constant dependent on temperature, and t is the degradation time. Rewriting the above equation as:

$$\ln(\frac{[\eta]_t}{[\eta]_o} \cdot e) \quad = \quad 1 - \alpha Kt \, ,$$

and applying this to the results of our viscosity measurements, leads to the plots shown in Fig. 2. From Fig. 2, the half-life for the polymer, $T_{1/2}$ (i.e. the time required to reduce its intrinsic viscosity $[\eta]_o$ to half) and the constant αK can be obtained, and related to the crystallinity of the PCE copolymer as shown in Fig. 3. It is clear that the degradation rate of the PCE copolymer increases upon reducing the crystallinity of the PCE copolymer.

The effect of enzyme on degradation rate of the PCE copolymer was tested by using lipase as the model enzyme. It was found that the effect of lipase on the degradation rate of the PCE copolymer is influenced by the composition of the copolymer. As shown in Table 2, the greater the PCL content, the greater is the degradation rate.

The drug release property of the copolymer was studied *in vitro* using 5-Fluorouracil as the model drug at pH 7.4 and at 37 °C. Figures 4 and 5 show that the 5-Fu release rate increases with increasing PEG content and with increasing segmental molecular weight of PEG. The trend is similar to that of the *in vitro* degradability of the PCE copolymer. The drug release rate of the PCE copolymer is thus also increased by reducing the crystallinity of the copolymer. Fig. 4 allows an estimate of time for 50% release of 5-Fu from its original capacity in the copolymer. The plot of T_{50} vs. crystallinity of the PCE copolymer is shown in Fig. 6. It is found that the T_{50} can be adjusted within a wide range from several hours to several months. This means that the drug release rate of the PCE copolymer can also be controlled by adjusting the composition of the copolymer.

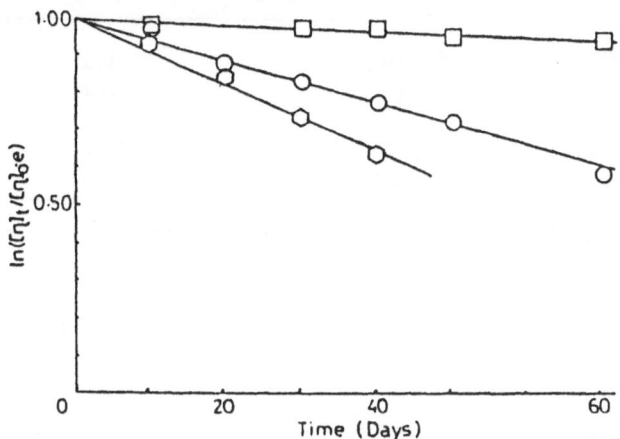

Fig. 2 : Dependence of degradability of the crystallinity of PCE copolymer (at pH 6.5 and 37 °C). Crystallinity (Xc%): -□- 49.2, -O- 46.3, - ⬡- 39.1

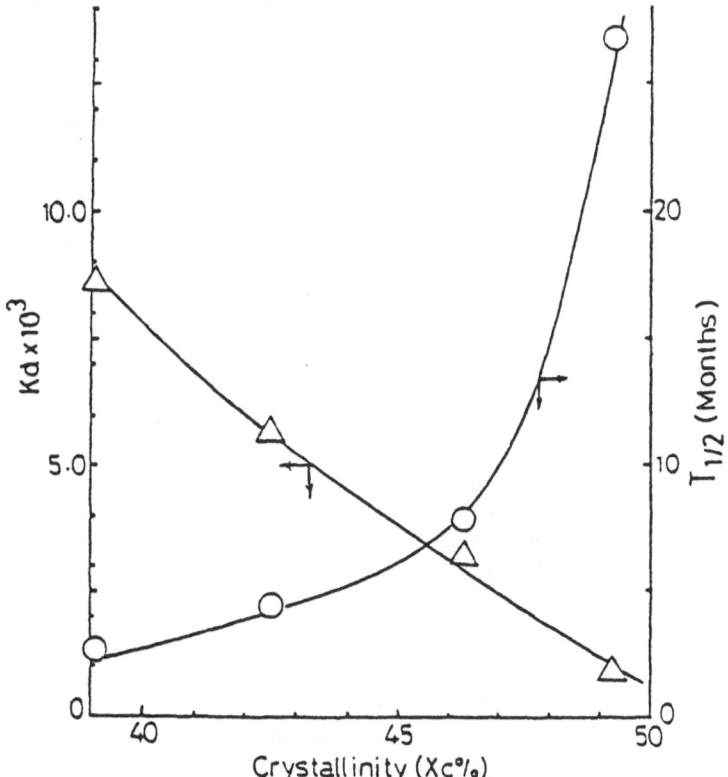

Fig. 3 : Dependence of degradation rate (T_w) and Kα on crystallinity of PCE copolymer (at pH 6.5, 37 °C)

Table 2 : Effect of lipase on the degradation rate of PCE copolymer (at pH
7.2 and 37 °C for 30 days)

Molar ratio of PCE (PCL/PEG)	Enzyme	$\Delta[\eta]\% = ([\eta]_o - [\eta]_t)/[\eta]_o \times 100$
100/0	Lipase	8.1
	none	6.4
75/25	Lipase	11.0
	none	10.8

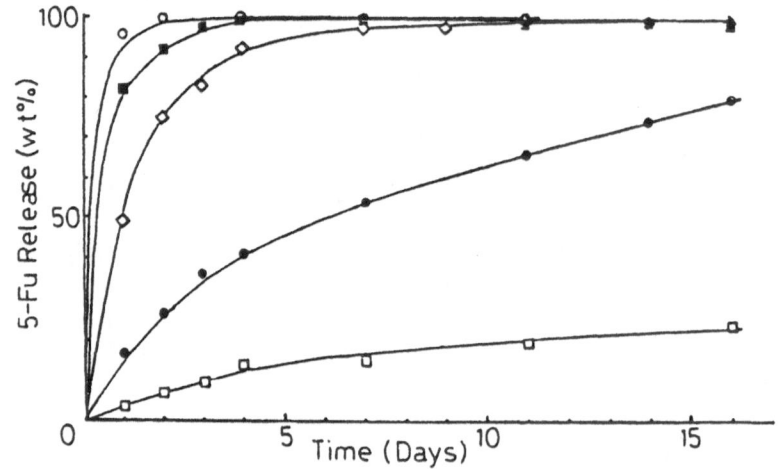

Fig. 4 : Drug release behaviour of PCE copolymer using 5-fluoro uracil as
model drug (at pH 7.4 and 37 °C). PCE composition (molar ratio of
CL/EG): -O- 100/0, -■- 80/20, -◊- 70/30, -●- 60/40, -□- 40/60

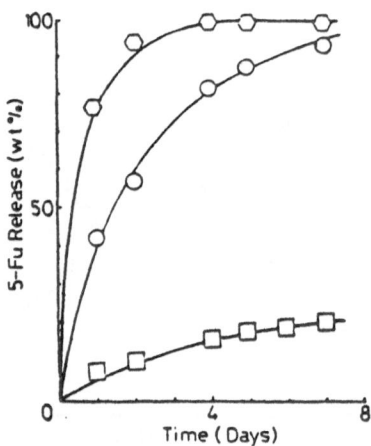

Fig. 5 : Dependence of 5-Fu release rate on molecular weight of PEG
component (at pH 7.4 and 37 °C). M_r of PEG: -□- 1,000, -O- 6,000,
-○- 10,000

296

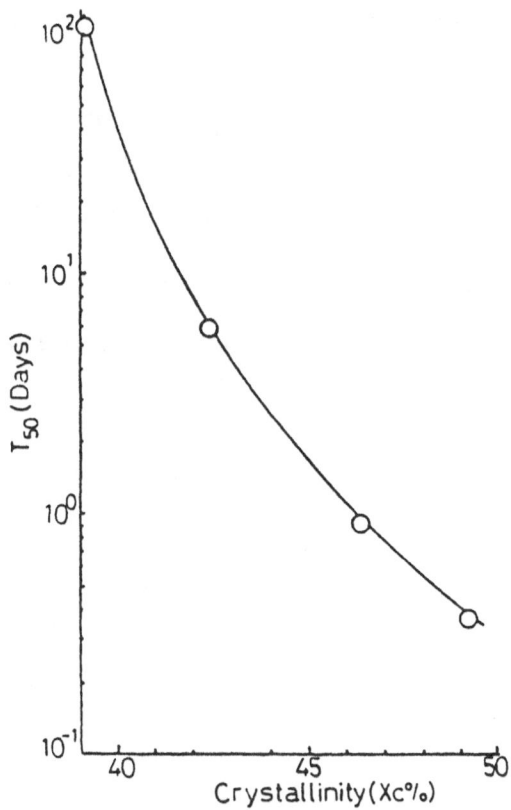

Fig. 6 : **Dependence of 5-Fu release rate on crystallinity of PCE copolymer (at pH 7.4 and 37 °C)**

On the other hand, with reducing crystallinity and hence increasing hydrophilicity of the PCE copolymer, the T_{50} decrease rate is much faster than the decrease rate of $T_{1/4}$. We attribute this to the high polyethylene glycol levels which help to hasten the solubility and diffusivity of 5-Fu from the matrix to the surrounding aqueous medium. For PCE copolymers containing lower PEG content where the structure is more rigid and the hydrophilicity is not so high, the effect of degradation rate of the copolymer itself on the rate of drug release could be an important factor. This aspect is being presently investigated in our laboratories.

References

1. R. Langer and W. Peppas, *J. Macro Sci.*, **C23(1)** (1983) 61.
2. E.J. Frazza and E.E. Schmitt, *J. Biomed. Mater. Res.*, **1** (1971) 43.
3. C.G. Pitt, A. Schindler and R.A. Zweidinger, In: *Drug Delivery Systems*, ed., H.L. Gabelnick, DHEW Publ. No. (NIH) 77: 1238 (1977), p.142.
4. A. Schindler, R. Jeffcoat, G.L. Kimmel, C.G. Pitt, M.E. Wall and R. Zweidinger, *Contemporary Topics in Polymer Science*, eds., E.M. Pearce,

J.R. Schaefgen, Plenum, New York, Vol. 2 (1977), p.251.

5. X.D. Feng, C.X. Song and W.Y. Chen, *J Polym. Sci.: Polym. Letts Edn.*, **21** (1983) 593.

6. D. Cohn and H. Younes, *Biomaterials*, **10** (1989) 466.

7. S.G. Wang and B. Qiu, 34th IUPAC International Symp. on Macromolecules, Book of Abstracts, 3-P11 (1992).

8. S.G. Wang and B. Qiu, *Polymer for Advanced Technologies*, 1993, p.4.

9. X.F. Li, S.G. Wang and Z.F. Li, *Gaofenzi Tongzum (Polymer Commun, China)*, **2** (1984) 95.

10. S.G. Wang, B. Qiu, J.W. Gao and Y.X. Duan, *ACTA Polymeric Sinica*, in press.

11. S.G. Wang, Y.D. Wang, Q.W. Mo, R. Liu and E.P. Xiao, *Gaofenzi Tongxun (Polymer Commun., China)*, **4** (1979) 246.

Main Group Elements and Their Compounds
V.G. Kumar Das (Ed)
Copyright © 1996 Narosa Publishing House, New Delhi, India

Frontiers of Main Group Metallacarborane Chemistry

Narayan S. Hosmane

Department of Chemistry, Southern Methodist University
Dallas, Texas 75275, USA

Heterocarboranes are those in which a polyhedral carborane (mixed hydrides of carbon and boron) cage is expanded through the incorporation of a heteroatom or a heteroatom group. When the heteroatom is a metal, they are referred to as metallacarboranes. Small carboranes, such as *nido*-dicarbahexaborane(8) and its C-trimethylsilyl-, alkyl-, and aryl- substituted derivatives, not only are important precursors to other *nido*-, arachno-, and *closo*-carboranes, *commo*-bis(metallacarborane) complexes, and di- and tetra-carbon metallacarboranes, they also can act as building blocks for the construction of multiple-decked sandwich complexes.[1] These extended sandwich systems are novel and they possess unusual spectroscopic properties. The carborane anions, $[R_2C_2B_4H_4]^{2-}$ and $[R_2C_2B_3H_3]^{4-}$ can be considered to be isolobal with $[C_5R_5]^-$ (R = alkyl, aryl or $SiMe_3$) anion in that they all have six electrons delocalized in II-type orbitals on an open pentagonal face. However, the systems differ in that metal orbitals tend to overlap more efficiently with carborane orbitals than with those of the cyclopentadienyl anion on formation of η^5-metal complexes.[2] The lower electronegativity and higher polarizability of boron versus carbon and higher ligand charge leads to stronger, more covalent interactions between the carborane ligands and the capping metal atom in metallacarborane systems than in the corresponding cyclopentadienyl metal complexes.[3] An added advantage is that nearly all of the characterized metallacarborane complexes are electrically neutral and are soluble in organic media.

Recently, there has been increased interest in the chemistry of main group metallacarboranes. This renewed interest has partly been the result of more structural data being available for these compounds and also due to the fact that "bare" main group metals and metalloids have been shown to insert, in their lowest and highest oxidation states, as integral members of single and double polyhedral carborane cages.[4] Recent developments in the area of carborane-nickel π-complexes have produced well-characterized multidecker species having up to ten decks, and a polydecker one which is semiconducting.[5] In comparing these compounds, as a class, with the metal/metalloid sandwich complexes derived from nido-C_2B_4-carboranes, it is hoped that the metallacarboranes in a similar extended system, particularly when the metals are in mixed valences, could produce novel materials of unusual physical and chemical properties.

Metallacarboranes are commonly synthesized by the reaction of a metal halide with either the mono- or dianion of a *nido*-carborane. There are two types of arrangements of the facial atoms, one in which the two carbons atoms are directly bonded to one another is called a "carbons adjacent" isomer, and another in which the carbons are separated by a boron atom is known as a "carbons apart" isomer. Since the "carbons adjacent" carborane precursors can be prepared more easily than the "carbons apart" isomers, most of the published results have been on metallacarboranes formed from the former ligands. However, recent results indicate that, not only are the latter carborane ligands just as effective as the former in bonding to metals, they also have the added advantage of being less susceptible to oxidative cage closure.

Results and Discussion

Metallacarboranes of Group 1 elements

The carborane monoanion can be obtained by the heterogeneous reaction of *nido*-$(CR)_2B_4H_6$ with NaH in THF (see Scheme I). The main point of interest in this reaction is that the stoichiometry is 1:1, even with the use of excess NaH or a stronger base, such as KH, and at elevated temperatures.[6] A possible explanation for this behavior was provided by the structures of the monosodium compounds, 1-Na(L)-2,3-$(SiMe_3)_2$-2,3-$C_2B_4H_5$, (L = THF or TMEDA), shown in Figures 1 - 3.[7] When L = THF the structure is that of an extended array of $Na_2(C_2B_4)_2$ dimers that are stacked on top of one another to give a series of - (carborane)$^-$- Na$^+$- (carborane)$^-$ - chains (see Figures 1 and 2), while with L = TMEDA the chain structure is broken but the ion cluster dimers remain (see Figure 3). These compounds are all fairly soluble in nonpolar and low dielectric constant solvents, indicating that the isolated ion clusters are quite stable in solution. Grimes and coworkers[8] have shown that the deprotonation of *nido*-2,3-RR'$C_2B_4H_6$ (where R = alkyl, arylmethyl and phenyl; R' = R, H) with NaH or KH in THF occurred at the surface of MH through the direct reaction of the bridged H with a H$^-$ ion in the hydride lattice.

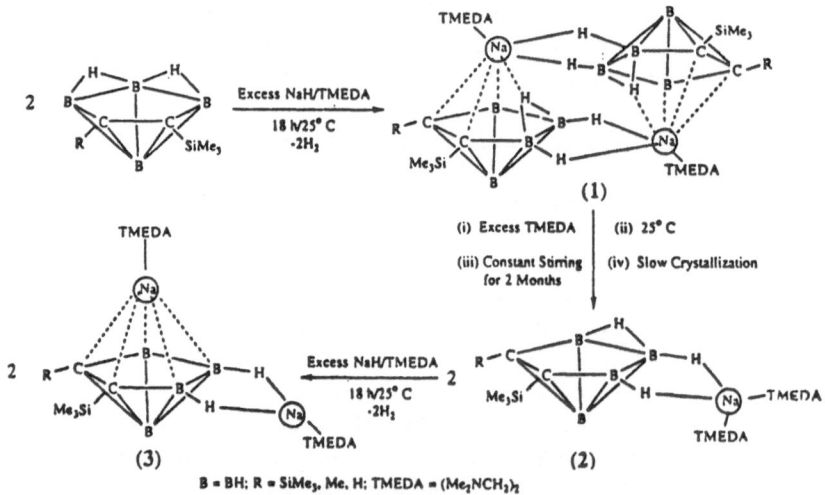

B = BH; R = SiMe₃, Me, H; TMEDA = (Me₂NCH₂)₂

Scheme I : Reactivity of *nido*-C₂B₄-carboranes with NaH

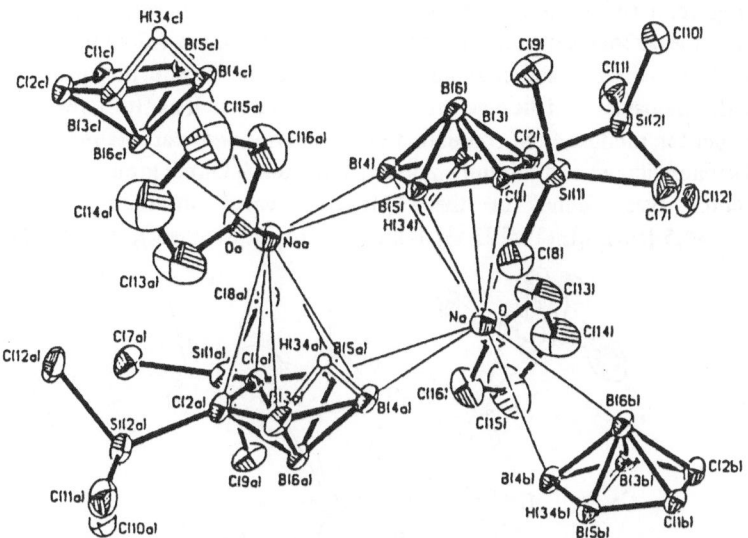

Fig. 1 : Perspective view of the dimeric form of "carbons adjacent" *nido*-natracarborane, 1-Na(THF)-2,3-(SiMe$_3$)$_2$-2,3-C$_2$B$_4$H$_5$
(Reprinted from ref. 7(b). Copyright 1993 American Chemical Society)

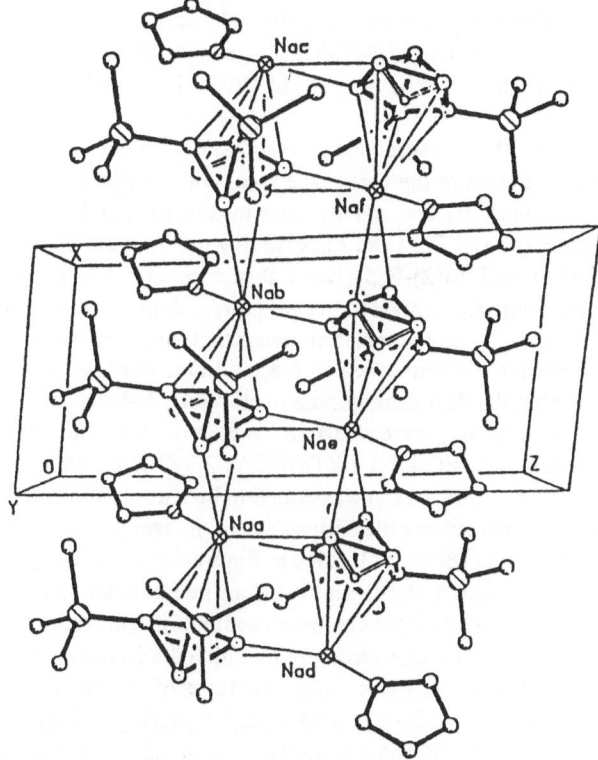

Fig. 2 : Packing diagram showing the extended interactions between the dimeric *nido*-natra-carborane clusters
(Reprinted from ref. 7(b). Copyright 1993 American Chemical Society).

From Figures 1 and 3 it is apparent that the second bridge hydrogen in an ion pair formed by the carborane monoanion and a solvated sodium ion would be effectively blocked from a direct reaction with a second hydride lattice site, thus preventing a second deprotonation. On the other hand, soluble bases, such as BuLi, react readily with either the monosodium compound or the neutral *nido*-carborane to form the mixed sodium/lithium or dilithium complexed dianion.[7] Direct group 1 metal-cage interaction was found in the TMEDA-solvated dilithium compound, *closo-exo*-4,5-[(μ-H)$_2$Li(TMEDA)]-1-Li(TMEDA)-2,3-(SiMe$_3$)$_2$-2,3-C$_2$B$_4$H$_4$.

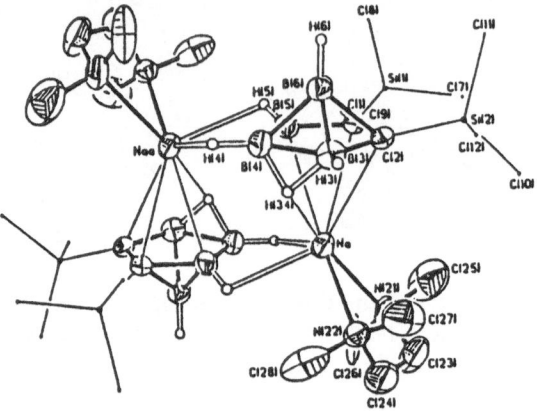

Fig. 3 : **Perspective view of a discrete dimeric unit of "carbons adjacent"**
nido-1-Na(TMEDA)-2,3-(SiMe$_3$)$_2$-2,3-C$_2$B$_4$H$_5$
(Reprinted from ref. 7(b). Copyright 1993 American Chemical Society).

The crystal structure of this compound shows that the two lithiums occupy quite different positions relative to the carborane face, with one lithium occupying an apical position above the C$_2$B$_3$ face, and the other located exopolyhedrally and about halfway between B(4) and B(5) and directed down below the plane of the C$_2$B$_3$ face[the Li(2)-B(4) and Li(2)-B(5) bond distances are 2.226 and 2.241 Å, respectively (Figure 4)]. In contrast to the structures of the sodium species shown in Figures 1 and 3, no extensive association exists between metallacarborane units. In an effort to acertain the extent to which these lithium arrangements are dictated by interactions within the dilithiacarborane itself rather than crystal lattice forces, MNDO-SCF calculations were carried out on the model compound *closo-exo*-4,5-[(μ-H)$_2$Li(TMEDA)]-1-Li(TMEDA)-2,3-C$_2$B$_4$H$_6$ which showed that the structure shown in Figure 4 can be assumed to be a minimum energy structure arising from interactions among the carborane and its two Li(TMEDA) groups.[7] The orientations of the TMEDA molecules in Figure 4 seem to be governed more by electrostatic rather than metal-ligand covalent interactions. Although crystal structures could not be obtained for the mixed sodium/lithium compounds, the ^{7}Li NMR spectrum is consistent with a structure in which the Li is *exo*-polyhedral and the Na occupies the apical position. The structures of the extensively solvated species, [Li(TMEDA)$_2$]$^+$[*nido*-2,3-(SiMe$_3$)$_2$-2,3-C$_2$B$_4$H$_5$]$^-$ (Figure 5) and *nido-exo*-4,5-[(μ-H)$_2$Na(TMEDA)$_2$]-2-(SiMe$_3$)-3-(Me)-2,3-C$_2$B$_4$H$_5$ (Figure 6), show that they are composed of discrete, well separated cation and anion units within the unit cell.[9] It is significant that these monoanions react readily with NaH to give either the mixed-lithium/sodium or disodium compounds of the dianion (see

302

Scheme I). Therefore, it seems that steric effects are as important as inherent acid/base strength in determining the reactivity of the *nido*-carborane anions.

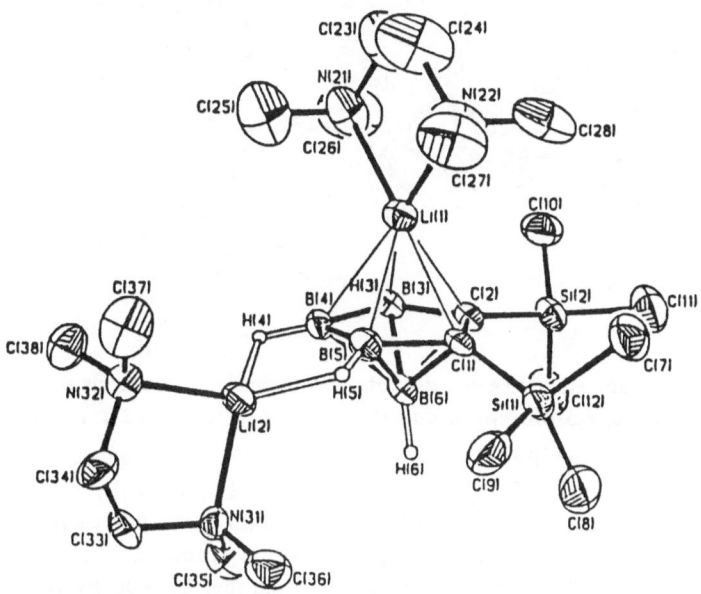

Fig. 4 : Perspective view of the zwitterionic, "carbons adjacent" *closo*-dilithiacarborane

(Reprinted from ref. 7(b). Copyright 1993 American Chemical Society.)

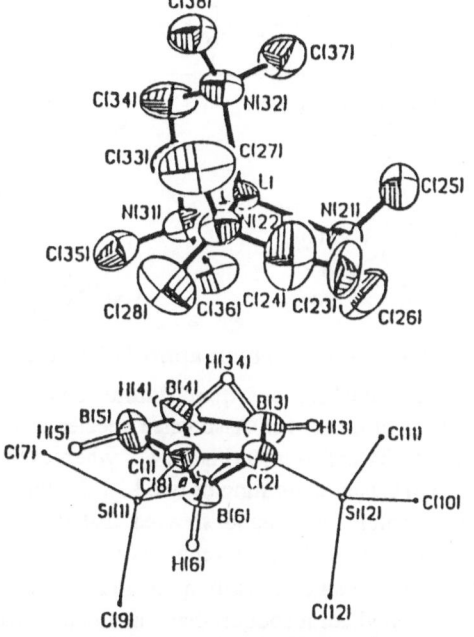

Fig. 5 : Perspective view of discrete [Li(TMEDA)₂]⁺ cationic and [*nido*- -2,3-(SiMe₃)₂-2,3-C₂B₄H₅]⁻ anionic units of a carborane salt

(Reprinted from ref. 9(a). Copyright 1993 American Chemical Society).

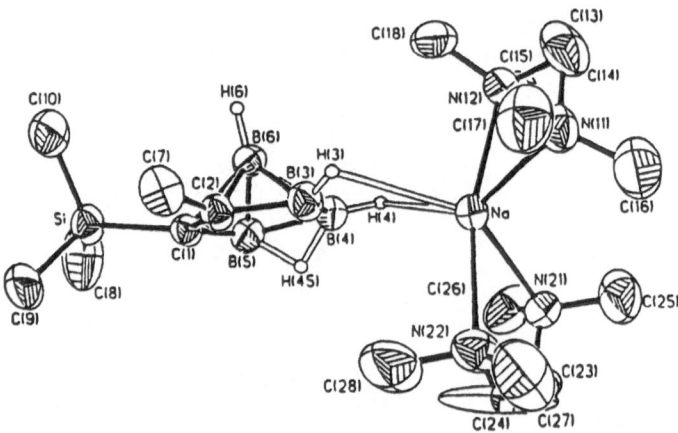

Fig. 6 : Crystal structure of a discrete, zwitterionic, monomeric monosodium salt of 2,3-dicarba-*nido*-hexaborate(1-) (Reprinted from ref. 9(b)).

The reaction of the dilithium complexed "carbons adjacent" dianion, [2-(SiMe₃)-3-(R)-2,3-C₂B₄H₄]²⁻ (R = SiMe₃, Me or H), with anhydrous NiCl₂ did not produce a nickelacarborane. Instead, it underwent an electron transfer reaction to yield Ni⁰ and the closed-cage oxidation product, *closo*-1-(SiMe₃)-2-(R)-1,2-C₂B₄H₄.[7b] This *closo*-carborane could then be reductively opened to give, ultimately, the group 1 metal complexed "carbons apart" dianion, [2-(SiMe₃)-4-(R)-2,4-C₂B₄H₄]²⁻, as outlined in Scheme II.[10]

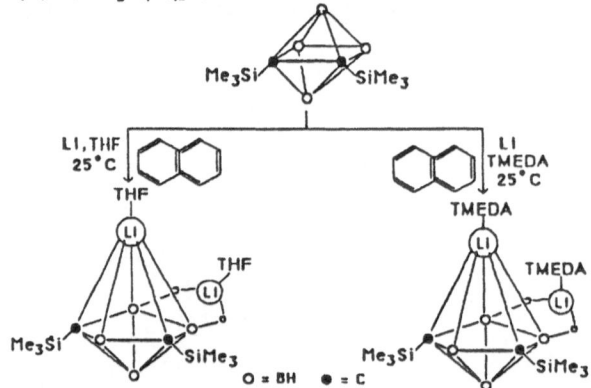

Scheme II : Synthesis of "carbons apart" *closo*-dilithiacarboranes

The X-ray analyses of the "carbons apart" dilithiacarboranes (Figures 7 and 8) show that the structure of the TMEDA-solvated species is almost identical to its corresponding "carbons adjacent" isomer (Figure 4)[7b] whereas the THF-solvated one exists as a fairly tight ion cluster consisting of two Li₂C₂B₄ units.[10a] The structure of the corresponding disodium compound showed that the compound existed as discrete dimers in which each carborane dianion was associated with an *endo*-sodium, which adopted an essentially η⁵-bonding posture with respect to the C₂B₃ face, and an *exo*-polyhedral sodium that was situated over a B₃ trigonal face formed by the apical boron and the two basal borons. Each sodium was also coordinated with sufficient THF molecules to give a somewhat tetrahedral arrangement of THF's and carboranes about the metal (see Figure 9).[10b] In contrast to the 2,3-C_2B_4-carborane system, the "carbons apart" ligand reacted with $NiCl_2$ to give the "carbons apart" *commo*-nickel complex.[10a]

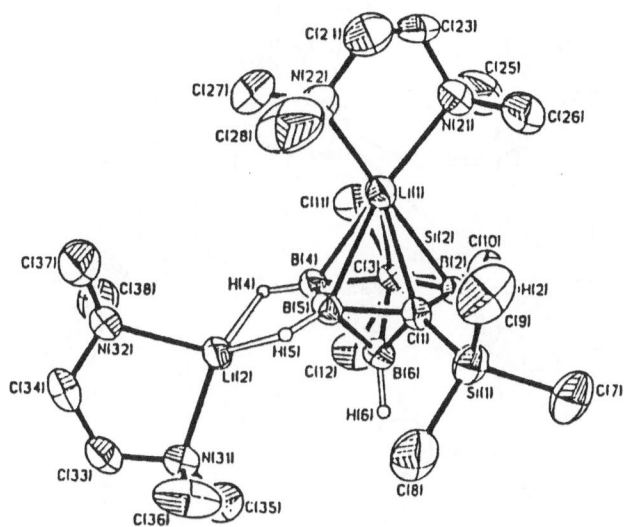

Fig. 7 : X-ray structure of a monomeric, TMEDA-solvated "carbons apart" *closo*-dilithiacarborane. (Reprinted from ref. 10(a). Copyright 1993 American Chemical Society.)

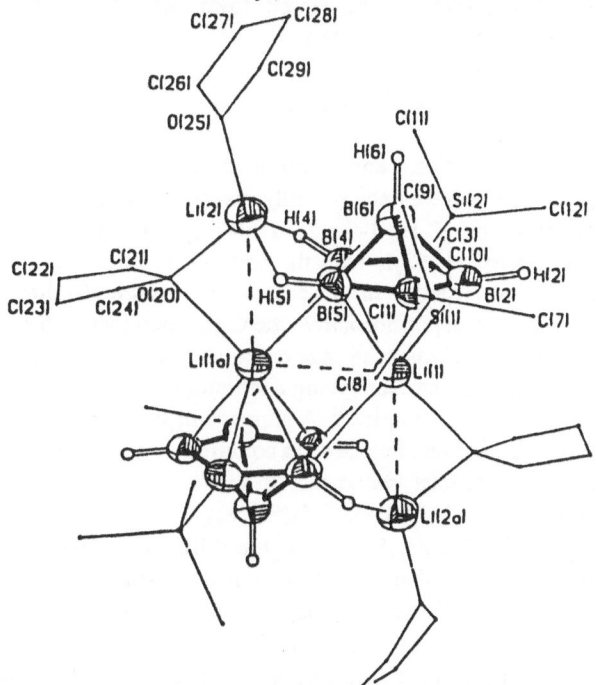

Fig. 8 : X-ray structure of a discrete dimeric unit of THF-solvated, "carbons apart" *closo*-dilithiacarborane. (Reprinted from ref. 10(a). Copyright 1993 American Chemical Society.)

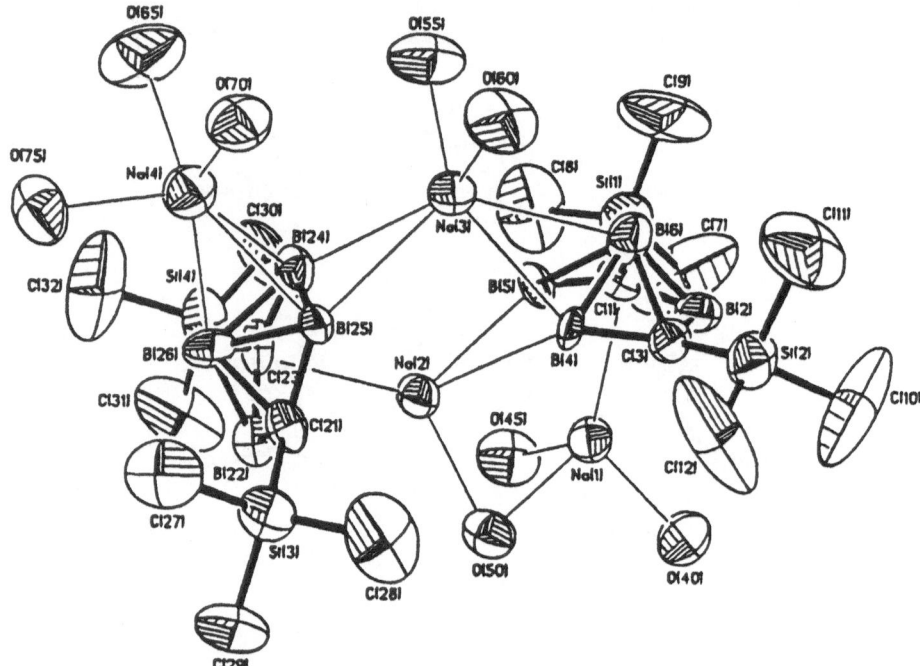

Fig. 9 : X-ray structure of the THF-solvated, "carbons apart" *closo*-dinatracarborane. (Reprinted from ref. 10(a). Copyright 1993 American Chemical Society.)

Metallacarboranes of Group 13 elements

The syntheses, reactivities, and structures of a number of group 13 metalla-C_2B_4-carboranes have been recently investigated. The "carbons adjacent" gallacarborane, *closo*-1-(t-C_4H_9-1-Ga-2,3-$(SiMe_3)_2$-2,3-$C_2B_4H_4$, was prepared by the reaction of $[Ga(t$-$C_4H_9)Cl_2]_2$ with the sodium lithium complexed carborane dianion in THF.[11] The structure of this compound (Figure 10) shows a t-C_4H_9Ga moiety occupying the apical position above the C_2B_3 open face of the carborane and the t-C_4H_9 group is oriented over the cage carbons. The Ga atom is slipped toward the boron side of the bonding face giving a structure very similar to that found for the 1-CH_3-1-Ga-2,3-$C_2B_4H_6$ complex.[12] A slippage of the capping metal toward the boron side of the carborane bonding face is a common structural feature found in the pentagonal bipyramidal stannacarboranes.[13] In the "carbons apart" system, a similar gallacarborane, *closo*-1-(t-C_4H_9)-1-Ga-2,4-$(SiMe_3)_2$-2,4-$C_2B_4H_4$, was prepared also by the reaction of $[Ga(t$-$C_4H_9)Cl_2]_2$ with the dilithium or disodium complexed "carbons apart" carborane dianion. Although its crystal structure could not be obtained, all spectroscopic data are consistent with a structure that is similar to that of its "carbons adjacent" analogue, shown in Figure 10.[4d,14] In both isomers, the Ga atom acts as a Lewis acid site and forms donor-acceptor complexes with a number of mono-, bi-, bis(bi) and tri-dentate Lewis bases. With multifunctional bases, such as 2,2'-bipyrimidine ($C_8H_6N_4$), the "carbons adjacent" *closo*-gallacarboranes form both bridged (Figure 11)[11] and unbridged (Figure 12)[15] adducts.

306

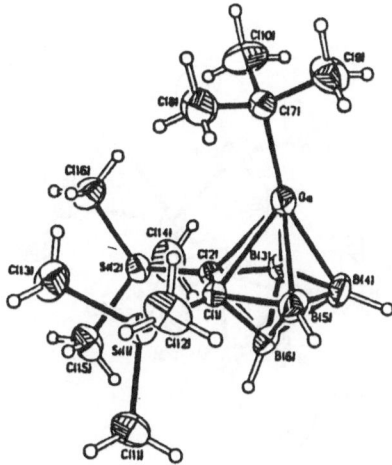

Fig. 10 : Crystal structure of the "carbons adjacent" *closo*-**gallacarborane.**
(Reprinted from ref. 11.   Copyright 1991 American Chemical Society.)

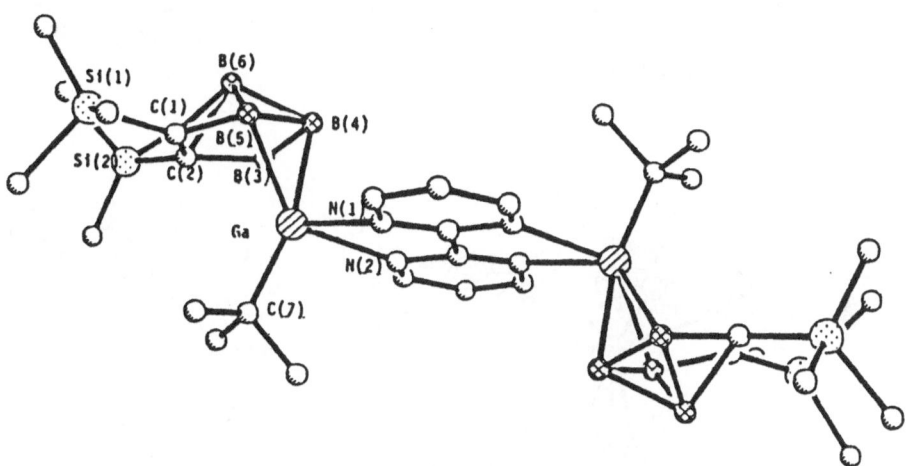

Fig. 11 : Crystal structure of a bridged complex of the "carbons adjacent"
gallacarborane,1,1'-(2,2'-C$_8$H$_6$N$_4$)[1-(*t*-C$_4$H$_9$)-1-Ga-2,4-(SiMe$_3$)$_2$-
2,4-C$_2$B$_4$H$_4$]$_2$. (Reprinted from ref. 11.   Copyright 1991 American
Chemical Society.)

On the other hand, only the unbridged adduct is formed when the "carbons apart"
closo-gallacarborane reacts with the 2,2'-bipyrimidine.[14]The crystal structure of this
complex (Figure 13) shows that there are two crystallographically independent
molecules within the unit cell and the Ga-bound 2,2'-bipyrimidine is opposite the
cage carbons in one molecule as in the unbridged gallacarborane analogue,
described above (Figure 12), and the reverse is true in the second.[14]

307

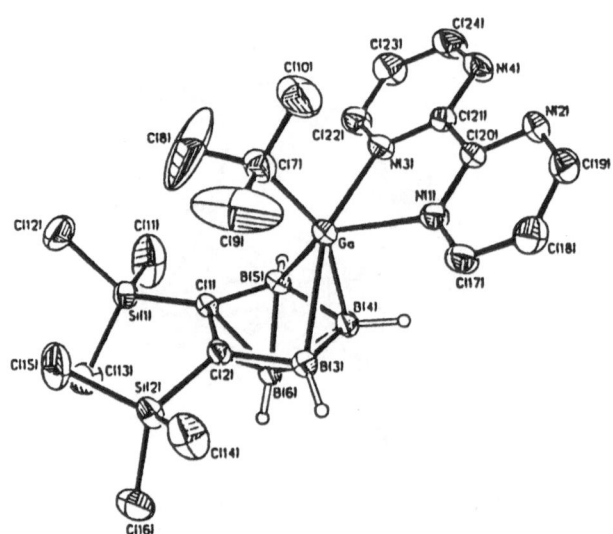

Fig. 12 : **Crystal structure of an unbridged complex of the "carbons adjacent" gallacarborane,1-(2,2'-C$_8$H$_6$N$_4$)-1-(t-C$_4$H$_9$)-1-Ga-2,3-(SiMe$_3$)$_2$-2,3-C$_2$B$_4$H$_4$.** (Reprinted from ref. 15. Copyright 1993 Plenum Publishing Corporation)

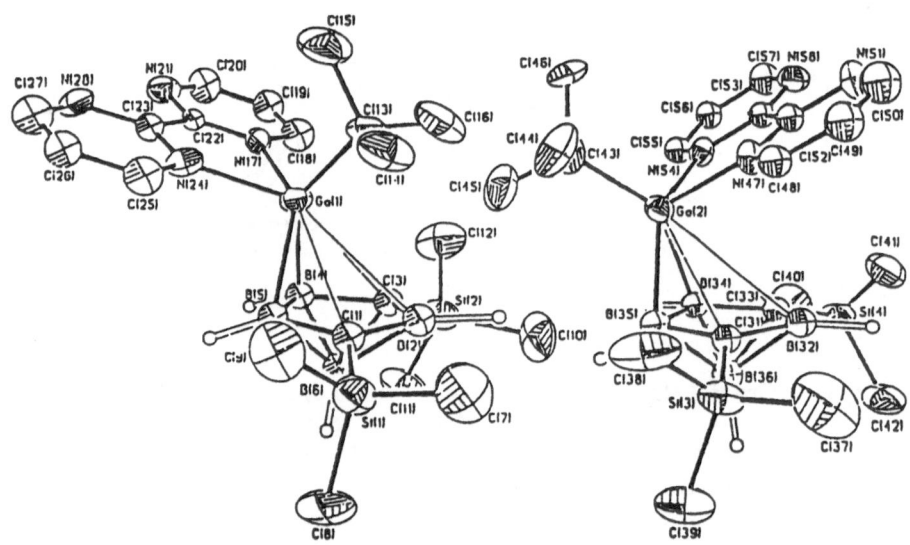

Fig. 13 : **Crystal structure of an unbridged complex of the "carbons apart" gallacarborane,1-(2,2'-C$_8$H$_6$N$_4$)-1-(t-C$_4$H$_9$)-1-Ga-2,4-(SiMe$_3$)$_2$-2,4-C$_2$B$_4$H$_4$.** (Reprinted from ref. 14.)

308

The structures of both the "carbons adjacent" and "carbons apart" gallacarborane-2,2'-bipyridine ($C_{10}H_8N_2$) adducts are shown in Figures 14 and 15.[4d,14] In most complexes, the base molecule is oriented away from the carbon atoms of the C_2B_3 carborane face. Similar increase in slip distortion and base orientation have been observed previously in the group 14 metallacarborane-Lewis base complexes.[4]

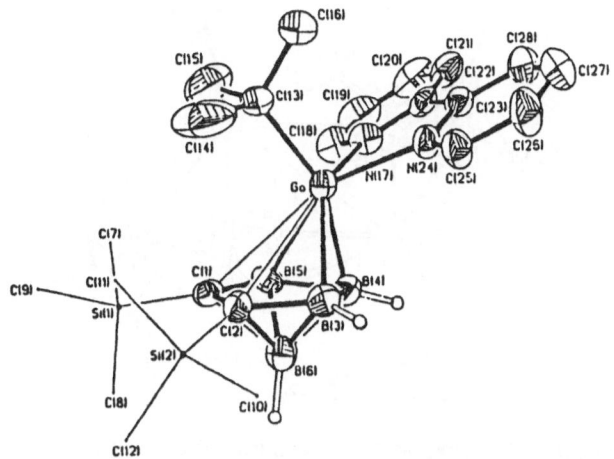

Fig. 14 : **X-ray structure of the "carbons adjacent" 1'-(2,2'-$C_{10}H_8N_2$)-1-(t-C_4H_9)-1-Ga-2,3-(SiMe$_3$)$_2$-2,3-$C_2B_4H_4$.** (Reprinted from ref. 14.)

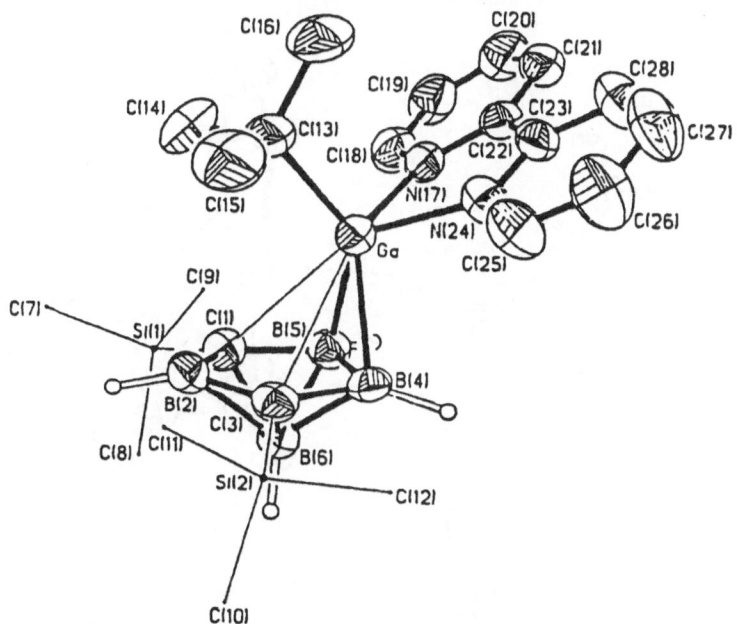

Fig. 15 : **X-ray structure of the "carbons apart" 1'-(2,2'-$C_{10}H_8N_2$)-1-(t-C_4H_9)-1-Ga-2,4-(SiMe$_3$)$_2$-2,4-$C_2B_4H_4$.** (Reprinted from ref. 14.)

Preliminary results on the synthesis and structure of the indacarborane, *closo*-1-(Me$_2$CH)In-2,3-(SiMe$_3$)$_2$-2,3-C$_2$B$_4$H$_4$, have also been described.[16] The solid state structure consists of dimeric clusters of opposed InC$_2$B$_4$ cages (Figure 16). The same structural feature has been observed for the "carbons apart" isomer, *closo*-1-(Me$_2$CH)In-2,4-(SiMe$_3$)$_2$-2,4-C$_2$B$_4$H$_4$ (Figure 17).[14]

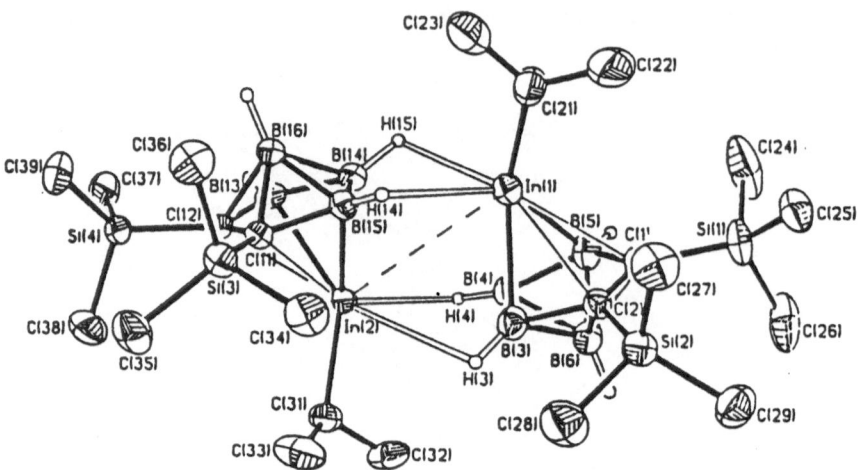

Fig. 16 : Crystal structure of the "carbons adjacent" *closo*-**indacarborane dimer, [1-(*i*-C$_3$H$_7$)-1-In^{III}-2,3-(SiMe$_3$)$_2$-2,3-C$_2$B$_4$H$_4$]$_2$.** (Reprinted from ref. 16(a). Copyright 1991 American Chemical Society.)

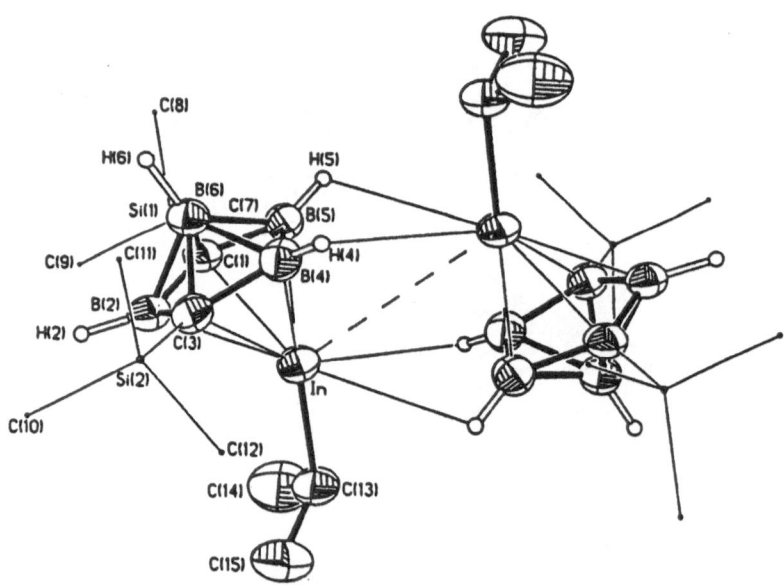

Fig. 17 : Crystal structure of the "carbons adjacent" *closo*-**indacarborane dimer, [1-(*i*-C$_3$H$_7$)-1-In^{III}-2,4-(SiMe$_3$)$_2$-2,4-C$_2$B$_4$H$_4$]$_2$.** (Reprinted from ref. 14.)

310

There seems to be a general tendency for the heavier members of the groups 13 and 14 metallacarboranes to form dimeric clusters, which may be the result of an increased ionic character in the metal-carborane bonds as one proceeds down these groups. As was the case with gallium, the indium atom acts as a Lewis acid site and adducts of the indacarboranes with the bases 2,2'-bipyrimidine and 2,2'-bipyridine have been prepared and well characterized.[14,16] Coordination of the indium by the base induces changes in the internal indacarborane geometry. The isopropyl-indium tilt angle increases from around 31° to 51°, and a comparison of the analogous In-carborane distances in the complex and its precursor shows that, on base coordination, In-C(cage) bond distances increase by 0.20 Å, while the In-B(basal) distances increase only slightly (~0.05 Å), with the In-B(unique) distance decreasing by about the same magnitude. The crystal structure of the "carbons adjacent" indacarborane-2,2'-bipyrimidine complex is shown in Figure 18.[16b]

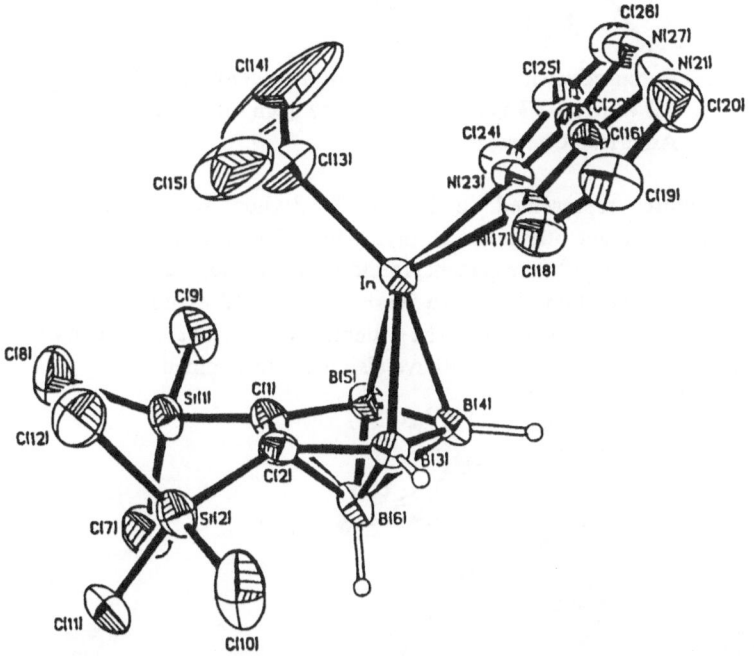

Fig. 18 : Crystal structure of the "carbons adjacent" *closo*-indacarborane dimer, 1-(2,2'-C$_8$H$_6$N$_4$)-1-(*i*-C$_3$H$_7$)-1-In^{III}-2,3-(SiMe$_3$)$_2$-2,3-C$_2$B$_4$H$_4$. (Reprinted from ref. 16(b).)

Metallacarboranes of Group 14 elements

The synthesis, structure and bonding of the 2,2'-bipyridine-"carbons adjacent" stannacarborane adducts have been extensively investigated (Figures 19 and 20).[4,13] These studies have been extended by an investigation of the reaction of *closo*-1-Sn-2-(SiMe$_3$)-3-(R)-2,3-C$_2$B$_4$H$_4$ (R = SiMe$_3$, Me, and H) with 2,2'-bipyrimidine to form the bridged donor-acceptor complex, 1,1'-(2,2'-C$_8$H$_6$N$_4$)[*closo*-1-Sn-2-(SiMe$_3$)-3-(R)-2,3-C$_2$B$_4$H$_4$]$_2$ in good yield.[17,18]

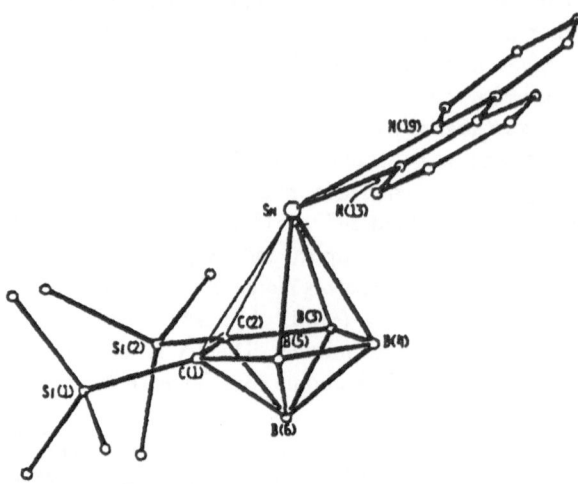

Fig. 19 : Perspective view of the "carbons adjacent" 1-(2,2'-C$_{10}$H$_8$N$_2$)-1-Sn-2,3-(SiMe$_3$)-2,3-C$_2$B$_4$H$_4$. (Reprinted from ref. 13(a). Compyright 1991 American Chemical Society.)

The structure, represented in Figure 21 (when R = SiMe$_3$), shows a base-stannacarborane orientation similar to the 2,2'-bipyridine system, except that the tin slippage is slightly less than that found for the analogous bipyridine complex. The *trans* configuration of the two "carbons adjacent" stannacarborane cages is probably dictated by steric factors. The structures of several *closo*-stannacarborane-monodentate Lewis base adducts have also been determined.

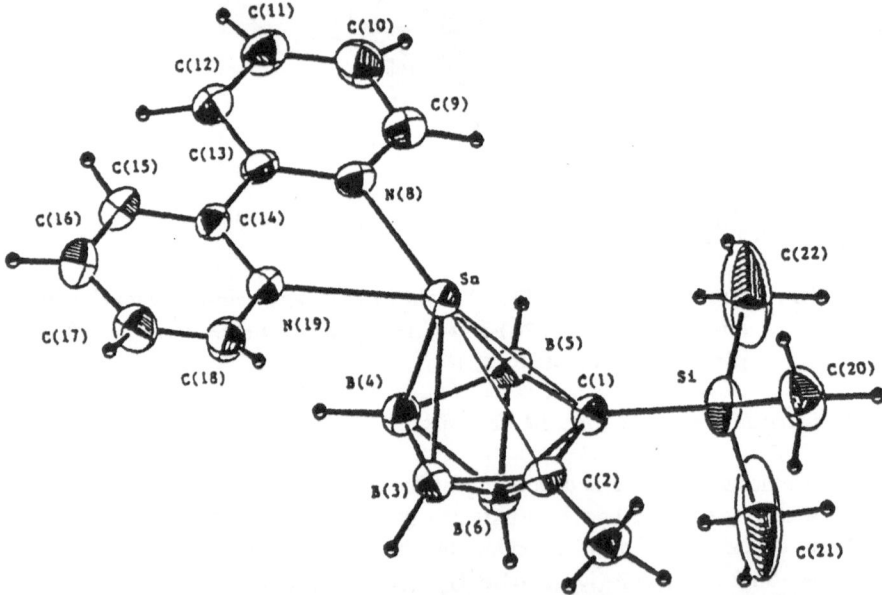

Fig. 20 : Perspective view of the "carbons adjacent" 1-(2,2'-C$_{10}$H$_8$N$_2$)-1-Sn-2-(SiMe$_3$)-3-(Me)-2,3-C$_2$B$_4$H$_4$. (Reprinted from ref. 13(b). Copyright 1987 International Union of Crystallography.)

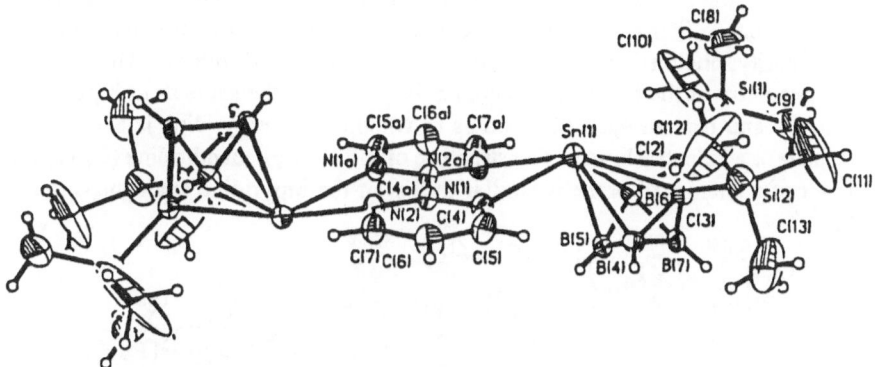

Fig. 21 : Crystal structure of a bridged "carbons adjacent" stannacarborane-2,2'-bipyrimidine complex. (Reprinted from ref. 18. Copyright 1989 American Chemical Society.)

The structure of 1-Sn[(η⁵-C₅H₅)Fe-(η⁵-C₅H₄CH₂(Me)₂N)]-2,3-(SiMe₃)₂-2,3-C₂B₄H₄, shown in Figure 22, exhibits less slip distortion than that found in the corresponding bipyridine complexes.[18] The studies have been extended to include tridentate Lewis base adducts of the "carbons adjacent" stannacarboranes.

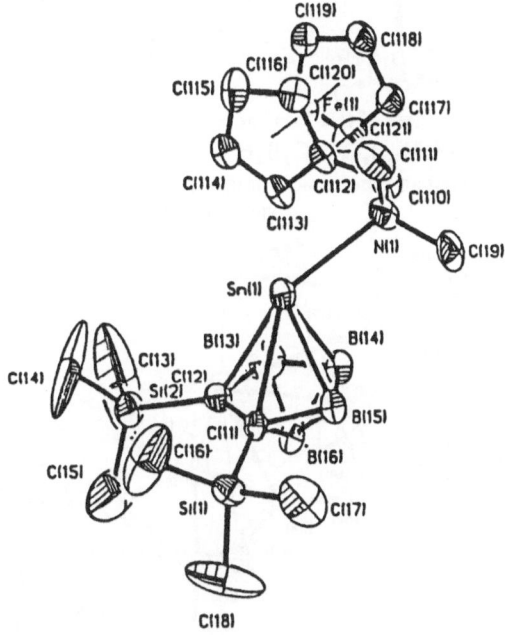

Fig. 22 : X-ray structure of the "carbons adjacent" 1-Sn[η⁵-C₅H₅)Fe(η⁵-C₅H₄CH₂(Me)₂N)]-2,3-(SiMe₃)₂-2,3-C₂B₄H₄. (Reprinted from ref. 18 Copyright 1989 American Chemical Society.)

Figures 23 and 24 show the structures of the terpyridine complex of 1-Sn-2-(SiMe$_3$)-3-(R)-2,3-C$_2$B$_4$H$_4$ (R = SiMe$_3$, and Me).[19,20] These compounds are of interest in that one of the end hexagonal rings of the terpyridine is oriented at an angle of about 13.9° from the plane of the other two rings. As a consequence of this nonplanarity, one Sn-N bond is significantly longer than the others . The SnC$_2$B$_4$ cage geometry of the terpyridine complex is essentially the same as its bipyridine analogue and it is an open question as to whether the terpyridine is acting as a tridentate or a bidentate ligand. The structure of the 1,10-phenanthroline (C$_{12}$H$_8$N$_2$) adduct of a stannacarborane is similar to that of the bipyridine analogue (Figure 25).[4f]

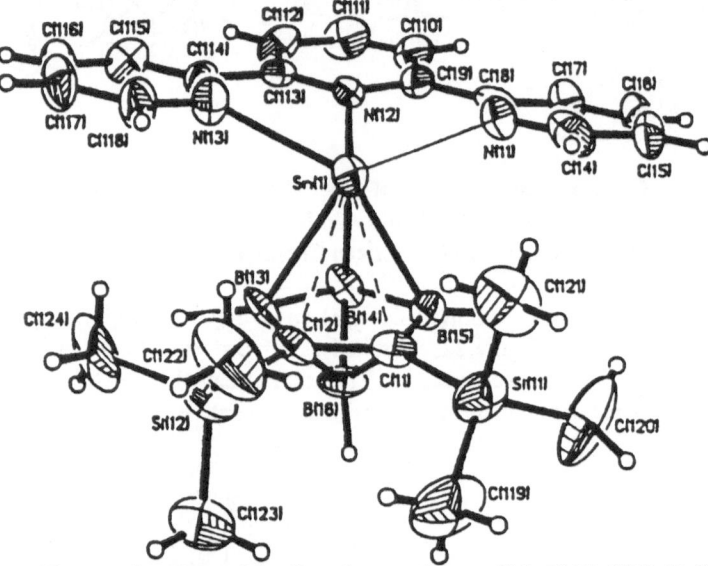

Fig. 23 : **Perspective view of the "carbons adjacent" 1-(2,2':6'2"-C$_{15}$H$_{11}$N$_3$)-1-Sn-2,3-(SiMe$_3$)$_2$-2,3-C$_2$B$_4$H$_4$.** (Reprinted from ref. 19 Copyright 1989 American Chemical Society.)

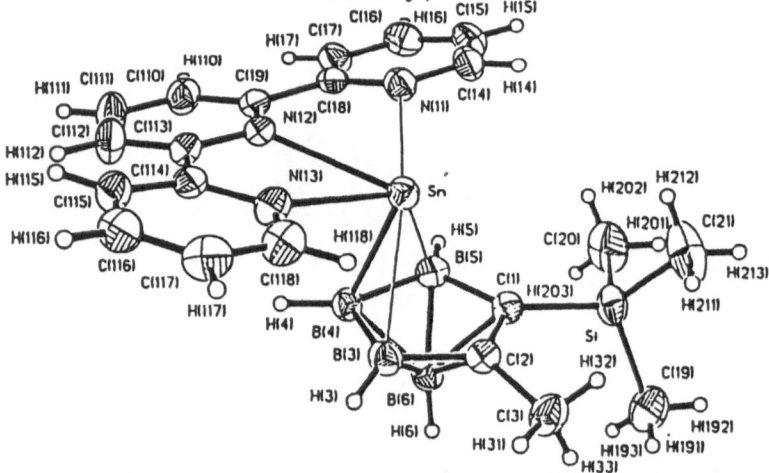

Fig. 24 : **Perspective view of the "carbons adjacent" 1-(2,2':6'2"-C$_{15}$H$_{11}$N$_3$)-1-Sn-2-(SiMe$_3$)-3-(Me)-2,3-C$_2$B$_4$H$_4$.** (Reprinted from ref. 20 Copyright 1988 International Union of Crystallography.)

314

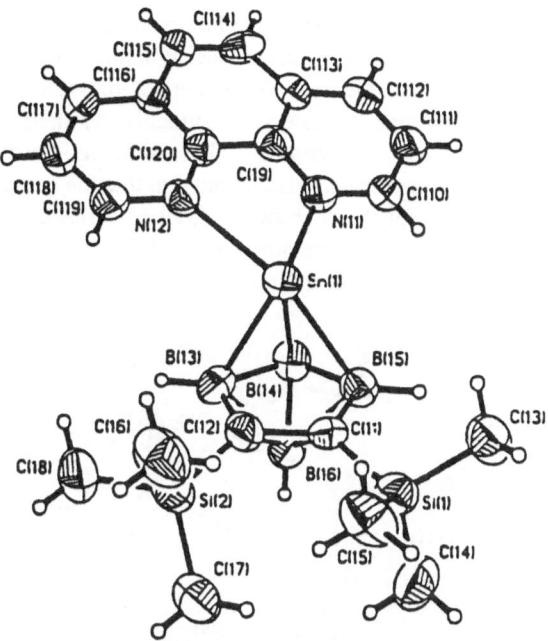

Fig. 25 : Perspective view of the "carbons adjacent" 1-(1,10-$C_{12}H_8N_2$)-1-Sn-2,3-(SiMe$_3$)$_2$-2,3-$C_2B_4H_4$. (Reprinted from ref. 4(f) Copyright 1990 VCH Publishers, Inc.)

Fenske-Hall and MNDO-SCF calculations were carried out on the "carbons adjacent" stannacarborane-bipyridine complexes.[21,22] Both calculations show that complexation with bipyridine gives rise to orbitals with antibonding tin-carbon (cage) interactions, which would encourage slippage. MNDO calculations indicate that the major bonding interactions between the bipyridine molecule and the stannacarborane are through tin orbitals oriented parallel to the C_2B_3 face of the carborane.[22] Hence, maximum tin-bipyridine bonding would be expected when the rings of the bipyridine molecule and the C_2B_3 face are essentially parallel. Repulsion between the two coordinating groups would prevent such an ideal alignment. An increased slip distortion of tin would tend to decrease ligand-ligand repulsion and yield a more favorable bipyridine orientation with stronger tin-bipyridine bonding. In general, one would expect a decrease in base-carborane dihedral angle and an increase in slip distortion on forming stronger tin-base adducts. This has been generally borne out by experiment. As described above, adducts with monodentate bases (Figure 22) show less slip distortion than found in the bipyridine complexes (Figures 19 and 20) as does the complex with the weaker bipyrimidine base (Figure 21).

The "carbons adjacent" *closo*-germacarboranes also form adducts with Lewis bases such as 2,2'-bipyridine.[23,24] The structure of 1-Ge^{II}($C_{10}H_8N_2$)-2,3-(SiMe$_3$)$_2$-2,3-$C_2B_4H_4$, shown in Figure 26, is similar to those of the bipyridine-stannacarborane adducts in that the germanium is slipped away from the cage carbons and the bipyridine is situated above the ring borons. However, the structure differs in that the apical germanium is twisted away from the carborane mirror plane so that the Ge-B(3) and Ge-B(5) bond distances are unequal (see Figure 26). The germanium can be considered to be η²-bonded to the unique boron and one basal boron. The two Ge-N bonds are also nonequivalent with one bond distance being 0.153 Å longer than the other. Since the 2,2'-bipyridine nitrogens are equivalent and the *closo*-germacarborane is presumably symmetric, there is no ready explanation for these distortions. The solution behavior of the bipyridine-germacarborane complexes is also unusual in that the room temperature proton-decoupled ¹¹B NMR spectrum shows single boron resonance, indicating fluxional behavior. It may be that the structure shown in Figure 26 represents only one of several structures that exist in solution.

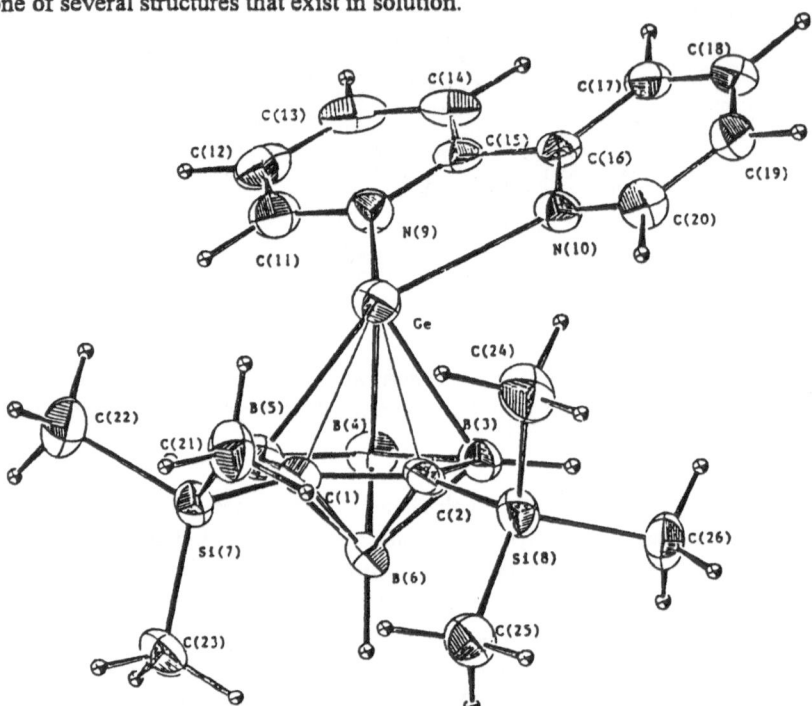

Fig. 26 : Crystal structure of the "carbons adjacent" $1-(2,2-C_{10}H_8N_2)-1-Ge-2,3-(SiMe_3)_2-2,3-C_2B_4H_4$. (Reprinted from ref. 24 Copyright 1988 American Chemical Society.)

The reaction of *closo*-1-Sn-2-(SiMe₃)-3-(R)-2,3-C₂B₄H₄ with an excess of GeCl₄ at 135°C in the absence of solvent produced the mixed valence "carbons adjacent" germacarborane, *closo*-Ge^II-2-(SiMe₃)-3-(R)-5-(Ge^IVCl₃)-2,3-C₂B₄H₃ in 44-60% yields.[25] The X-ray crystal structures [Figures 27 (R = SiMe₃) and 28 (R = H)][25,26] show that the Ge^II is η⁵-bonded to the open pentagonal face of the carborane and is symmetrically situated above the face. The other germanium, in a + 4 oxidation state, is involved in an exopolyhedral GeCl₃ group bonded to the unique boron of the cage *via* a Ge-B sigma bond. This is one of the few cases of a main group *closo*-metallacarboranes that is not slip distorted. Theoretical studies[22] on the stannacarboranes indicate that electron withdrawing groups on the unique boron should favour a more centroidal location of the capping heteroatom which may be the case for the mixed valence germacarborane.

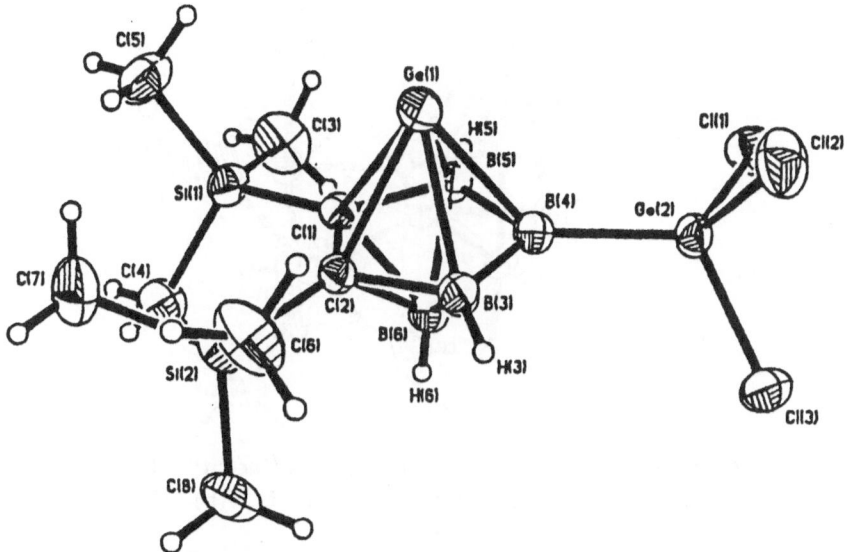

Fig. 27 : **X-ray structure of the mixed valence "carbon adjacent"** *closo*-**digermacarborane (R = SiMe₃).** (Reprinted from ref. 25. Copyright 1988 Americal Chemical Society.)

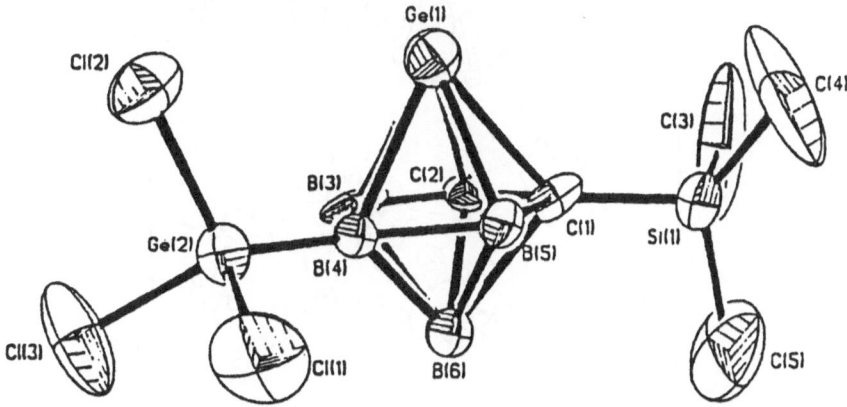

Fig. 28 : **X-ray structure of the mixed valence "carbons adjacent"** *closo*-**digermacarborane (R = H).** (Reprinted from ref. 26.)

Of all the group 14 metallacarboranes the least studied have been the plumbacarboranes. The X-ray crystal structure of 1-Pb-2,3-(SiMe₃)₂-2,3-C₂B₄H₄, shown in Figure 29, confirms the *closo*-geometry in that it shows the Pb to be essentially symmetrically bonded above the C_2B_3 face.[27] The bond distances indicate that, if any distortion exists, it is one in which the Pb is slipped toward B(5), a basal boron that is bonded to a cage carbon. The crystal packing diagram (Figure 30) shows that the solid consists of closely associated [Pb(SiMe₃)₂C₂B₄H₄]₂ dimers with a crystallographic center of symmetry halfway between the two Pb atoms. This dimeric structure could be due to an increased ionic character in the Pb-carborane bonds in the plumbacarboranes.[27]

317

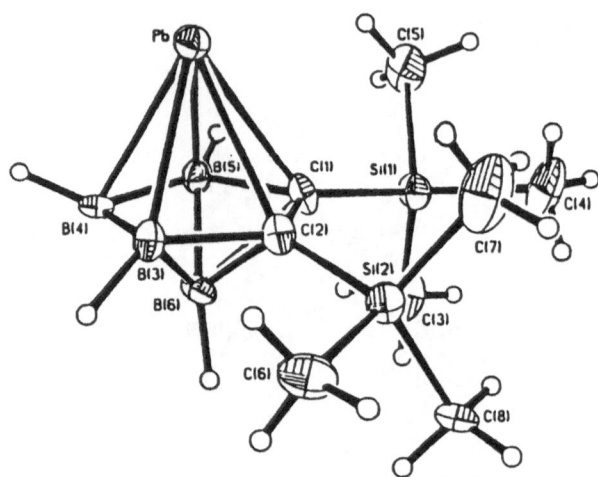

Fig. 29 : Crystal structure of the "carbons adjacent" closo-1-Pb-2,3-(SiMe₃)₂-2,3-C₂B₄H₄. (Reprinted from ref. 27(b). Copyright 1990 American Chemical Society).

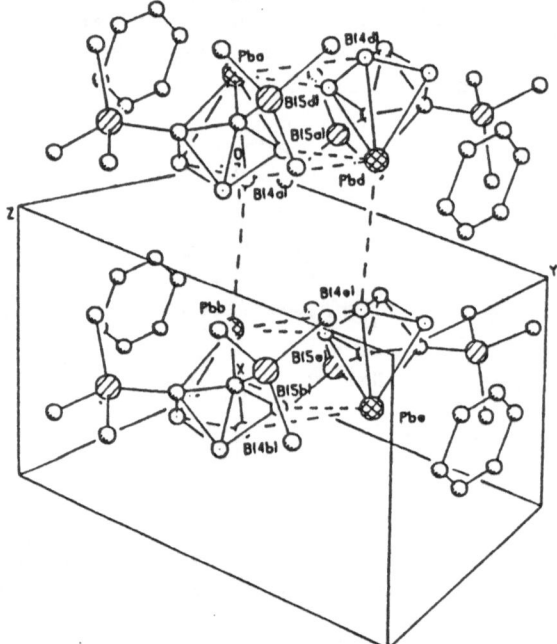

Fig. 30 : Crystal packing diagram showing the extended network of dimeric *closo*-plumbacarboranes with benzene molecules of crystallization. (Reprinted from ref. 27(b). Copyright 1990 American Chemical Society.)

As in the cases of *closo*-stanna- and germacarboranes, the *closo*-plumbacarboranes also behave as Lewis acids in forming donor-acceptor complexes, despite the presence of a lone pair of electrons on the metal.[27b,c] The crystal structure of 1-Pb(C₁₀H₈N₂)-2,3-(SiMe₃)₂-2,3-C₂B₄H₄, shown in Figure 31, also exhibits a significant slippage of the apical Pb toward the borons of the C₂B₃ face.[27b]

318

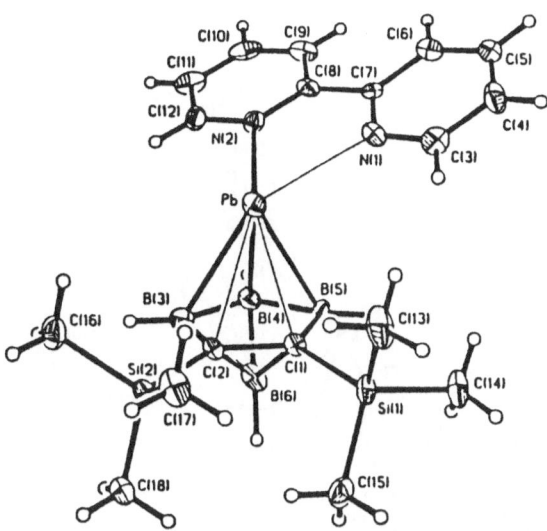

Fig. 31 : Crystal structure of the "carbons adjacent" 1-Pb(2,2'-$C_{10}H_8N_2$)-2,3-(SiMe$_3$)$_2$-2,3-$C_2B_4H_4$. (Reprinted from ref. 27(b). Copyright 1990 American Chemical Society.)

Although the *closo*-plumbacarboranes form the expected bridged donor-acceptor complexes 1,1'-(2,2'-$C_8H_6N_4$)[*closo*-1-Pb-2-(SiMe$_3$)-3-(R)-2,3-$C_2B_4H_4$]$_2$ with 2,2'-bipyrimidine, the *closo*-germacarboranes produce the unbridged species 1-Ge-(2,2'-$C_8H_6N_4$)-2-(SiMe$_3$)-(R)-2,3-$C_2B_4H_4$ even when excess of 2,2'-bipyrimidine was used.[28] Crystal structure of the germanium complex (Figure 32) with one molecule of cocrystallized 2,2'-bipyrimidine shows that the complex consists of only one distorted germacarborane unit, and a bipyrimidine molecule, that acts unusually as a monodentate ligand, is bonded to the apical Ge atom. This Ge-bound 2,2'-bipyrimidine base is not situated exactly opposite the C-C(cage) bond, rather it is tilted toward the two basal borons above the C_2B_3 face. The nonbonded contacts between the atoms of the complex and the cocrystallized 2,2'-bipyrimidine exceed the sum of their van der Waals radii, and consequently, the geometry of the complex was unaffected by the presence of an extra molecule of the ligand in the crystal lattice.[28] On the other hand, the crystal structure of the lead complex [Figure 33 (R = SiMe$_3$)] is similar to that of the tin analogue[17,18] and shows that the apical lead is slipped significantly from the centroidal position above the C_2B_3 face in a distorted pentagonal-bipyramidal geometry. The direct relationship between the crystal packing and the ligand-ligand dihedral angle was also evident in these structures.[28]

Donor-acceptor complexes of *closo*-germa and *closo*-plumbacarboranes with the ferrocenyl-methyl-*N,N*-dimethylamine (ferrocene amine) have also been prepared and their structures (Figures 34 and 35) determined, recently.[29] A common structural feature found in all the group 14 ferrocene amine complexes is that the ferrocenyl group is not in a position of minimum steric interaction with the metallacarborane moiety, but is rotated in a manner that some of the ferrocene atoms and the metal are within their van der Waals radii.[4,18,29]

319

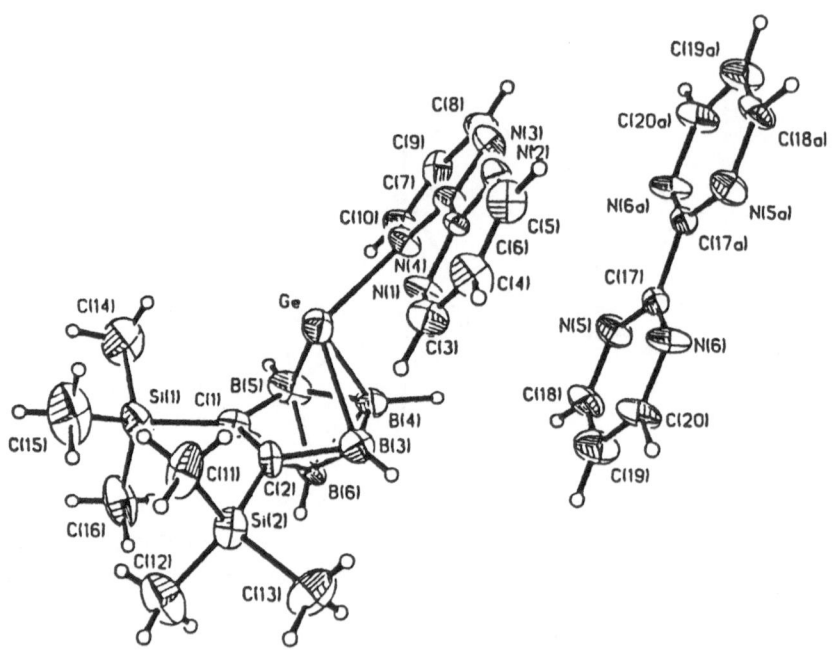

Fig. 32 : Crystal structure of the unbridged "carbons adjacent" 1-Ge(2,2'-$C_8H_6N_4$)-2,3-$(SiMe_3)_2$-2,3-$C_2B_4H_4$ with the co-crystallized 2,2'-bipyrimidine. (Reprinted from ref. 28. Copyright 1990 American Chemical Society.)

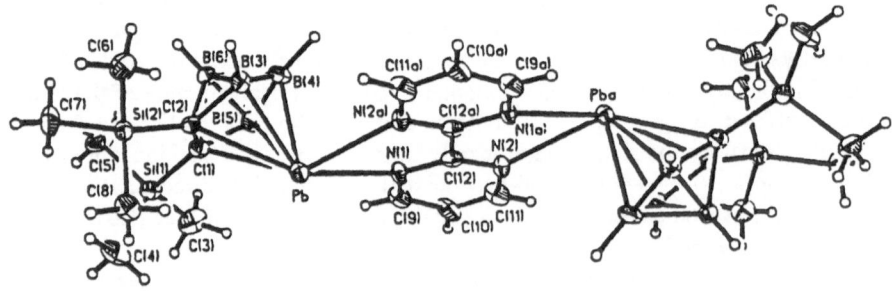

Fig. 33 : Perspective view of the bridged " carbons adjacent" 1,1'-(2,2'-$C_8H_6N_4$)[1-Pb-2,3-$(SiMe_3)_2$-2,3-$C_2B_4H_4$]$_2$ showing the *trans* configuration of the plumbacarboranes. (Reprinted from ref. 28. Copyright 1990 American Chemical Society.)

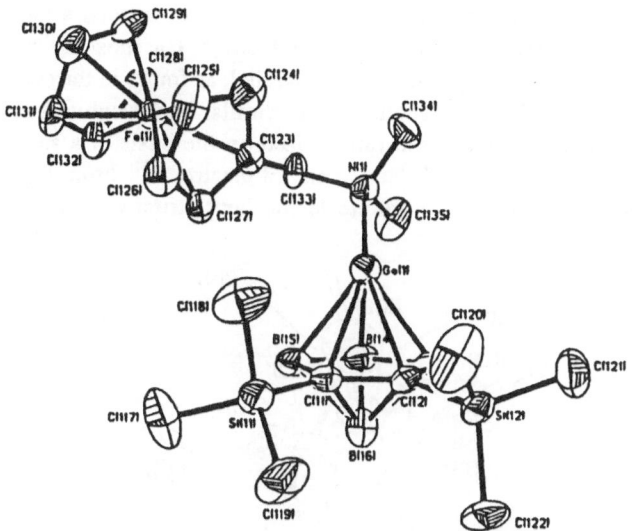

Fig. 34 : Perspective view of the "carbons adjacent" 1-Ge[η⁵-C₅H₅)Fe(η⁵-C₅H₄CH₂(Me)₂N)]-2,3-(SiMe₃)₂-2,3-C₂B₄H₄. (Reprinted from ref. 29. Copyright 1992 American Chemical Society.)

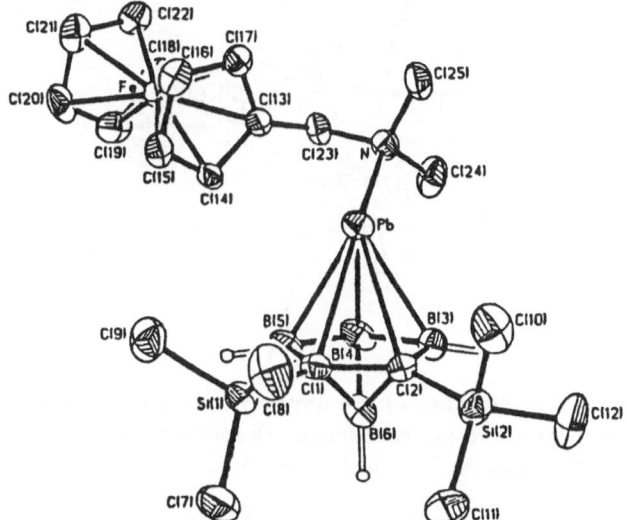

Fig. 35 : Perspective view of the "carbons adjacent" 1-Pb(η⁵-C₅H₅)Fe(η⁵-C₅H₄CH₂(Me)₂N)]-2,3-(SiMe₃)₂-2,3-C₂B₄H₄. (Reprinted from ref. 29. Copyright 1992 American Chemical Society.)

The reaction of a *closo*-stannacarborane with TiCl₄ [eq. (1)]

$$2\ [1\text{-}Sn^{II}\text{-}2\text{-}(SiMe_3)\text{-}3\text{-}(Me)\text{-}2,3\text{-}C_2B_4H_4] + 2\ TiCl_4 \xrightarrow{\ \ THF/benzene/25°C\ \ }$$

$$2\ TiCl_3(THF)_3 + SnCl_2 + commo\text{-}1,1'\text{-}Sn^{IV}[2\text{-}(SiMe_3)\text{-}3\text{-}(Me)\text{-}2,3\text{-}C_2B_4H_4]_2$$

(1)

produced the novel Sn^{IV} bent-sandwich, whose structure (Figure 36) shows that the Sn, in a formal +4 state, is sandwiched between two carborane ligands and is slipped toward the boron sides of the bonding faces.[30] The most striking aspect of this structure is that the C_2B_4 cages are not parallel, as found in the corresponding Si^{IV} and Ge^{IV} sandwiches,[4] but slightly bent. Similar bent-sandwich structures in group14 metallocenes have been rationalized on the basis of a stereochemical influence exerted by the metal's "lone pair" of electrons.[31] From Figure 36 it is apparent that other factors contribute to the bent structures of the group 14 sandwich complexes.

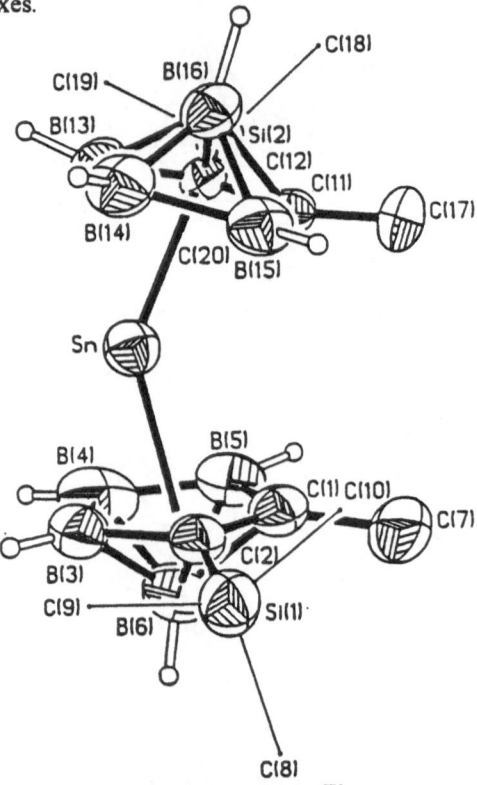

Fig. 36 : **Crystal structure of the novel, Sn^{IV} bent-sandwiched, "carbons adjacent" stannacarborane.** (Reprinted from ref. 30. Copyright 1992 American Chemical Society.)

Recently, a number of "carbons apart" group 14 metallacarboranes have also been prepared.[4d,32] The *closo*-stanna- and plumbacarboranes show a reactivity pattern with mono-, bi-, bis(bi)-, and tridentate Lewis bases similar to that observed in "carbons adjacent" isomers. The X-ray structures show that each complex has a distorted pentagonal bipyramidal geometry with the capping metal that coordinates to one or two nitrogen atom(s) of the respective Lewis base in a manner similar to that found in the corresponding "carbons adjacent" isomers (Figures 37 - 40).[32,33] However, unlike the "carbons adjacent" species, 2,2'-bipyrimidine unbridges the stannacarborane (Figure 38) and the ferrocenyl carbons do not interact with the apical tin atom in the ferrocene amine complex (Figure 39), but resides away from it with the least steric repulsion.[32]

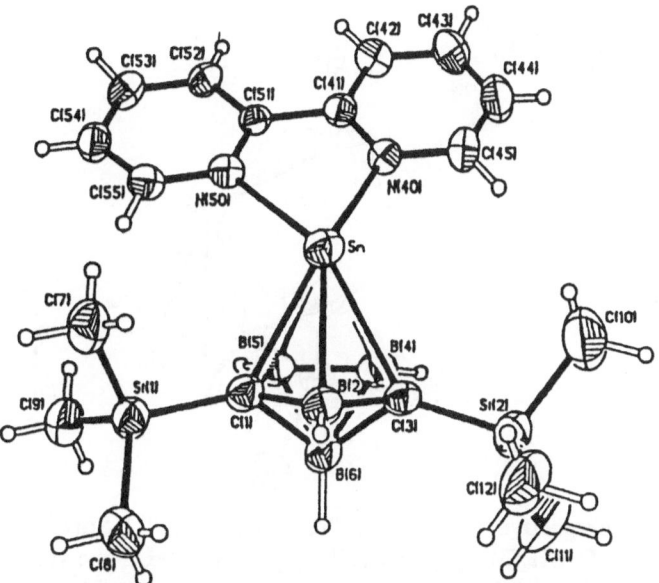

Fig. 37 : Perspective view of the "carbons apart" 1-Sn(2,2'-C$_{10}$H$_8$N$_2$)-2,4-(SiMe$_3$)$_2$-2,4-C$_2$B$_4$H$_4$. (Reprinted from ref. 32.)

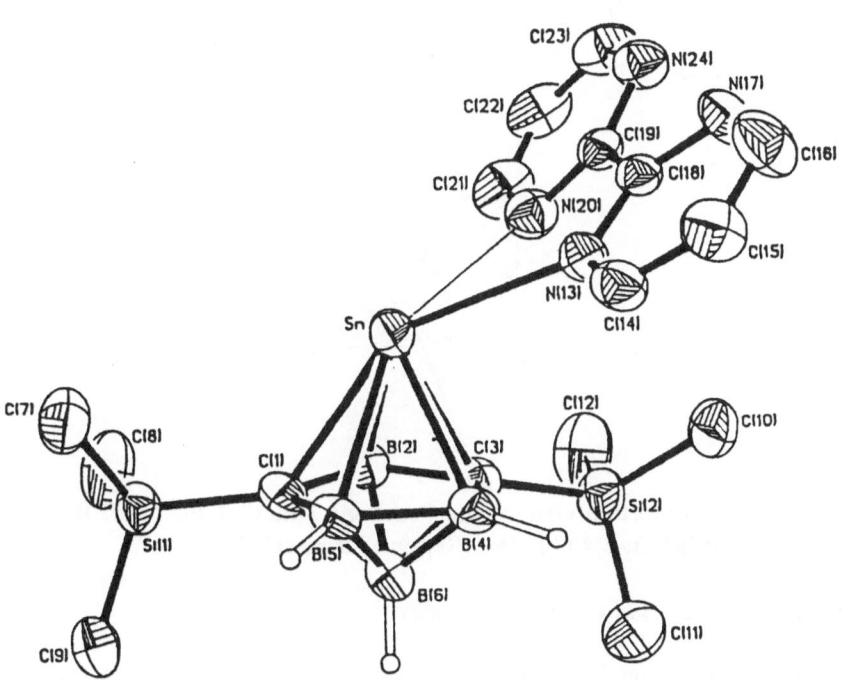

Fig. 38 : Perspective view of the unbridged "carbons apart" 1-Sn(2,2'-C$_8$H$_6$N$_4$)-2,4-(SiMe$_3$)$_2$-2,4-C$_2$B$_4$H$_4$. (Reprinted from ref. 32.)

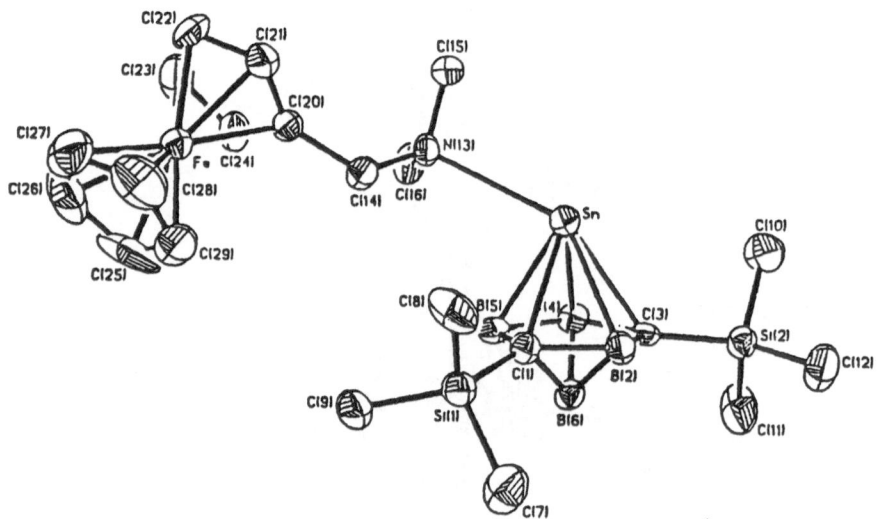

Fig. 39 : Perspective view of the "carbons apart" 1-Sn(η^5-C$_5$H$_5$)Fe(η^5-C$_5$H$_4$CH$_2$(Me)$_2$N)]-2,4-(SiMe$_3$)$_2$-2,4-C$_2$B$_4$H$_4$ showing an ideal position of the ferrocene amine ligand. (Reprinted from ref. 32.)

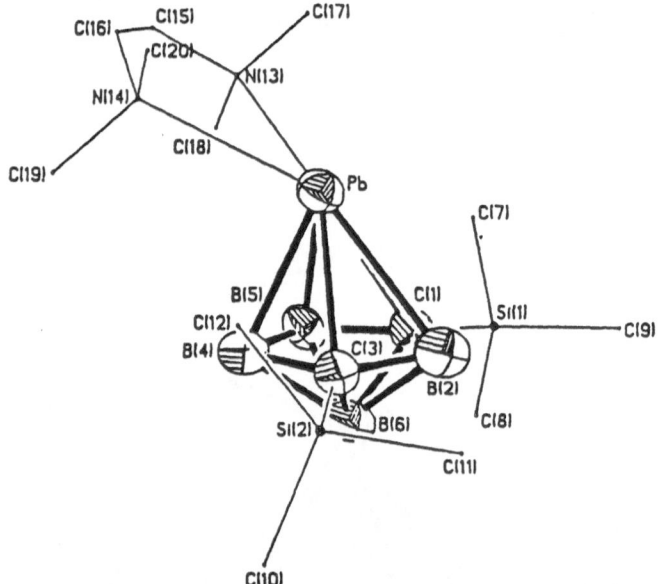

Fig. 40 : Perspective view of the "carbons apart" 1-Pb^{II}(TMEDA)-2,4-(SiMe$_3$)$_2$-2,4-C$_2$B$_4$H$_4$. (Reprinted from ref. 33.)

Heterocarboranes of Group 15 elements

A number of group 15 heterocarboranes was synthesized and their reactivity studies explored.[34] The reactivity of these heterocarboranes toward trialkylaluminum in forming donor-acceptor complexes is identical to those of icosahedral *closo*-phospha- and *closo*-arsacarboranes.[35]

Summary

Recent findings in our laboratory (described above) and elsewhere clearly demonstrate that C_2B_4-carborane-based organometallic chemistry is an emerging area of very considerable scope and versatility. Although the metal-C_2B_4 carborane ligand interactions are similar to those of the 12-vertex (icosahedral) metallacarboranes, and some of the reactivities are likewise related, the stereochemistry is markedly different. Moreover, oxidative face-to-face ligand fusion, construction of multidecker metal sandwich complexes and formation and isolation of stable species such as Ge(IV) and Sn(IV) sandwich complexes have no known parallel in the icosahedral systems.[1-5] Thus the chemistry of C_2B_4 main group metallacarboranes has evolved with distinct characteristics, even though with obvious relationships to other areas. Consequently, new frontiers of organometallic chemistry have begun to be investigated.

Acknowledgement

This work was supported, in parts, by grants from the National Science Foundation (CHE-9100048), the Robert A. Welch Foundation (N-1016), the donors of the Petroleum Research Fund, administered by the American Chemical Society, and the University Lecture Series Fund for Faculty Excellence from Southern Methodist University. The perseverance of numerous undergraduate students, postdoctoral associates, and other co-workers in many of these studies is gratefully acknowledged.

References

1. (a) R.N. Grimes, *Chem. Rev.*, **92** (1992) 251. (b) R.N. Grimes, *Carboranes*, Academic Press, N.Y., 1970; In: *Comprehensive Organometallic Chemistry*, eds., G. Wilkinson, F.G.A. Stone, E.W. Abel, Pergamon Press, Oxford, Vol. 1, 1982 and references therein. (c) In: *Metal Interactions with Boron Clusters*, ed., R.N. Grimes, Plenum: N.Y., 1982. (d) R.N. Grimes, *Pure and Appl. Chem.*, **63** (1991) 369. (e) Chapters In: *Electron Deficient Boron and Carbon Clusters*, eds., G.A. Olah, K. Wade, R.E. Williams, Wiley, New York, 1991.

2. R.N. Grimes, In: *Advances in Boron and the Boranes*, eds., A. Greenberg, J.F. Liebman, R.E. Williams, VCH Publishers, New York, 1988, Chapter 11, pp. 235-263.

3. M.J. Calhorda and D.M.P. Mingos, *J. Organomet. Chem.*, **229** (1982) 229.

4. (a) N.S. Hosmane and J.A. Maguire, Recent Advances in the Chemistry of Main Group Heterocarboranes, In: *Advances in Boron and The Boranes*, eds., A. Greenberg, J.F. Liebman, R.E. Williams, VCH Publishers, New York, 1988, Chapter 14, p. 297. (b) N.S. Hosmane and J.A. Maguire, *Adv. Organomet. Chem.*, **30** (1990) 99. (c) N.S. Hosmane, *Pure and Appl. Chem.*, **63** (1991) 375. (d) A.K. Saxena, J.A. Maguire, J.J. Banewicz and N.S. Hosmane, *Main Group Chem. News*, **1(2)** (1993) 14. (e) J.A. Maguire, J.S.

Fagner, U. Siriwardane, J.J. Banewicz and N.S. Hosmane, *Struct. Chem.*, **1** (1990) 583.

5. W. Siebert, *Angew. Chem.*, **97** (1985) 924; *Angew. Chem. Int. Ed. Engl.*, **24** (1985) 943; W. Siebert, *Pure Appl. Chem.*, **59** (1987) 947.

6. (a) T. Onak and G.B. Dunks, *Inorg. Chem.*, **5** (1966) 439. (b) C.G. Savory and M.G.H. Wallbridge, *J. Chem. Soc., Dalton Trans.*, (1974) 880.

7. (a) N.S. Hosmane, U. Siriwardane, G. Zhang, H. Zhu and J.A. Maguire, *J. Chem. Soc.,Chem. Commun.*, (1989) 1128. (b) N.S. Hosmane, A.K. Saxena, R.D. Barreto, H. Zhang, J.A. Maguire, L. Jia, Y. Wang, A.R. Oki, K.V. Grover, S.J. Whitten, K. Dawson, M.A. Tolle, U. Siriwardane, T. Demissie and J.S. Fagner, *Organometallics*, **12** (1993) 3001.

8. M.E. Fessler, T. Whelan, J.T. Spencer and R.N. Grimes, *J. Am. Chem. Soc.*, **109** (1987) 7416.

9. (a) Y. Wang, H. Zhang, J.A. Maguire and N.S. Hosmane, *Organometallics*, **12** (1993) 3781. (b) N.S. Hosmane, Y. Wang, A.K. Saxena, H. Zhang, J.A. Maguire and L. Jia, *Organometallics*, to be submitted for publication.

10. (a) H. Zhang, Y. Wang, A.K. Saxena, A.R. Oki, J.A. Maguire and N.S. Hosmane, *Organometallics*, **12** (1993) 3933. (b) N.S. Hosmane, L. Jia, H. Zhang, J.W. Bausch, G.K.S. Prakash, R.E. Williams and T.P. Onak, *Inorg. Chem.*, **30** (1991) 3793.

11. N.S. Hosmane, K.-J. Lu, H. Zhang, L. Jia, A.H. Cowley and M.A. Mardones, *Organometallics*, **10** (1991) 963.

12. R.N. Grimes, W.J. Rademaker, M.L. Denniston, R.F. Bryan and P.T. Greene, *J. Am. Chem. Soc.*, **94** (1972) 1865.

13. (a) N.S. Hosmane, P. de Meester, N.N. Maldar, S.B. Potts, S.S.C. Chu and R.H. Herber, *Organometallics*, **5** (1986) 772. (b) U. Siriwardane, N.S. Hosmane and S.S.C. Chu, *Acta Crystallogr., Cryst. Struct. Commun.*, **C43** (1987) 1067.

14. N.S. Hosmane, A.K. Saxena, H. Zhang, J.A. Maguire, A.H. Cowley, M.A. Mardones, D. Atwood and V. Atwood, unpublished results.

15. N.S. Hosmane, H. Zhang, K.-J. Lu, J.A. Maguire, A.H. Cowley and M.A. Mardones, *Struct. Chem.*, **3** (1992) 183.

16. (a) N.S. Hosmane, K.-J. Lu, H. Zhang, A.H. Cowley and M.A. Mardones, *Organometallics*, **10** (1991) 392. (b) N.S. Hosmane, K.-J. Lu, H. Zhang, J.A. Maguire, A.K. Saxena, A.H. Cowley, M.A. Mardones, *Organometallics*, in press.

17. N.S. Hosmane, M.S. Islam, U. Siriwardane, J.A. Maguire and C.F. Campana, *Organometallics*, **6** (1987) 2447.

18. N.S. Hosmane, J.S. Fagner, H. Zhu, U. Siriwardane, J.A. Maguire, G. Zhang and B.S. Pinkston, *Organometallics*, **8** (1989) 1769.

19. U. Siriwardane, J.A. Maguire, J.J. Banewicz and N.S. Hosmane, *Organometallics*, **8** (1989) 2792.

20. U. Siriwardane and N.S. Hosmane, *Acta Cryst. Cryst. Struct. Commun.*, **C44** (1988) 1572.

21. R.D. Barreto, T.P. Fehlner and N.S. Hosmane, *Inorg. Chem.*, **27** (1988) 453.

22. J.A. Maguire, G.P. Ford and N.S. Hosmane, *Inorg. Chem.*, **27** (1988) 3354.

23. N.S. Hosmane, U. Siriwardane, M.S. Islam, J.A. Maguire and S.S.C. Chu, *Inorg. Chem.*, **26** (1987) 3428.

24. N.S. Hosmane, M.S. Islam, B.S. Pinkston, U. Siriwardane, J.J. Banewicz and J.A. Maguire, *Organometallics*, 7 (1988) 2340.

25. U. Siriwardane, M.S. Islam, J.A. Maguire and N.S. Hosmane, *Organometallics*, 7 (1988) 1893.

26. N.S. Hosmane, K.-J. Lu, U. Siriwardane, H. Zhang and J.A. Maguire, to be submitted for publication.

27. (a) N. S. Hosmane, U. Siriwardane, H. Zhu, G. Zhang and J.A. Maguire, *Organometallics*, 8 (1989) 566. (b) N.S. Hosmane, K.-J. Lu, H. Zhu, U. Siriwardane, M.S. Shet and J.A. Maguire, *Organometallics*, 9 (1990) 808. (c) U. Siriwardane, K.-J. Lu and N. S. Hosmane, *Acta Crystallogr., Cryst. Struct. Commun.*, C46 (1990) 1391.

28. N.S. Hosmane, K.-J. Lu, U. Siriwardane and M.S. Shet, *Organometallics*, 9 (1990) 2798.

29. N.S. Hosmane, K.-J. Lu, H. Zhang, J.A. Maguire, L. Jia and R.D. Barreto, *Organometallics*, 11 (1992) 2458.

30. L. Jia, H. Zhang and N. S. Hosmane, *Organometallics*, 11 (1992) 2957.

31. P. Jutzi, Adv. *Organomet. Chem.*, 1986, 26, 217; *Chem. Rev.*, 86 (1986) 983.

32. N.S. Hosmane, L. Jia, H. Zhang and J.A. Maguire, *Organometallics*, submitted for publication.

33. N.S. Hosmane, L. Jia, H. Zhang, J.A. Maguire and A.K. Saxena, unpublished results.

34. N.S. Hosmane, K.-J. Lu, A.H. Cowley and M.A. Mardones, *Inorg. Chem.*, 30 (1991) 1325.

35. P. Jutzi, D. Wegener and M. Hursthouse, *J. Organomet. Chem.*, 418 (1991) 277.

Main Group Elements and Their Compounds
V.G. Kumar Das (Ed)
Copyright © 1996 Narosa Publishing House, New Delhi, India

Aspects of the Chemistry of Subvalent Organic Compounds of Germanium, Tin and Lead; Synthesis and Structures of the Bis{bis(trimethylsilyl)amido}tin(IV) Chalcogenides

Peter B. Hitchcock, Eunseok Jang and **Michael F. Lappert**

*School of Chemistry and Molecular Sciences, University of Sussex,
Brighton, BN1 9QJ, United Kingdom*

Until about 20 years ago bi- or trivalent non-polymeric compounds of the Group 14 elements germanium, tin and lead were unknown. The reason for this was then believed to be electronic, in that covalent derivatives of these elements were expected to obey the 8-electron rule. This situation has changed drastically in the intervening period. The success in overcoming this electronic paradigm was based on aiming for kinetically stabilizing target molecules by careful ligand selection. Thus, numerous mononuclear homoleptic compounds of the formula MX_2 (M = Ge, Sn or Pb) or $M'X'_2$ (M' = Ge or Sn) were made in which the ligand X^- or X'^-, although monohapto, was bulky and preferably was free from β-hydrogen atoms. Among the thermally stable, but highly reactive, compounds of this class to have been made were those in which $X = CHR_2$, NR_2, OAr or SAr (R = $SiMe_3$, Ar = $C_6H_2Bu^t_3$-2,4,6) and $X' = CHR_2$ or NR_2. These aspects have been reviewed.[1]

A more recent development concerns the first stable heteroleptic compounds MXX' (X and X' are different). An interesting feature of such a molecule, and indeed of the homoleptic analogues MX_2, is their structures which, in principle, could be linear corresponding to a triplet ground state, or V-shaped characteristic of a singlet electronic ground state. Thus far, only the latter type has been found. A further point concerning a heteroleptic compound MXX' is that it is prochiral at the metal M and an addition reaction with a reagent A-B can lead to a chiral tetravalent compound of formula MX(X')(A)B. One paper has been published;[2] it included the preparation and crystallographic characterization of $Sn(NR_2)(OC_6H_2Bu^t_2$2,6-Me-4), while other aspects have been briefly reviewed.[3]

Extensions of the MX_2 studies have been carried out to bi- or trinuclear tin(II) compounds, in which either each of the M atoms is two-coordinate, or a compound is obtained having two 2- and one 4-coordinate tin(II) centres.[4] Among such compounds to have been prepared are compounds 1-4. However, in this paper, we shall restrict ourselves to discussing a third new development, namely the synthesis and structures of the compounds $[Sn(NR_2)_2(\mu\text{-E})]_2$ [R = $SiMe_3$ and E = S (5), Se (6) or Te (7)] and $[(R_2N)_2Ge(\mu\text{-Te})_2Sn(NR_2)_2]$ (8). Each of **5 - 7** (see also ref. 5) was obtained by oxidative addition of $Sn(NR_2)_2$ + E, or for **8** from $Ge(NR_2)_2 + Sn(NR_2)_2$ + Te.

Table 1 : Selected published geometric data for some 4-membered ring compounds: bond length (Å) and angles (°), with estimated standard deviations in parentheses

	[SnBu'$_2$(μ-S)]$_2$[a] [ref. 6]		SnBu'$_2$-(μ-Se)]$_2$[b] [ref. 6]	[SnBu'$_2$(μ-Te)]$_2$[b] [ref. 6]	[Ge(NR$_2$)$_2$(μ-Te)]$_2$[b] (9) [ref. 1]
l_1	2.442(4)	2.410(4)	2.553(2)	2.758(1)	2.599(2)
l_2	-	-	-	-	2.596(2)
l_3	2.419(4)	2.434(4)	2.549(2)	2.754(1)	2.592(2)
l_4	-	-	-	-	2.595(2)
l_5	2.19(1)	2.21(2)	2.18(2)	2.20(1)	1.867(12)
l_6	2.23(1)	2 19(2)	2.17(2)	2.21(1)	1.869(10)
l_7	-	-	-	-	1.873(12)
l_8	-	-	-	-	1.858(13)
a	93.8(1)	93.8(1)	97.5(1)	100.0(1)	94.28(6)
b	-	-	-	-	94.47(6)
c	-	-	-	-	85.59(6)
d	86.2(1)	86.2(1)	82.5(1)	80.0(1)	85.58(6)
e	117.0(6)	118.0(6)	115.4(8)	117.0(4)	108.1(5)
f	-	-	-	-	109.0(6)

[a][SnBu'$_2$(μ-S)]$_2$ occurs in two different forms; data in the left hand column are for the monoclinic crystal, and those in the right hand column are for the triclinic crystal. [b]Monoclinic crystal.

Bis{bis(trimethylsilyl)amido}germanium(IV) and -tin(IV) chalcogenides $\{M(NR_2)_2(\mu\text{-}E)_n$ (M = Ge or Sn; E = S, Se or Te; R = SiMe$_3$) had previously been synthesized at Sussex by a reaction between bis{bis(trimethylsilyl)amido}-germanium(II) or -tin(II) and an elementary chalcogen.[5] Di-t-butyltin chalcogenides $[SnBu^t_2(\mu\text{-}E)]_2$ and di-t-amyltin chalcogenides $[SnAm^t_2(\mu\text{-}E)]_2$ (E = S, Se or Te) had been prepared from Na$_2$E and di-t-butyltin(IV) chloride and di-t-amyltin(IV) chloride, respectively.[6] Of these, $[SnBu^t_2(\mu\text{-}E)]_2$ (E = S, Se or Te)[6] and $[Ge(NR_2)_2(\mu\text{-}Te)]_2$[5] were characterized by single crystal X-ray diffraction, each of the four compounds having a planar four-membered ring geometry. Table 1 shows selected comparative structural data for these four compounds.

Experimental

Synthesis of $[Sn(NR_2)_2(\mu\text{-}S)]_2$ (R = SiMe$_3$) (5) by reflux

Sublimed sulfur powder (0.17 g, 5.3 mmol) was placed into a THF solution (*ca.* 50 cm^3) of bis{bis(trimethylsilyl)amido{tin(II)7 (2.3 g, 5.2 mmol). The mixture was refluxed for *ca.* 8 h. The pale yellow reaction mixture was cooled to room temperature and the small amount of unreacted sulfur and other insolubles were filtered off yielding a yellow filtrate, which was concentrated *in vacuo*. A colourless-to-opaque crystalline solid was obtained, from which the volatiles were removed *in vacuo*. The residual solid was extracted into n-C$_5$H$_{12}$ (*ca.* 30 cm^3). The extract was concentrated to *ca.* 10 cm^3 and was placed at -30 °C. After a week colourless crystals of the title product 5 (1.54 g, 63%) were obtained, which were dried slowly *in vacuo*.

Synthesis of $[Sn(NR_2)_2(\mu\text{-}S)]_2$ (R = SiMe$_3$) (5) by sonication

Sublimed sulfur (0.35 g, 11.0 mmol) was placed into a THF solution (*ca.* 50 cm^3) of bis{bis(trimethylsilyl)amido}tin(II) (4.55 g, 10.3 mmol). The vessel was immersed in an ultrasonic cleaning bath which was preset at 25 °C. An almost instant colour change to a more intense yellow was observed. The reaction mxiture was removed from the bath after 1h of sonication, then filtered free of the very slight preciptiate. Volatiles were removed from the filtrate yielding a crystalline solid which was extracted into n-C$_5$H$_{12}$ (*ca.* 30 cm^3). The title product 5 (4.49 g, 91%) was obtained using the workup procedure described above.

Synthesis of $[Sn(NR_2)_2(\mu\text{-}Se)]_2$ (R = SiMe$_3$) (6) by reflux

Selenium powder (0.4 g, 5.1 mmol) was added to a THF (*ca.* 50 cm^3) of bis{bis(trimethylsilyl)amido}tin(II) (2.22 g, 5.0 mmol). The mixture was refluxed for *ca.* 14 h; it had become dark brown. After cooling to room temperature, the small amount of unreacted selenium and other insolubles were filtered off yielding a red filtrate, which was concentrated *in vacuo* yielding red needles. Volatiles were removed and the residual solid was dried *in vacuo*, then extracted into n-C$_5$H$_{12}$ (*ca.* 30 cm^3). The extract was placed at -30 °C for a week. Volatiles were removed slowly *in vacuo* to give red needles of the title product 6 (1.73 g, 67%).

Synthesis of [Sn(NR₂)₂(μ-Se)]₂ (R = SiMe₃) (6) by sonication

Selenium powder (0.64 g, 8.1 mmol) wasa added to a THF solution (*ca.* 50 cm³) of bis{bis(trimethylsilyl)amido}tin(II) (3.6 g, 8.1 mmol). The vessel was immersed in an ultrasonic cleaning bath preset at 25 °C. The red reaction mixture was removed from the bath after 4 h of sonication, then filtered following cooling to room temperature to give a red filtrate from which volatiles were removed slowly *in vacuo* yielding red needles. The needles were then extracted into *n*-C₅H₁₂ (*ca.* 30 cm³) to give a bright red solution, which was placed at -30 °C for a week. Volatiles were removed slowly *in vacuo* and the residual red crystals of the title product 6 (2.86 g, 92%) were dried *in vacuo*.

Synthesis of [Sn(NR₂)₂(μ-Te)]₂ (R = SiMe₃) (7) by reflux

Tellurium powder (0.56 g, 4.4 mmol) was added to a THF solution (*ca.* 50 cm³) of bis{bis(trimethylsilyl)amido}tin(II) (1.86 g, 4.2 mmol). The mixture was refluxed for *ca.* 17 h. The dark brown reaction mixture was allowed to cool to room temperature, and the small amount of unreacted tellurium and other insolubles were filtered off, to give a dark red filtrate. Volatiles were removed from the filtrate *in vacuo* yielding red needles, which were extracted into *n*-C₅H₁₂ (*ca.* 30 cm³) yileding a red solution, which was placed at -30 °C for a week. Volatiles were slowly removed from the latter *in vacuo* to provide red needles of the title product 7 (1.62 g, 68%) which were dried *in vacuo*.

Synthesis of [Sn(NR₂)₂(μ-Te)]₂ (R = SiMe₃) (7) by sonication

Tellurium powder (1.41 g, 11.1 mmol) was added to THF soolution (ca. 50 cm³) of bis{bis(trimethylsilyl)amido}tin(II) (4.86 g, 11.0 mmol). The vessel was immersed in an ultrasonic bath preset at 25 °C, and 4 h of sonication was carried out yielding a dark red reaction mixture. After cooling to room temperature, the small amount of unreacted tellurium powder and other insolubles were filtered off giving a red filtrate from which volatiles were removed *in vacuo* affording red needles. These were extracted into *n*-C₅H₁₂ (*ca.* 30 cm³) yielding a bright red solution, which was placed at -30 °C for a week. Volatiles were removed very slowly from this *in vacuo* and the residual red crystals of the title product 7 (5.75 g, 92%) were dried *in vacuo*.

Synthesis of [(NR₂)₂Ge(μ-Te)₂Sn(NR₂)](R = SiMe₃) (8)

Tellurium powder (1.95 g, 15.3 mmol) was added to a THF solution (*ca.* 50 cm³) containing Ge(NR₂)₂ (1.96 g, 5 mmol) and Sn(NR₂)₂ (4.47 g, 10 mmol); the mixture was refluxed for *ca.* 10 h. The cooled dark brown reaction mixture was filtered through a cannula, yielding a dark red filtrate and a black powder. Red needles were obtained upon removal of volatiles from the filtrate *in vacuo*, which were then extracted into *n*-C₅H₁₂ (*ca.* 30 cm³) to give a red solution. Slowly concentrating the solution *in vacuo* and placing at -30 °C for a week yielded a mixture of mainly orange-red with a small amount of red needles. Further removal of volatiles was carried out *in vacuo*, and a mixture (2.33 g) of orange-red (8) and red needles

{[Ge(NR$_2$)$_2$(μ-Te)]}$_2$ and 7} was obtained. Compound **8** (0.65 g, 12%) was later separated manually.

Crystallographic data for the structural analysis of compounds 5 - 8

The data are summarized in Table 2. The diffractometer was an Enraf-Nonius CAD4. Further details may be obtained from the authors.

Compound **8** was obtained as orange-red, triclinic needles in 68% yield, m.p. 145 °C (decomp.); Anal: found: C, 26.4; H, 6.45; N, 4.94%. C$_{24}$H$_{72}$GeN$_4$Si$_8$SnTe$_2$ calcd: C, 25.7; H, 6.42; N, 5.00%.

Results and Discussion

Synthesis of [Sn(NR$_2$)$_2$(μ-E)]$_2$ [R = SiMe$_3$ and E = S (5), Se (6), or Te (7) by oxidative addition of chalogen (E) to Sn(NR$_2$)$_2$ activated by sonication

Improved yields[5] of each of the compounds **5 - 7** was obtained by prolonged reflux in THF (a in Scheme 1). By use of ultrasound, almost quantitative yields were obtained at ambient temperature in tetrahydrofuran (THF), after a short reaction time (b in Scheme 1). The new data on compounds **5 - 7** are summarized in Tables 3 and 4.

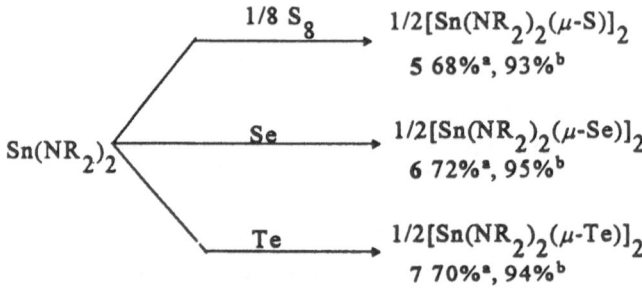

Scheme 1 : New data on the synthesis of [Sn(NR$_2$)$_2$(μ-E)]$_2$ [R = SiMe$_3$ and E = S (5), Se (6) or Te (7)]: (a) reflux for 8 (5), 14 (6) or 17 (7) h using THF as solvent; (b) by sonication at 25 °C for 1 (5), 4 (6) or 4 (7) h using THF as solvent. Yields were obtained by integration of the δ(^{119}Sn) signals, using ^{119}Sn{^{1}H} NMR spectroscopy. Crystals were obtained by recrystallisation from *n*-C$_5$H$_{12}$.

Colourless (**5**) and red (**6** or **7**) needles were readily obtained by recrystallization from *n*-C$_5$H$_{12}$. The melting point of each of the compounds **5 - 7** was consistent with that reported in the earlier work.[5] Precipitation of chalcogen powder for each of the compounds **5 - 7** in *n*-C$_5$H$_{12}$ solution (or separation in the solid) was observed; precipitation of a white (**5**) or black (**6** or **7**) powder was noted after the mixture had been kept at -30 °C under argon for *ca.* 5 day. When any one of the compounds **5 - 7** was kept under argon at room temperature and exposed to light, a faster decomposition rate was found, with the chalcogen precipitating in *ca.* 2 days.

Table 2 : Crystal data for 5 - 8

Compound		5	6	7	8
Formula		$C_{24}H_{72}N_4S_2Si_8Sn_2$	$C_{24}H_{72}N_4Se_2Si_8Sn_2$	$C_{24}H_{72}N_4Si_8Sn_2Te_2$	$C_{24}H_{72}GeN_4Si_8Sn_2Te_2$
Formula weight		943.1	1036.9	1134.1	1088
Crystal size (mm)		0.2 x 0.2 x 0.2	0.25 x 0.25 x 0.25	0.3 x 0.15 x 0.1	0.3 x 0.3 x 0.2
Crystal system		triclinic	triclinic	triclinic	triclinic
Cell dimensions:	a(Å)	9.444(2)	8.905(4)	8.754(2)	8.709(2)
	b(Å)	11.310(4)	11.291(7)	11.357(1)	11.095(7)
	c(Å)	23.727(7)	12.925(4)	26.402(3)	26.098(14)
	α(°)	79.49(3)	107.02(4)	79.07(1)	78.52(5)
	β(°)	88.12(3)	93.41(3)	81.82(1)	80.55(3)
	γ(°)	68.80(3)	109.79(4)	73.24(1)	73.37(4)
Cell volume (Å^3)		2321.5	1151.0	2457.2	2352.4
Z		2	1	2	2
Space group		$P\bar{1}$ (No. 2)	$P\bar{1}$ (No. 2)	$P\bar{1}$ (No. 2)	$P\bar{1}$ (No. 2)
D_{calc} (g cm^{-3})		1.35	1.5	1.53	1.54
μ (cm^{-1})		13.9	28.8	24.0	26.03
Number of observed reflections, $|F^2| > 2\sigma$ (F^2)		5027	365	8642	4688
σ(F^2)		$\{\sigma^2(I) + (0.04I)^2\}^{1/2}/L_p$	$\{\sigma^2(I) + (0.04I)^2\}^{1/2}/L_p$	$\{\sigma^2(I) + (0.04I)^2\}^{1/2}/L_p$	$\{\sigma^2(I) + (0.04I)^2\}^{1/2}/L_p$
R		0.085	0.025	0.048	0.04
Rw		0.114	0.034	0.065	0.045

333

Table 3 : Summary of new characterising (excluding detailed X-ray) data obtained on $[Sn(NR_2)_2(\mu\text{-}E)]_2$ [R = SiMe$_3$ and E = S (5), Se (6) or Te (7)] and $[(R_2N)_2Ge(\mu\text{-}Te)_2Sn(NR_2)_2]$ (8)

Crystal appearance	Yield (%)	Proton-decoupled NMR spectral data[c]				
		^{29}Si{^{1}H}	^{77}Se{^{1}H}	^{119}Sn{^{1}H}	^{125}Te{^{1}H}	INEPT ^{15}N
5 Colourless needles, triclinic	63[a], 91[b]	$\delta = +7.4$, $^2J(^{29}\text{Si-}^{119}\text{Sn}) = 13$ Hz		$\delta = -106.6$, $^2J(^{119}\text{Sn-}^{117}\text{Sn}) = 609$ Hz		$\delta = 321.1$, $^2J(^{15}\text{N-}^{119}\text{Sn}) = 28$ Hz
6 Red needles, triclinic	67[a], 92[b]	$\delta = +7.3$	$\delta = -640.1$, $^1J(^{77}\text{Se-}^{119}\text{Sn}) = 1129\text{Hz}$, $^1J(^{77}\text{Se-}^{117}\text{Sn}) = 1028$ Hz	$\delta = -382.6$, $^1J(^{119}\text{Sn-}^{77}\text{Si}) = 1129\text{Hz}$, $^2J(^{119}\text{Sn-}^{117}\text{Sn}) = 786$ Hz		$\delta = -322.2$ $^1J(^{15}\text{N-}^{119}\text{Sn}) = 5$ Hz,
7 Red needles, triclinic	68[a], 92[b]	$\delta = +7.3$		$\delta = -988.8$, $^1J(^{119}\text{Sn-}^{125}\text{Te}) = 2711$ Hz, $^2J(^{119}\text{Sn-}^{117}\text{Sn}) = 567$ Hz	$\delta = +760.2$, $^1J(^{125}\text{Te-}^{119}\text{Sn}) = 2711$ Hz, $^1J(^{125}\text{Te-}^{117}\text{Sn}) = 2589$ Hz	$\delta = -326.7$, $^1J(^{15}\text{N-}^{119}\text{Sn}) = 5$ Hz
8 Orange-red needles, triclinic	68[a]	$\delta = +5.47$ (Ge), $+8.13$ (Sn)		$\delta = -1023.9$, $^1J(^{119}\text{Sn-}^{125}\text{Te}) = 2778$ Hz,	$\delta = +964.5$, $^1J(^{125}\text{Te-}^{119}\text{Sn}) = 2761$ Hz, $^1J(^{77}\text{Se-}^{117}\text{Sn}) = 1028$ Hz	$\delta = -317.1$ (Ge), -332.1 (Sn)

[a]Isolated yield obtained by method a of Scheme 1. [b]Isolated yield obtained by method b of Scheme 1. [c]The multinuclear NMR spectra were recorded at 305 K in C$_6$D$_6$ on a Bruker AMX500 spectrometer, chemical shifts are relative to those for SiMe$_4$ (^{29}Si), SeMe$_2$ (^{77}Se), SnMe$_4$ (^{119}Sn), TeMe$_2$ (^{125}Te) and MeNO$_2$ (^{15}N). [d]$\delta(^1\text{H}) = x + 0.48$; $\delta[^{13}\text{C}\{^1\dot{\text{H}}\}] = +7.3$. [e]Yield based on integral of ^{119}Sn{^{1}H} NMR spectral signals

Table 4 : Selected structural parameters obtained by single crystal X-ray diffraction at 173 K for compounds $[Sn(NR_2)_2(\mu\text{-}E)]_2$ [R = SiMe$_3$ and E = (5), Se (6) or Te (7)] and $[(R_2N)_2Ge(\mu\text{-}Te)_2Sn(NR_2)_2]$ (8)

Compounds	5	6[a]	7	8
r(Sn-E)/Å	2.416(5)	2.544(1)	2.752(1)	2.728(1)
	2.413(6)	2.538(1)	2.757(1)	2.721(1)
	2.427(6)		2.751(1)	
	2.409(5)		2.755(1)	
r(Ge-Te)/Å				2.615(1)
				2.621(1)
r(Sn-N)/Å	2.034(13)	2.050(3)	2.066(6)	2.042(6)
	2.080(2)	2.047(3)	2.062(7)	2.032(7)
	2.061(12)		2.054(8)	
	2.070(2)		2.060(7)	
r(Ge-N)/Å				1.873(7)
				1.872(6)
r(Sn···Sn)/Å	3.321(1)	3.436(2)	3.677(8)	
r(Sn···Ge)/Å				3.583(6)
R(E···E)/Å	3.517(6)	3.744(3)	4.101(9)	3.957(8)
E-Sn-E(°)	93.5(2)	94.9(1)	96.34(3)	93.14(3)
	93.3(2)		96.12(3)	
Te-Ge-Te(°)				98.18(3)
Sn-E-Sn(°)	86.4(2)	85.09(1)	83.71(3)	
	86.9(2)		83.74(3)	
Sn-Te-Ge(°)				84.27(3)
N-Sn-N(°)	112.7(6)	110.1(1)	109.2(3)	107.1(3)
	114.6(6)		107.7(3)	
N-Ge-N(°)				109.0(3)

[a]Compound **6** has an inversion centre in the mid-point of the ring plane.

Multinuclear NMR spectroscopic studies of compounds $[Sn(NR_2)_2(\mu\text{-}E)]_2$ [R = SiMe$_3$ and E = S (5), Se (6) or Te (7)], $[(R_2N)_2Ge(\mu\text{-}Te)_2Sn(NR_2)_2]$ (8), and some comparative data

Proton-decoupled multinuclear NMR spectroscopic studies using ^{13}C, ^{15}N, ^{29}Si, ^{77}Se, ^{119}Sn and ^{125}Te nuclei were carried out. Data were obtained for each of the compounds **5 - 8** in C$_6$D$_6$ using a Bruker AMX500 spectrometer at ambient temperature, and are summarized in Table 3. In addition to the fact that compounds **5 - 7** have numerous NMR-active spin-½ nuclei, most of them are located next to one another. Hence by determining one-bond coupling constants between chalcogen (^{77}Se and ^{125}Te) and tin (^{117}Sn and ^{119}Sn), and also the transannular ^{119}Sn-^{117}Sn coupling constant for each of the compounds **5 - 7**, it was possible to deduce their central ring structures. From chemical shifts and coupling constants based on the ^{1}H, ^{13}C, ^{15}N and ^{29}Si nuclei, it was possible to establish that the bis{bis(trimethylsilyl)amido} ligand was intact. Figure 1 shows this NMR-friendly environment for compounds **5 - 7**; parameters *a* and *c - e* are also appropriate for **8**. From the ^{119}Sn{^{1}H} NMR spectra, it was evident that a small amount of unreacted Sn(NR$_2$)$_2$ (δ = +776 ppm) was present in each of the three reaction mixtures [Sn(NR$_2$)$_2$ + E] or [Ge(NR$_2$)$_2$ + Sn(NR$_2$)$_2$ + Te].

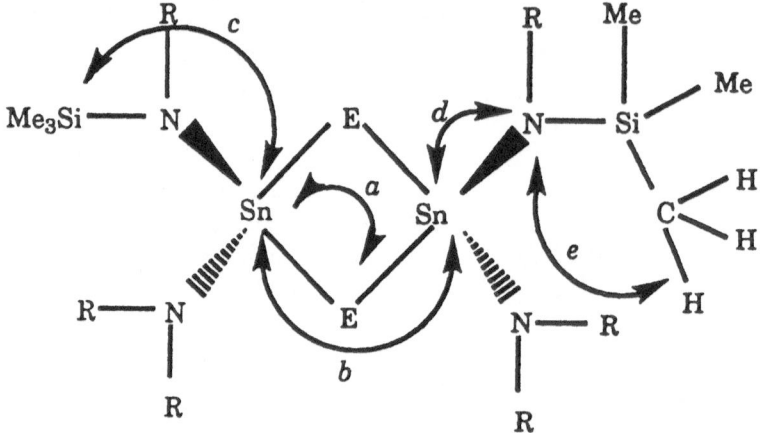

Fig. 1 : **Some selected NMR spectral coupling information available for [Sn(NR$_2$)$_2$(μ-E)]$_2$ [R = SiMe$_3$ and E = S (5), Se (6) or Te (7)]: (a) $^1J(^{119}$Sn-^{77}Se) and $^1J(^{117}$Sn-^{77}Se) for 6 and $^1J(^{119}$Sn-^{125}Te) and $^1J(^{117}$Sn-^{125}Te) for 7; (b) $^2J(^{119}$Sn-^{117}Sn); (c) $^2J(^{119}$Sn-^{29}Si), but this was obtained only for 5; (d) $^1J(^{15}$N-^{119}Sn) by the refocussed INEPT technique, obtained by polarisation transfer via a three bond N-H coupling [the optimal results were obtained by assuming (e) $^3J(^{15}$N-^{1}H) = *ca.* 1.8 Hz].**

Tin has two spin-½ nuclei and selenium and tellurium one each, and the natural abundance of each is between 7 and 8%. The ^{119}Sn{^{1}H} spectrum of the selenium analogue **6** showed a single signal flanked by two sets of satellites; these are attributed to $^1J(^{119}$Sn-^{77}Se) and $^2J(^{119}$Sn-^{117}Sn). The former had twice the intensity of the latter, consistent with each tin nucleus being coupled to two equivalent selenium nuclei. Comparable conclusions are drawn from the ^{119}Sn{^{1}H} spectrum of the tellurium analogue **7**. Each of the ^{77}Se{^{1}H} (for **6**) and ^{125}Te{^{1}H} (for **7**) spectra showed a single peak with closely-spaced pairs of satellites due to both ^{119}Sn and ^{117}Sn nuclei. Initially (prior to the X-ray analysis) the structure of [(R$_2$N)Ge(μ-Te)$_2$Sn(R$_2$N)$_2$] (**8**) was deduced from the rich NMR spectral information, Table 3.

The insensitive nucleus enhanced by polarisation transfer (INEPT) technique was used to obtain spectra from the insensitive ^{15}N nucleus for each of **5 - 8**. Polarisation transfer was from the trimethylsilyl protons *via* a three-bond N-H coupling, $^3J(^1$H-^{15}N). In order to obtain the transfer, the delays within the INEPT pulse had to be lengthened for signal through relaxation. Interestingly $^1J(^{15}$N-^{119}Sn) for compound **5** was substantial 28.4 Hz, but for compounds **6** (4.9 Hz) and **7** (4.8 Hz) the values were small. The observed ^{1}H and ^{13}C{^{1}H} NMR spectral data for **5 - 7** were consistent with those of our earlier work.[7] For **8**, $\delta(^1$H) = +0.48 and

$\delta[^{13}C\{^1H\}] = +7.3$. From these and the $^{29}Si\{^1H\}$ NMR data it is evident that there is no substantial change in these parameters for the four compounds.

A linear relationship was found between the tin chemical shifts for compounds **5 - 7** and the Pauling electronegativity of the appropriate chalcogen. Similarly, a plot of $\delta[^{125}Te\{^1H\}]$ against the Pauling electronegativity of the transannular Group 14 elements for $[Ge(NR_2)_2(\mu\text{-}Te)]_2{}^6$ ($\delta = +1184$), **3** and **4** showed linearity.

The X-ray structures of $[Sn(NR_2)_2(\mu\text{-}E)]_2$ [R = SiMe$_3$ and E = S (5), Se (6) or Te (7)], $[R_2N)_2Ge(\mu\text{-}Te)_2Sn(NR_2)_2]$ (8) and some comparative data

For each of the compounds **5 - 8**, X-ray quality single crystals were obtained from $n\text{-}C_5H_{12}$ solution. X-ray crystal data were collected at 173 K on an Enraf-Nonius CAD4 diffractometer using monochromated Mo-K$_\alpha$ radiation. Key data are summarized in Table 4 and are available for comparison with published results (Table 1). Figure 2 and Table 5 illustrate the molecular structure and the atom labelling scheme for **5**, and are typical of the four compounds. More detailed data are available from the authors.

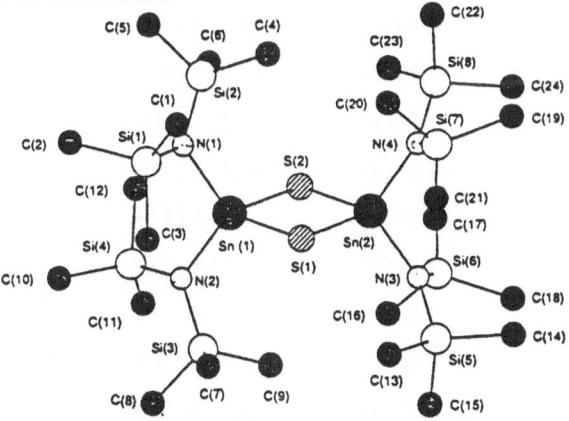

Fig. 2 : The molecular structure of $[Sn(NR_2)_2(\mu\text{-}S)]_2$ (R = SiMe$_3$) (5), showing the atom labelling scheme

Each of the four molecules was found to contain a planar four-membered ring core of alternating tin and chalcogen atoms, the sum of the internal angles being close to 360°. Two bulky N(SiMe$_3$)$_2$ groups were attached to each tin atom (or for **8** also Ge) with one above and the other below the ring plane. Compounds **5** and **7** showed slightly distorted ring planes, while compound **6** was found to have an inversion centre in the mid-point of the ring; **8** also had a planar ring core. Bond lengths between tin and chalcogen increased with atomic number of E. The Sn-N bond lengths for compound **5 - 7** were reduced by *ca.* 0.025 - 0.061 Å, and the N-Sn-N angles were increased by *ca.* 3.01 - 9.94°, compared with the values for the tin(II) starting material Sn(NR$_2$)$_2$ [Sn-N = 2.096(1) and 2.088(6) Å; N-Sn-N = 104.7(2)°].[8] This is attributed mainly to the radius of Sn^{4+} being smaller than of Sn^{2+}, but a steric effect may have been a contributory factor. The sum of angles at N in each of the four compounds **5 - 8** was approximately 360° (consistent with a trigonal planar environment at nitrogen).

Table 5 : Non-hydrogen intramolecular distances (Å) and angles (°) with estimated standard deviations in parentheses for $[Sn(R_2N)_2(\mu\text{-}S)]_2$ (R = $SiMe_3$) (5)

(a) Bonds			
Sn(1)-S(1)	2.461(5)	Sn(1)-S(2)	2.413(6)
Sn(1)-N(1)	2.034(13)	Sn(1)-N(2)	2.08(2)
Sn(2)-S(1)	2.427(6)	Sn(2)-S(2)	2.409(5)
Sn(2)-N(3)	2.061(12)	Sn(2)-N(4)	2.07(2)
Si(1)-N(1)	1.74(2)	Si(1)-C(1)	1.94(3)
Si(1)-C(2)	1.86(3)	Si(1)-C(3)	1.86(3)
Si(2)-N(1)	1.78(2)	Si(2)-C(4)	1.85(2)
Si(2)-C(5)	1.86(2)	Si(2)-C(6)	1.95(2)
Si(3)-N(2)	1.74(2)	Si(3)-C(7)	1.95(2)
Si(3)-C(8)	1.82(2)	Si(3)-C(9)	1.85(2)
Si(4)-N(2)	1.72(2)	Si(4)-C(10)	1.86(2)
Si(4)-C(11)	1.84(2)	Si(4)-C(12)	1.86(2)
Si(5)-N(3)	1.76(2)	Si(5)-C(13)	1.86(2)
Si(5)-C(14)	1.88(3)	Si(5)-C(15)	1.87(2)
Si(6)-N(3)	1.74(2)	Si(6)-C(16)	1.88(2)
Si(6)-C(17)	1.86(3)	Si(6)-C(18)	1.87(2)
Si(7)-N(4)	1.70(2)	Si(7)-C(19)	1.87(2)
Si(7)-C(20)	1.83(2)	Si(7)-C(21)	1.88(2)
Si(8)-N(4)	1.78(2)	Si(8)-C(22)	1.84(2)
Si(8)-C(23)	1.94(2)	Si(8)-C(24)	1.89(3)
(b) Angles			
S(1)-Sn(1)-S(2)	93.5(2)	S(1)-Sn(1)-N(1)	110.9(5)
S(1)-Sn(1)-N(2)	115.1(4)	S(2)-Sn(1)-N(1)	114.8(5)
S(2)-Sn(1)-N(2)	108.6(5)	N(1)-Sn(1)-N(2)	112.7(6)
S(1)-Sn(2)-S(2)	93.3(2)	S(1)-Sn(2)-N(3)	114.6(5)
S(1)-Sn(2)-N(4)	108.2(4)	S(2)-Sn(2)-N(3)	109.2(4)
S(2)-Sn(2)-N(4)	115.3(4)	N(3)-Sn(2)-N(4)	114.6(6)
Sn(1)-S(1)-Sn(2)	86.4(2)	Sn(1)-S(2)-Sn(2)	86.9(2)

Table 5 (cont'd)

N(1)-Si(1)-C(1)	113(1)	N(1)-Si(1)-C(2)	109(1)
N(1)-Si(1)-C(3)	111(1)	C(1)-Si(1)-C(2)	113(1)
C(1)-Si(1)-C(3)	101(1)	C(2)-Si(1)-C(3)	110(2)
N(1)-Si(2)-C(4)	111(1)	N(1)-Si(2)-C(5)	112(1)
N(1)-Si(2)-C(6)	111.8(9)	C(4)-Si(2)-C(5)	106(1)
C(4)-Si(2)-C(6)	112(1)	C(5)-Si(2)-C(6)	104(1)
N(2)-Si(3)-C(7)	113.8(8)	N(2)-Si(3)-C(8)	110.8(9)
N(2)-Si(3)-C(9)	111(1)	C(7)-Si(3)-C(8)	105(1)
C(7)-Si(3)-C(9)	109(1)	C(8)-Si(3)-C(9)	107(1)
N(2)-Si(4)-C(10)	111(1)	N(2)-Si(4)-C(11)	109.5(9)
N(2)-Si(4)-C(12)	111(1)	C(10)-Si(4)-C(11)	111(1)
C(10)-Si(4)-C(12)	106(1)	C(11)-Si(4)-C(12)	108(1)
N(3)-Si(5)-C(13)	116.1(8)	N(3)-Si(5)-C(14)	110(1)
N(3)-Si(5)-C(15)	112(9)	C(13)-Si(5)-C(14)	108(1)
C(13)-Si(5)-C(15)	105(1)	C(14)-Si(5)-C(15)	105(1)
N(3)-Si(6)-C(16)	113(1)	N(3)-Si(6)-C(17)	112.5(8)
N(3)-Si(6)-C(18)	112(1)	C(16)-Si(6)-C(17)	104(1)
C(16)-Si(6)-C(18)	107(1)	C(17)-Si(6)-C(18)	107(1)
N(4)-Si(7)-C(19)	111(1)	N(4)-Si(7)-C(20)	112.4(9)
N(4)-Si(7)-C(21)	111.7(9)	C(19)-Si(7)-C(20)	110(1)
C(19)-Si(7)-C(21)	105(1)	C(20)-Si(7)-C(21)	106(1)
N(4)-Si(8)-C(22)	110.7(9)	N(4)-Si(8)-C(23)	115.8(9)
N(4)-Si(8)-C(24)	110.3(9)	C(22)-Si(8)-C(23)	105.0(9)
C(22)-Si(8)-C(24)	106(1)	C(23)-Si(8)-C(24)	108(1)
Sn(1)-N(1)-Si(1)	117.9(9)	Sn(1)-N(1)-Si(2)	119(1)
Si(1)-N(1)-Si(2)	121.0(8)	Sn(1)-N(2)-Si(3)	119.3(8)
Sn(1)-N(2)-Si(4)	114.3(9)	Si(3)-N(2)-Si(4)	122.6(9)
Sn(2)-N(3)-Si(5)	124.0(8)	Sn(2)-N(3)-Si(6)	117.3(8)
Si(5)-N(3)-Si(6)	118.7(7)	Sn(2)-N(4)-Si(7)	117.5(9)
Sn(2)-N(4)-Si(8)	123.4(8)	Si(7)-N(4)-Si(8)	119.0(9)

In compound **7**, the Sn-Te and Sn-N bonds were longer by *ca.* 0.16 and 0.2 Å, respectively than the Ge-Te and Ge-N bonds in $[Ge(NR_2)_2(\mu\text{-}Te)]_2$ (**9**) (*c.f.* Table 1); likewise, the Te-M-Te angles were wider and the M-Te-M angles narrower in compound **7** (M = Sn) than in the Ge analogue.[5] As far as the endocyclic bond angles and distances are concerned, the Se parameters are similar in **6** and **9**. The compounds $[SnBu^t_2(\mu\text{-}E)]_2$ (E = S, Se or Te) had longer Sn-C bonds than the Sn-N bonds of the appropriate compounds **5** - **7** by *ca.* 0.128 - 0.197 Å, and the C-Sn-C angles in the former were also wider than the N-Sn-N in the latter by *ca.* 2.38 -

10.33°.[6] The Sn-E bonds in the two series of compounds were of similar dimensions. However, whereas the endocyclic angles and distances for $[SnBu^t_2(\mu\text{-}S)]_2$ were similar[6] to those of compound **5**, for $[SnBu^t_2(\mu\text{-}Se)]_2$ the Sn-Se bonds were shorter by *ca.* 0.1 Å than in compound **6**, and for $[SnBu^t_2(\mu\text{-}Te)]_2$ the Te-Sn-Te angles were wider by *ca.* 3.6 - 3.7° (and hence the Sn-Te-Sn angles narrower by a similar amount).

The Ge-Te bonds in **8** were longer by *ca.* 0.02 Å compared to those in **9**[5], while the Sn-Te bonds were *ca.* 0.03 Å shorter than in **7**. The Te-Ge-Te angle in **8** was *ca.* 4° wider than in **9**, while the Te-Sn-Te angle was *ca.* 3° narrower than in **7**.

Acknowledgements

We are grateful to Lucky-DC Silicone Co., Ltd. (South Korea) for a grant for E. Jang and S.E.R.C. for support.

References

1.   M.F. Lappert, *Silicon, Germanium, Tin and Lead Compounds*, **9** (1986) 129; M.F. Lappert and R.S. Rowe, *Coord. Chem. Rev.*, **100** (1990) 267; M.F. Lappert, Ch. 12, in *Chemistry and Technology of Silicon and Tin*, eds., V.G. Kumar Das, Ng Seik Weng and M. Gielen, Oxford University Press, New York, 1992.

2.   H. Braunschweig, R.W. Chorley, P.B. Hitchcock and M.F. Lappert, *J. Chem.Soc.Chem. Commun.*, (1992) 1311.

3.   M.F. Lappert, in *Frontiers of Organogermanium -Tin and -Lead Chemistry*, eds., E. Lukevics and L. Ignatovich, Latvian Institute of Organic Synthesis, Riga, pp 8-27; M.F. Lappert, *Silicon, Germanium, Tin and Lead Compounds*, in press.

4.   H. Braunschweig, P.B. Hitchcock, M.F. Lappert and L-J. Pierssens, unpublished work cited in ref. 3.

5.   P.B. Hitchcock, H.A. Jasim, M.F. Lappert, W-P. Leung, A.K. Rai and R.E. Taylor, *Polyhedron*, **10** (1991) 1203.

6.   H. Puff, G. Bertram, B. Ebeling, M. Franken, R. Gattermayer, R. Hundt, W. Schuh and R. Zimmer, *J. Organomet. Chem.*, **379** (1989) 235.

7.   M.J.S. Gynane, D.H. Harris, M.F. Lappert, P.P. Power, P. Riviére and M. Riviére-Baudet, *J. Chem. Soc., Dalton Trans.*, (1977) 2004.

8.   R.W. Chorley, P.B. Hitchcock, M.F. Lappert, W-P. Leung, P.P. Power and M.M. Olmstead, *Inorg. Chim. Acta.*, **198-200** (1992) 203.

Main Group Elements and Their Compounds
V.G. Kumar Das (Ed)
Copyright © 1996 Narosa Publishing House, New Delhi, India

Organotin Compounds : Studies on Sn-C, Sn-Sn, C-C and C-O-C Bond-forming Reactions in Aqueous Media

Daniele Marton and **Giuseppe Tagliavini**

Dipartimento di Chimica Inorganica, Metallorganica ed Analitica, Università di Padova, Via Marzolo, 1, I-35131-Padova (Italy)

Recently, the use of water as a medium, either alone or with a cosolvent, to perform organic reactions under heterogeneous conditions has been receiving much attention.[1-13] Water affords considerable advantages with respect to the more conventional organic solvents, being inexpensive and non-flammable, and, in many cases, allowing one-pot procedures. In addition, it clearly helps to reduce the environmental impact of the chemical work-up by reducing the volume of toxic organic solvents. With regard to its effects, the use of water or water-cosolvent media, instead of hydrocarbons, leads to increased reaction rates,[1-9] and the promotion of different stereochemical reaction courses.[13]

Organometallics have increasingly featured as reactants in water-borne reactions. These include a number of C-C[14-17], C-O-C[18,19] Sn-C[20-24] and Sn-Sn[24] bond-forming reactions. So far, the majority of the examined reactions is restricted to the sole use of silicon and tin derivatives bearing allyl or allyl-like groups. The reason for this resides in the fact that allyl derivatives undergo reactions with many electrophiles at rates which are significantly higher than protonolysis in aqueous media.

This paper reviews the following topics: (1) preparation of tri- and di-organotin derivatives by direct synthesis, (2) preparation of allylstannanes by zinc-mediated coupling of allyl- or allenyl-bromides with R_3SnCl and R_2SnCl_2 in H_2O(salt)/cosolvent media, (3) preparation of hexaalkyl- and hexaaryl-distannanes by means of a Wurtz-type coupling of triorganotin derivatives, (4) preparation of piperidines *via* an aminomethano desilylation-cyclization- and amines *via* aminomethano destannylation process, (5) aldol addition of silyl enol ethers with carbonyl compounds, (6) reactions of allyl- and allenyl-tin chlorides with carbonyl compounds, and, (7) acetalizations mediated by organotin halides.

Sn-C bond-forming reactions

Direct synthesis of organotin compounds

An early report by K. Sisido dealt with the direct synthesis of organotin halides in water.[20] Benzyl chloride - and its ring-substituted derivatives - reacted with tin

341

powder suspended in water to give tribenzyltin chloride according to the reaction:[20]

$$3 \ C_6H_5CH_2Cl + 2 \ Sn \xrightarrow[\text{autoclave - stirring}]{H_2O, \ 104 - 108 \ °C} (C_6H_5CH_2)_3SnCl + SnCl_2 \quad (1)$$
$$(97\% \ \text{yield})$$

Later, Sisido *et al.*[21] speculated that bis(dibenzylchlorotin)oxide, formed as an intermediate in the above reaction, was responsible for the conversion to tribenzyltin chloride (Scheme 1).

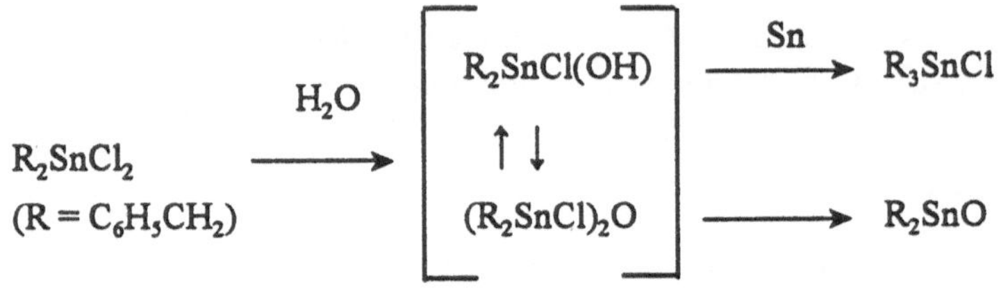

$$R_2SnCl_2 \quad (R = C_6H_5CH_2) \xrightarrow{H_2O} \left[\begin{array}{c} R_2SnCl(OH) \\ \uparrow \downarrow \\ (R_2SnCl)_2O \end{array} \right] \begin{array}{c} \xrightarrow{Sn} R_3SnCl \\ \\ \longrightarrow R_2SnO \end{array}$$

Scheme 1

In support of this, they advanced their findings that both dibenzyltin dichloride and bis(dibenzylchlorotin)oxide were converted into tribenzyltin chloride and tin(II) chloride when treated with metallic tin in water.[21] However, the proposed mechanistic pathway depicted in Scheme 1 does not take into account the stoichiometry of reaction 1. A more plausible pathway that takes into account the acceleration of this reaction when it is performed in water, its radical nature,[20,25] the transient formation[21] of monobenzyltin monochloride [BzSnCl] and the well-known hydrolytic equilibria of organotin chlorides is that depicted in Scheme 2.

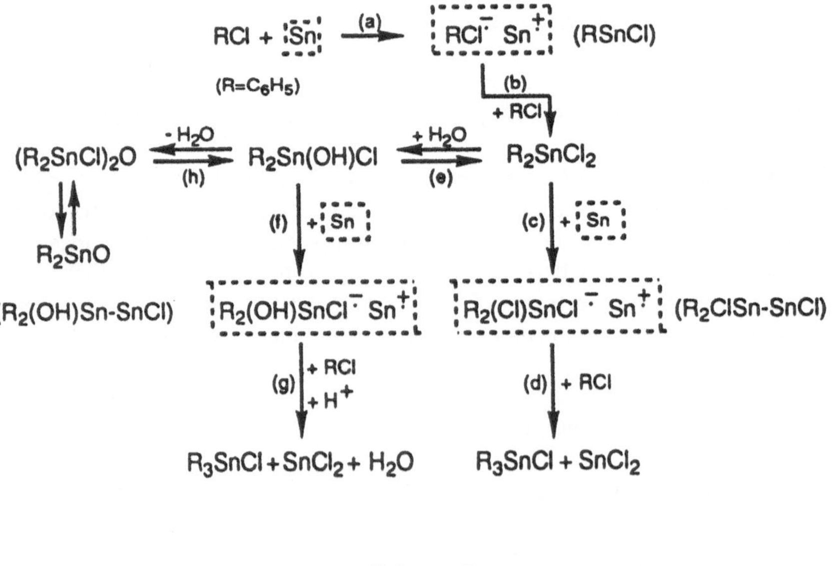

Scheme 2

342

It is to be noted that this reaction occurs in a stirred emulsified heterogeneous medium comprising water, benzyl chloride,[a] and tin powder. Following current suggestions about these heterogeneous systems,[1-13,24] we feel that tin and benzyl chloride particles are forced to interact under the high cohesive energy density[b] developed by the surrounding water phase[13] in such a way that the generated radical ions, adsorbed on the tin surface, will be trapped, step by step, $b \to c \to d$, by the benzyl chloride to form tribenzyltin chloride as the final product. In such reactions, dibenzyltin dichloride represents an intermediate, which is subjected to hydrolytic equilibria to form oxy- or hydroxy-species. These latter species (*e.g.* $R_2Sn(OH)Cl$) are able to interact with tin, affording tribenzyltin chloride through steps *f* and *g*. In point of fact, the radical ions represented in Scheme 2, RCl^-Sn^+, and $R_2(Cl)SnCl^-Sn^+$, and $R_2(OH)SnCl^-Sn^+$ are formally equivalent to the presumed transient intermediate [RSnCl] (monobenzyltin monochloride), as well as to $R_2(Cl)Sn-SnCl$, and $R_2(OH)Sn-SnCl$, respectively. They take into account the oxidation of tin from Sn(0) to Sn(II) and Sn(IV). Even dialkyltin dichlorides (alkyl = Et, Pr, and Bu) are converted into trialkyltin chlorides under the mediation of tin or other metals in water[22] at 160 °C.

Preparation of triorganoallyl- and diorganoallyl-stannanes

Triorganoallyl- and diorganoallyl-stannanes are easily obtained in an one-pot synthesis in the presence of water and air at room temperature. A Wurtz-type reductive coupling reaction mediated by zinc powder, as depicted in eq 2, forms the basis of this novel preparative method.[23,24]

$$n\,RBr \;+\; R'_{4-n}SnCl_n \;\xrightarrow[\text{r.t., stirring}]{\text{cosolvent/H}_2O(NH_4Cl)/Zn}\; R'_{4-n}SnR_n \quad (2)$$

(R = Allyl or allyl-like group; R' = Bu, Pr,Et, Me, Ph; n = 1, 2)

Twenty-one allylstannanes have been prepared *via* this simple procedure using various allyl bromides, RBr (R:$CH_2CH=CH_2$, $CH_2=C(CH_3)CH$, $CH_3CH=CHCH_2$,

$(CH_3)_2C=CHCH_2$, $CH\equiv CCH_2$, $C_6H_5CH=CHCH_2$, $CH_2CH_2CH=CHCH$,

$CH_2CH_2CH_2CH=CHCH$, $CH_2CH_2CH_2CH_2CH=CHCH$), and organotin derivatives such as R_3SnX (R=Me, Et, Pr, Bu and Ph; X=Cl, I, and OH), Bu_2SnCl_2 and $(Bu_2SnCl)_2O$, with yields from 50 to 90%.

Among the various cosolvents used (cyclohexane, toluene, tetrahydrofuran, acetonitrile, 2-propanol and pyridine), the best results were obtained with

---

a   Benzyl chloride is insoluble in water. Moreover, Sisido *et al.* appear not to have taken into account the fact that this compound is easily hydrolyzed to benzyl alcohol in water at about 100 °C.

b   Water is thought to have a cohesive energy density of 22 Kbars (See ref. 13) cyclohexane (see Table 1), where the reaction between Bu_3SnCl and $(C_4H_7)Br$ ($C_4H_7=\alpha$-methylallyl-, *trans*- and *cis*-crotyl) yields with high stereoselectivity $Bu_3SnCH(CH_3)CH=CH_2$ (α-isomer) as the sole product.

Table 1 : Effect of cosolvents on the stereochemical course of the reaction[a]

$$Bu_3SnCl \ + \ (C_4H_7)Br \ \xrightarrow[\text{r.t., stirring}]{\text{cosolvent/H}_2\text{O(NH}_4\text{Cl)/Zn}} \ Bu_3Sn(C_4H_7)$$

Cosolvent	$Bu_3Sn(C_4H_7)$	Isomeric composition isolated mixture, %			Degree of isomerization	*trans:cis* ratio
		CH:CHCH (Me)SnBu$_3$	CH$_3$CH:CH CH$_2$SnBu$_3$	(Z)-CH$_3$CH: CHCH$_2$SnBu$_3$		
cyclohexane	85	100	0	0	0	-
toluene	74	80	8	12	0.20	40:60
THF	83	70	11	19	0.30	37:63
2-propanol	81	45	20	35	0.55	36:64
acetonitrile	78	35	26	39	0.65	40:60
pyridine	58	10	36	54	0.90	40:60

[a]The experimental runs were all performed under atmospheric conditions with the following amounts of components: Bu_3SnCl, 10 g (30.7 mmol); $(C_4H_7)Br$, 8.35 g (61.5 mmol), consisting of a mixture of *trans*-crotyl-(76%), *cis*-crotyl-(9%) and *a*-methylallyl-bromide (15%); Zn, 4 g (61.5 mmol); cosolvent, 25 mL and $H_2O(NH_4Cl$ sat.), 50 mL.

The stereochemical course of these reactions has also been extensively studied with regard to their dependence on the R groups of the R_3SnCl substrate (R=Me, Et, Pr, Bu and Ph), and the salts (NH$_4$Cl, LiCl, NH=C(NH$_2$)$_2$·HCl) employed.

Two reactions are involved in the overall process: (a) the coupling reaction, which gives rise stereoselectively to $R_3SnCH(CH_3)CH=CH_2$ (α-isomer) as the sole product, and (b) the subsequent isomerization of the α-isomer furnishing mixtures of (α, *trans*, *cis*)-isomers. The balance between the two reactions depends upon the nature either of the R group or the cosolvent employed. In cyclohexane, the α-isomer is exclusively obtained with R= Bu, while with R=Me, Et and Pr it is found as a major component of the ternary isomeric mixture. In tetrahydrofuran, 2-propanol, acetonitrile and pyridine, the isomerization occurs to an extent which is dependent on the polarity and the coordinating ability of the cosolvent itself. The observed stereoselectivity has been hypothesized (Scheme 3) to occur via a one-electron transfer from the zinc metal to the $(C_4H_7)Br$ (path 1) to form stereoselectively an adsorbed $CH_2=CHCH(CH_3)Br^-Zn^+$ radical ion that is trapped by the R_3SnCl reactant to form the α-isomer, $R_3Sn(CH(CH_3)CH=CH_2$ (path 2).

The electron transfer, promoted by the high interfacial energy developed by water,[13] occurs under kinetic control. Among the several transition states, those that are more compact will be favoured; these will have more negative values for $\Delta V^{\ddagger}$.[26] Thus, the transition state containing the branched allyl group will have a smaller volume than those containing *cis*- and *trans*-linear allyl groups. The branched radical ion adsorbed on the zinc particle surface will be trapped by the R_3SnX species (path 2) to form the sole $R_3SnCH(CH_3)CH=CH_2$ isomer which, in some particular cosolvent, will undergo isomerization *via* the metallotropic equilibrium (3), mediated either by some unreacted R_3SnX or the cosolvent itself.

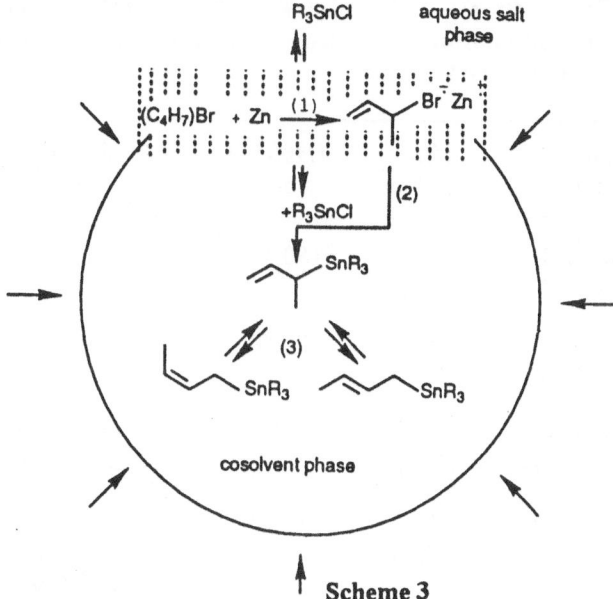

Scheme 3

Sn-Sn bond-forming reactions

Preparation of hexaaryl- and hexaalkyl-ditin compounds

Zinc, as condensing agent, has shown to be able to promote, the coupling of the Ph_3Sn groups to form hexaphenyldistannane. The distannane, in fact, is recovered as by-product together with the main triphenylallylstannane product when allyl bromides are allowed to react with Ph_3SnX compounds (X=Cl, I, OH).

Thus, hexaaryldistannanes can be easily prepared in a $THF/H_2O(NH_4Cl)/Zn$ medium using the sole triaryltin derivatives as starting reactants. Table 2 shows the results achieved for the coupling reactions performed with R_3SnX substrates (R = Ph, p-$CH_3C_6H_4$, m-$CH_3C_6H_4$, X = Cl, I, OH).

Preliminary investigations indicate that this procedure can be generalized to prepare hexaalkyldistannanes, provided that the work-up is performed under a nitrogen atmosphere. Bu_6Sn_2 has been isolated (54% yield) by reacting the sole Bu_3SnCl in $THF/H_2O(NH_4Cl)/Zn$ in a nitrogen atmosphere (Table 2).

The ready synthesis of Sn-Sn bonds by this procedure is to be contrasted with the more tedious well-known preparative methods of ditin compounds based on the reaction of the same substrates with sodium in organic solvent or liquid ammonia, or of triaryltinlithium with triaryltin halides.[27,28]

With regard to the mechanistic aspects of this Wurtz-type reaction, it has been

345

assumed[24] that $R_3SnCl^-Zn^+$ radical ions are produced by a one-electron transfer from the metal to the electrophilic R_3SnCl at a favourable rate, which enables the ions to be trapped by other R_3SnCl molecules to form ditin compounds. The acceleration of the reaction brought about in the water-cosolvent medium parallels that observed for reactions subjected to sonification. In fact, trialkyltin halides (R=Me, Bu) give good yields of Me_6Sn_2 (60%) and Bu_6Sn_2 (94%) with lithium wire in THF under sonification.[29]

Table 2 : Preparation of some hexaaryldistannanes[a] and hexabutyldistannane[b] from triorganotin derivatives in a THF/H_2O(NH$_4$Cl)/Zn medium

Organotin g (mmol)	Zn g (mmol)	THF/H_2O mL	Product g (yield %)
Ph_3SnCl 10.0 (25.9)	2.9 (44.3)	150/150	$Ph_3SnSnPh_3$ 6.14 (67)
Ph_3SnI 4.0 (8.4)	1.0 (15.3)	50/50	$Ph_3SnSnPh_3$ 1.76 (60)
Ph_3SnOH 5.5 (15.0)	1.5 (22.7)	60/60	$Ph_3SnSnPh_3$ 3.46 (66)
$(p\text{-}C_6H_4)_3SnCl$ 4.0 (7.7)	1.0 (15.3)	25/25	$(p\text{-}C_6H_4)_3SnSn(p\text{-}C_6H_4)_3$ 1.09 (36)
$(m\text{-}C_6H_4)_3SnCl$ 4.0 (7.7)	1.0 (15.3)	25/25	$(m\text{-}C_6H_4)_3SnSn(m\text{-}C_6H_4)_3$ 0.81 (27)
Bu_3SnCl 10.0 (30.7)	4.0 (61.4)	25/50	$Bu_3SnSnBu_3$ 4.84 (54)

[a]At 20 °C in air, [b]At 20 °C under nitrogen

C-C bond-forming reactions

Piperidines via an aminomethano desilylation-cyclization process and amines via an aminomethano destannylation process

Grieco and coworkers[30,31] have reported aqueous conditions for the hydrolytically-sensitive allyl silanes. As shown in Scheme 4, aminomethylation followed by cyclization to give piperidine derivatives proceeds faster than the protodesilylation normally observed with allyl silanes in acidic medium. The generation of iminium ions (**A**) and (**B**) is followed by inter- and intra-molecular addition to the corresponding C-C double bond. The final carbenium ion captures a nucleophile (HO⁻) to give a 4-hydroxypiperidine derivative.

$$R - \overset{\oplus}{NH_3}\ \overset{\ominus}{TFA} + O{=}CH_2 \xrightarrow[\text{35 °C, 24 h}]{H_2O} \left[\begin{array}{c} H \\ R \end{array} \!\!\!\overset{\oplus}{N}{=}CH_2\ \overset{\ominus}{TFA} \right]$$
(A)

$$\underset{R}{\overset{H}{\diagdown}} N \diagdown \diagup \diagup{=} \xleftarrow{\diagup\diagup{-}SiMe_3}$$

$$\downarrow O{=}CH_2 \cdot H^+$$

$$\left[\begin{array}{c} CH_2 \\ \overset{\oplus}{\underset{R}{N}} \diagdown \diagup \diagup{=}\ \overset{\ominus}{TFA} \end{array} \right] \longrightarrow \left[\begin{array}{c} R \\ N \\ \overset{\oplus}{\underset{\ominus}{}} \\ TFA \end{array} \right] \xrightarrow{H_2O} \begin{array}{c} R \\ N \\ \\ OH \end{array}$$
(B)   (81% yield)

(R = C₆H₅CH₂, TFA = OOCCF₃)

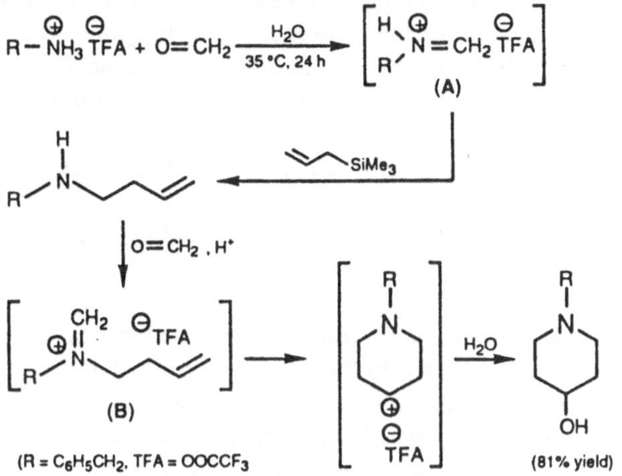

Scheme 4

(R = Bz; TFA = CF₃COO)

Several piperidine derivatives have been prepared *via* this simple procedure, through reactions of allyl silanes with iminium salts in water[30] or with immonium ions[31] generated *in situ*. Allylstannanes do not form piperidine under these conditions,, instead bis-homoallyl amines are formed (eq. 3):[32]

$$RNH_3 \cdot TFA \xrightarrow[\substack{HCHO(H_2O)/MeOH/CHCl_3, \\ (R = Bz; TFA = CF_3COO)}]{Bu_3SnCH_2CH{=}CH_2} RN(CH_2CH_2CH{=}CH_2)_2 \quad (3)$$
(97% yield)

Reactions performed with secondary amines lead only to products of aminomethano desilylation (eq. 4)

$$\underset{(R = Bz)}{RNHMe \cdot TFA} \xrightarrow[\text{HCHO(H}_2\text{O)/ 50 °C, 68 h}]{Me_3SiCH_2CH{=}CH_2} RN(Me)CH_2CH_2CH{=}CH_2 \quad (4)$$
(76% yield)

Under identical conditions, the reaction of *N*-methyl-*N*-benzylammonium trifluoroacetate with aqueous formaldehyde and allyltributylstannane (eq. 5) yields the same tertiary amine of eq. 4.

$$\underset{(R = Bz)}{RNHMe \cdot TFA} \xrightarrow[\text{HCHO(H}_2\text{O)/MeOH/CHCl}_3\text{, r.t., 2 h}]{Bu_3SnCH_2CH{=}CH_2} RN(Me)CH_2CH_2CH{=}CH_2 \quad (5)$$
(~100% yield)

It is of interest to note that the reaction with allyltributylstannane, as depicted in eq. 5, proceeds to completion after 2 h at ambient temperature, while use of allyltrimethylsilane (eq. 4) requires 68 h at 50 °C to realize a 76% yield of the same tertiary amine. This marked difference in rate is explicable in terms of the greater nucleophilicity of the double bond of an allylstannane compared to that of an allylsilane.[33] It is surprising, however, that no protodestannylation[34] is observed for the allyltributylstannane. In this case, as in others (*vide infra*)C-C bond-formation occurs at a rate much higher than that of protodestannylation.

Aldol addition of silyl enol ethers with carbonyl compounds

Data on the diastereoselectivity of the aldol addition of 1-trimethylsiloxy-cyclohexene with benzaldehyde performed under different conditions[13,35] are given in Scheme 5.

Under Mukayama's conditions (entry 1),[36] 1-trimethylsiloxycyclohexene and benzaldehyde reacted to yield a 25:75 *syn:anti* mixture of the two aldol adducts. No reaction occurs at atmospheric pressure in the absence of a Lewis acid catalyst such as $TiCl_4$, but a 90% yield of the product is obtained when the reaction is operated under a pressure of 10 Kbar at 60 °C (entry 2). In this case the diastereoisomeric ratio is reversed, since the more compact transition state $(\Delta V^{\ddagger})^{26}$ leads to the *syn* form as main product. Similar *syn:anti* ratios are registered when the reaction is performed at room temperature in water (entry 3) or in water-tetrahydrofuran medium (entry 4). Owing to the hydrolysis of the silyl enol ether in both cases of entries 3 and 4, yields are moderate. A smooth 1,4-addition of siloxycyclohexene has been found for α,β-unsaturated ketones, while saturated ketones do not react under these conditions.

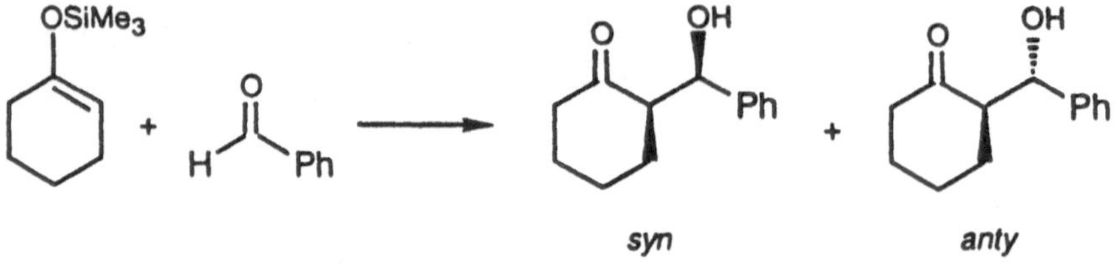

Entry	Medium	Temp (°C)	Time	Yield (%)	syn:anti
1	$CH_2Cl_2/TiCl_4$	20	2 h	82	25:75
2	$CH_2Cl_2/10$ Kbar	60	9 d	90	75:25
3	H_2O	20	5 d	23	85:15
4	H_2O/THF (1:1)	20	5 d	45	74:26

Scheme 5

The addition of 1-trimethylsiloxycyclohexene and benzaldehyde represents an example where the role of water is essential either in accelerating the reaction rate or in determining a specific stereochemical course, as has been found in other cases[3,24,36] β-Acetylenic carbinols can be readly prepared in a one-pot reaction, in the presence of water, by the addition of allenyldibutyltin chloride to carbonyl compounds (eq. 6).[14]

$$Bu_2(CH_2=C=CH)SnCl + RCOR' + 1/2\ H_2O \xrightarrow[\text{5-60 min}]{\text{r.t., stirring}}$$

$$RC(OH)R'CH_2\text{-}C\equiv CH + RC(OH)R'CH=C=CH_2 + 1/2\ (Bu_2SnCl)_2O \quad (6)$$

(90 - 95% yield)   (10-5% yield)

$(R=C_2H_5, (CH_3)_2CH, (CH_3)_3C, CH_3CH=CH, R' = H; R = R' = CH_3)$

Organotin chlorides containing allenic- as well as allyl-tin bonds are activated reagents, the carbonyl species being able to coordinate to the tin center[37] in a process that facilitates the overall reaction as depicted in Scheme 6.

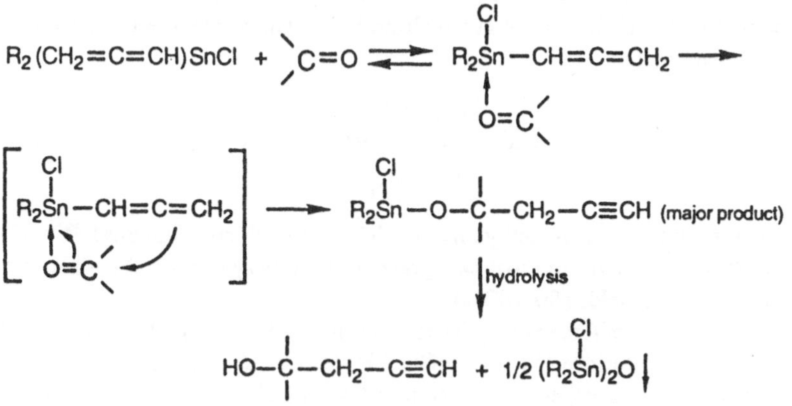

$$R_2(CH_2=C=CH)SnCl + \underset{/}{\overset{\backslash}{C}}=O \rightleftharpoons \begin{matrix} Cl \\ | \\ R_2Sn-CH=C=CH_2 \\ | \\ O=C\overset{/}{\underset{\backslash}{}} \end{matrix} \longrightarrow$$

$$\left[\begin{matrix} Cl \\ | \\ R_2Sn-CH=C=CH_2 \\ \big\downarrow\big\uparrow \\ O=C\overset{/}{\underset{\backslash}{}} \end{matrix} \right] \longrightarrow \begin{matrix} Cl \\ | \\ R_2Sn-O-\overset{|}{\underset{|}{C}}-CH_2-C\equiv CH \end{matrix} \text{(major product)}$$

$$\Big\downarrow \text{hydrolysis}$$

$$HO-\overset{|}{\underset{|}{C}}-CH_2-C\equiv CH + 1/2\ (R_2\overset{\overset{Cl}{|}}{Sn})_2O \big\downarrow$$

Scheme 6

Since the reaction involves an allenic rearrangement, the major product being the propargylic carbinol, a cyclic mechanism (S_Ei') has been proposed. This new route to known propargylic carbinols is simple and convenient and the use of organotin substrates greatly improves this method, which has advantages over the previous methods.[38]

Addition of a chemically inert mixture of $Bu_3SnCH=C=CH_2$ and RCHO (R= CH_3, C_2H_5, $(CH_3)_2CH$, $(CH_3)_3C$) to Bu_2SnCl_2 in the presence of water affords isomeric mixtures of α-allenic and β-acetylenic alcohols with the α-allenic isomer predominating (~ 75%). The overall process is represented in Scheme 7.

When the mixture of $Bu_3SnCH=C=CH_2$ and RCHO is added to the scrambling reagent Bu_2SnCl_2, mixed $Bu_2(CH\equiv CCH_2)SnCl$ is rapidly formed (step 1). This can add to the aldehyde (step 2) and, if the rate of this addidition is greater than that of the isomerization (step 3), the α-allenic alcohol is formed in major amount;

otherwise the β-acetylenic alcohol is produced *via* step 4. β-Acetylenic alcohols predominate only when HCHO and α,β-unsaturated aldehydes (R = CH_2=CH, CH_2=C(CH_3), CH_3CH=CH, C_3H_7CH=CH) are used: the stereochemical course of the reactions appears to be dependent upon the addition and isomerization rates (step 2 and 3 respectively) of the Bu_2(CH≡CCH_2)SnCl intermediate.

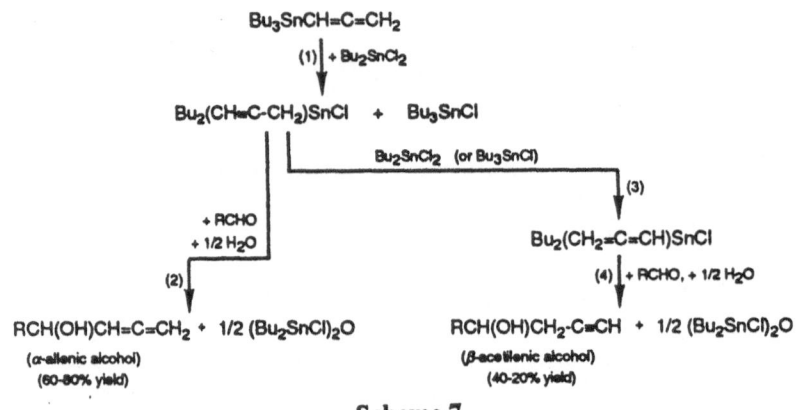

Scheme 7

Allyl-, crotyl-, 1-methylallyl-, cyclohex-2-enyl- and cinnamyl-stannations of carbonyl compounds have been also realized.[14,16,17] The procedure is based on the general reaction (eq. 7):

$$Bu_2RSnCl + R'COR'' + 1/2\ H_2O \xrightarrow[\text{stirring}]{\text{water, 20 °C}} R(OH)CR'R'' + 1/2\ (Bu_2SnCl)_2O\ (7)$$

where R = allyl, crotyl, 1-methylallyl, cyclohex-2-enyl, and cinnamyl, R' = H or alkyl, R'' ≠ R' = alkyl. In most cases reactions are very rapid and carbinols are obtained in high yields (80-100%).

Allylstannations of three aldehydes RCHO (R=C_2H_5, (CH_3)$_2$CH, and (CH_3)$_3$C) with crotyltin chlorides $Bu_{3-n}Cl_nSnCH_2$CH=CHCH_3 (n=1,2,and 3) have been carried out in the presence of aqueous 0.1-4.0 M $HClO_4$. The stereochemical course of these reactions to give the *threo/erythro* alcohols RCH(OH)CH(CH_3)CH=CH_2 and/or *E/Z* alcohols RCH(OH)CH_2CH=CHCH_3 depends on the number, n, of the chlorine atoms bonded to tin and on the concentration of perchloric acid. With increasing acid concentration, the *threo/erythro* pair is gradually replaced by the *E/Z* isomer pair, which latter is exclusively obtained in 4 M $HClO_4$. However, in most of the cases examined, the ratio *anti:syn* = (*threo* + *E*):(*erythro* + *Z*) remains constant for any given aldehyde and crotyl compound. Stereoselectivity, which is fairly good in some cases, is determined by the nature of the aldehyde: *syn*-convergence occurs with pivalaldehyde, and *anti*-convergence with isobutyraldehyde. No stereoselection is found in the case of propionaldehyde.

Table 3 collects some relevant results: comparison is made between the data obtained in water and those in 4 M aqueous $HClO_4$ for the three allylating organotin reagents. Figures in bold indicate the cases in which there is fairly good stereoselection; the major isomer is present in the range 75-83%.

Table 3 : Stereoselection of allylation of aldehydes in the presence of water or in 4 M aqueous HClO$_4$ using Bu$_{3-n}$Cl$_n$SnCH$_2$CH=CHCH$_3$ substrates (n = 1, 2 and 3)

RCHO	Bu$_2$ClSnCH$_2$CH=CHCH$_3$		BuCl$_2$SnCH$_2$CH=CHCH$_3$		Cl$_3$SnCH$_2$CH=CHCH$_3$	
R	water t:e[a]	4 M HClO$_4$ E:Z[b]	water t:e[a]	4 M HClO$_4$ E:Z[b]	water t:e[a]	4 M HClO$_4$ E:Z[b]
C$_2$H$_5$	52:48	52:48	49:51	54:46	44:56	47:53
(CH$_3$)$_2$CH	66:34[c]	59:41	76:24[c]	76:24[d]	75:25[c]	83:17[d]
(CH$_3$)$_3$C	46:54	18:82[e]	45:55	33:67[e]	22:78[f]	24:76[e]

[a]t:e = threo-:erythro-β-methylallylic alcohols. [b]E:Z = E-:Z-crotyl alcohols.
[c]Threo-convergence. [d]E-convergence. [e]Z-convergence. [f]Erythro-convergence.

C-O-C bond-forming reactions

Acetalizations mediated by BuSnCl$_3$ in water

Butyltin trichloride, owing to its ability to act as an acid and as a dehydrating agent, has been largely used for acetalizations of diols and polyols.[39,40] With this kind of catalyst precursor, the presence of water can be tolerated[18,19] so that even aqueous commercial solutions of aldehydes can be used, as the following example indicates[18] (eq. 8).

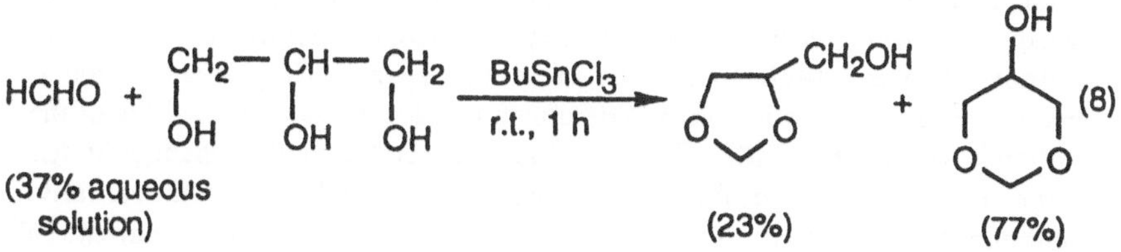

$$\text{HCHO} + \begin{matrix} \text{CH}_2-\text{CH}-\text{CH}_2 \\ | \quad\quad | \quad\quad | \\ \text{OH} \quad \text{OH} \quad \text{OH} \end{matrix} \xrightarrow[\text{r.t., 1 h}]{\text{BuSnCl}_3} \quad (8)$$

(37% aqueous solution)   (23%)   (77%)

Of special interest is the synthesis promoted by organotins as shown in Scheme 8, where R = CH$_2$=CHCH$_2$. The diol is easily prepared by the allylstannation[41] of an aqueous solution of glyoxal using allyldibutyltin chloride (step 1). Its acetalization yields the cyclic acetal as a mixture of two isomers A and B (step 2), where the ratio $A:B$ = 63:37 is the same as that of the starting meso-:racemate-diol.[19]

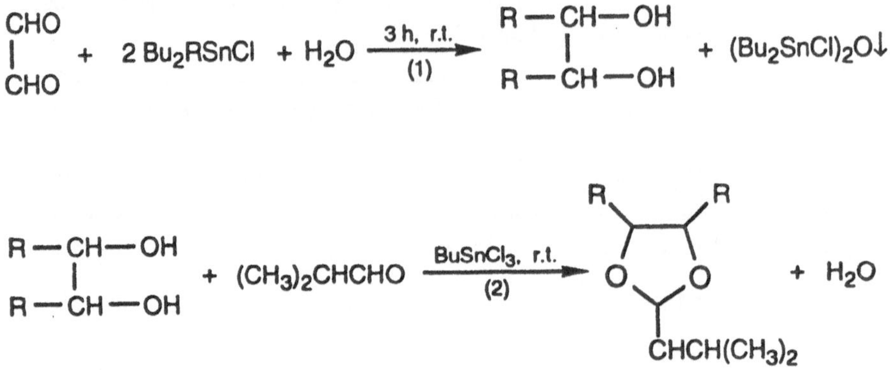

CHO
| + 2 Bu$_2$RSnCl + H$_2$O $\xrightarrow[\text{(1)}]{\text{3 h, r.t.}}$ R—CH—OH
CHO | + (Bu$_2$SnCl)$_2$O↓
                                                                      R—CH—OH

R—CH—OH
| + (CH$_3$)$_2$CHCHO $\xrightarrow[\text{(2)}]{\text{BuSnCl}_3,\ \text{r.t.}}$ + H$_2$O
R—CH—OH

(five-membered dioxolane ring with two R groups, O atoms, and CHCH(CH$_3$)$_2$ substituent)

Scheme 8

Concluding remarks

The present report is essentially devoted to the preparation of organotins in the presence of water or in water-cosolvent media. A description of their use in aqueous media for organic synthesis has been given together with that of some related organosilicon compounds.

With regard to the literature dealing with the C-C bond formation in aqueous media for non-organometallic reactions, .*i.e.*, Diels-Alder reactions, Claisen rearrangement reactions, Barbier-type reactions, main-group metal mediated reactions and transition-metal-catalyzed reactions, the reader is directed to references 1-11, 42.

From the present report, the following important points can be emphasized: (1), so far, most of the work has been restricted to only silicon and tin derivatives bearing allyl and allyl-like groups; (2) reactions can be performed in a one-pot synthesis and are accelerated in the presence of water, (3) stereoselection is encountered in aqueous media as is demonstrated for the stereoselectivity of the synthesis of the Bu$_3$SnCH(CH$_3$)CH=CH$_2$ isomer[24] and the change in the stereochemical courses of the aldol reaction of silyl enol ethers with carbonyl compounds[15] (5), results in water have some analogies with sonofication procedures.[43]

Acknowledgments

The authors gratefully acknowledge support of this work by the Consiglio Nazionale delle Ricerche (CNR, Rome), Progetto Finalizzato Chimica Fine II and the Ministero dell' Università e della Ricerca Scientifica e Tecnologica (MURST, Rome).

References

1. J. Multzer, H.J. Altenbach, M. Braun, K. Krohn and H.U. Reissig, In: *Organic Synthesis Highlights* ,VCH; Weinheim, Germany,1991, p. 71.
2. P.A. Grieco, *Aldrichimica Acta*, **24** (1991) 59.
3. D.C. Rideout and R. Breslow, *J. Am. Chem. Soc.*, **102** (1980) 7816.
4. R. Breslow, U. Maitra and D.C. Rideout, *Tetrahedron Lett.*, **24** (1983) 1901.
5. R. Breslow and U. Maitra, *Tetrahedron Lett.*, **25** (1984) 1239.
6. R. Breslow and T. Guo, *J. Am. Chem. Soc.*, **110** (1988) 5613.
7. P.A. Grieco, P. Garner and Z. He, *Tetrahedron Lett.*, **24** (1983) 1897.
8. R. Braun, F. Schuster and J. Sauer, *Tetrahedron Lett.*, **27** (1986) 1285.
9. P.A. Grieco, K. Yoshida and P. Garner, *J. Org. Chem.*, **48** (1983) 3137.
10. T. Kunz and H.U. Reissig, *Liebigs Ann. Chem.*, (1989) 891.
11. H. Mattes and C. Benezra, *Tetrahedron Lett.*, **26** (1985) 5697.
12. C. Petrier and J.L. Luche, *J. Org. Chem.*, **50** (1985) 910.
13. A. Lubineau and E. Meyer, *Tetrahedron*, **44** (1988) 6065.
14. A. Boaretto, D. Marton, G. Tagliavini and A. Gambaro, *J. Organomet. Chem.*, **286** (1985) 9.
15. A. Boaretto, D. Marton and G. Tagliavini, *J. Organomet. Chem.*, **297** (1985) 149.
16. D. Furlani, D. Marton, G. Tagliavini and M. Zordan, *J. Organomet. Chem.*, **341** (1988) 345.
17. D. Marton, G. Tagliavini and N. Vanzan, *J. Organomet. Chem.*, **376** (1989) 269.
18. D. Marton, P. Slaviero, G. Tagliavini, N. Vanzan and M. Zordan, In: *Chemistry and Technology of Silicon and Tin*, eds., V.G. Kumar Das, Ng Seik Weng and M. Gielen, Oxford University Press, Oxford, 1992, p. 277.
19. G. Tagliavini, *J. Organomet. Chem.*, **437** (1992) 15.
20. K. Sisido, Y. Takeda and Z. Kinugawa, *J. Am. Chem. Soc.*, **83** (1961) 538.
21. K. Sisido, S. Kozima and T. Hanada, *J. Organomet. Chem.*, **9** (1967) 99.
22. K. Sisido and S. Kozima, S. *J. Organomet. Chem.*, **11** (1968) 503.
23. T. Carofiglio, D. Marton and G. Tagliavini, *Organometallics*, **11** (1992) 2961.
24. E. Brescacin, F. von Gyldenfeldt, D. Marton and G. Tagliavini, *Organometallics*, submitted for publication.
25 K. Sisido and Y. Takeda, *J. Org. Chem.*, **26** (1961) 2301.
26 M.R.J. Dack, *Chem. Soc. Rev.*, **4** (1975) 211.
27 R.K. Ingham, S.D. Rosenberg and H. Gilman, *Chem. Rev.*, **60** (1960) 459.
28 P. Harrison, in *Chemistry of Tin*, Blackie, London, 1959.
29 P. Boudjouk and B. Hee-Han, *Tetrahedron Lett.*, **22** (1981) 3813.
30. S.D. Larsen, P.A. Grieco and W.F. Fobare, *J.Am. Chem. Soc.*, **108** (1986) 3512.
31. P.A. Grieco and W.F. Fobare, *Tetrahedron Lett.*, **27** (1986) 5067.
32. P.A. Grieco and A. Bahsas, *J. Org. Chem.*, **52** (1987) 1378.
33. J.A. Mangravite, J.A. Verdone and H.G. Kuivila, *J. Organomet. Chem.*, **104** (1976) 303.

34. M. Pereyre, J.P. Quintard and A. Rahm, In: *Tin in Organic Synthesis*, Butterworths, London, 1987.

35. A. Lubineau, *J. Org. Chem.*, **51** (1986) 2142.

36. H.J. Schneider and N.K. Saugwan, *J. Chem. Soc., Chem. Comm.*, (1986) 1787.

37. G. Tagliavini, V. Peruzzo, G. Plazzogna and D. Marton, *Inorg. Chim. Acta,,* **24** (1977) L47.

38. See ref 16 in 14.

39. D. Marton, G. Tagliavini and P. Slaviero, *Gazz. Chim. Ital.*, **119** (1989) 359.

40. D. Marton and G. Tagliavini,. *Main Group Metal Chem.*, **13** (1990) 363.

41. G. Tagliavini and D. Marton, *Gazz. Chim. Ital.*, **118** (1988) 483.

42. Chao-Jun Li *Chem. Rev.*, **93** (1993) 2023.

43. R. F. Abdulla, *Aldrichimica Acta*, **21** (1988) 31.

Main Group Elements and Their Compounds
V.G. Kumar Das (Ed)

Supramolecular Associations in Organotin, Organoantimony and Other Main Group Organometallic Compounds

Ionel Haiduc[a] and Cristian Silvestru[b]

[a]Instituto de Quimica, Universidad Nacional Autonoma de Mexico, Ciudad Universitaria, 04510 Mexico D.F., Mexico, and [b]Facultatea de Chimie, Universitatea Babes-Bolyai, RO-3400 Cluj-Napoca, Roumania

Supramolecular chemistry is defined as "the chemistry of molecular assemblies and of the intermolecular bond"[1] and deals with "organized entities of higher complexity that result from the association of two or more chemical species held together by intermolecular forces".[2] Much attention thus far has been given to exploring the connectivity between the area termed supramolecular chemistry and traditional coordination chemistry, and from this preoccupation has emerged the conclusion that this new area of chemical science can be regarded as a "generalized coordination chemistry".[3] It seems that the potential interconnections between supramolecular chemistry and organometallic compounds have, so far, received less consideration, although many organometallic compounds form intermolecular associations of various types which fall within the frame of the definitions cited above.

It is the purpose of this paper to demonstrate a possible link between supramolecular and organometallic chemistry, and to survey the types of supramolecular associations in organometallic compounds, using illustrations predominatly from our own work. X-ray structural studies have revealed many examples of organometallic molecules which, in principle at least, could exist as definite, separate entities, yet form oligomeric or polymeric structures. These have been described in various ways in the literature, but the term "supramolecular association" was very seldom, if ever, used.

Coordination and supramolecular association

Let us consider, by way of example, firstly organotin(IV) compounds. The fully organosubstituted tetraorganotin compounds are four-coordinate, tetrahedral compounds, 1. The functional derivatives (X = electronegative substituent) may maintain the four-coordinate state as shown in 2 and 3, or they may exhibit the tendency to expand the coordinate number to five or six, and to achieve trigonal bipyramidal (sometimes square pyramidal) and octahedral geometry, respectively. Monodentate (monoconnective) ligands, e.g. halogens, hydroxo, alkoxo, thiolato and other electronegative ligands, may increase the coordination number by coordination of additional units to the metal, to form anionic species 4 and 5, or by polymerization (or oligomerization) to form polynuclear bridged species 6 and 7 in triorganotin and diorganotin compounds, respectively. The first type of compounds formed by coordination of additional ligands belong to the area of traditional coordination chemistry. The polymers can be described as supramolecular associations, and belong to the new area of chemical science, supramolecular chemistry. Depending upon the nature of R and X, a tri- or diorganotin compound may maintain its four-coordinate, monomeric state, or may undergo supramolecular association with expansion of the coordination number. In the supramolecular associations the monodentate (monoconnective) ligand X becomes doubly monodentate (biconnective), if denticity is defined as the number of coordination sites occupied by a ligand around a metal coordination centre.

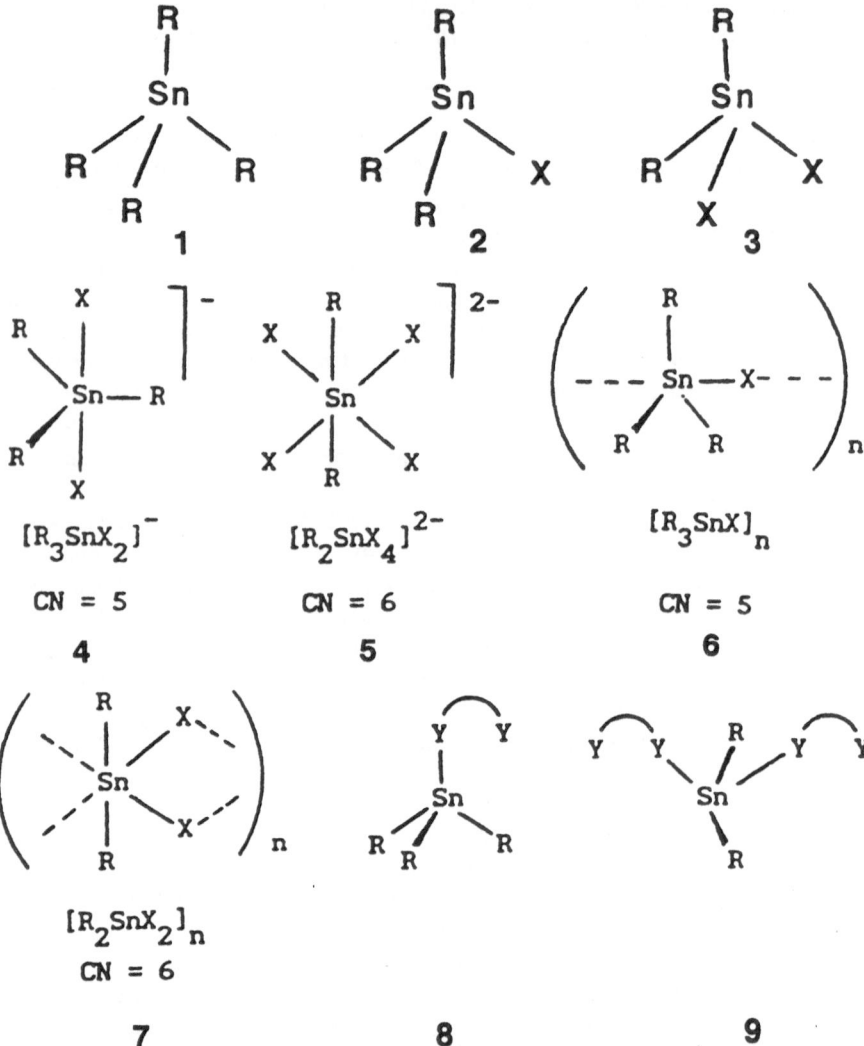

Ambient ligands possessing two potential coordinating sites, e.g. carboxylato, phosphinato, dithiophosphinato or dithiophosphato, dithicarbamato, xanthato and the like, can act as monodentate (monoconnective), bidentate (symmetrical or unsymmetrical, i.e. isobidentate and anisobidentate) or bridging (bimetallic biconnective or triconnective)[4] ligands. Thus, if tin is used again as an example, the metal can maintain its coordination number four and the tetrahedral geometry, like in **8** and **9**, or can expand it intramolecularly (chelating ring formation, biconnective coordination), e.g. in **10** or **11**, or intermolecularly, to form supramolecular associations as shown in **12** and **13**. Thus, a tri- and diorganotin derivative has the choice to remain molecular and monomeric, without or with expansion of the coordination number, or to form supramolecular associations. The choice will be decided by factors which are not discussed here in any detail, including the nature, size and shape of the organic group R.

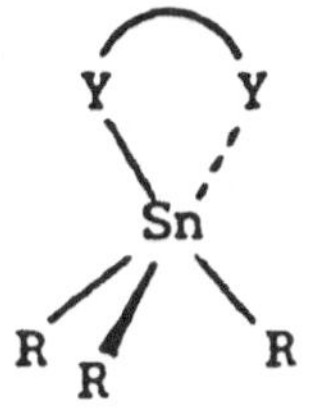

10

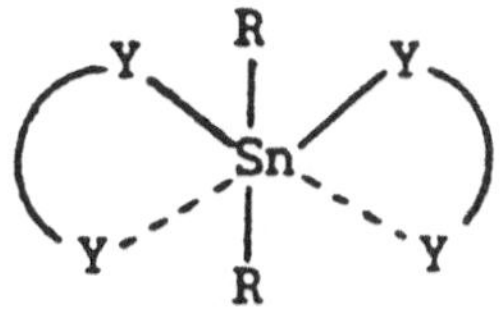

11

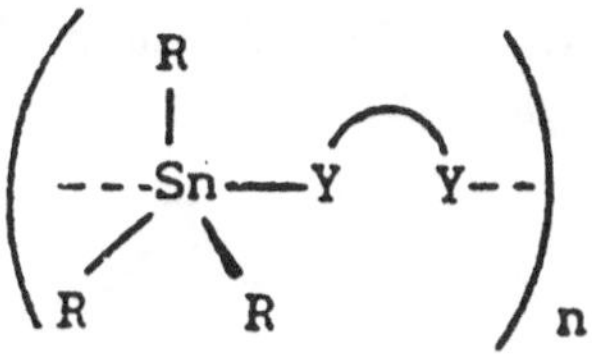

12

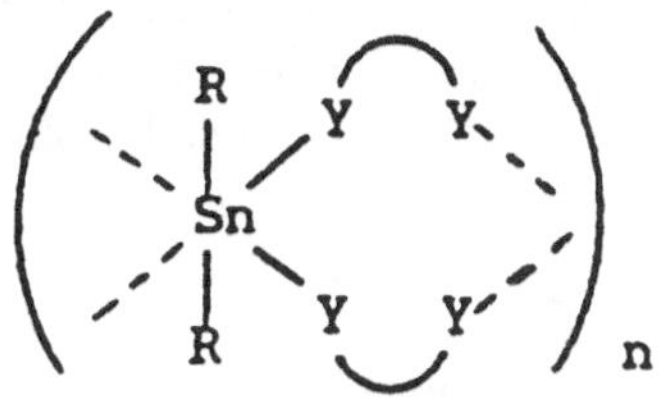

13

A given ambient ligand may behave in various ways, depending on its nature. A good example is offered by the dithiophosphorus ligands, i.e. dithiophosphinates, $R_2PS_2^-$, and dithiophosphates, $(RO)_2PS_2^-$ which exhibit a variety of coordination patterns,[4a] and by other dithiolato ligands (dithiocarbamates, xanthates, etc.).[4b] These ligands are seldom monodentate; they tend to become aniso- or isobidentate (chelating) or bridging. In many cases, the anisobidentate chelating (bimetallic biconnective) ligand may engage in further bonding, becoming bimetallic triconnective, and thus leading to supramolecular associations. For example, triphenyltin diethyldithiophosphate is molecular, monomeric with tetracoordinate tin (**14**), while trimethyltin dimethyldithiocarbamate contains an anisobidentate, chelating ligand, which makes the tin five-coordinate, with a weaker intramolecular secondary Sn...S bond, as shown in 15.[6]

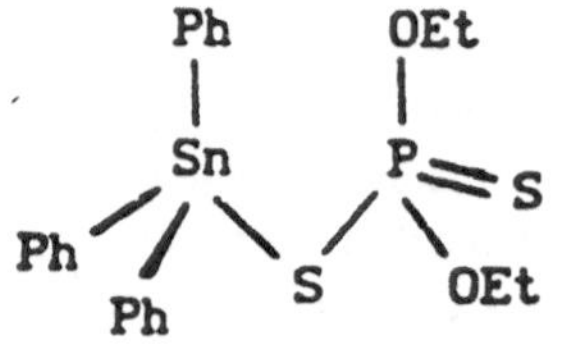

Ph   OEt

Sn   P==S

Ph / \ S / OEt
   Ph

14

Sn–S   2.458 Å

P–S   2.054 Å

P=S   1.931 Å

Me   Me
 \ /
  N
  |
  C
 / \\
S   S
 \ ⋮
  Sn
 / \
Me   Me

15

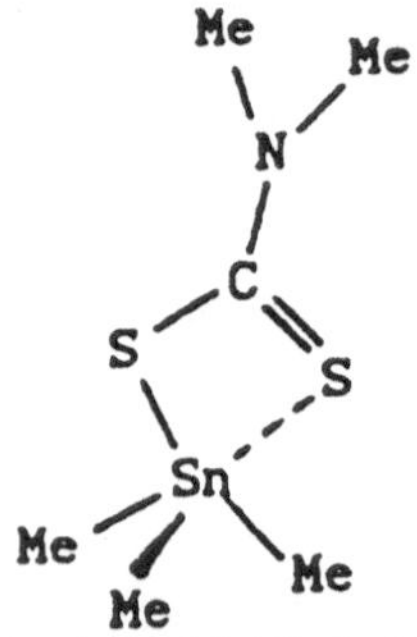

Sn–S   2.47 Å

Sn...S   3.16 Å

In diphenyltin bis(diethyldithiophosphate) the ligand is anisobidentate (16),[7a] and similarly also in dimethyltin bis(dimethyldithiophosphinate)[7b], -bis(diethyldithiophosphinate)[7c], and -bis(dimethyldithioarsinate)[7d], but in diphenyltin bis(diisopropyldithiophosphate) (17) the ligand is isobidentate. This compound is the only known example of a 1,1-dithiolato ligand displaying isobidentate coordination towards tin, but it seems that chelating ligands with larger bite angles are able to coordinate in this way without difficulty, like in bis(tetramethyldithioimidodiphosphinato)dimethyltin (18).[9]

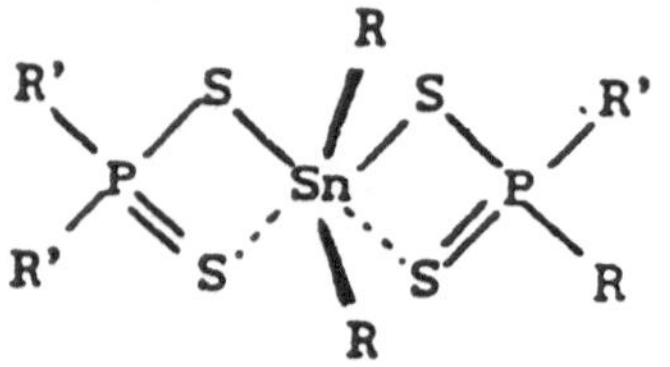

16

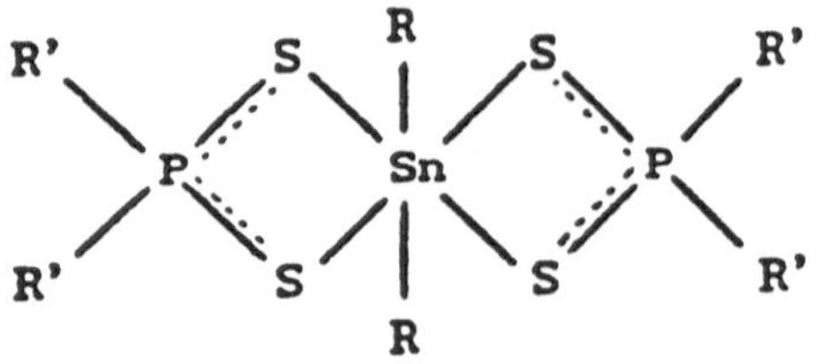

17

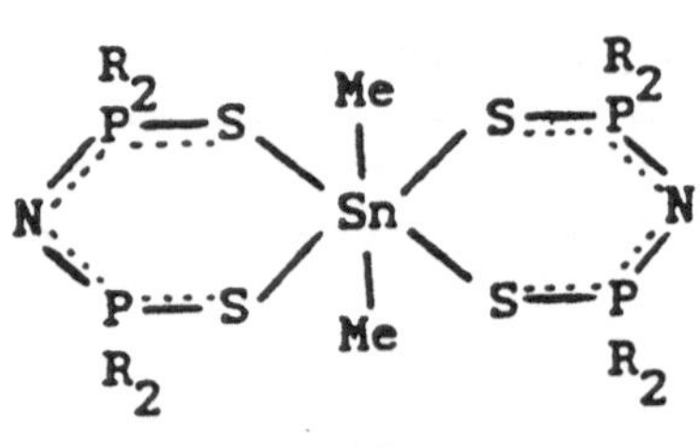

18

Bridging coordinations of the same type of ligand, leading to supramolecular associations, are discussed later in the text.

Bond types in supramolecular associations

Supramolecular associations can be formed through various types of intermolecular bonds. These include:

(a) normal coordination, i.e. donor-acceptor two electron bonds
(b) "semibonding" or secondary bonding - longer than covalent bonds but shorter than van der Waals contacts
(c) metal-metal secondary interactions
(d) hydrogen bonds
(e) pi-bonds
(f) ionic interactions

Dimeric, tetrameric, hexameric and polymeric supramolecular associations can be formed through these types of interactions.

Normal coordinate bonds

Frequently, supramolecular association is achieved through normal, two electron, donor-acceptor bonds, so common in coordination chemistry. These are formed when the metal is coordinatively unsaturated, possesses at least one vacant orbital in the valence shell and is attached to a functional group, possessing in turn lone electron pairs. The most common are halogen-, hydroxy-, alkoxo- and thiolato-bridged polynuclear species, but polyatomic bridges are also possible. The interatomic distances for these bonds are of the same order of magnitude as for the normal covalent bonds, and are usually slightly longer than, but in the range of estimated sums of covalent radii.

Numerous examples can be provided for this type of supramolecular structures. The most typical are the dimeric hydroxotin derivatives, e.g. $[EtSnCl_2OH(H_2O)]_2$,[10] dimeric distannoxanes like $[ClMe_2SnOSnMe_2Cl]_2$[11a] or $[ClPri_2SnOSnPri_2]_2$[11b] and many dimeric organotin carboxylates,[12] as well as nitrogen-tin compounds, e.g. $[Me_2SnS_2N_2]_2$.[13] *

With organophosphorus ligands, examples include the hydroxy-(diphenylthiophosphato)diphenyltin(IV) $\{Ph_2Sn(OH)[OSP(OPh)_2]\}_2$[14] and hydroxy(diphenylthiophosphinato)diphenyltin, $[Ph_2Sn(OH)OSPPh_2]_2$ (19)[15]. We found a unique tetrameric structure for trimethyltin(IV) diphenylphosphinate, $[Me_3SnO_2PPh_2]_4$ (20).[16] The compound contains a 16-membered ring, formed from linear O-Sn-O units (O-Sn-O 176.2°, Sn-O 2.243 and 2.245 Å) connected through phosphinato ligands. The bending of the chain leading to cyclization, occurs at the oxygen (Sn-O-P) and phosphorus (O-P-O) atoms.

Cyclic hexameric structures have been reported for triphenyltin diphenylphosphate, $[Ph_3SnO_2P(OPh)_2]_6$[17a] and triphenyltin methyl(methylphosphonate), $[Ph_3SnO_2P(OMe)Me]_6$[17b] (21), as well as for triphenyltin N-phthaloylglycinate.[18]

---

*Editor's note: A number of organotin stannates derived from dicarboxylic acids also possess supramolecular structures (see Chapter 32)].

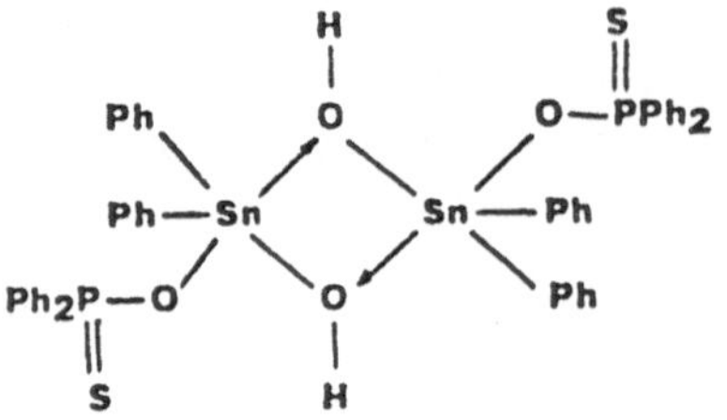

19

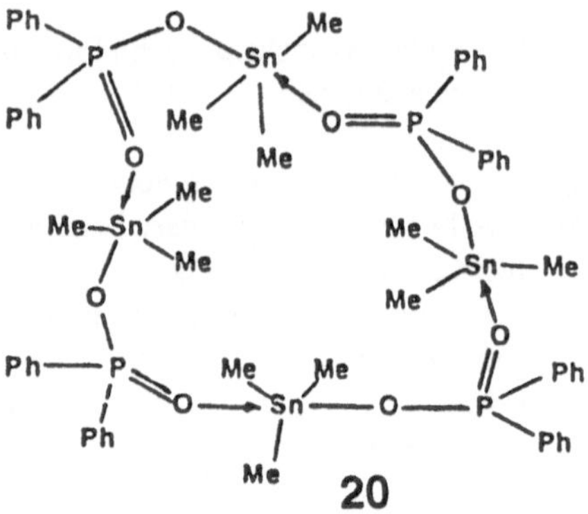

20

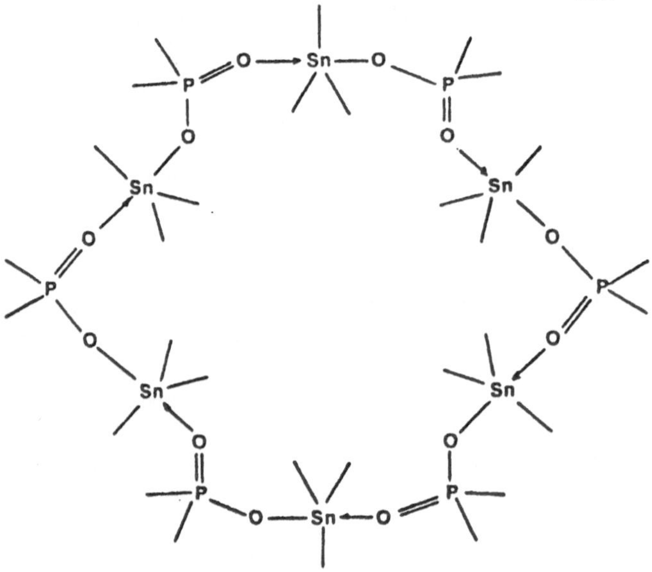

21

The polymeric structure of type **6** is much more numerous and include helical or zig-zag chains of triorganotin functional derivatives [R$_3$Sn-X...], e.g. trimethyltin fluoride;[19] triphenyltin fluoride,[20] triphenyltin hydroxide,[21] trimethyltin methoxide,[22] trimethyltin cyanide,[23] benzyldimethyltin hydroxide,[24] trimethyltin methylsulfinate,[25] triphenyltin phenylseleninate,[26] and triorganotin alkanoates.[12] Many diorganoantimony compounds are similarly polymeric, e.g. diphenylantimony(III) fluoride and dimethylantimony(III) azide.[27]

The derivatives of oxo- and thiophosphorus ligands also yield polymeric supramolecular structures. Thus, trimethyltin dimethylthiophosphinate, [Me$_3$SnOP(S)Me$_2$] was found to be a polymer with thiophosphinato bridges.[28] We found that dimethyltin(IV) bis(diphenylthiophosphinate), [Me$_2$Sn(OSPPh$_2$)$_2$] is a helical polymer[29] (**22**), in contrast to dimethyltin(IV) bis(diethylthiophosphinate), [Me$_2$Sn(OSPEt$_2$)$_2$], which is monomeric.[30]

Among organoantimony(III) compounds polymeric structures have been found for diphenylantimony(III) diphenylphosphinate, [Ph$_2$SbO$_2$PPh$_2$] (**23**), diphenylantimony(III) diphenylthiophosphinate, [Ph$_2$SbOSPPh$_2$] (**24**)[31] and diphenylantimony(III) dimethyldithioarsinate, $[Ph_2SbS_2AsMe_2]^{32}$ (**25**).

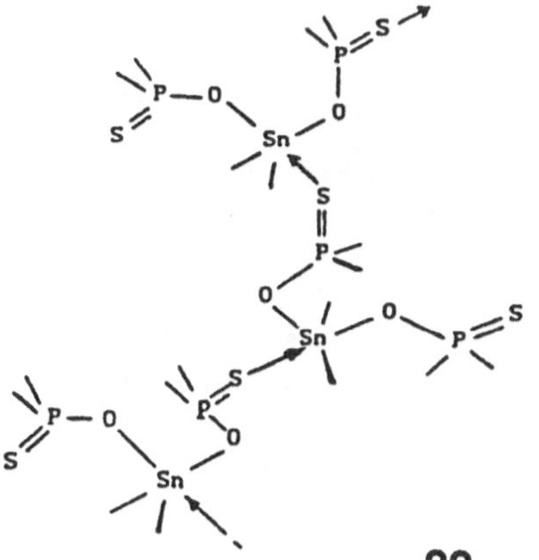

22

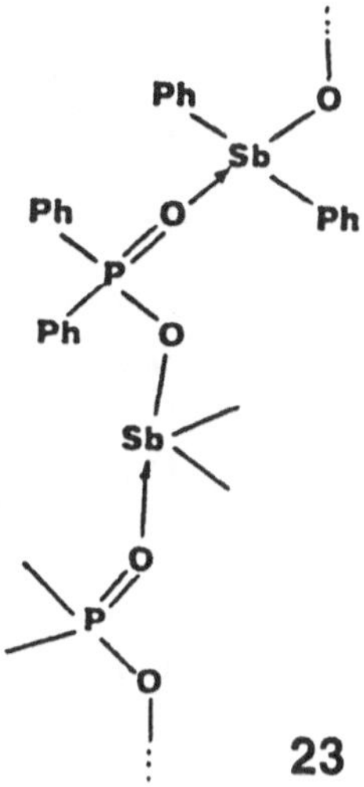

23

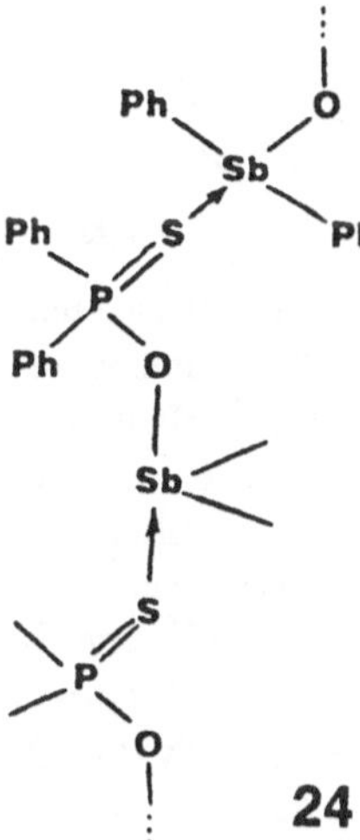

24

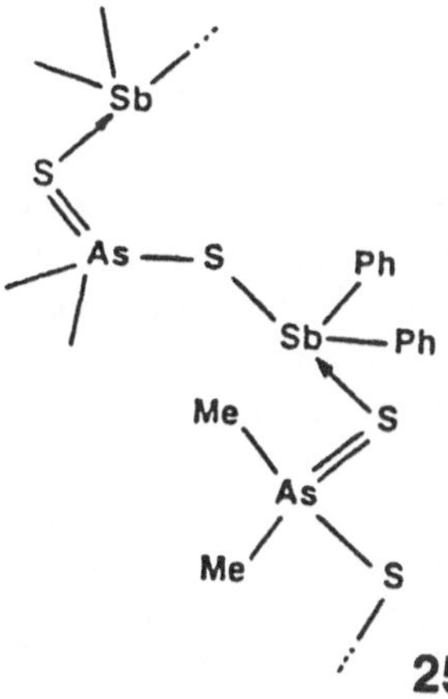

25

Secondary bonding ("semibonding") interactions

Interatomic distances longer than covalent bonds but shorter than the sum of van der Waals radii fall into the category of "secondary interactions".[33] Such interactions occur frequently in main group metal compounds, both intra- and intermolecularly. The intramolecular secondary interactions are usually observed when anisobidentate chelating ligands with small bite angles are coordinated to main group metals. In structures **10** and **11**, these are indicated by dotted lines. Intermolecular secondary bonds are shown in **6, 7, 12** and **13**; this leads to supramolecular association of the monomeric units. The intermolecular secondary bonds may not differ significantly in their length from the intramolecular secondary bonds. Thus, for example in **26** the intramolecular Sb...S secondary bond is 3.440 Å, while the intermolecular Sb...S distance within the dimer is 2.490 Å.[34] A similar structure is observed for the dithioarsinato dimer **27**.[34]

The compound **28** was formulated as a monocyclic species because the intermolecular Sb...S distance (3.556 Å) is shorter than the transannular Sb...S distance (4.013 Å, nonbonding).[35]

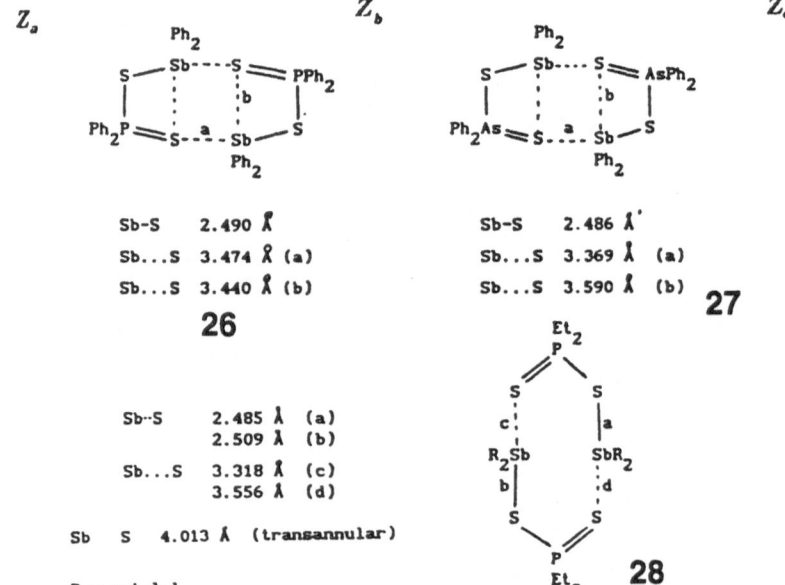

Z_a Z_b Z_c

Sb-S	2.490 Å		Sb-S	2.486 Å
Sb...S	3.474 Å (a)		Sb...S	3.369 Å (a)
Sb...S	3.440 Å (b)		Sb...S	3.590 Å (b)

26

27

Sb-S	2.485 Å (a)
	2.509 Å (b)
Sb...S	3.318 Å (c)
	3.556 Å (d)
Sb S	4.013 Å (transannular)

R = p-tolyl

28

Dimeric association through Sb...S secondary bonding has also been observed in methylantimony(III) diethyldithiocarbamate,[36] methylantimony(III) ethylxanthate[37] and phenylantimony(III) methylxanthate[38]. Bismuth compounds are also known to form Bi...S secondary bond supramolecular dimers, e.g. methylbismuth(III) diethyldithiocarbamate,[39] phenylbismuth(III) diethyldithiocarbamate[40] and phenylbismuth(III) methylxanthate[41].

In our hands, bismuth(III) diphenyldithiophosphinate[42] and bismuth(III) dimethyldithiophosphinate[43] have also been shown to be dimeric (structure **29**) but the mesityl derivative MesitylBi(S$_2$PPh$_2$)$_2$ is a monomer (with Bi-S 2.662 Å and intramolecular Bi...S 3.112 Å).[44]

29

362

Polymeric structures formed through secondary bonding have been identified in trimethyltin chloride[45a,b] (structural type 6), dimethyltin dichloride,[46] bis(chloromethyltin) dichloride[47] (structural type 7), cyclopentadienyltin chloride (double chain),[48] trimethyltin thiocyanate[49] and triphenyltin thiocyanate[50] (structural type 6), dialkyldithiastanolanes[51] and in phenyldithiastibolane[52].

With dithiophosphorus ligands, polymeric supramolecular associations, formed through Sb...S secondary bonds, were observed in diphenylantimony(III) diisopropyldithiophosphate, $[Ph_2SbS_2P(OPri)_2]$[53] and in dimethylantimony(III) dimethyldithiophosphinate, $[Me_2SbS_2PMe_2]$ (30).

In tellurium dithiophosphates, $ArTeS_2P(OR)_2$, the association is achieved through biconnective bridges as shown in 31,[55,56] but a new type of bridging for a dithiophosphorus ligand has been found in the dithiophosphinate $[PhTeS_2PPh_2]$ (32).[57]

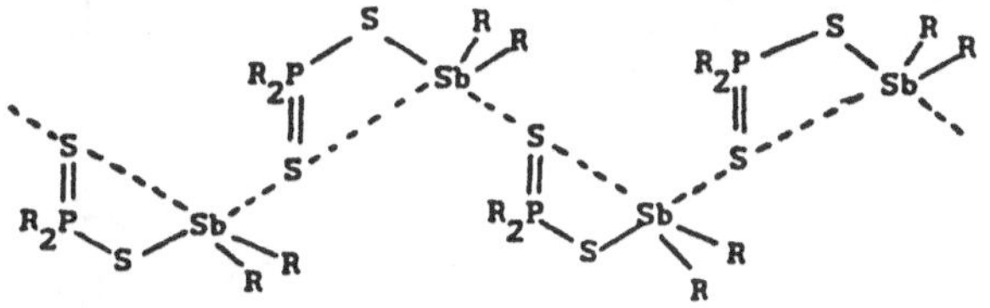

R = Me   Sb–S   2.555 Å
          Sb...S 3.158 Å

30

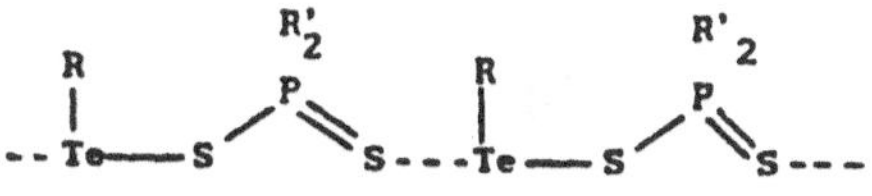

31

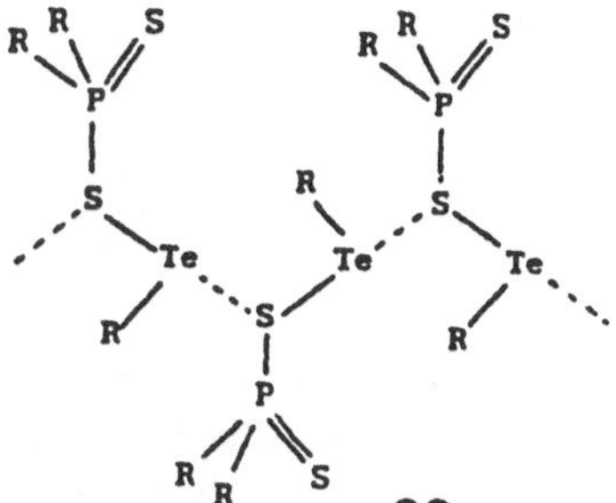

32

Metal-metal secondary interactions

The compounds containing Sb–Sb bond, e.g. stibacyclopentadienes[58] are known to form supramolecular associations through Sb...Sb secondary bonds,[58] but the dimeric structure 33 of a tellurium(I) diphenyltinphosphinate, $[PhTeS_2PPh_2]_2$ was totally unexpected.[59]

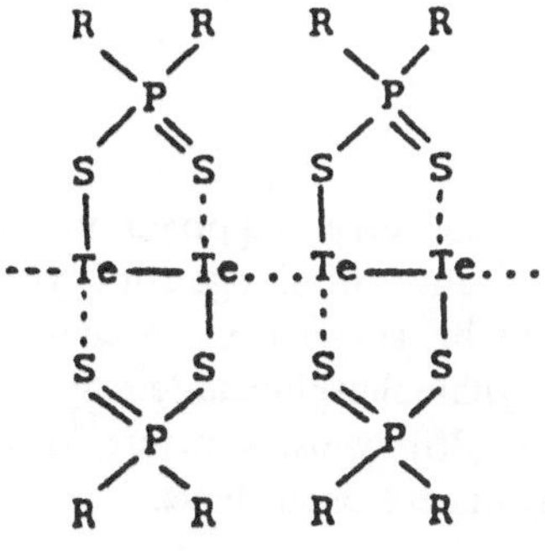

R = Ph

Te–S   2.493 Å
Te...S   3.064 Å

Te–Te   2.723 Å
Te...Te 3.668 Å

33

Hydrogen bonding interactions

Hydrogen bonding plays an important role in the supramolecular association of many organic compounds, but in organometallic derivatives it has been much less frequently identified. Some examples can, however, be cited. Thus, dimeric hydrogen-bonded cyclic structures were observed in [{(Me$_3$Si)$_3$C}PhSi(OH)$_2$}]$_2$,[60] [Zn(CRR'SiMe$_2$OH)$_2$]$_2$[61] and [Me$_2$AsO(OH)]$_2$,[62]; cyclic tetramers in [Ph$_3$Si-OH]$_4$,[63] and [Ph$_3$Ge-OH]$_4$,[64]; cage-like hexamers, for example, in (Me$_3$Si)$_3$CSi(OH)$_3$,[65]. Polymeric structures with skeletal bonds are displayed by diphenylsilanediol,[66] tetra-t-butyldisilanediol-1,2[67] and by di-t-butyl-germanediol[68].

An interesting supramolecular structure formed through skeletal hydrogen bonding was found[69] in hydroxotriphenylantimony(III) diphenylphosphinate, [HO-SbPh$_3$OP(O)Ph$_2$] (**34**).

34

Inter-chain hydrogen bonds lead to double helical supramolecular associations, e.g. in Me$_3$Sn(O$_3$SPh).H$_2$O[70] or in Me$_3$SnN$_3$.Me$_3$PbOH[71]. A very interesting structure which deserves special mention is that of [Me$_3$Sn]+ [Ph(HO)P(O)OSnMe$_3$OP(O)(OH)Ph]$^-$, in which a system of interchain hydrogen bonds connects the helical chain in a complex supramolecular structure.[72]

π-Bonding interactions

This type of bonding, which will not be elaborated in this paper, is observed in main group metal cyclopentadienyls, several of which are known to be polymeric, e.g. Pb(η^5-C$_5$-H$_5$)$_2$;[73]

Ionic interactions

Ionic or electrostatic interactions are very important in supramolecular chemistry. Many lithium and sodium derivatives of nitrogen or oxygen compounds (amines, alcohols, etc) are known to be associated. Among the "salts" of organophosphorus acids, sodium diphenylthiophosphinate-tetrahydrofuran solvate, [Na(OSPPh$_2$).2THF]$_2$, is a cyclic dimer (**35**), whose structure[74] is reminiscent of seemingly unrelated diorganoantimony dimers, cited above.

35

Acknowledgements

The new data presented in this review are the result of fruitful international collaboration with many colleagues from other countries, whose names are mentioned in the reference list. We are grateful to them for their cooperation and for sharing with us the pleasure of exploring this area. We are also grateful to UNAM and CONACYT, Mexico, for visiting research grants to the authors of this paper, to the British Council for the support of some of the research reported, and to our home University for leave of absence.

References

1. J.M. Lehn, *Pure Appl. Chem.*, **50** (1978) 871.
2. J.M. Lehn, *Angew. Chem.*, **100** (1988) 91; *Angew. Chem. Int. Ed. Engl.*, **27** (1988) 89.
3. J.M. Lehn, In: *Perspectives in Coordination Chemistry*, eds., A.F. Williams, C. Floriani and A.E. Mehrbach, Verlag Helvetica Chimica Acta, Basel and VCH Weinheim, 1992, p. 447.
4. (a) I. Haiduc, *Revs. Inorg. Chem.*, **3** (1981) 353. (b) E.R.T. Tiekink, *Main Group Metal Chem.*, **15** (1992) 161.
5. K.C. Molloy, M.B. Hossain, D. Van der Helm, J.J. Zuckerman and I. Haiduc, *Inorg. Chem.*, **18** (1979) 3507.
6. G.M. Sheldrick and W.S. Sheldrick, *J. Chem. Soc.*, **A** (1970) 490.
7. (a) B. Liebich and M. Tomassini, *Acta Cryst.*, **B34** (1978) 944. (b) K.C. Molloy, M.B. Hossain, D. Van der Helm, J.J. Zuckerman and F.P. Mullins, *Inorg. Chem.*, **20** (1981) 2172. (c) C. Silvestru, I. Haiduc, S. Klima, U. Thewalt, M. Gielen and J.J. Zuckerman, *Organomet. Chem.*, **327** (1987) 181. (d) L. Silagahi-Dumitrescu, I. Haiduc and J. Weiss, *J. Organomet. Chem.*, **263** (1984) 159.
8. K.C. Molloy, M.B. Hossain, D. Van der Helm, J.J. Zuckerman and I. Haiduc, *Inorg. Chem.*, **19** (1980) 2041.
9. I. Haiduc, C. Silvestru, H.W. Roesky, H.G. Schmidt and M. Noltemeyer, *Polyhedron*, **12** (1993) 69.
10. C. Lecomte, M. Protas and M. Devaud, *Acta Cryst.*, **B32** (1976) 923.

11. (a) R. Graziani, U. Casellato and G. Plazzogna, *Acta Cryst.*, **C39** (1983) 1188. (b) H. Puff, I. Bung, E. Friedrichs and J. Jansen, *J. Organomet. Chem.*, **254** (1983) 23.

12. E.R.T. Tiekink, *Appl. Organomet. Chem.*, **5** (1991) 1.

13. H. Roesky and H. Wiezer, *Angew. Chem.*, **87** (1975) 254; Angew. Chem. Int. Ed. Engl., **14** (1975) 258.

14. F.A.K. Nasser, M.B. Hossain, D. Van der Helm and J.J. Zuckerman, **22** (1983) 3107.

15. F. Caruso, M. Rossi, C. Silvestru and I. Haiduc, 1993 (unpublished data).

16. M.G. Newton, I. Haiduc, R.B. King and A. Sivlestru, *J. Chem. Soc., Chem. Commun.*, (1993) 1229.

17. (a) K.C. Molloy, F.A.K. Nasser, C.L. Barnes, D. Van der Helm and J.J. Zuckerman, *Inorg. Chem.*, **21** (1982) 960; (b) J.G. Masters, F.A.K. Nasser, D. Van der Helm and J.J. Zuckerman, *J. Organomet. Chem.*, **385** (1990) 39.

18. S.W. Ng, V.G. Kumar Das, G. Pelizzi and F. Vitali, *Heteroatom. Chem.*, **1** (1990) 433.

19. H.C. Clark, R.J. O'Brien and J. Trotter, *J. Chem. Soc.*, (1964) 2332.

20. D. Tudela, E. Gutierez-Puebla and A. Monge, *J. Chem. Soc., Dalton Trans.*, (1992) 1069.

21. C. Gildewell and D.C. Liles, *Acta Cryst.*, **B34** (1978) 129.

22. A.M. Domingos and G.M. Sheldrick, *Acta Cryst.*, **B30** (1974) 519.

23. E.O. Schlemper and D. Britton, *Inorg. Chem.*, **5** (1966) 507.

24. U. Wannagat, V. Damrath, V. Huch, M. Veith and U. Harder, *Organomet. Chem.*, **443** (1993) 153.

25. R. Hengel, U. Kunze and D. Strahle, *Z. Anorg. Allg. Chem.*, **423** (1976) 35; G.M. Sheldrick and R. Taylor, *Acta Cryst.*, **B33** (1977) 135.

26. V. Chandrasekar, M.G. Muralidhara, K.R.J. Thomas and E.R.T. Tiekink, *Inorg. Chem.*, **31** (1992) 4707.

27. D.B. Sowerby, In: *The Chemistry of Metal-Carbon Bond*, eds., F.R. Hartley and S. Patai, John Wiley & Sons, N.Y., 1985.

28. A.F. Shihada, I.A.A. Jasim and F. Weller, *J. Organomet. Chem.*, **268** (1984) 125.

29. C. Silvestru, I. Haiduc, F. Caruso, M. Rossi, B. Mahieu and M. Gielen, *J. Organomet. Chem.*, **448** (1993) 75.

30. I. Haiduc, C. Silvestru, F. Caruso, M. Rossi and M. Gielen, *Rev. Roumaine Chim.*, in press.

31. M.J. Begley, D.B. Sowerby, D.M. Wesolek, C. Silvestru and I. Haiduc, *J. Organomet. Chem.*, **316** (1986) 281.

32. D.B. Sowerby, M.J. Begley, L. Silaghi-Dumitrescu, I. Silaghi-Dumitrescu and I. Haiduc, to be published.

33. N.W. Alcock, *Advan. Inorg. Chem.*, **15** (1972) 1.

34. C. Silvestru, L. Silaghi-Dumitrescu, I. Haiduc, M.J. Begley, M. Nunn and D.B. Sowerby, *J. Chem. Soc., Dalton Trans.*, (1986) 1031.

35. K.H. Ebert, H.J. Breunig, C. Silvestru and I. Haiduc, *Polyhedron*, in press.

36. M. Wieber, D. Wirth, J. Metter and C. Burschka, *Z. Anorg. Allg. Chem.*, **520** (1985) 65.

37. M. Wieber, D. Wirth and C. Burschka, *Z. Anorg. Allg. Chem.*, **505** (1983) 141.
38. M. Hall, D.B. Sowerby and C.P. Falshaw, *J. Organomet. Chem.*, **315** (1986) 321.
39. M. Wieber, D. Wirth, J. Metter and C. Burschka, *Z. Anorg. Allg. Chem.*, **520** (1985) 65.
40. M. Ali, W.R. McWhinnie, A.A. West and T.A. Harmor, *J. Chem. Soc., Dalton Trans.*, (1990) 899.
41. C. Burschka, *Z. Anorg. Allg. Chem.*, **485** (1982) 217.
42. M.J. Begley, D.B. Sowerby and I. Haiduc, *J. Chem. Soc., Dalton Trans.*, (1987) 145.
43. F.T. Edelmann, F. Noltemeyer, C. Silvestru, I. Haiduc and R. Cea-Olivares, to be published.
44. R. Schultz, K.H. Ebert, H.J. Breunig, C. Silvestru and I. Haiduc, to be published.
45. (a) M.B. Hossain, J.L. Lefferts, K.C. Molloy, D. Van der Helm and J.J. Zuckerman, *Inorg. Chim. Acta*, **36** (1979) L409. (b) J.L. Lefferts, K.C. Molloy, M.B. Hossain, D. Van der Helm and J.J. Zuckerman, *J. Organomet. Chem.*, **240** (19822) 349.
46. A.G. Davies, H.J. Milledge, D.C. Puxley and P.J. Smith, *J. Chem. Soc. A*, (1970) 2862.
47. N.G. Bokii, Yu, T. Struchkov and A.K. Prokofiev, *Zhur. Strukt. Khim.*, **13** (1972) 665.
48. K.D. Bos, E.J. Bulten, K.J.G. Noltes and A.L. Speck, *J. Organomet. Chem.*, **99** (1975) 71.
49. R.A. Forder and G.M. Sheldrick, *J. Organomet. Chem.*, **221** (1970) 115.
50. A.M. Domingos and G.M. Sheldrick, *J. Organomet. Chem.*, **67** (1974) 257.
51. A.G. Davies, S.D. Slater, D.C. Povey and G.W. Smith, *J. Organomet. Chem.*, **352** (1988) 283.
52. H.M. Hoffmann and M. Dräger, *J. Organomet. Chem.*, **329** (1987) 51.
53. C. Silvestru, I. Haiduc, M.J. Begley and D.B. Sowerby, *J. Organomet. Chem.*, **426** (1992) 49.
54. C. Silvestru, I. Haiduc, K.H. Ebert and H.J. Breunig, (1993) unpublished results.
55. S. Husebye, K. Maartmann-Moe and O. Mikalsen, *Acta Chem. Scand.*, **43** (1989) 868.
56. S. Husebye, K. Maartmann-Moe and O. Mikalsen, *Acta Chem. Scand.*, **44** (1990) 464.
57. C. Silvestru, I. Haiduc and K.H. Ebert, to be published.
58. A.J. Ashe III, W. Butler and T.R. Diephouse, *J. Amer. Chem. Soc.*, **103** (1981) 207.
59. M.G. Newton, R.B. King, I. Haiduc and A. Silvestru, *Inorg. Chem.*, in press.
60. S.S. Al-Juaid, N.B. Buttrus, R.I. Damja, Y. Derouiche, C. Eaborn, P.B. Hitchcock and P.D. Lickiss, *J. Organomet. Chem.*, **371** (1981) 287.
61. F.I. Aigbirhio, S.S. Al-Juaid, C. Eaborn, A. Habtemariam, P.B. Hitchcock and J.D. Smith, *J. Organomet. Chem.*, **405** (1991) 149.
62. J. Trotter and T. Zobel, *J. Chem. Soc.*, (1965) 4466.
63. H. Puff, K. Braun and H. Reuter, *J. Organomet. Chem.*, **409** (1991) 119.

64. G. Fergusson, J.F. Gallagher, D. Murphy, T.R. Spalding, C. Glidewell and H.D. Holden, *Acta Cryst.*, **C48** (1992) 1228.

65. C. Eaborn, *J. Organomet. Chem.*, **371** (1989) 287.

66. P.E. Tomlins, J.E. Lydon, D. Akrigg and B.M. Sheldrick, *Acta Cryst.*, **C41** (1985) 941.

67. R. West and E.K. Phan, *J. Organomet. Chem.*, **403** (1991) 43.

68. H. Puff, S. Franken, W. Schuh and W. Schwab, *J. Organomet. Chem.*, **254** (1983) 33.

69. K.H. Ebert, H.J. Breunig, C. Silvestru and I. Haiduc, unpublished results.

70. P.G. Harrison, R.C. Phillips and J.A. Richards, *J. Organomet. Chem.*, **114** (1976) 47.

71. N. Beriazzi, A. Alonzo, F. Di Bianco and G.C. Stocco, *Inorg. Chim. Acta*, **12** (1978) 123.

72. K.C. Molloy, M.B. Hossain, D. Van der Helm, D. Cunningham and J.J. Zuckermman, *Inorg. Chim.*, **20** (1981) 2402.

73. I. Panotton, G. Bombieri and U. Croato, *Acta Cryst.*, **21** (1966) 823.

74. F.T. Edelmann, M. Noltemeyer, C. Silvestru and I. Haiduc, unpublished results.

Main Group Elements and Their Compounds
V.G. Kumar Das (Ed)

Synthesis, Structure and Bipotency of Tin(IV) Organyls

V.G. Kumar Das[a], Ng Seik Weng[b] and Lo Kong Mun[a]

*[a]Department of Chemistry and [b]Institute of Advanced Studies,
University of Malaya, 59100 Kuala Lumpur, Malaysia*

Organotin(IV) compounds are among the most actively studied groups of organometallics and possess a wide range of established and potential applications based on their chemical and biological properties.[1]

In an earlier article,[2] we had outlined the essential thrust of our work in organotin chemistry which has been directed at discovering new or improved properties of organotins for application as crop protectants and wood preservatives. To a large extent, therefore, our programme on the synthesis and structural elucidation of organotin compounds has been conducted in tandem with structure-activity relationship studies on insects and fungal pathogens of economic importance.

Two major approaches in this regard have been the application of the strategies of

i)   the introduction of suitable functional moieties in the non-carbon bonded anionic X group of a triorganotin biocide, R_3SnX, and

ii)   the introduction of one or more ring substituents selectively into only one phenyl ring of Ph_3SnX.

In respect of (i), focus was given in the first place to the synthesis of compounds containing functionalised acetate as the "triorganotin carrier group", X. These have the general formula $R_3SnOC(O)Y$. Among the compounds synthesised and tested for their biological activity and selectivity against fungi and insects were those in which the substituent Y group was $CH_2SC(S)NR'_2$ (R' = alkyl and cycloalkyl);[3] $CH_2SC(S)(OR')$ (R' = alkyl, terpenyl, steroid or sugar residue);[4] $CHR'NC(O)C_6H_4N.H_2O$, $SSnPh_3$, $CH_2C(O)Ph$, etc.[2,5]

The results revealed that the anionic X group on R_3SnX can and does exert a moderately strong influence on the overall toxicity of the organotin compounds, contrary to previous generalisations attesting to their relatively marginal influence in comparison with the organic R groups.[6]

In respect of (ii), compounds of the type $(4\text{-}ClC_6H_4)SnPh_2OH$ and $(3,5\text{-}Cl_2C_6H_3)SnPh_2OH$ were synthesised and shown to be superior in their antifungal activity compared to Ph_3SnOH or $(4\text{-}ClC_6H_4)_3SnOH$.[2] The presence of halogen (in contrast to electron-donating ring substituents) tended also to lower the phytotoxicity of the compounds relative to either Ph_3SnOH or Ph_3SnCl.[7]

Further developments on these leads are briefly discussed in this paper which include consideration of synthetic and structural aspects of a selected number of new compounds representative of the range studied.

Functionalised organotin esters, $R_3SnOC(O)Y$

Extended studies on these derivatives were prompted by two appealing features associated with this class of compounds, namely, the ease of introducing a variety of functional moieties in the ester unit as the pendant Y group, including those which are inherently biologically active, and the general tendency for organotin ester compounds to exist as crystalline products which permits the possibility of absolute structure determinations by X-ray diffraction.

Organotin carboxylates containing N-C(S)O or N-C(O)S substituents in the ester group

The biologically active pendant groups, *O*-monothiocarbamyl and *S*-monothiocarbamyl, constitute obvious extensions to our earlier work involving the dithiocarbamyl group. Their attachment to tin as carbamylacetates may be achieved in a relatively straightforward manner (Schemes 1 and 2). The organic acids were prepared by established methods[8,9]

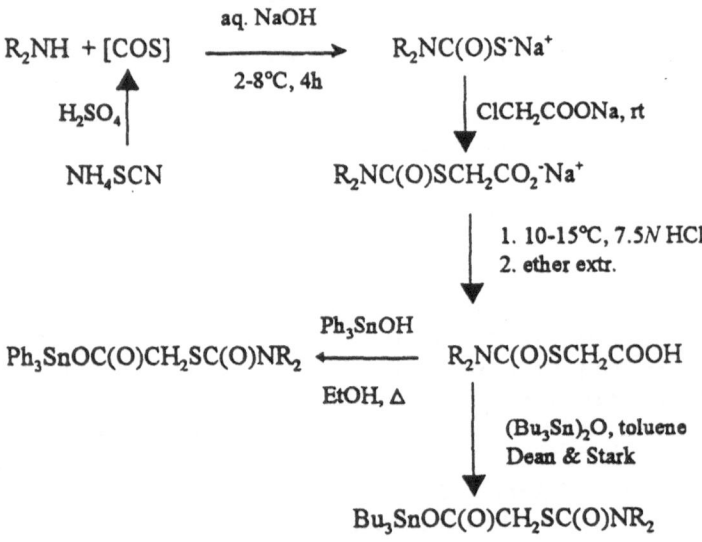

[R_2N = piperidinyl, morpholinyl, pyrrolidinyl]

Scheme 1

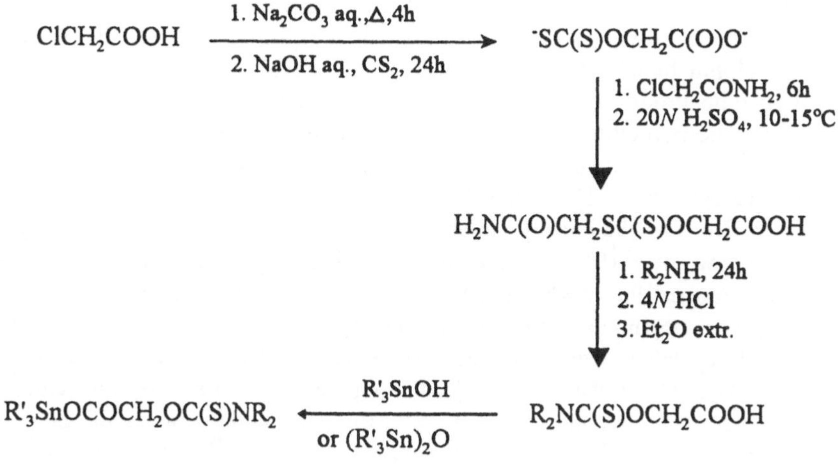

$$\text{ClCH}_2\text{COOH} \xrightarrow[\text{2. NaOH aq., CS}_2\text{, 24h}]{\text{1. Na}_2\text{CO}_3\text{ aq.,}\Delta\text{,4h}} \text{}^-\text{SC(S)OCH}_2\text{C(O)O}^-$$

1. ClCH$_2$CONH$_2$, 6h
2. 20N H$_2$SO$_4$, 10-15°C

↓

$$\text{H}_2\text{NC(O)CH}_2\text{SC(S)OCH}_2\text{COOH}$$

1. R$_2$NH, 24h
2. 4N HCl
3. Et$_2$O extr.

↓

$$\text{R}'_3\text{SnOCOCH}_2\text{OC(S)NR}_2 \xleftarrow[\text{or (R}'_3\text{Sn)}_2\text{O}]{\text{R}'_3\text{SnOH}} \text{R}_2\text{NC(S)OCH}_2\text{COOH}$$

Scheme 2

The triorganotin S-monothiocarbamylacetates have been compared with their dithiocarbamato analogues in tests on the 4th instar larvae of *Plutella xylostella* (Linnaeus), more commonly known as the diamond-back moth (DBM), a major destructive pest of cruciferous vegetables.[10,11] Some selected data on the toxicity (by topical application) of the compounds (largely triphenyltins) towards the larvae as well as on the anti-acetylcholinesterase activity of the compounds using *in vitro* preparations of the enzyme derived from a highly tolerant strain of DBM are shown in Table 1.

Table 1: Toxicity by topical application on 4th instar DBM (R-strain) larvae and acetylcholinesterase inhibitory trends

Compound	LC$_{50}$ (μg/μL)	Molar LC$_{50}$ (mmoles/L)	I$_{50}$ (moles/L)
Ph$_3$SnOC(O)CH$_2$SC(S)NMe$_2$	1.70 ± 0.29	3.23	16.9
(4-ClC$_6$H$_4$)Ph$_2$SnOC(O)CH$_2$SC(S)NMe$_2$	1.21 ± 0.14	2.36	3.0 x 10^6
(Cyh)$_3$SnOC(O)CH$_2$SC(S)NMe$_2$	0.81 ± 0.10	1.78	>10^6
Ph$_3$SnOC(O)CH$_2$SC(S)OR' (R' = menthyl)	2.32 ± 0.50	3.62	-
Ph$_3$SnOC(O)CH$_2$SC(O)NMe$_2$	0.91 ± 0.31	1.80	1.3 x 10^{-2}
Ph$_3$SnOC(O)CH$_2$SC(O)NHPh	0.85 ± 0.15	1.50	-
Ph$_3$SnOC(O)CH$_2$SC(O)NH(4-ClC$_6$H$_4$)	0.53 ± 0.19	0.90	9.0x10^{-5}
Ph$_3$SnOCOCH$_3$	2.47 ± 0.31	6.06	>10^6
Malathion (Gold Coin^R, 84% a.i.)	14.08	5.69	-
Methomyl (Lannate^R, 90% a.i.)	2.32	2.3x10^{-3}	-

The acute toxicity data (LC$_{50}$ values) reveal that the dithiocarbamylacetate, Ph$_3$SnOC(O)CH$_2$SC(S)NMe$_2$ is more toxic than Ph$_3$SnOAc; replacement of the thione sulphur by oxygen further improves the activity, but replacement of the dialkylamino group by an alkoxide (xanthylacetate) results in a lowering of activity. The monothiocarbamylacetates also show better acetylcholinesterase (AChE) inhibitory properties (expressed in the Table as I$_{50}$ values) compared to the dithiocarbamylacetates. Thus, Ph$_3$SnOC(O)CH$_2$SC(O)NMe$_2$ is seen to be 10^3-fold more active than the corresponding dithiocarbamylacetate, but surprisingly a dramatic drop (over 10^6-fold) in activity attends the replacement of a tin-bound phenyl group in Ph$_3$SnOC(O)CH$_2$SC(S)NMe$_2$ by a p-chlorophenyl group. In general, most organotin compounds are totally ineffective against AChE, with estimated I$_{50}$ values well exceeding 10^6 moles/L. The inhibitory potency of Ph$_3$SnOC(O)CH$_2$SC(O)NH(4-ClC$_6$H$_4$) is the highest that we have recorded among the organotin compounds screened. At 1.0x10^{-4}M concentration of the test compounds, the trend in enzyme inhibitory potencies is as shown in Table 2.[11]

In the solution state, the S- and O-monothiocarbamylacetates, like the dithiocarbamylacetates, are essentially 4-coordinated structures as deduced from NMR (^{119}Sn, ^{13}C) studies. In the solid state, tin-119m Mössbauer spectral data (80K) on these compounds are in accord with pentacoordinated structures with $trans$-trigonal bipyramidal (tbp) geometries at tin in nearly all cases. X-ray structural investigations of Bu$_3$SnOC(O)CH$_2$SC(S)NMePh,[12] Ph$_3$SnOC(O)CH$_2$OC(S)NMe$_2$[13] and Ph$_3$SnOC(O)CH$_2$SC(O)N(CH$_2$)$_4$CH$_2$[13]

confirm the pentacoordination at tin and the carboxylate-bridged structural feature which is indeed typical[14] of triorganotin alkanoates.

Table 2 : Relative inhibition of AChE by selected organotin compounds of concentration 1.0x10^{-4} M

Compound	% inhibition of AChE
Ph$_3$SnOC(O)CH$_2$SC(O)NH(4-ClC$_6$H$_4$)	74.6
Ph$_3$SnOC(O)CH$_2$SC(O)NHPh	31.1
Ph$_3$SnOC(O)CH$_2$SC(O)N(CH$_2$)$_4$CH$_2$	24.4
Ph$_3$SnOC(O)CH$_2$SC(O)NH(2,4-Cl$_2$C$_6$H$_3$)	13.4
Ph$_3$SnSC(O)N(CH$_2$)$_4$CH$_2$	14.8
(4-ClC$_6$H$_4$)Ph$_2$SnOC(O)CH$_2$SC(O)NHPh	15.8
Ph$_3$SnOC(O)CH$_2$SC(O)NMe$_2$	42.1
Ph$_3$SnOC(O)CH$_2$SC(S)NMe$_2$	31.8

Figure 1 shows the molecular structure of Ph$_3$SnOC(O)CH$_2$SC(O)N(CH$_2$)$_4$CH$_2$;

Figure 1 shows the molecular structure of $Ph_3SnOC(O)CH_2SC(O)\underline{N(CH_2)_4CH_2}$;

the intermolecular Sn-O' distance [2.322(6) Å] is only slightly shorter than that in
$Bu_3SnOC(O)CH_2SC(S)NMePh$ [2.393(5) Å][12]

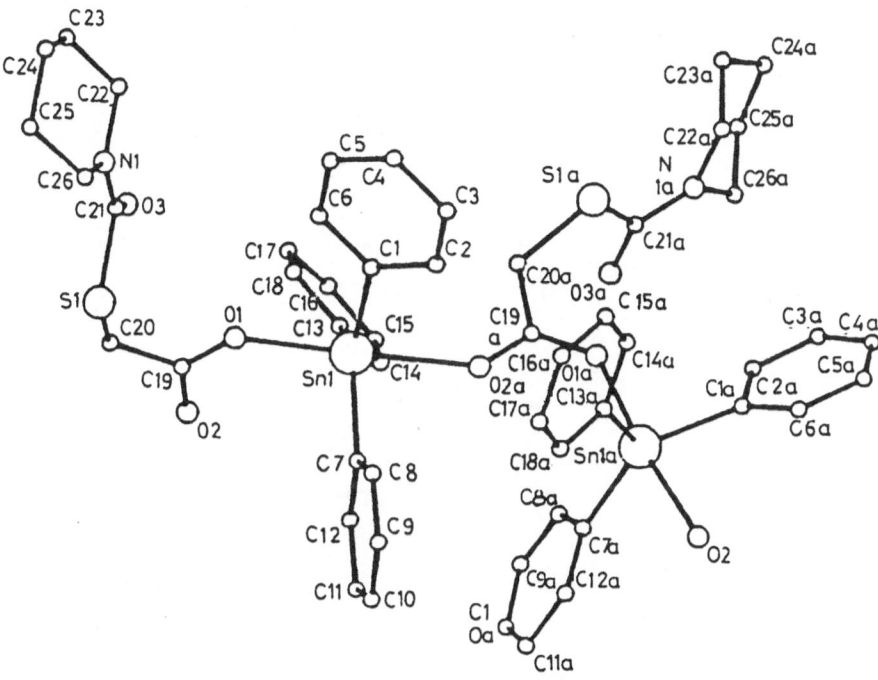

Fig. 1 : Molecular structure of $Ph_3SnOC(O)CH_2SC(O)\underline{N(CH_2)_4CH_2}$

An odd result is the essentially 4-coordinated structure of $Ph_3SnOC(O)CH_2SC(S)\underline{N(CH_2)_3CH_2}$ (IS 1.47 mms^{-1}, QS 2.92 mm s^{-1}), whose

crystal structure is displayed in Fig.2; the non-bonded Sn...O' contact is 2.906(7) Å.

Alternative bonding modes to carboxyl bridging are not often observed, even when other potential donor sites are present in the ester function. Recent examples from our laboratories where such alternative bonding modes have been encountered include the ureido-oxygen bridged (Sn-O' = 2.352 Å) $Ph_3SnOC(O)(CH_2)_2NHCONH_2$,[15] the phosphate-bridged (Sn-O' = 2.420 Å) $Ph_3SnOC(O)(CH_2)_nP(O)(OEt)_2$ (n = 1,2),[16] the sulphonyl-oxygen bridged (Sn-O' = 2.8697 Å) $Cyh_3SnOC(O)CH_2SO_2CH_2Ph$[17] and the sulphonato-oxygen bridged (Sn-O' = 2.575 Å) $[Cyh_2NH_2][R_3SnO_2CC_6H_4-2-SO_3]$[18,19] stannates. Even rarer is the chelating mode adopted by the carboxyl group. This has been observed recently[20] in $[Cyh_2NH_2]_2[Bu_2Sn(O_2CC_6H_4-2-SO_3)_2]$ wherein the diorganostannate anion has a 6-coordinated skewed trapezoidal bipyramidal (STB) geometry [C-Sn-C 133.5(4)°; Sn-O 2.083(4), 2.070(4) Å; Sn-O 2.648(5), 2.700(5) Å]; the sulphonato groups merely engage the ammonium cations in hydrogen-bonding to give an overall tight network structure.

373

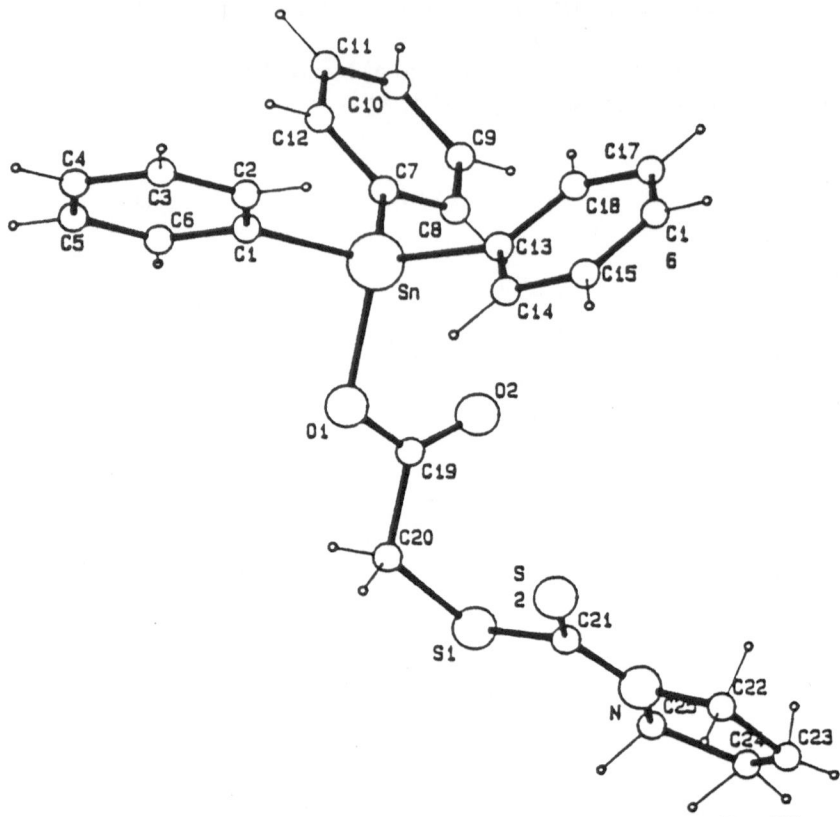

Fig. 2 : Molecular structure of $Ph_3SnOC(O)CH_2SC(S)\underline{N(CH_2)_3CH_2}$

Organotin carboxylates with olefinic functionalities in the esteryl unit

As examples of this category, we have chosen some essential oil components with known antifungal properties as the base ligands, namely cinnamaldehyde, citral and citronellal. *A priori* conversion of the precursor ligands to their corresponding acids (Scheme 3) provided a convenient means for attachment of the triorganostannyl moiety.[21] The terpenic esters were obtained by adding $(Bu_3Sn)_2O$ or Ph_3SnOH in the appropriate stoichiometry to the terpenic acids in petroleum ether in the presence of freshly activated molecular sieves (4Å), with slight warming for 0.5 h, followed by filtration and work-up. For triphenyltin cinnamates, ethanol was used as the solvent medium, but for tributyltin cinnamates, toluene was the preferred medium for the condensation reaction.

The triorganotin alkenoates exhibited NMR spectral characteristics typical of 4-coordinated tin species in solution ($CDCl_3$). For the triorganotin 3,7-dimethyl-2,6-octadieonates, ^{119}Sn NMR spectra revealed, however, two near-coincident tin peaks in the intensity ratio of $3:1$, which may be rationalised in terms of *cis* and *trans*-isomerism about the $C_2=C_3$ bond and the relative orientation of the CO_2Sn fragment with respect to the substituents on C_3. As shown in Fig. 3, the configurations Z_a and Z_c may be considered to be less likely on steric grounds so that the E configurations, presumably equivalent on the ^{119}Sn NMR time scale, are statistically dominant.

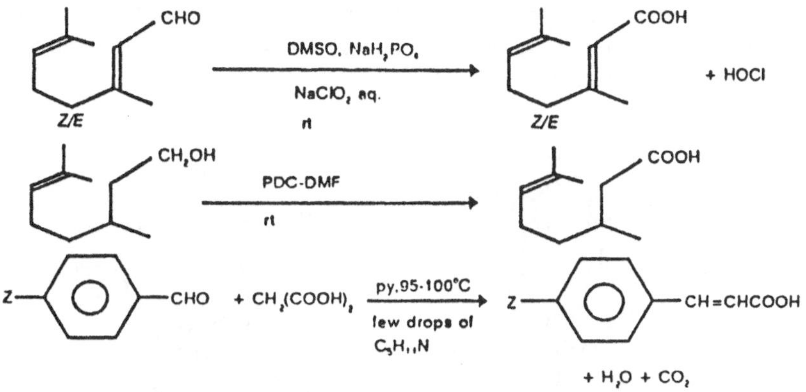

DMSO, NaH, PO₄

NaClO, aq.

CHO → COOH Z/E Z/E + HOCl

CH₂OH → COOH (PDC-DMF, rt)

$$Z-\langle\bigcirc\rangle-CHO + CH_2(COOH)_2 \xrightarrow[\text{few drops of } C_5H_{11}N]{\text{py. 95-100°C}} Z-\langle\bigcirc\rangle-CH=CHCOOH$$

+ H₂O + CO₂

PDC-DMF = pyridinium dichromate-dimethylformamide

Scheme 3

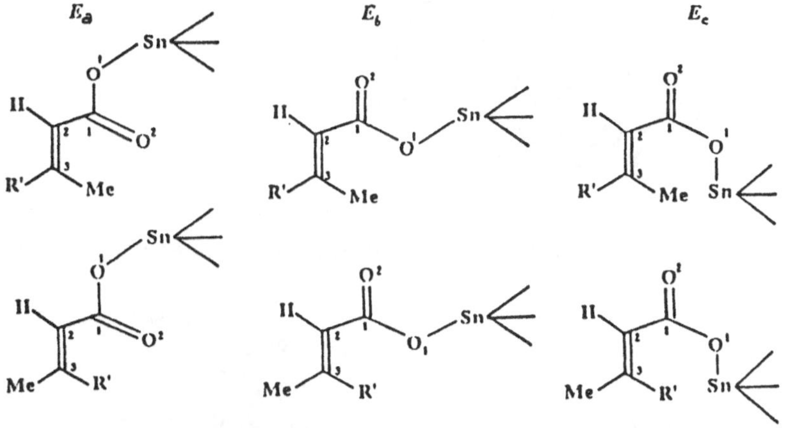

E_a E_b E_c

R₃SnO...

Fig. 3 : Z/E Configurations of $R_3SnOCOCH:C(Me)CH_2CH_2CH:CMe_2$
$\delta(^{119}Sn)$: +97.7/+96.6 (R=Bu); -125.1/-124.6(R=Ph); -129.7/-129.3 (R=p-ClC₆H₄)

The duality of peaks was also seen in the ^{13}C spectrum for C_1 and for each of the terpenic skeletal carbon atoms, C_2 to C_7 and C_{10}; only one set of butyltin and phenyltin peaks, however, could be discerned in the spectra. Solid state structural analysis by tin-119m Mössbauer spectroscopy gave IS and QS values typical of 5-coordinate structures for all the triorganotin esters save the triaryltin 4-nitrocinnamates for which the QS values lie in the range 2.35-2.49 mm s^{-1}, with corresponding ρ (QS/IS) values < 2.1. This is strongly indicative of 4-coordination.[22] Confirmation of this has come from comparison crystal structure studies[21] on triphenyltin cinnamate (Fig. 4a), -p-chlorocinnamate (figure not presented) and -p-nitrocinnamate (Fig. 4b). The carboxyl oxygen atom points into one face of the tetrahedral polyhedron around the tin in the p-nitrocinnamate molecule at a distance of 2.699(3) Å. The p-nitrocinnamato anion of the parent acid has the nitro group participating in *through-resonance* with the negatively-charged carboxyl oxygen atom, providing extra stabilization of the anion. This stabilization probably reduces the Lewis basicity of the carboxyl oxygen atom and favours a monomeric structure for the triphenyltin derivative. The Sn–O bond distance (2.067(2) Å) compares favourably with that (2.076(4), 2.086(4) Å) found in the two independent molecules in the crystal structure of the 4-coordinate compound $Ph_3SnOC(O)C_6H_4$-2-NO_2.[23]

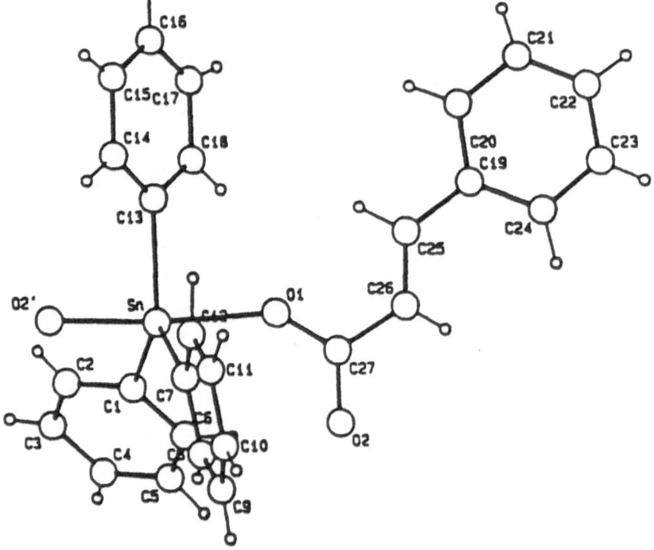

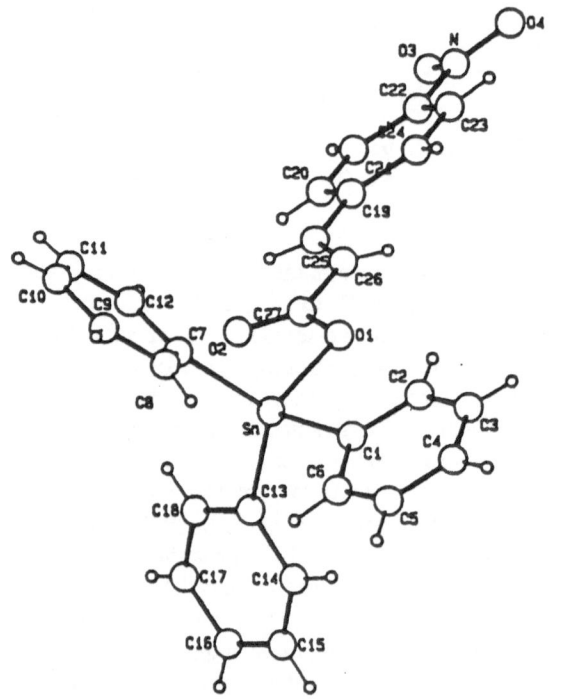

Fig. 4a : Molecular geometry of triphenyltin cinnamate

Fig. 4b : Molecular geometry of triphenyltin *p*-nitrocinnamate

376

A wood-block test evaluation of tributyltin cinnamate and its p-methoxy- and p-nitro- analogues as well as of the tributyltin esters of the two terpenic acids for their antifungal activity against *Coriolus versicolor* (white rot) and *Gloephyllum trabeum* (brown rot) was undertaken. Rubberwood, a medium hardwood which finds much use in the local furniture industry was used for this purpose. The results, summarised in Table 3, show that the compounds afforded generally better protection against *G. trabeum*, and that one of the compounds, $Bu_3SnOC(O)CH:C(Me)(CH_2)_2CH:CMe_2$, showed the highest activity against both the decay fungi.[24]

Table 3 : Toxic limits[a] of some tributyltin compounds towards wood decaying fungi in Rubberwood (*Hevea brasiliensis*)

$Bu_3SnOC(O)R'$	Toxic limits (kg m^{-3})	
R' =	*C. versicolor*	*G. trabeum*
$CH:CHC_6H_5$	0.32-0.60	< 0.16
$CH:CHC_6H_4$-*p*-NO_2	0.65-1.33	< 0.16
$CH:CHC_6H_4$-*p*-OCH_3	0.75-1.24	< 0.16
$CH:C(CH_3)(CH_2)_2CH:C(CH_3)_2$	0.22-0.45	< 0.12
$CH_2CH(CH_3)CH_2CH_2CH:C(CH_3)_2$	0.18-0.33	< 0.16
$(Bu_3Sn)_2O$	0.42-0.85	<0.11

[a]The toxic limit of a test compound is defined as the interval between that concentration of the toxicant and its associated loading in wood which permits significant decay and the next highest in the series which prevents decay. An average weight loss greater than or equal to 3% in each set of replicates or a loss of 5% or more of its initial mass in any one wood block constitutes significant decay.

Organotin stannates derived from dicarboxylic acids

More recently, our choice of Y groups has been extended to include $(CH_2)_xC(O)O^-$ moieties, in which the dicyclohexylammonium ion has featured strongly as the counterion. The products have been thoroughly studied by X-ray diffraction and several novel structural features have been observed. Structural complexity appears to be the rule rather than the exception here, with simple stannates being a rarity. An example of the latter is dicyclohexylammonium triphenyltin succinate, obtained by reacting Ph_3SnOH in ethanol with equimolar amounts of succinic acid and dicyclohexylamine. The succinato ligand is non-chelating but axially links planar Ph_3Sn^+ cations into chains that run parallel to the *c*-axis of the unit cell; a distorted *trans*-C_3SnO_2 tbp geometry results at tin.[25] The dicyclohexylammonium cations surround the polymeric chains and are H-bonded to the acyl oxygens of the succinato ligand.

A similar attempt using oxalic acid in place of succinic acid yielded $[Cyh_2NH_2]^+[Ph_3Snox]^-$, whose geometry at tin has been assigned as *cis*-C_3SnO_2

on Mössbauer evidence (IS 1.02, QS 2.15 mm s⁻¹). In seeking crystal structure confirmation for this, we undertook the synthesis of the related [Me₄N]⁺[Ph₃Snox]⁻ (prepared as in equation 1), but instead a trinuclear stannate resulted whose composition was that of a 1:1 adduct of the expected stannate with bis(triphenyltin) oxalate.[26]

$$[Me_4N]Cl + AgOC(O)C(O)OAg + Ph_3SnCl \xrightarrow{\text{MeOH}} [Me_4N]^+[Ph_3Sn^1ox]^-$$
$$+$$
$$[Ph_3Sn^2OC(O)C(O)OSn^3Ph_3]$$

The geometry at Sn¹ (the stannate portion) is *cis*-tbp (Fig 5), while the geometries at the Sn² and Sn³ atoms are the ubiquitous *trans*-tbp.

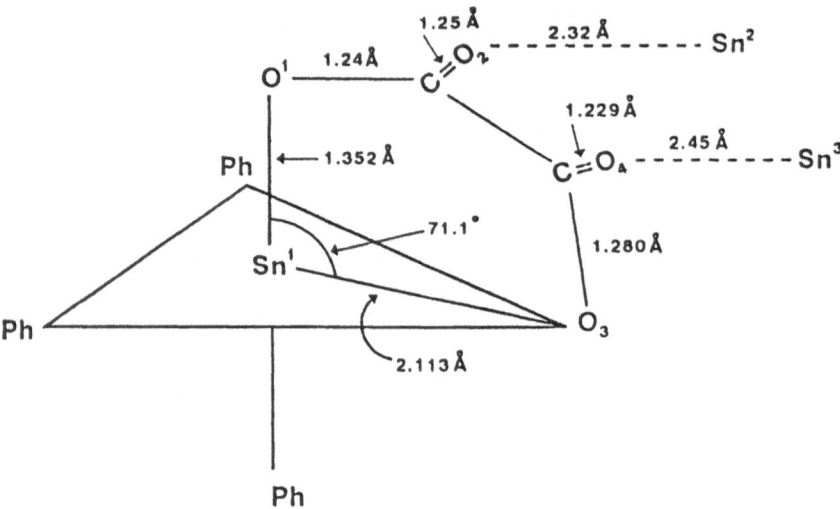

Fig. 5: The coordination geometry of the tin atom in [Ph₃Snox]⁻

This structure is an unusual example of a triorganotin compound having within the same molecule two 5-coordinate tin sites in distinct *cis*- and *trans*-tbp geometries. Yet another example is the compound μ-oxalatobis(tricyclohexyltin) (IS/QS: 1.44/2.96; 1.54/3.87 mm s⁻¹) whose unit cell contains two symmetry-independent molecules (a and b).[27] In either molecule, anisobidentate chelation (Fig. 6) of a Cyh₃Sn group by the oxalato unit renders this tin atom 5-coordinate in a *cis*-tbp configuration [molecule a: Sn-O$_{ax}$ = 2.48(1), Sn-O$_{eq}$ = 2.119(9) Å, O$_{ax}$-Sn-O$_{eq}$ = 68.5(4)°, molecule b: Sn-O$_{ax}$ = 2.471(8), Sn-O$_{eq}$ = 2.123(10) Å, O$_{ax}$-Sn-O$_{eq}$ = 69.2(3)°]. Chelation permits the remaining two oxygen atoms of the oxalato ligand to bind axially to two adjacent planar Cyh₃Sn girdles, one of which is the molecule's own half and the other the corresponding half of the symmetry-independent partner (molecule a: Sn-O 2.22(1), Sn-O = 2.40(1) Å, O-Sn-O = 176.0(4)°; molecule b: Sn-O = 2.242(8), Sn-O = 2.410(9) Å, O-Sn-O = 176.2(4)°]. This leads to a linear chain structure parallel to the *a*-axis.

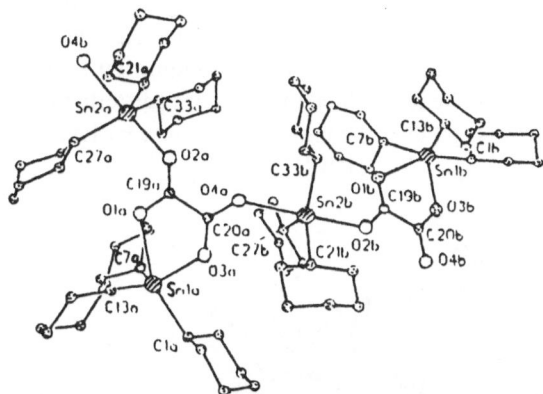

Fig. 6 : Molecular structure of μ-oxalatobis(tricyclohexyltin)
(Reprinted from Ref. 27. Copyright 1994 - Elsevier Sequoia)

A ^{13}C NMR study of the compound in chloroform indicates the presence of a molecule having two magnetically-equivalent 4-coordinate tin centres, as suggested by the observation of only one set of tin-carbon couplings [$^1J(^{119}Sn-^{13}C) = 328.0$ Hz] in the NMR.

The chelating mode of the oxalato anion has also been observed in the di- and mono-organotin compounds, [Cyh$_2$NH$_2$]$_2$$^+$[Ph$_2$Sn(ox)$_2$$^{2-}$][28] and [Cyh$_2NH_2$]$_2$$^+$ [BzSn(ox)$_2$OC(O)Ph]$^{2-,29}$ which have six-fold (octahedral) coordination at tin, and network structures as a consequence H-bonding interactions involving the ammonium cations. Interestingly, the monobenzyltin compound was a serendipitous product that we obtained upon reacting what we presumed was Bz$_2$SnO (formed when Bz$_3$SnOH was left exposed to air) with [Cyh$_2$NH$_2$]$^+$[Hox]$^-$ in ethanolic medium. The X-ray analysis, in this case, provided the evidence that the actual organotin starting material involved was benzyloxotin benzoate, BzSnO(OCOPh) (equation 2).

$$Bz_3SnOH \xrightarrow{\text{air}} BzSnO(OCOPh) \xrightarrow[\text{ether}]{[Cyh_2NH_2]^+[HOx]^-} [Cyh_2NH_2]^+[BzSn(ox)_2OC(O)Ph]^- \quad ....(2)$$

Complex structures are invariably obtained with tributyltins in their interactions with dicarboxylic acids, especially in alcoholic media. Thus with oxalic acid, TBTO yielded the products indicated in equations (3) and (4).

$$(Bu_3Sn)_2O + 2\ H_2Ox.2H_2O + 2Cyh_2NH \xrightarrow{\text{toluene}} 2\ [Bu_3Snox]^-[Cyh_2NH_2]^+ \quad ....(3)$$

$$\downarrow \text{ethanol}$$

[Cyh$_2$NH$_2$]$_2$$^+$[Bu$_3$SnOC(O)C(O)OSnBu$_3$].[Bu$_3$SnOC(O)C(O)O.EtOH]$_2$$^-$ ...(4)

In the crystal structure of the tetranuclear product, the [Cyh$_2$NH$_2$]$^+$ cations form hydrogen bonds to the dianionic chain, which is composed of two ethanol-coordinated Bu_3Snox^- anions coordinated to a $(Bu_3Sn)_2ox$ molecule;[30] a three-dimensional network structure is the result. As shown in Fig. 7, the centrosymmetric dianion consists of four Bu_3Sn units linked through three oxalate groups into a chain; the inner Sn atoms have less distorted *trans*-tbp geometries [*trans* angle 175.5(1)°] than the terminal ethanol-coordinated tins [$Sn-O_{ethanol} =$ 2.465(4) Å; *trans*-angle 172.3(10)°].

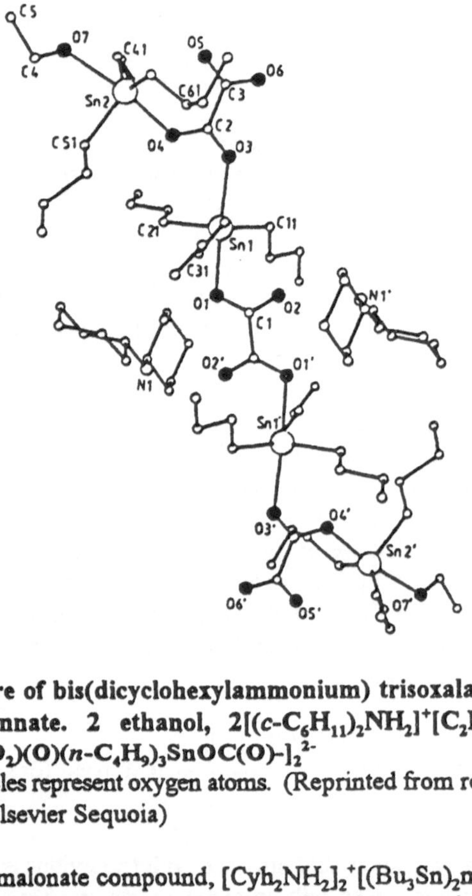

Fig. 7 : Structure of bis(dicyclohexylammonium) trisoxalatotetrakis(tri-*n*-butylstannate. 2 ethanol, $2[(c-C_6H_{11})_2NH_2]^+[C_2H_5OH(n-C_4H_9)_3-SnO(CO_2)(O)(n-C_4H_9)_3SnOC(O)-]_2^{2-}$
Dark circles represent oxygen atoms. (Reprinted from ref. 30. Copyright 1990 - Elsevier Sequoia)

In the related malonate compound, $[Cyh_2NH_2]_2^+[(Bu_3Sn)_2mal][Bu_3Snmal]_2^-$, which contains no ethanol, the two $[Bu_3Snmal]^-$ anions are similarly linked to a $[(Bu_3Sn)_2mal]$ molecule; however, a 3-dimensional carboxylate-bridged network is formed because both $[Bu_3Snmal]^-$ anions are additionally linked to other $[(Bu_3Sn)_2mal]$ molecules.[31]

Unlike the alkanoic dicarboxylic acids, 2,6-pyridinedicarboxylic acid behaves as a doubled-up monocarboxylic. In the crystal structure of $[Cyh_2NH_2][Bu_3Sn(O_2C)_2C_5H_3N]$, each carboxylate moiety is involved in coordination to a tin atom *via* one O atom only, that is, the ligand functions as a bidentate bridging ligand.[32] The pyridyl nitrogen is not involved in coordination. By way of contrast, in triphenyltin nicotinate[33] and isonicotinate,[34] the pyridyl nitrogen participates in intermolecular bonding (Sn-N:2.56-2.58 Å).

We have also synthesised several diorganostannate esters (equations 5 & 6).[35]

$$2Cyh_2NH + 2H_2ox.2H_2O + Bu_2SnO \xrightarrow{\text{ethanol}} [Cyh_2NH_2]_2^+[Bu_2Snox_2]^{2-} \quad (5)$$
$$\mathbf{1}$$

$$2Cyh_2NH + 3H_2ox.2H_2O \xrightarrow{\text{ethanol}} [Cyh_2NH_2]^+[\{Bu_2Snox(H_2O)\}_2ox]^{2-} \quad (6)$$
$$+ 2Bu_2SnO \quad \mathbf{2}$$

A skewed trapezoidal bipyramidal geometry exists for compound **1**, which is obtained as a monomeric species (C–Sn–C 146.7(3)°). The dinuclear species **2** has each of its tin atoms in a distorted pentagonal bipyramidal geometry with apically disposed butyl groups [C–Sn–C 172.2(3)°; Sn–OH$_2$ 2.422(4) Å]. As shown in Fig.8, there are 2 independent oxalate ligands in the structure - a bidentate and a tetradentate. Both **1** and **2** feature extended H-bonded networks involving the oxygen atoms of the dianion and the *N*-bound hydrogen atoms. Against the tumours MCF-7 and WiDr, compounds **1** and **2** showed moderate activity, with significantly lower ID$_{50}$ values compared to cisplatin[35]: MCF-7: 1 76, 2 95; WiDr: 1 323, 2 353 μg/mL.

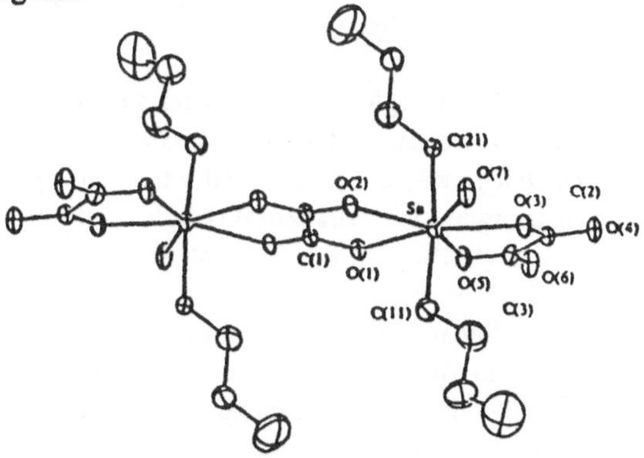

Fig. 8 : Molecular structure of dimeric dianion in [(c-C$_6$H$_{11}$)$_2$NH$_2$]$_2$[{(n-C$_4$H$_9$)$_2$Sn(O$_2$CCO$_2$)(H$_2$O)}$_2$(O$_2$CCO$_2$)]
(Reprinted from ref. 35. Copyright 1992 - John Wiley and Sons)

Substituted aryl- and heteroaryltin compounds

An increasing focus in our organotin synthetic programme has been on compounds of the above category, often containing mixed organic groups on the metal.

Substituted aryl tins

A number of tetra- and triphenyltin compounds containing a monosubstituted ring of formula (Z-C$_6$H$_4$)SnPh$_3$ and (Z-C$_6$H$_4$)SnPh$_2$X, respectively, where X = OH, Cl or OCOR (R = (CH$_2$)$_2$COPh, CH$_2$SC(S)NMe$_2$; Z = CHO, COMe or the carbonyl protectant 1,3-dioxolane) have been synthesised, along with stable oxime and semicarbazone derivatives of the ring-borne carbonyl functional group. Scheme 4 summarises the synthetic steps for the mono-ring substituted tetraphenyltins.

Conversions of the tetraorganotins to the triorganotin derivatives were achieved using appropriate electrophilic reagents.

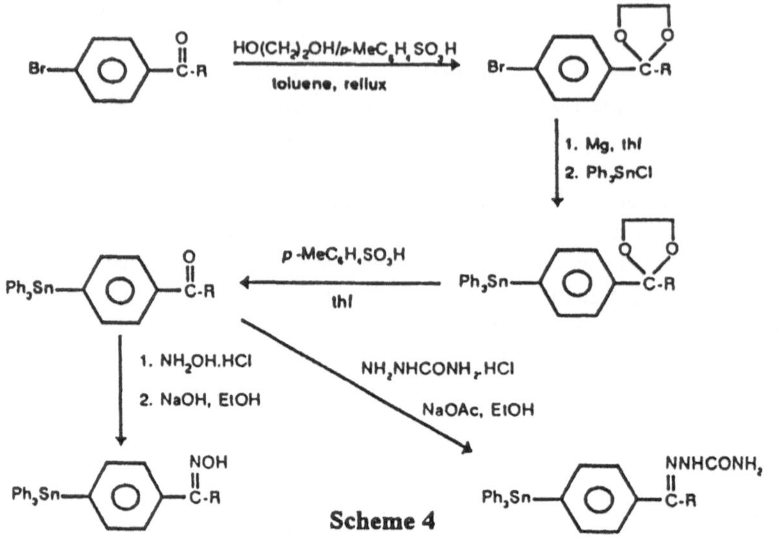

Scheme 4

The tetraorganotins all yielded single line Mössbauer spectra (IS:1.20-1.24 mm s^{-1}) attesting to their 4-coordinated tetrahedral structures in the solid state; their tin-119 chemical shifts (CDCl$_3$) lie in the range -130.1 to -131.3 ppm; *c.f.* Ph$_4$Sn - 128.8 ppm. Among the triorganotins, tetrahedral structures in the solid state exist for the chlorides, with ρ (QS/IS) ~1.86, but not for the hydroxides or ester derivatives ($\rho > 2.1$).

We were also attracted to the use of the 2-oxazolinyl substituent in our synthetic work because of its relative inertness toward a variety of synthetic manipulations, and the availability of methods for its conversion under mild conditions to the carboxylic acid function.

As shown in scheme 5, stannylation of 2-phenyl-2-oxazolinyl lithium is readily achieved with both tri- and diorganotin dihalides, with the latter yielding directly triorganotin halides in 40-60% yields.

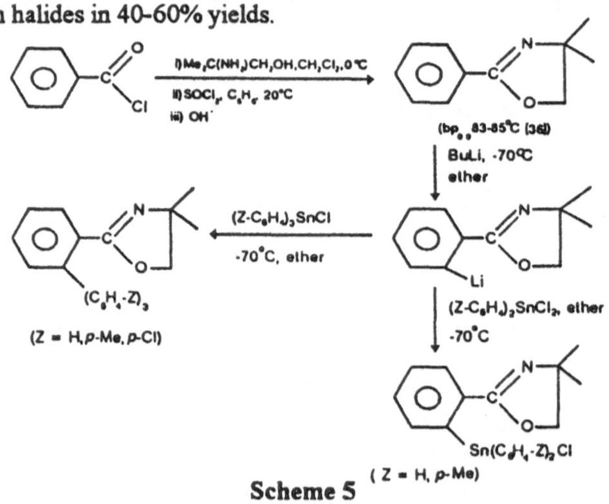

Scheme 5

Extension of the above work to stannylated thiophenes has also been carried out (Scheme 6) to yield both tetra- and triorganotin derivatives.

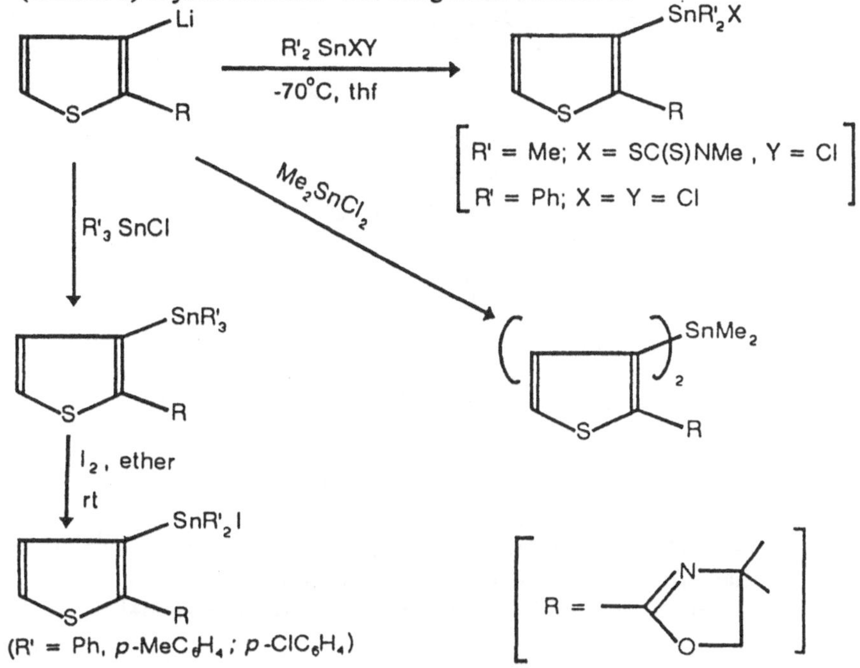

$$
\left[
\begin{array}{l}
R' = Me;\ X = SC(S)NMe,\ Y = Cl \\
R' = Ph;\ X = Y = Cl
\end{array}
\right]
$$

$$
R = \left[
\begin{array}{c}
\text{oxazoline structure}
\end{array}
\right]
$$

(R' = Ph, *p*-MeC$_6$H$_4$; *p*-ClC$_6$H$_4$)

Scheme 6

Whereas the lithio derivative of 2-oxazolinylthiophene (LH) failed to react with Me$_3$Sn(S$_2$CNMe$_2$)$_2$, and with Me$_2$SnCl$_2$ yielded the unsymmetrical tetraorganotin product, the use of Me$_2$SnCl(S$_2$CNMe$_2$) and Ph$_2$SnCl$_2$, on the other hand, resulted largely in monodehalogenated triorganotin products Me$_2$SnL(S$_2$CNMe$_2$) and Ph$_2$SnLCl.[37] With [2-(4,4-dimethyl-2-oxazolinyl)-3-thienyl]triaryltins, selective mono-aryl cleavage was accomplished with iodine in ether. The cleavage of the tin-aryl bond over the relatively more labile tin-heteroaryl bond is attributed to the coordinative interaction at tin by the oxazolinyl substituent, and the relief of steric strain in the molecule accompanying the loss of a bulky aryl group. Crystal structure analyses of LSnMe$_2$(S$_2$CNMe$_2$) (Fig. 9) and LSnPh$_2$Cl (figure not presented) indicate that the oxazolinyl group is intramolecularly coordinated to tin through its N atom [Sn-N 2.720(3) Å]. This yields a trigonal bipyramidal geometry at tin in which the more electronegative S and N atoms occupy apical positions.[37] By way of contrast, the crystal structure study of the tetraorganotin LSn(*p*-tolyl)$_3$ reveals a tetrahedral geometry at tin which is slightly distorted as a result of the orientation of oxygen (rather than nitrogen) of the oxazolinyl group (Sn-O non bonded contact is 2.977(3) Å) (Fig. 10) towards one face of the polyhedron around tin.[38]

Further studies on other stannylated heterocycles are currently in progress.

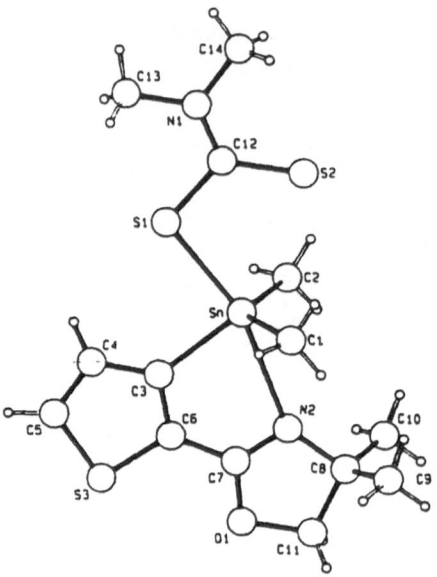

Fig. 9 : Molecular structure of [2-(4,4-dimethyl-2-oxazolinyl)-3-thienyl]dimethyltin *N,N*-dimethyldithiocarbamate
(Reprinted from ref. 37. Copyright 1992 - Elsevier Sequoia)

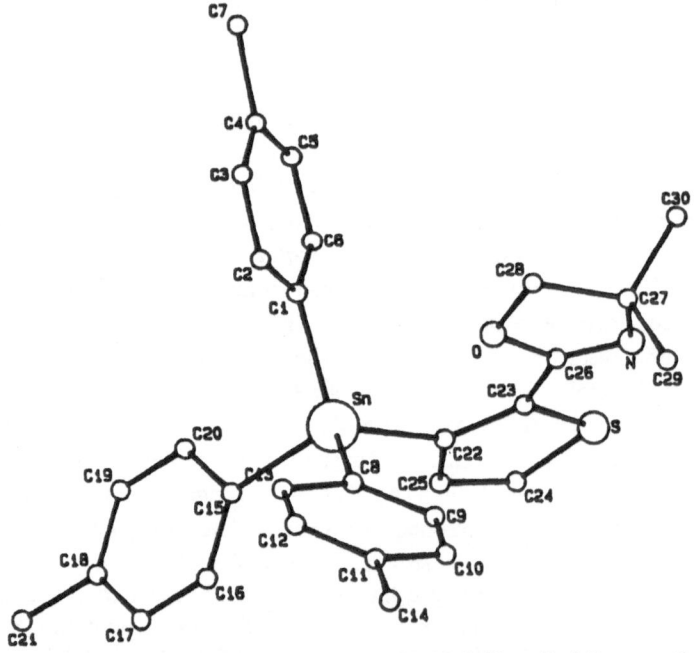

Fig. 10 : Molecular structure of [2-(4,4-dimethyl-2-oxazolinyl)-3-thienyl]tri(*p*-tolyl)tin
(Reprinted from ref. 38. Copyright 1994 - Elsevier Sequoia)

Acknowledgements

Acknowledgement is made to the National Science Council for R&D (Grant No. 2-07-04-06) for the generous support of this research programme and to the following postgraduate students who have contributed immensely to the work described herein: Selvasothi Selvaratnam, L.-F. Siah, Nazni Wasi Ahmad, Normawati Samsodin and Ng Mei Peng.

References

1. V.G. Kumar Das, S.W. Ng and M. Gielen, eds, *Chemistry and Technology of Silicon and Tin*, Oxford University Press, London, 1992.
2. V.G. Kumar Das, A.J. Kuthubutheen, S. Balabaskaran and S.W. Ng, *Main Gp. Met. Chem.* **12** (1989) 389.
3. V.G. Kumar Das, A.J. Kuthubutheen, S.W. Ng and W.K. Ng, Malaysian Patent, MY-101623-A, Dec. 1991.
4. S.W. Ng, A. Zainudin, A.J. Kuthubutheen and V.G. Kumar Das, *Organotin xanthylacetates: tin-119m Mössbauer and in vitro fungitoxicity study*, in Ref. 1, pp. 340-348.
5. S.W. Ng, V.G. Kumar Das, G. Pelizzi and F. Vitali, *Heteroatom Chem.* **1** (1990) 433.
6. G.J.M. van der Kerk and J.G.A. Luijten, *J. Appl. Chem.* **4** (1954) 314; **6** (1954) 56.
7. S. Balabaskaran, K. Tilakavati and V.G. Kumar Das, *Appl. Organomet. Chem.* **1** (1987) 347.
8. B. Holmberg, *J. Prakt. Chem.* **84** (1911) 634.
9. G. Hilgetag and A. Martini, eds, *Preparative Organic Chemistry*, Wiley, Berlin, 1972; H. Tilles and P.E. Hoch, U.S. Patent 4,147,715 (April 1979).
10. Nazni W. Ahmad, Sofian Azirun Mohd., S. Balabaskaran and V.G. Kumar Das, *Appl. Organomet. Chem.* **7** (1993) 583.
11. Nazni W. Ahmad, Tay Siew Huang, S. Balabaskaran, K.M. Lo and V.G. Kumar Das, *Metal-Based Drugs* **1** (1994) 1.
12. S.W. Ng, Chen Wei, V.G. Kumar Das, C.K. Yap and R.J. Butcher, *Crystal structure of tri-n-butyltin(IV) N-methyl-N-phenyldithiocarbamylacetate, (n-$C_4H_9)_3SnOC(O)CH_2SC(S)N(CH_3)(C_6H_5)$*, in Ref. 1., pp 565-571.
13. Ng Mei Peng, M. Phil dissertation, University of Malaya (1993).
14. S.W. Ng, Chen Wei and V.G. Kumar Das, *J. Organomet. Chem.* **345** (1988) 59.
15. K.M. Lo, V.G. Kumar Das, W.-H. Yip and T.C.W. Mak, *J. Organomet. Chem.* **412** (1992) 21.
16. S.W. Ng and V.G. Kumar Das, *J. Chem. Crystallogr.*, **24** (1994) 337.
17. K.M. Lo, S. Vijayalakshmi and V.G. Kumar Das, unpublished results.
18. S.W. Ng, V.G. Kumar Das and E.R.T. Tieknik, *J. Organomet. Chem.* **411** (1991) 121.
19. S.W. Ng and V.G. Kumar Das, *J. Cryst. & Spectros. Res.* **22** (1992) 507.
20. S.W. Ng, V.G. Kumar Das and M. Gielen, *Appl. Organomet. Chem.* **6** (1992) 489.
21. L.-F. Siah, M. Phil dissertation, University of Malaya (1992).
22. R.H. Herber, H.A. Stöckler and W.T. Reichle, *J. Chem. Phys.* **42** (1965) 2447.

23. S.W. Ng, V.G. Kumar Das and W.T. Robinson, *Malays. J. Sci.* **12** (1990) 57.

24. L.-F. Siah, Salamah Selamat and V.G. Kumar Das, *Studies on the structure and wood-preserving properties of some tributyltin alkenoates*, in "Frontiers of Organogermanium, -tin and -lead Chemistry" eds, E. Lukevics and L. Ignatovich, Latvian Inst. of Organic Synthesis, 1993, pp. 301-318.

25. S.W. Ng, V.G. Kumar Das, G. Xiao, Dick van der Helm, J. Holeček and A. Lyčka, *Heteroatom Chem.* **2** (1991) 495.

26. S.W. Ng and V.G. Kumar Das, *J. Organomet. Chem.*, 456 (1993) 175.

27. S.W. Ng, V.G. Kumar Das, S.-L. Li and T.C.W. Mak, J. Organomet. Chem., **467** (1994) 47.

28. S.W. Ng and V.G. Kumar Das, *Main Gp. Metal Chem.* **16** (1993) 87.

29. S.W. Ng and V.G. Kumar Das, *Main Gp. Metal Chem.* **16** (1993) 95.

30. S.W. Ng, V.G. Kumar Das, M.B. Hossain, F. Goerlitz and Dick van der Helm, *J. Organomet. Chem.* **390** (1990) 19.

31. S.W. Ng, V.G. Kumar Das, B.W. Skelton and A.H. White, *J. Organomet. Chem.* **430** (1992) 139.

32. S.W. Ng, V.G. Kumar Das and E.R.T. Tieknik, *J. Organomet. Chem.* **403** (1991) 111.

33. S.W. Ng, V.G. Kumar Das, F. van Meurs, J.D. Schagen and L.H. Straver, *Acta Crystallogr.* **C45** (1989) 570.

34. S.W. Ng and V.G. Kumar Das, *J. Cryst. & Spectros. Res.* **22** (1992) 371.

35. S.W. Ng, V.G. Kumar Das, M. Gielen and E.R.T. Tieknik, *Appl. Organomet. Chem.* **6** (1992) 19.

36. P.D. Pansegrau, E.F. Reiker and A.I. Meyers, *J. Am. Chem. Soc.* **110** (1988) 7178.

37. K.M. Lo, S. Selvaratnam, S.W. Ng, Chen Wei and V.G. Kumar Das, *J. Organomet. Chem.* **430** (1992) 149.

38. S. Selvaratnam, K.M. Lo and V.G. Kumar Das, *J. Organomet. Chem.* **464** (1994) 143.

Main Group Elements and Their Compounds
V.G. Kumar Das (Ed)

Selenium : the Element with Two Faces

Robert J. Magee

Department of Chemistry, La Trobe University
Bundoora, Vic. 3083, Australia

During his travels in China in the year 1295, Marco Polo commented on the herbage growing in Turkestan and Mongolia, which he suggested was the cause of the severe disorientation in the grazing animals. In later centuries, this became known as the "blind staggers". The cause, however, was still unknown, but it was established that the most poisonous of the herbage producing the staggers was milk vetch (*Astragalus*), also known to cowboys as *locoweed.*[1]

The element selenium was not discovered until 1817, but it was not until 1960 that it was established that the blind staggers in grazing animals was due to high concentrations of selenium in *Astragalus*, which has a remarkable property of concentrating selenium from the soil. For example, a specimen in Wyoming was found to contain 1.4% Se. These accumulator plants, as they are called, suffer no ill effects, but are poisonous to grazing animals. It was such events that gave selenium a bad reputation: it is not surprising that it was classified as a toxic element.

Dr. Klaus Schwarz working in Heidelberg, Germany in the 1940-50's was baffled. Liver necrosis, produced in animals on diets low in protein, could be arrested by **Factor-3** - a substance of unknown composition present in about a milligram of material, obtained by a laborious processing of a ton of porcine kidneys. However, in May 1957, he got a vital clue - a smell of garlic from one of his preparations. Ultimately, it was shown that the vital component in **Factor-3** was selenium. Here is the case of selenium showing its other face and behaving as a beneficial element.

In the 1970's, the Jekyll and Hyde behaviour of selenium was again revealed in studies on two mysterious diseases of long-standing. The first concerns the inhabitants of Hubei Province, China, where an endemic disease took a heavy toll of their lives. Thus, in the period 1961-64, almost 50% of the inhabitants died from the disease. The most common signs were loss of hair and nails. In areas of high incidence, skin lesions, tooth decay, abnormalities of the nervous system and finally death were common. Selenium was found to be the cause. Extremely high concentrations of selenium were found in the stony coal bed of the region and in the soil. The disease was finally named **selenosis**. Dietary intake of selenium (from foods grown in the soil) averaged 4.99 mg (3.20 - 6.69 mg)/d. Hair and blood levels averaged 32.2 μg/g and 3.2 μg/ml, respectively.

By contrast, in Keshan County, China, a disease known as **Keshan Disease** showed a severe outbreak. It resulted in heart failure not only in the middle-aged and elderly, but also among children. One type of heart failure, sometimes acute, with a mortality rate of up to 50% in children, was common. The puzzle was that Keshan Disease occurred spasmodically among the peasant population in a wide area stretching from north-east to south-west China. Its distribution was irregular, with areas of high incidence surrounded by regions relatively free from the disease.

In 1974, a large scale study was carried out in Mianning Country in South China. The study clearly revealed that selenium deficiency was the cause - confirmed by measurements on soil, foods, hair, blood and urine. The daily intake was found to be 7 - 10 µg Se, figures which are well below those for USA and Europe (see Table 1). Over a two year period, inhabitants were supplemented with selenium in their diet with sodium selenite. By 1977, Keshan Disease had been eliminated. These are further examples of selenium showing its two faces.[2]

Table 1 : Some world-wide dietary intakes

Keshan Region (China)	7 µg/d
Huber Province (China)	4,900 µg/d
Finland	25 - 30 µg/d
New Zealand	20 - 25 µg/d
USA	50 - 200 µg/d
Japan	130 µg/d
Germany	45 µg/d

Selenium Chemistry[3-5]

Selenium has properties somewhat similar to those of its group 16 congeners: sulphur (a non-metal); tellurium (a metalloid); polonium (a metal). There are six stable isotopes and several unstable ones, of which the most useful is ^{75}Se ($t_{1/2}$ = 120.4d).

The three elements show a great variety of allotropy. There are three allotropic forms of selenium, interest in which has been largely occasioned by the use of Se in electronic devices.

1. Metallic hexagonal : crystalline, stable, grey to black in colour; high electrical conductivity when exposed to light.

2. Red selenium : monoclinic crystals from cooling molten Se-consists of Se_8 molecules. Due to different packings of Se_8 rings, three red forms exist (α, β, γ-monoclinic).

3. Amorphous : can exist as black, amorphous red or colloidal Se. This form and the red forms are poor conductors of electricity.

Electronic configuration : [Ar] $3d^{10}$ $4s^2$ $4p^4$. Se 4d orbitals are available for bonding.

In its compounds Se shows the following valencies: -2, -1, 0, 2, 4, 6. Only a few unstable compounds exist in the +2 oxidation state. In the +4 oxidation state, Se shows both reducing and oxidizing properties; in the +6 state only oxidizing properties are manifested. Se resists oxidation to its highest oxidation state.

Se exists in the -2 state only in important organo-selenium compounds. The +6 oxidation state is most widely encountered in inorganic compounds.

Selenious acid (H_2SeO_3)

This gives rise to selenites. In soluble form, selenites are highly toxic. Affinity for iron and ease of reduction to elemental Se makes this form quite unavailable to plants; and the ready activity of selenites with cysteine of glutathione is of considerable biological interest.

Selenic acid (H_2SeO_4)

Alkaline and oxidising conditions favour formation and stability of +6 oxidation state, i.e., selenates. Most selenates are quite soluble and highly toxic. In soil, selenates are easily leached and are available to plants.

Selenides

Water soluble Se compounds can be reduced to selenides under acidic conditions or in biological systems. Of interest are hydrogen selenide (H_2Se), dimethylselenide and trimethylselenonium. Dimethyl selenide has a marked garlic odour.

Organoselenium compounds

Many organoselenium compounds have been synthesised in the search for anti-metabolites for cancer therapy and bacteriostatic compounds, e.g., 6-amino-8-selenopurine and selenoguanine.

Distribution

Selenium occurs in the earth's crust with an average abundance of *ca.* 0.09 ppm. Most soils contain 0.1 - 2.0 ppm, although there is an extremely wide variation.

Both chemical form and total content determine whether Se gets incorporated into vegetation. In alkaline soils, Se is usually present as water-soluble selenate and is thus available to plants; whereas in acidic soil it is often found as basic ferric selenite of very low solubility. There is a wide variation in food products. As shown in Table 2, vegetables and fruits are generally low in selenium content (< 0.01 μg/g), compared to grain and cereal products, meat and seafood.[2]

The species *Astragalus* and *Stanleya* can accumulate selenium to quite spectacular levels and have been used as "selenium indicators" for the underlying soils.

389

Table 2 : Selenium content of some foods

Fruits/Vegetables	<0.01 ppm
Cereals	0.01 - 0.7 ppm
Meats	0.03 ppm
Seafood	0.5 ppm

Methods of analysis

Many techniques have been used for the determination of selenium in biological and environmental samples. Fluorometry is popular and has often been used as a reference method.[6] Selenium(IV) reacts with 3,3'-diaminobenzidine (DAB) and with 2,3-diaminonaphthalene (DAN) to give strongly fluorescent products, which provide a sensitive method for the detection of selenium. It has been claimed that 2,3-DAN gives greater sensitivity and better extractability into organic solvents from acidic media. Fluorometry is, however, subject to interferences from co-extracted species. Detection limit for the fluorometric method is about 2×10^{-8} g/l.

While it has been claimed that conventional flame AAS is not sensitive enough for Se determination, an Ar - H_2 flame improves sensitivity, but interferences increase. Flameless AAS offers high sensitivity but is not free from interferences. Recently, there has been a marked increase in the use of hydride-generation AAS and ETAAS.[7] Due to its high sensitivity (2×10^{-8} to 10^{-9} g Se), simplicity and non-destructive nature, Neutron Activation Analysis (NAA), in particular, thermal neutron activation, is widely used[8], especially for biological samples. Radioactive Se nuclides are formed by (n, γ) reactions. Many inter-laboratory studies have been reported. A sub-committee of IUPAC carried out a study on the determination of total Se in lyophilized human blood serum involving 27 participating laboratories. Techniques used were fluorometry, ETAAS, NAA (instrumental or with radiochemical separation), hydride-generation AAS, isotope dilution MS and XRF spectrometry. ETAAS and hydride-generation AAS elicited the largest systematic differences in concentration.

In the period 1985 - 1989, the European Bureau of Reference certified 7 biological materials for Se with concentrations in the range 0.03 - 10 mg/kg. Certification analyses were carried out using fluorometry, hydride-generation combined with ICP, AAS, ETAAS, instrumental or radiochemical NAA. For Se concentrations < 0.3 mg/kg, fluorometry was adjudged as the best and preferred method; for concentrations above this level, the use of instrumental NAA was recommended.

In a comparison of several techniques for the determination of Se in biological fluids, three analytical techniques, fluorometry, hydride-generation AAS and ETAAS, were used to determine Se in plasma samples, while hydride-generation AAS and fluorometry were employed for whole blood, urine, haemo-dialysis fluid and water. The overall correlation coefficient for hydride-generation AAS *vs* fluorometry was 0.97. All methods gave accurate results for certified materials.

390

More recently, gas chromatography[9] and electroanalytical methods have grown in popularity, including DDP, catalytic polarography, DPCSV and DAASV.[10,11]

Uses of selenium[12-15]

The largest use of selenium is in the electronics and related industries. Its important use in xerography is due to the photoconductivity of the element. Semi-conductors, photoelectric cells, infra-red optic materials all involve selenium. Other major uses are in the manufacture of pink and red glasses, heat resistant red pigments in plastics, enamels, paints, inks (cadmium sulphoselenide). Organo-selenium compounds are important synthetic reagents.

Selenium illustrates more than any other element the dichotomy between essentiality and toxicity. Thus, the public image of selenium has, over the past 60 years, passed through a number of phases. Selenium, as a component of *glutathione peroxidase* (GPx), is known to reduce lipid peroxides. Thus, *glutathione peroxidase* (in the form of selenocysteine) with *catalase* and *superoxide dismutase*, is part of the cellular anti-oxidant defence system against free radicles. However, GPx is not the only selenoprotein in tissues.

The importance of selenium is that it affects all components of the immune system and is necessary for its optimum performance. In general, selenium deficiency results in immune suppression, while low dosage supplementation results in augmentation and/or restoration of immunologic functions.

The Biochemical Role of Selenium[16-19]

The biochemical role of selenium in mammals has been clearly established as part of the active site of the peroxide-destroying enzyme *glutathione peroxidase*. This enzyme, present in the cytosol and mitochondrial matrix of cells as well as in body fluids, is formed by four identical sub-units, each containing one atom of selenium as selenocysteine.

The metabolic functions of the selenoenzyme are vital for cells, as it is part of the mechanism responsible for the metabolism and detoxification of oxygen. Reduction of oxygen, in the form of active species such as superoxide, hydroxyl free radicals or organic peroxides, is brought about by various metalloenzymes - Cu, Zn or Mn-containing *superoxide dismutases*, iron-containing *catalase* and *glutathione peroxidase*. These have specific functions and different sub-cellular localisations: they act in a complementary manner, but cannot substitute for one another.

Homeostasis is maintained by the activity of the metalloenzymes, as well as of scavengers of free radicals, e.g., Vitamins E and C and reduced glutathione. Thus, selenium deficiency, for example, will lead to a loss of GPx activity with biochemical consequences which, if not compensated by other defence mechanisms, will result in clinically-identifiable diseases - peroxidative, oxidative or 'free radical' disease. In addition to this essential activity, other important biochemical roles for selenium are known. Selenium deficiency exacerbates the toxicity of drugs, insecticides or halogenated hydrocarbons that act by the production of toxic oxygen derivatives in the organism. Similarly, selenium supplementation alleviates the toxic effects of drugs and antibiotics as well as many chemical carcinogens.

Furthermore, the element has been shown to be very effective in the prevention of toxic manifestations from Cd, Hg, Pb, As, *cis*-Pt derivatives. One mechanism suggested is the formation of biologically-inactive selenides (Cd, Pb, Pt, Ag, Hg), which accumulate as granules in some organs. Selenium may also divert toxic metals away from binding with some vital proteins to binding with less important ones due to the formation of active selenotrisulphide centres (Hg, Cd), or it may act by metabolic interference (As). Such interactions can occur with selenium at physiological levels and are, therefore, significant as defence mechanisms against some heavy metals.

Other roles for selenium have been identified recently in some metabolic pathways or physiological functions. For example, a selenoproptein of 10,000 daltons has been characterised in human muscle and is thought to be connected with muscular dystrophy occurring in selenium-deficient subjects.

Keratinoid selenoproteins have also been identified in human spermatozoa. Their responsibility in maintaining the integrity of flagella has been proved in cases of infertility due to selenium deficiency.

Marginal selenium deficiency states are more widespread than previously thought, although pathological implications sometimes remain unclear. Thus, hemolytic anaemia and immune dysfunction have been found to be manifestations of selenium deficiency in protein-calorie malnutrition occurring in African countries.

Low selenium status has been found in some neurological disorders and in ageing. The element may actively prevent the progressive failure of the immune system and the accumulation of peroxides and free radicals that are thought to be common features of human ageing.

Selenium deficiency

In humans, the link between Se-deficiency and endemic cardiomyopathy resulting from low selenium intake from food, has been established. The condition is responsive to selenium. In a case control study in eastern Finland (a low selenium country) and an area with an exceptionally high mortality from cardiovascular disease, an association between cardiovascular death, myocardial infarction and low levels of selenium was established.

Furthermore, in recent work, a correlation has been found between human dietary selenium in 27 countries with the incidence of cancer at 17 major sites.

A significant negative correlation has been observed between selenium intakes and cancer of the large intestine, rectum, prostate, ovary, lungs and leukaemia.[20,21]

Generally, it appears that selenium manifests itself in lower tumour incidence, reduction of tumour yield and a longer latency period. However, it would appear that the level of selenium required to demonstrate a cancer preventative effect is greatly in excess of nutritional needs of 100-200 µg/d (Table 3). Outside of this optimal level, deficiency or toxicity can occur.

Table 3 : Intake of selenium in relation to health

	Humans	Animals	Plants
Deficient	8 µg/d	0.8 µg/d	20 µg/d
Normal	100-200 µg/d	0.3 -4 µg/d	1000 µg/d
Toxic	500 µg/d	4 µg/d	1000 - 2000 µg/d

Selenium Toxicity

Animals

Syndromes are well known for farm animals in seleniferous regions. (i) *Acute selenosis* from the consumption of highly seleniferous plants, producing symptoms and death in a few hours; (ii) *Chronic selenosis* from moderately toxic amounts of these plants, eaten over a period of time - "blind staggers"; (iii) *Chronic selenosis* where forages containing less, but still excessive levels of Se (5-30 ppm) are eaten over a period of weeks or months - "alkali disease"; which includes hair loss, malformed hooves, lameness, emaciation.

Oral or parenteral administration of selenium compounds is widespread and has established that there is a narrow margin of safety between therapeutic and toxic doses.

Human

Industrial accidents with selenium have revealed many cases of acute and chronic selenosis in humans by consumption or by inhalation. Apart from the Hubei Province episode in China, there was an unusual case reported in USA.[22] Twelve individuals showed symptoms of selenosis due to the ingestion of overly-potent selenium tablets. The tablets ingested were supposed to contain 0.273 mg Se; however, analysis later showed a content of 27.3 mg Se per tablet. The individuals were found to have ingested between 27 and 2310 mg Se. The most common symptoms were nausea, vomiting, nail changes, irritability. About half experienced hair loss, loss of nails, diarrhoea, abdominal cramps, garlic breath odour. All recovered in a few days. Exhaustive tests showed no changes or abnormalities in blood chemistry. Liver and kidney functions were also normal.[23,24]

It should be noted that the recommended level of selenium intake by the Food and Nutrition Board, USA, is 50-200 µg/day for adults.[25]

Selenium Status

Dietary intake

Estimates of dietary intake of selenium worldwide covers more than 3 orders of magnitude. Most extreme values come from areas of human Se deficiency, e.g., Keshan Region, China (7 µg/d) and at the other extreme, the chronic selenosis areas, e.g., Hubei Province, China (4900 µg/d) (see Table 1). Less extreme values of low dietary selenium intakes have been reported, e.g., 25-30 µg/d in Finland and New Zealand (both low selenium countries). USA, Japan, Venezuela (50-200 µg/d) and Germany (38-47 µg/d). Dietary intake in a country often varies from region to region. The selenium status of some countries has only recently been investigated. A Hungarian study (1993) reported that rock-soil systems were low in selenium, as were blood serum levels.

New data (1993) on the selenium content of soils, cereal crops, human blood serum and hair from different regions in former Yugoslavia indicates a very low Se-status of the population. It is claimed that the Se content of grains, blood serum, etc., approach those of the low selenium regions of China.[26]

Blood, Tissue

Selenium levels in blood and blood components are widely used as indicators of selenium status in humans since these are sensitive to dietary intake and complement studies on dietary intake. Table 4 lists Se level in human blood serum of several European communities; the average level is estimated at 78 - 83 µg/L.

Table 4 : Selenium status (blood serum)

Average 10 European countries	78 - 83 µg/L
Greece	63 ± 14 µg/L
England	109 ± 14 µg/L
Germany	
- men	98 ± 19 µg/L
- women	89 ± 17 µg/L
Italy	87 - 93 µg/L
Yugoslavia (former)	69 ± 18 µg/L
Croatia	64 ± 11 µg/L

Recently, however, a case has been made for toenail clippings which, it claimed, reflects the dietary intake of selenium better than intakes calculated from dietary data. Nevertheless, a significant correlation exists between dietary selenium intake and whole blood selenium.[27] In some countries, where dietary intake is low, selenium supplementation is used. For example, in Finland, commercial fertilizers have been supplemented with sodium selenate, since 1st July 1984: for hay and fodder (6 mg/kg Se); for cereals (16 mg/kg Se).[28] The effect of supplementation was first observed in dairy products in June 1985 and in wheat, flour, beef, bovine liver from the beginning of August 1985. A study involving 108 students was systematically documented from November 1985, when blood serum selenium levels were 1.05 µmol/L . A steady increase was observed until November 1989, when a mean level of 1.6 µmol/L was reached.

In another study, changes in plasma selenium levels in New Zealand were examined over the period 1981-1992. Mean plasma selenium levels ranged between 46-54 µg/L until 1987, after which (1988-1991) there was a dramatic sustained increase to between 66-70 µg/L. The increase closely followed the deregulation of the New Zealand wheat market, after which New Zealanders began to be exposed to higher selenium levels from imported Australian and U.S. wheat.   Prior to 1988, flour was made from the low selenium New Zealand wheat.

Tissue and Organ Distribution of Selenium

Selenium is widely distributed in organs and tissues with the highest concentrations in the liver, kidneys, spleen, pancreas, heart and lungs. Selenium is excreted within two weeks. The chemical form of selenium is important. Retention of inorganic selenium is limited, whereas organically-bound selenium has a longer lasting effect.   In this connection, it has been claimed that **selenomethionine** is more effective in raising the blood selenium concentration than selenate.

The selenium content of 10 different organs from German traffic accident victims showed that kidneys had the highest content (771 ± 169 ng/g, wet weight). This was more than twice the levels found in the liver (291 ± 78 ng/g); skeletal muscle (111 ± 17 ng/g), brain (110 ± 21 ng/g) and other organs such as lungs, spleen and prostate.[29]

Essentiality and Toxicity

There is a strong relationship between essentiality and abundance of an element in the earth's crust and in sea water (although there are some exceptions) since, in the process of natural selection, the rarer elements were not so freely available to the dependent organism. Thus, the most toxic elements are much rarer than the essential elements. Life evolved using these elements, which were both abundant and accessible, then became dependent on them.

It is also true that some relatively common elements are not used by living organisms, because similar roles are played by even more common elements. For example, any role that might be played by Br (10th most abundant element) has already been taken by Cl (10^3 times more abundant).

Each essential or non-essential element has a spectrum of biological action (behaviour). Fig. 1 shows schematically the biological response dependence on the organ concentration of an **essential** element and a **toxic** element. There are two important aspects to this: (i) almost any element in excess ultimately becomes harmful; (ii) even with a toxic element, the organism can cope with some small amount (threshold) before toxic effects become apparent.

Animals and humans maintain a concentration, which is within the optimal concentration range by a complex set of physiological reactions known as **homeostasis**. Thus, the concentrations of all essential elements are under homeostatic control.

As our civilization developed, the concentration of heavy metals in the environment steadily increased and organisms had to adapt to their presence.

Thus, evolution and adaptation permitted elements to traverse the following pathway:

toxic elements → tolerable impurities → useful elements → essential elements

On this basis, the seven elements, Si, V, Cr, Se, Br, Sn and F, are now considered to be essential for health. Thus, vanadium, formerly thought to be non-essential, has been found in *Ascidians* (sea squirts); the element is concentrated from seawater and is involved in a respiratory pigment.

There are, at the present time, still some elements thought to be important, if not yet essential, but specific functions are, as yet, unknown, e.g, arsenic.

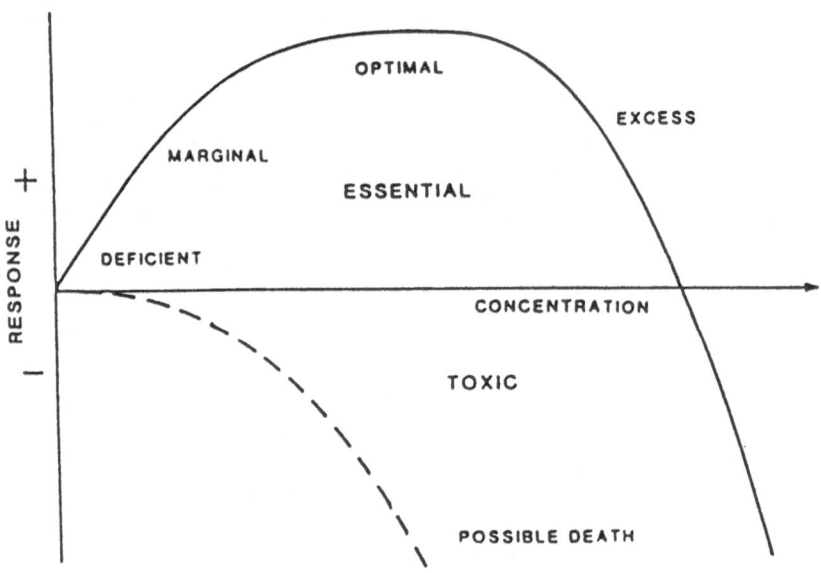

Fig. 1 : **Biological response dependence on tissue concentration of an essential nutrient (solid curve) and of a deleterious substance (dashed curve). The relative position of the two curves on the concentration axis is arbitrary and one of convenience.**

References

1. Marco Polo, in *The Travels of Macro Polo*, p.81, Chapter 43, Livenight, New York, 1926.
2. R.J. Magee and B.D. James, "Selenium", in *Handbook on Metals in Clinical Chemistry*. H.G. Seiler, A. Sigel and H. Sigel (eds), Marcel Dekker, New York, 1993.
3. R.A. Zingaro and W.C. Cooper (eds), *Selenium*, Van Nostrand-Reinhold Co., New York, 1974.
4. N.N. Greenwood and A. Earnshaw, *Chemistry of the Elements*, Pergamon Press, Oxford, Chapter 16, 1984.
5. J. Alexander, J. Hogberg, Y. Thomassen and J. Aaseth, "Selenium", in *Handbook of Toxicity of Inorganic Compounds*. H.G. Seiler, H. Sigel (eds), Marcel Dekker, New York, Chapter 53, 1988.
6. T.-S. Koh, *J. Assoc. Off. Anal. Chem.*, **70** (1987) 664.
7. M. Verlinden, H. Deelstra and E. Adriaenssens, *Talanta* **28** (1981) 637.
8. E.M. Bem, *Environ. Health Perspect*, **37** (1981) 183.
9. G. Kolbl, K. Kalcher, K.J. Irgolic and R.J. Magee, *Chromatographia* (1993), in press.
10. Z. Lu, M. Mao and J. Wei, *Environ. Inorg. Chem.* (1983), 373.
11. Li Chiang, B.D. James and R.J. Magee, *Sci. Int.* (Lahore) , **4** (1992) 25.
12. G. Lockitch, *Crit. Rev. Clin. Lab. Sci.*, **27** (1989) 483.
13. K.W. Franke, *J. Nutr.*, **8** (1984) 597.
14. L. Fishbein, *Toxicol. and Environ. Chem.*, **12** (1986) 1.
15. A.C. Mason, *Food Sci. Tec.*, **28** (1988) 325.
16. J. Neire, *Experentia*, **47** (1991) 187.
17. C.K. Chow and A.L. Tappel, *J. Nutr.*, **104** (1987) 441.
18. L.J. Machlin and A. Bendich, *FASEB J.*, **1** (1987) 441.
19. L. Kiremidjian-Schumacher and G. Stotzky, *Environ. Res.*, **42** (1987) 227.
20. H.J. Cohen, M. Brown, J. Lyons, N. Avissar, D. Hamilton and P. Liegey, *Essential and Toxic Trace Elements in Human Health and Disease*, pp. 201-210, Alar R. Less, New York, 1983.
21. P.J. Collipp and S.Y. Chan, *New Engl. J. Med.*, **304** (1981) 1210.
22. L. Fishbein, *Adv. Mod. Toxicol.*, **2** (1977) 191.
23. G.Q. Yang, S. Wang, R. Zhou and S. Sun, *Am. J. Clin. Nutri.*, **37** (1983) 872.
24. G.Q. Yang, S.W. Wang, R. Zhou and S. Sun, *Acta Acad. Med. Sinica* (Suppl. 2), **3** (1981) 1.
25. Recommeded Dietary Allowances, 9th Rev. Edn, National Academy of Sciences, Washington, DC, 1980, p.162ff.
26. O.A. Lavender, *Ann. Rev. Nutr.*, **7** (1987) 227.
27. G.N. Schrauzer and D.A. White, *Bioinorg. Chem.*, **8** (1978) 303.
28. J.H. Watkinson, *Am. J. Clin. Nutri.*, **34** (1981) 936.
29. O. Oster, G. Schmiedel and W. Prellwitz, *Biol. Trace Element Res.*, **15** (1988) 23.

Main Group Elements and Their Compounds
V.G. Kumar Das (Ed)
Copyright © 1996 Narosa Publishing House, New Delhi, India

Tin in Relation to Toxicity : Myths and Facts

P.J. Smith[a] and V.G. Kumar Das[b]

[a]*International Tin Research Institute, Kingston Lane, Uxbridge UB8 3PJ,
Middlesex, UK,. and [b]Department of Chemistry, University of Malaya,
59100 Kuala Lumpur, Malaysia*

World consumption of tin chemicals has increased steadily since 1950. Organotin compounds, although known one hundred years prior to this date, were not introduced extensively into industry until the mid- to late-1950's. By 1980, their annual production had risen to 30-35,000 tonnes and, in 1992, it stood at about 50,000 tonnes.[1] Inorganic tin compounds, which have been used for many hundreds of years in coloured pigments and glazes (primarily in pottery and in artists' paints), have experienced a similar steady growth worldwide over the last 40 years, with some 16,000 tonnes currently being utilized in industry.[2] The estimated consumption of tin metal for chemicals in 1990 was 24,000 tonnes (Table 1), representing 12% of the total tin consumption.[3] The reasons for this growth in tin chemicals are the diversity of their applications coupled, in many cases, with their generally low toxicity.

Table 1 : **Western world tin chemicals consumption (1990) by end use (tonnes of Sn)[3]**

PVC Stabilizers	10,000
Ceramic Industry	2,500
Biocides	3,500
Glass Industry	2,500
Plating Chemicals	2,000
Catalysts	1,500
Other	2,000
Total :	24,000

In recent years, certain organotin compounds and, to a lesser extent, tin and its inorganic derivatives, have been subjected to some degree of unfair criticism, *e.g.* unjustified comparisons have been made between tin and the toxic heavy metals, cadmium, lead and mercury. It is the purpose of this article, therefore, to place tin and its compounds in the correct perspective and to correct any myths which have arisen with regard to their toxicity.

Tin and its Inorganic derivatives

Tin is a non-toxic metal [4] and is reported [5] to be beneficial, even if not yet proved to be an essential element. Inhalation of elemental tin does not produce any effects in man, whereas extended exposure to stannic oxide dust and fumes can produce a benign pneumoconiosis.[6] The generally accepted maximum allowable concentration of tin in the air of work-rooms is significantly higher than the corresponding values for the toxic heavy metals, lead, cadmium and mercury [7] (Table 2).

Table 2 : ACGIH threshold limit values (TLV) (1993/1994)[7]

	TLV-TWA (mg metal/ m^3 air)
Pb (inorganic dusts and fumes)	0.15
Cd (metal and compounds)	0.01
Hg (inorganic compounds)	0.10
Sn (metal and inorganic compounds)	2.00

Tin is not included in the tables of inorganic chemicals of health significance in drinking water in the new WHO Guidelines for Drinking Water Quality[8] (Table 3), or in the list of toxic substances whose concentrations are restricted in UK mineral waters [9] (Table 4), or in the list of inorganic primary contaminants in the US EPA National Primary Drinking Water Standards[10] (Table 5), whereas cadmium, mercury and lead appear in all of these. In the UK, there is a legal general maximum limit of 200 mg/kg for tin in food [11] and a much lower value of 1 mg/kg for lead[12].

Table 3 : WHO guideline values (GV) for inorganic chemicals of health significance in drinking water (April, 1993)

Contaminant	GV (mg L^{-1})
Sb	0.005 [a]
Ba	0.70
Cd	0.003
Cr	0.05[a]
Cu	2.00[a]
Pb	0.02
Mn	0.50[a]
Hg	0.001
Mo	0.07
Ni	0.02

[a]Provisional value

399

Table 4 :   Maximum limits (ML) of toxic substances allowed in UK natural mineral waters[9]

Contaminant	ML (mg L^{-1})
Cd	0.005
Cr	0.05
Hg	0.001
Ni	0.05
Sb	0.01
Pb	0.01

Table 5 :   US EPA national primary drinking water standards (Jan., 1987)

Contaminant	MCL^a (mg L^{-1})
Ba	2.0
Cd	0.005
Cr	0.1
Pb	0.015
Hg	0.002
Ag	0.05

^{a}MCL = maximum concentration limit

Tin, in the form of an alloy, may have a role to play in fuel economics. Over the past five years, tin alloy-based fuel saving devices which, in the form of porous metal pellets, are inserted into the fuel line or, less commonly, into the fuel tank, of either spark ignition or diesel engines, have become commercially available (Fig. 1). Typical fuel savings reported[13-15] for motor cars and heavy vehicles are in the 5-10% range and some manufacturers are offering products which are claimed to be suitable for marine engines (*vide infra*).

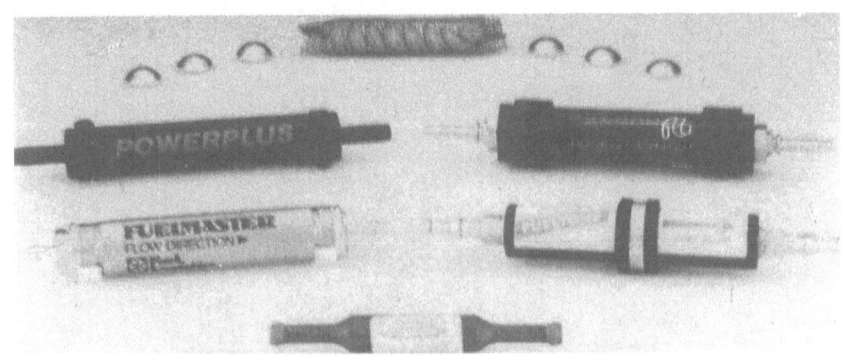

Fig. 1 : Tin alloy-based fuel saving devices

The acute oral toxicities of a series of inorganic tin(II) and tin(IV) compounds in rats are summarised in Table 6[16-20] and it can be seen that these are generally very low (>700 mg/kg).

The lower LD_{50} values observed for the two tin(II) fluoride salts reflect the higher toxicity of the anion, *c.f.* NaCl and NaF in Table 6. A two-year feeding study with stannous chloride in rats and mice has shown that, at 2,000 ppm, this compound was not carcinogenic in the test animals.[21] Similarly tin(IV) chloride and tin(II) chloride display a low biological activity towards marine and freshwater algae, when compared to the highly active compound tributyltin chloride [22,23] (Table 7).

Table 6 : Acute oral toxicities of inorganic tin(IV) and tin(II) compounds (rats)

Compound	LD_{50} (mg/kg)	Ref.
$ZnSn(OH)_6$	> 5,000	16
$ZnSnO_3$	> 5,000	16
$SnOCH_2CH_2O$	>10,000	17
$Sn(O.CO.CHEtBu)_2$	5,870	17
$Sn(O.CO.CO.O)$	3,400	17
SnO	>10,000	17
$SnCl_2$	700	18
SnF_2	188.2	19
$NaSn_2F_5$	221	19
[NaF]	[22]	20
[NaCl]	[3,000]	20

Table 7 : Toxicity of $SnCl_4$ and $SnCl_2$ to algae[22,23]

Algae	IC_{50} (mg L^{-1})		
	$SnCl_4$	$SnCl_2$	Bu_3SnCl
A. falcatus	12	14	0.02
S. quadricauda	>50	>50	0.016
A. flos-aquae	>5	>5	0.013
Lake Ontario algae	>50	>50	0.003
S. costatum	-	0.325*	0.00036*
T. pseudonana	-	0.316*	-

*EC_{50} values

401

The low toxicity of inorganic tin salts has given rise to a number of pharmaceutical applications, *e.g.* in toothpastes $(Sn_2P_2O_7)$[24], in dentifrices (SnF_2)[5] and in radiopharmaceutical technetium-99 *m* scanning agents (SnF_2)[5]. Potential new drug uses being developed include the treatment of hyperbilirubinaemia or neonatal jaundice {dichloro(protoporphyrin IX)tin(IV)}[25] and hypertension $(SnCl_2)$[26] and the photodynamic therapy of cancer {various tin(IV) porphyrin derivatives}.[25]

Methylation of inorganic tin and methyltin species in aquatic environments appears to be a slow process [6,27] and Aldridge has suggested[4] that it is chemically unlikely that tin could be biologically methylated significantly to give toxicologically significant amounts of trimethyltin derivatives. In this context, it is important to note that, while trimethyltin compounds are highly toxic to mammal[4], their biological activities against a range of aquatic organisms appear to be of a generally low order of magnitude, as are those of the other methyltins, Me_nSnX_{4-n}, where n = 1-4 (Table 8).[22,23,28-31]

Table 8 : Aquatic toxicity of methyltin compounds[22,23,28-31]

Compound	Concentration (mg L^{-1})				
	Red killifish[a] (O. latipes)	Mud crab larva[b] (R. harrisii)	Water flea[c] (D. magna)	Diatom[d] (S. costatum)	Alga[e] (A. falcatus)
Me_4Sn	6.4	-	40	> 0.5	-
$MeSnCl_3$	-	-	90	0.078	23
Me_2SnCl_2	6.0	20	88	> 0.5	21
Me_3SnCl	5.6	0.11(0.37)$^{1-f}$	0.47	0.214	5.5
Bu_3SnCl	0.05	0.02	0.013	0.00036	0.02

[a]48 h LC$_{50}$, [b]LC$_{50}$, [c]24 h EC$_{50}$, [d]EC$_{50}$, [e]IC$_{50}$, [f]Fiddler crab larva (*U. pugilator*)

Organotin Compounds

Unlike their inorganic tin counterparts, the organotin compounds display a wide range of biological properties, with the most active compounds being the triorganotins, R_3SnX. These derivatives find applications primarily as fungicides in wood preservation (R = Bu), and as agrochemical fungicides (R = Ph) and miticides (R = Cy or $PhMe_2CCH_2$). The tributyl- and triphenyl-tin compounds have been in use for thirty years in many countries and there have been very few undesirable effects reported [4,32,33] in humans. Only one case of a possible fatality through occupational exposure to these two classes of triorganotin compounds - involving a chemical slurry of triphenyltin chloride, diphenyltin dichloride, hexane and other unidentified compounds - has been observed,[33] although the agent responsible could not be determined from the available data.

In Table 9, the five commercial triorganotin agrochemicals are compared to the world's ten largest (in sales [34]) agrochemical products in terms of their acute oral LD_{50} values in the rat.[35,36] It may be seen that the acute oral LD_{50} range for the triorganotin agrochemicals (110-2631 mg/kg) is very similar to that of the ten leading non-tin pesticides, with the exception of trifluralin (>10,000 mg/kg). No carcinogenic effects have been found in animal studies involving the triorganotin agrochemicals, nor in any of the other tin chemicals which have been tested to date (Table 10).[21,36-43]

Table 9 : Comparison of major agrochemical products with their triorganotin counterparts

Agrochemical (common name)	Sales[34] ($ M)	LD_{50} oral, rat[35,36] (mg/kg)
Glyphosate (H)	1,000	5,600
Chlorpyriphos (I)	300 - 500	135 - 163
Paraquat (H)	300 - 500	150
Atrazine (H)	300 - 500	1,869 - 3,080
Dimethoate (I)	300 - 500	291 - 325
2,4-D (H)	250 - 300	375
Metolachlor (H)	250 - 300	2,780
Bentazone (H)	250 - 300	1,710
Trifluralin (H)	250 - 300	> 10,000
Alachlor (H)	250 - 300	930 - 1,350
Fenbutatin oxide (M)	-	2,630
Azocyclotin (M)	-	631
Cyhexatin (M)	-	540
Fentin hydroxide (F)	-	110 - 171
Fentin acetate (F)	-	140 - 298

H = herbicide, I = insecticide, F = fungicide, M = miticide

In 1987, Dow Chemicals voluntarily cancelled their US tricyclohexyltin hydroxide (Plictran) registration after studies with this miticide in rabbits showed a positive teratology response.[44] This investigation was refuted in later work by other companies [45] and, although tricyclohexyltin hydroxide is not approved in the UK[46], it is currently registered for use in many European[47] and South American countries. Elf Atochem North America has maintained import tolerances - pesticide residue levels allowable on foods imported in the US - with the US EPA.[45]

Table 10 : Carcinogenicity studies on tin chemicals

Compound	Species	Type of test	Results
SnCl$_2$	rats & mice (♂ & ♀)	105 wk feeding	negative[21]
Bu$_2$Sn(O.CO.Me)$_2$	mice (♂ & ♀) rats(♂)	78 wk feeding	no conclusive evidence for carcinogenicity[37]
(Bu$_3$Sn)$_2$O	rats (♂ & ♀)	106 wk feeding	no compound related tumours found[38]
Bu$_3$SnF	mice (♂)	6 mth dermal	negative[39]
Ph$_3$SnOH	rats & mice (♂ & ♀)	78 wk feeding	negative[40]
Ph$_3$SnOCOMe	mice (♂ & ♀)	78 wk feeding	negative[41]
Cy$_3$SnOH	rats (♂ & ♀)	2 yr feeding	negative[42]
1-Cy$_3$Sn(1,2,4-triazole)	rats	2 yr feeding	no positive effects[36]
{(PhMe$_2$CCH$_2$)$_3$Sn}$_2$O	mice (♂ & ♀)	18 mth feeding	negative[43]

The long history of safe use for the triorganotin pesticides has ensured continuing developments in this area. These involve the evolution of novel, active compounds, such as the sterically hindered triorganotin miticide, (Me$_3$SiCH$_2$)(PhMe$_2$CCH$_2$)$_2$SnCl,[48] and the extension of existing agrochemicals, such as the fungicides, triphenyltin acetate and triphenyltin hydroxide, to new areas, namely the control of fungal diseases of yams[49,50] and rice[51]. Several new-generation triorganotin biocides with superior fungicidal and/or larvicidal properties compared to triphenyltin acetate or hydroxide as well as manifesting improved environmental properties have also been recently reported.[52,53]

Finally, it is of interest to note in the agrochemical field that, in the UK, cadmium compounds and certain mercury derivatives are not allowed as pesticides,[46] and that a common feature of the triorganotin agrochemicals is their ability to break down rapidly, under the influence of sunlight and microorganisms, to non-toxic inorganic tin(IV) species.[35]

Tributyltins in marine antifouling paints

Although the use of triorganotin compounds in antifouling paints for small boats (typically <25m length) has been prohibited or phased out in most countries, the tributyltin methacrylate/methyl methacrylate co-polymer paint formulations are widely employed on large ocean-going vessels as antifouling self-polishing coatings. In November, 1990, the International Maritime Organisation (IMO)[54] recommended that the average release rates of these paints do not exceed 4 μg organotin/cm^2/day.

Recent studies of the half-lives of tributyltin species in aquatic environments summarised in the more recent literature indicate that these are generally between a few days and a few weeks whilst, in sediments, the breakdown rates appear to be significantly slower.[55] Dealkylation of the triorganotin moiety in water is accompanied by a progressive reduction in toxicity towards aquatic organisms, *e.g.* Table 11.[56,57] It is interesting to note in this context that the trialkyltin derivative,

404

trimethyltin chloride, shows a higher No-Observable-Effect Concentration (NOEC) than the other triorganotins.

It has been estimated that the current economic benefits to the marine industry, including fuel savings, from the use of tributyltin (TBT) co-polymer paints, amount to about \$ 3,000 M annually and that this drop in world fleet fuel consumption gives rise to a significant reduction in CO_2 and SO_2 emissions into the environment (Table 12).[54]

Table 11 : No-observable-effect concentrations for di- and tri-organotins (fish)

Compound	Species	NOEC (mg L^{-1})	Ref.
Bu_3SnCl	rainbow trout (*O. mykiss*)	0.039×10^{-3}	56
$(Bu_3Sn)_2O$	Japanese medaka (*O. latipes*)	0.32×10^{-3}	57
Ph_3SnCl	rainbow trout (*O. mykiss*)	0.046×10^{-3}	56
Cy_3SnCl	rainbow trout (*O. mykiss*)	$< 1.21 \times 10^{-3}$	56
Me_3SnCl	rainbow trout (*O. mykiss*)	$> 14.6 \times 10^{-3}$	56
Ph_2SnCl_2	rainbow trout (*O. mykiss*)	55.0×10^{-3}	56
Bu_2SnCl_2	rainbow trout (*O. mykiss*)	48.6×10^{-3}	56
Bu_2SnCl_2	Japanese medaka (*O. latipes*)	$< 320 \times 10^{-3}$	57

Table 12 : Benefitis of using TBT copolymer antifouling paints[54]

A. Economic benefits	
	\$ M
Fuel savings	500
Extended drydocking	400
Reduced maintenance	800
Indirect savings	1,000
Total :	2,700

B. Environmental benefits

Reduction of world fleet fuel consumption by 7 metric tonnes per annum (4%), resulting in 22 metric tonnes of CO_2 and 0.6 metric tonnes of SO_2 not being emitted into the environment.

Tributyltin fungicides for wood preservation

Tributyltin compounds have been used as fungicides in wood preservative formulations - mostly in Europe and America - for over 25 years and, during this time, there have been no reports [58,59] of cases of acute systemic poisoning or of long-term adverse toxic effects in workers. In the UK, for example, bis(tributyltin) oxide may be employed in industrial use products for application by vacuum/pressure impregnation or enclosed immersion, the maximum ready-for-use strength being 1.5% w/w, and tributyltin naphthenate in industrial use products at 2.0% w/w. Tributyltin fungicides are not allowed in home (do-it-yourself) or remedial treatment formulations in the UK. The legislative situation in a number of other countries has been summarised recently by Schweinfurth and Ventur.[59]

Increasingly, tributyltin fungicides are finding use for the protection of tropical timbers, such as Rubberwood (*Hevea brasiliensis*), from fungal attack[60-62] and, presently, applications for approval of tributyltin naphthenate in a number of mainly ASEAN countries are being evaluated.[59]

Recent studies in Malaysia[60,62] and in Indonesia[63] have indicated that certain tributyltin carboxylates are active against blue stain fungi in Rubberwood and in Ramin wood at levels of 0.3% w/w. Since the use of sodium pentachlorophenoxide (NaPCP) as an anti-blue stain chemical is being phased out on environmental and toxicological grounds, it is an interesting possibility that the tributyltin fungicides may find use as alternative treatments in this field.

Dialkyltin compounds

The major use of tin chemicals is as heat and light stabilizers for PVC (Table 1). Di- and mono-/di- mixtures of methyl-, and octyl-tin compounds, primarily as *iso*-octyl-thioglycollate dervatives, have been approved in many countries for 20-30 years as non-toxic additives for PVC food packaging and drink bottles; certain dimethyltin stabiliser formulations are allowed by the US National Sanitation Foundation (NSF) for potable (drinking) water PVC pipe. The US maximum permissible levels for metals extracted from PVC potable water piping are shown in Table 13 [64] and it may be seen that the value for Pb (0.02 mg/l) exceeds the new WHO drinking water Guideline Value in Table 3 (0.01 mg/l; formerly 0.05 mg/l) for this metal. Since many countries outside the US are still using lead stabilizers in PVC drinking water pipe, it may be expected that this new drinking water Guideline Value will facilitate a move by the stabilizer manufacturers to organotin and other non-toxic metal stabilizer additives for this application. This is already taking place in Japan.

Table 13 : US maximum permissible levels (MPL) for metals extracted from PVC potable water piping[64]

Metal	MPL (mg L^{-1})
Sn	0.05
Pb	0.02
Cd	0.005
Hg	0.002

The acute oral toxicities of the common organotin stabilizers are listed in Table 14[65] and it may been seen that these are generally quite low. For the series of *iso*-octyl-thioglycollates, the LD_{50} values for the trialkyltin by-products [65] have been included for comparison and it is apparent that the trimethyltin derivative is highly toxic and, therefore, stringent precautions are taken by the stabilizer manufacturers to prevent any significant formation of this compound in the product.

Table 14 : Acute oral toxicity of organotin PVC stabilizers[65]

Compound	LD_{50} rats (mg/kg)
$MeSn(SCH_2CO_2Oct-i)_3$	920 - 1700
$Me_2Sn(SCH_2CO_2Oct-i)_2$	620 - 1380
$BuSn(SCH_2CO_2Oct-i)_3$	1063
$Bu_2Sn(SCH_2CO_2Oct)_2$	510 - 1037
$OctSn(SCH_2CO_2Oct-i)_3$	3400 - >4000
$Oct_2Sn(SCH_2CO_2Oct-i)_2$	1200 - 2100
$Oct_2Sn(OCOCH{:}CHCO_2)$	4500
$Me_3SnSCH_2CO_2Oct-i$	20.4
$Bu_3SnSCH_2CO_2Oct-i$	1350
$Oct_3SnSCH_2CO_2Oct-i$	26550

Tin chemicals tested to date have not shown any carcinogenic properties. In fact, over the last 10-15 years, certain diorganotin complexes have been subject to intensive research efforts in the field of anti-cancer drugs, although no compound has yet shown a sufficiently high activity to justify its inclusion in a human clinical trial. The anti-tumour activity is primarily governed by the alkyl groups of the diorganotin moiety, rather than by the coordinated anionic ligands, the function of the latter being to aid transport of the active R_2Sn species to the site of action in the cell, where it is then released by hydrolysis[66,67]. Some typical anti-tumour activity results are summarised in Table 15[66] (see also accompanying article by M. Gielen in this book).

Table 15 : Anti-tumour activities of diorganotin compounds against P388 leukaemia in mice[66]

Compound	Dose (mg/kg)	T/C (%)
$Et_2SnBr_2{\cdot}phen$	25	176
$Et_2SnCl_2{\cdot}phen$	25	177
$Et_2SnI_2{\cdot}phen$	200	184
Ph_2SnF_2	-	196
$Ph_2SnCl_2{\cdot}2py$	-	180
Bu_2SnCl_2	3.0	186[a]

[a]Renal carcinoma i.p.

407

Conclusions

1. Tin and its inorganic compounds are essentially non-toxic and a number of these find uses in the pharmaceutical field.

2. The tin chemicals studied to date are not carcinogenic and, in fact, certain of these show anti-tumour properties.

3. Organotin compounds display a wide range of toxicities, depending on the nature and number of the organic groups. The triorganotin biocides have a long history of safe use in industry and cases of fatalities are rare.

4. Many dialkyltin derivatives are non-toxic and are used as PVC stabilizers in food packaging, drink bottles and in potable water piping.

Acknowledgements

The authors are grateful to Ms S.E. Ziegler, American Conference of Governmental Industrial Hygienists, Inc., Dr H. Galal-Gorchev, World Health Organisation, and Mr M. Phillips, Wood Mackenzie Ltd., for permission to include their data; to Dr D.B. Russell, Elf Atochem North America, Inc., Dr A. Suzuki, Hokko Chemical Industry Co. Ltd. and Mr R.F. Bennett of Canterbury, England, for valuable comments on the manuscript; to ITRI, Uxbridge, for permission to publish this paper; and to the Royal Society of Chemistry, London for financial support.

References

1. P.J. Smith, *Appl. Organomet. Chem.*, **6** (1992) 1.
2. Anon, *ITRI Ann. Rept.*, 1992, p.12 (ITRI Publicn. No.731).
3. P.K. Frame, *Proc. 5th Internat. Tinplate Conf.*, ITRI, London, 1992, p.10 (ITRI Publicn. No.727).
4. W.N. Aldridge, in : *Chemistry and Technology of Silicon and Tin*, eds V.G. Kumar Das, S.W. Ng and M. Gielen, Oxford University Press, Oxford, 1992, p.78.
5. J.M. Tsangaris and D.R. Williams, *Appl..Organomet. Chem.*, **6** (1992) 3.
6. Anon, *Tin and Organotin Compounds : A Preliminary Review*, Environ. Health Criteria 15, WHO, Geneva, 1980.
7. Anon, *1993-1994 Threshold Limit Values for Chemical Substances and Physical Agents and Biological Exposure Indices*, American Conference of Governmental Industrial Hygienists (ACGIH), Cincinnati, 1993.
8. Anon, *Guidelines for Drinking-Water Quality*, Draft Manuscript, WHO, Geneva, April, 1993.
9. Anon, *The Natural Mineral Waters Regulations 1985*, SI No.71, HMSO, London, 1985.

10. Anon, *National Primary Drinking Water Standards*, US EPA, Washington DC, January, 1987.

11. Anon, *The Tin in Food Regulations 1992*, SI No.496, HMSO, London, 1992.

12. Anon, *The Lead in Food Regulations 1979*, SI No.1254, HMSO, London, 1979.

13. I. Kuah, *Motor. Leisure*, Sept., 1989, p.11.

14. Anon, *Asiaweek*, 7 Sept., 1990, p.74.

15. Wribro Ltd., *Eur. Pat. Applicn.*, 0 399 801 A1, 28.11.90.

16. P.A. Cusack and S. Karpel, *Tin and its Uses*, **165** (1991) 1.

17. M.H. Gitlitz and M.K. Moran, *Kirk-Othmer Encyclopaedia of Chemical Technology*, **23** (1983) 42.

18. H.O. Calvery, *Food Res.*, **7** (1942) 313.

19. J.K. Lim, G.J. Renaldo and P. Chapman, *Caries Res.*, **12** (1978) 177.

20. Anon, *Registry of Toxic Effects of Chemical Substances*, 1985-1986 Edn., US Nat. Inst. Occup. Saf. Health, Washington DC, April, 1987.

21. Anon, *Nat. Tox. Prog. Tech. Rept.* NTP-81-33, US NIH Publicn. No.82-1787, June, 1982.

22. P.T.S. Wong, Y.K. Chau, O. Kramer and G.A. Bengert, *Can. J. Fish. Aquat. Sci.*, **39** (1982) 483.

23. G.E. Walsh, L.L. McLaughlan, E.M. Lores, M.K. Louie and C.H. Deans, *Chemosphere*, **14** (1985) 383.

24. Unilever N.V., *Eur. Pat. Applicn.*, 0 514 966 A2, 25.11.92.

25. A.J. Crowe, *Tin and its Uses*, **162** (1990) 4; **163** (1990) 16.

26. D. Sacerdoti, B. Escalante, N.G. Abraham, J.C. McGiff, R.D. Levere and M.L. Schwartzman, *Science*, **243** (1989) 388.

27. R.J. Maguire, *Water Poll. Res. J. Canada*, **26** (1991) 243.

28. D.A. Wright and W.H. Roosenburg, *Arch. Environ. Contam. Toxicol.*, **11** (1982) 491.

29. R.B. Laughlin, Jr., R.B. Johannesen, W. French, H. Guard, and F.E. Brinckman, *Environ. Toxicol. Chem.*, **4** (1985) 343.

30. H. Nagase, T. Hamasaki, T. Sato, H. Kito, Y. Yoshioka and Y. Ose, *Appl. Organomet. Chem.*, **5** (1991) 91.

31. M. Vighi and D. Calamari, *Chemosphere*, **14** (1985) 1925.

32. Anon, *Tributyltin Compounds*, Environ. Health Criteria 116, WHO, Geneva, 1990.

33. Anon, *Occupational Exposure to Organotin Compounds*, US Nat. Inst. Occup. Saf. Health, Washington DC, November, 1976.

34. M. Phillips, Wood MacKenzie Agrochemical Service, Edinburgh, *Personal Communication*, 1993.

35. B. Sugavanam, *Tin and its Uses*, **126** (1980) 4.

36. Anon, *The Pesticide Manual*, British Crop Protection Council, Farnham, 1991.

37. Anon, *US Nat. Cancer Inst. Carcinogen. Tech. Rep. Ser.*, 1979, No.183.

38. P.W. Wester, E.I. Krajnc, F.X.R. van Leeuwen, J.G. Loeber, C.A. van der Heijden, H.A.M.G. Vaessen and P.W. Helleman, *Food Chem. Toxicol.*, **28** (1990) 179.

39. A.W. Sheldon, *J. Paint. Technol.*, **47** (1975) 54.

40. Anon, *US Nat. Cancer Inst. Carcinogen. Tech. Rep. Ser.*, 1979, No.139.

41. J.R.M. Innes, B.M. Ulland, M.G. Valerio, L. Petrucelli, L. Fishbein, E.R. Hart, A.J. Palotta, R.R. Bates, H.L. Falk, J.J. Gart, M. Klein, I. Mitchell and J. Peters, *J. Nat. Cancer Inst.*, **42** (1969) 1101.

42. Anon, *1970 Evaluations of some Pesticide Residues in Food*, FAO/WHO, Rome, 1971, p.527.

43. Anon, *1977 Evaluations of some Pesticide Residues in Food*, FAO/WHO, Rome, 1977, p.232.

44. Anon, *Chem. Eng. News*, **65**(39) (1987) 19.

45. D.B. Russell, Elf Atochem North America, *Personal Communication*, 1993.

46. Anon, *Pesticides 1993*, Ref. Book 500, HMSO, London, 1993.

47. Anon, *European Directory of Agrochemical Products*, Roy. Soc. Chem., London, 1990, Vol.3.

48. Nitto Kasei Co., *Eur. Pat. Applicn.*, 0 361 323 A2, 4.4.90.

49. U.S. Attahiru, T.T. Iyaniwura, A.D. Adaudi and J.J. Bonire, *Vet. Hum. Toxicol.*, **33** (1991) 554.

50. P.F. Olurinola, J.O. Ehinmidu and J.J. Bonire, *Appl. Environ. Microbiol.*, **58** (1992) 758.

51. P.P. Pablico and K. Moody, *Crop Protection*, **10** (1991) 45.

52. Nazni W. Ahmad, Sofian-Azirun Mohd., S. Balabaskaran and V.G. Kumar Das, *Appl. Organomet. Chem.*, 7 (1993) 583.

53. Nazni W. Ahmad, Tay Siew Huang, S. Balabaskaran, K.M. Lo and V.G. Kumar Das, *Metal-based Drugs*, 1 (1994) 1.

54. Anon, *TBT-Copolymer Antifouling Paints:the Facts*, ORTEP Association/Marine Painting Forum, July, 1992.

55. C. Stewart and S.J. de Mora, *Environ. Technol.*, 11 (1990) 565.

56. H de Vries, A.H. Penninks, N.J. Snoeij and W. Seinen, *Sci. Total. Environ.*, 103 (1991) 229.

57. P.W. Wester, J.H. Canton, A.A.J. van Iersel, E.I. Krajnc and H.A.M.G. Vessen, *Aquat. Toxicol.*, 16 (1990) 53.

58. F.I. Gilbert, Jr., C.E. Minn, R.C. Duncan and J. Wilkinson, *Arch. Environ. Contam. Toxicol.*, 19 (1990) 603.

59. H. Schweinfurth and D. Ventur, in : *The Chemistry of Wood Preservation*, ed., R. Thompson, Roy. Soc. Chem., London, 1991, p.192.

60. A.J. Kuthubutheen, Y. Salahudin and V.G. Kumar Das, in: *Chemistry and Technology of Silicon and Tin*, eds., V.G. Kumar Das, S.W. Ng and M. Gielen, Oxford University Press, Oxford, 1992, p.289.

61. C.K. Yap, L.T. Hong, R. Hill, K.M. Lo, S.W. Ng and V.G. Kumar Das, *J. Troical for Sci.*, 1 (1990) 390.

62. L.F. Siah, Salamah Selamat and V.G. Kumar Das, in: *Frontiers of Organogermanium, -tin and -lead Chemistry*, eds., E. Lukevics and L. Ignatovich, Latvian Institute of Organic Synthesis, 1993, p. 301.

63. A. Martawijaya, in: *Properties and Uses of Timber Species of Industrial Plantation Forests*, eds., A. Martawijaya, S. Widarmana, S. Karnasudirdja and M.A.S. Bulharman, Forest Products R&D Centre, Bogor, Indonesia, 1 (1992) 12.

64. Anon, *Plastics Piping Components and Related Materials*, ANSI/NSF Internat. Stand., 14-1990, Nat. Sanit. Foundn., Ann Arbor., 1990, p.9.

65. P.J. Smith, *Toxicological Data on Organotin Compounds*, ITRI Publicn. No.538, 1978.

66. A.J. Crowe, in: *Tin-based Antitumour Drugs*, ed., M. Gielen, Springer-Verlag, Berlin, 1990, NATO ASI Series, Vol. H 37, p.69.

67. A.J. Crowe, in: *Metal Complexes in Cancer Chemotherapy*, ed., B.K. Keppler, VCH, Weinheim, 1993, p.370.

Main Group Elements and Their Compounds
V.G. Kumar Das (Ed)

Interaction of Organotin Compounds with Model and Biological Membranes

Enrico Bertoli, Fabio Tanfani, Annarina Ambrosini
and Giovanna Zolese

*Institute of Biochemistry, Medical School, University of Ancona,
Via Ranieri, 60131 Ancona (Italy)*

Organotin compounds are used in a variety of technical applications:[1,2] as polyvinyl chloride (PVC) stabilizers, industrial catalysts, industrial and agricultural biocides, surface disinfectans, wood preserving and marine antifouling agents. In general, the toxicity of organotin compounds decreases from tri-to mono-organotins and it depends on the type of alkyl or aryl ligands. Trimethyltins, triethyltins and tri-*n*-propyltins are most toxic to insects, mammals, and Gram-negative bacteria respectively, whereas tri-n-butyltins show maximal toxicity to Gram-positive bacteria, yeast, fungi and fish;[3]; triphenyltin compounds are particularly toxic to fungi and phytoplankton[4]. Among trialkyltin compounds, the class of tri-*n*-butyltins has found wide use commercially as a biocide; these compounds find applications as surface disinfectants, as slimicides in cooling water and as antifouling agents in a variety of formulations for marine paints. As a consequence, tributyltin compounds and their degradation products (dibutyl- and monobutyltins) are often found at relatively high concentrations in areas of heavy boating or shipping traffic and in harbours.

Because of the biocidal activity and widespread usage of organotin compounds, the concern about their potential enviromental and health effects has increased in the last decade. Tributyltin chloride can change the permeability of erythrocytes and liposomes, and it can pass into the outer mitochondrial membranes where it acts as a chloride/hydroxide exchanger.[5,6] Trialkyltin compounds are also potent inhibitors of oxidative phosphorylation[7] and the major effect is the inhibition of the mitochondrial F_oF_1 ATP synthase complex as a result of an interaction with the F_o component.[8,9] Tributyltin compounds bind to a limited number of tissue proteins.[10] The higher toxicity of trialkyltin compounds with respect to their lower analogs may find a possible explanation in their high hydrophobicity. For instance, tributyltins are soluble in a variety of low or non polar solvents including biological phospholipid membranes. In this context, plasma membrane damage has been proposed to explain the cytotoxic effect of tributyltin compounds in a variety of isolated cells.[5] In this paper, we present data on the effect of organotin conpounds on model biological membranes and on the mitochondrial ATP synthase complex in natural membranes and in the isolated form.

Materials and Methods
Tri-*n*-butyltin chloride (TBTC) and tri-*n*-butyltin acetate (TBTA) were obtained from Aldrich Chemical Company. Hepes, Tricine, NADH, Dipalmitoyl

phosphatidylcholine (DPPC), Dimyristoyl phosphatidylcholine (DMPC) and Dimyristoyl phosphatidylglycerol (DMPG) were purchased from Sigma. 1,6-Diphenyl-1,3,5-hexatriene (DPH) and its charged derivative 1-(4-trimethylaminophenyl)-6-phenyl-1,3,5-hexatriene (TMA-DPH) were obtained from Molecular Probes (Eugene, OR). The fluorescent 5-coordinate organotin-flavone complex of 3-hydroxyflavone (Hof) $Bu_2SnBr(of)$ was a kind gift of Prof. Griffiths (Chemistry Deptartment, Warwick University, U.K.).

Preparation of F_oF_1 ATP synthase

ATP synthase complex (from beef heart mitochondria) was prepared from Mg-ATP submitochondrial particles[11] as originally described by Berden and Voorn-Brouwer.[12]

Preparation of liposomes for DSC and infrared analysis

TBTC and TBTA (in ethanol) were mixed with DPPC (in methanol/chloroform, 1/3, v/v) with DPPC/organotin molar ratios (R) within 10 to 200 range. Multilamellar liposomes were obtained by hydrating the lipid (after solvent removal by nitrogen drying) in a 100 mM Hepes, 100 mM KCl pH 7.4 buffer. A DPPC/buffer of weight ratio 1/3 was used.

Preparation of liposomes for fluorescence measurements

Liposome suspensions with a final lipid concentration corresponding to 0.3 mM Pi were used. DPH or TMA-DPH was added to the phospholipid suspensions up to a final fluorescent probe/lipid Pi molar ratio of 1/800.

Analytical procedures

ATP hydrolitic activity was measured at 37 °C by monitoring at 340 nm the oxidation of NADH in a coupled lactate dehydrogenase-pyruvate kinase regenerating system.[11]

Protein concentration was determined according to a literature methhod.[13]

Analysis of ATP synthase residual phospholipids content was performed after extraction with chloroform/methanol.[14]

DSC measurements were made using a Perkin -Elmer DSC-2 calorimeter. Heating scans were performed at 2.5 K/min

Infrared spectra were recorded on a Perkin-Elmer 1760-x Fourier-tranform infrared spectrometer at 2 cm^{-1} resolution by using a Beer-Norton apodisation function. The detector was deuterated triglycine sulfate (DTGS).

Fluorescence measurements were performed using a Perkin-Elmer MPF-66 spectrofluorometer. The degree of DPH and TMA-DPH polarization (P) was obtained using the OBEY program POLM available from Perkin-Elmer. The excitation and the emission wavelengths were set at 360 and 430 nm, respectively.

For the experiments with the fluorescent organotin compound Bu$_2$SnBr(of), the excitation and the emission wavelengths were set at 395 and 450 nm respectively.

Results and Discussion

Molecular interactions of organotin compounds with artificial membrane lipids

The hydrophobic character of tributyltin compounds prompted the present investigations on the effects of insertion of these molecules into lipid bilayers. Tri-*n*-butyltin acetate (TBTA) and tri-*n*-butyltin chloride were chosen to study the possible role of the anion moiety in perturbing the physico-chemical properties of liposomes. The thermotropic characteristic of DPPC multilamellar liposomes containing TBTA and TBTC were studied by FT-IR, DSC and fluorescence spectroscopy. Liposomes made of DMPC and DMPG were also studied using this last technique.

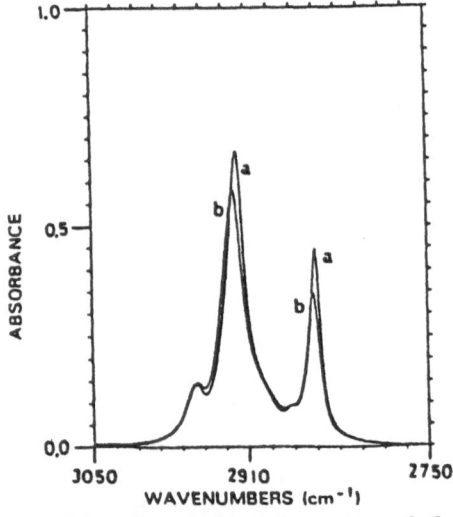

Fig. 1 :   Temperature-dependent infrared spectra of the C-H   stretching region of DPPC multilamellar liposomes.
Spectra (a) and (b) were recorded before 41 °C and after 43 °C the main DDPC phase transition.   An increase in temperature leads to an increase in the CH$_2$ symmetric and asymmetric stretching frequencies and a decrease in band intensity.

The infrared spectrum of hydrated phospholipids in the 3000-2700 cm^{-1} region (see Fig. 1) displays three well-resolved peaks located at 2960 (CH$_3$ asymmetric stretching), 2920 (CH$_2$ asymmetric stretching) and 2850 cm^{-1} (CH$_2$ symmetric stretching). The CH$_2$ stretching frequencies correspond primarily to conformational disorder and increase with the introduction of gauche conformers (as a consequence of increase in temperature).[15,16] These peaks and in particular that at 2850 cm^{-1} are useful for the monitoring of the temperature-induced phase transition of liposomes. DPPC liposomes show a main phase transition (T$_m$) at about 41.5 °C. The presence of TBTA induces a low decrease of T$_m$ even at a DPPC/organotin molar ratio (R) of 10. At this R value, the range of temperatures within which the main phase

414

transition occurs is higher with respect to the control (data not shown). The effect of TBTC is markedly different: it considerably decreases T_m at all TBTC concentrations (about 4 °C at R equal to 200 and 100 and about 9 °C at R equal to 25). Moreover, in liposomes with R values of 200 and 100, the lipid phase transition encompasses a 10 °C temperature range, whereas in liposomes with an R value equal to 25, the main transition occurs in a 2 °C range, which is slightly higher with respect to that of control DPPC (data not shown). These results are confirmed by DSC data shown in Fig. 2. The curve (c) of panel (B) indicates the coexistence of different phases.

Infrared and DSC data are useful for studying pesticides in the hydrophobic core of the membrane. The greater effect of TBTC as compared with TBTA on the thermotropic characteristics of DPPC liposomes may find one possible explanation in the common hydrolysis product, tributyltin hydroxide TBT(OH).[17] If TBT(OH) is the most active species on DPPC liposomes, it should be produced in a large quantity from TBTC, since the latter affects the DPPC bilayers to a great extent. The hydrolysis process probably is modulated by the presence of liposomes that avoid direct contact of pesticides with water; this should further differentiate the different rate of hydrolysis of TBTC and TBTA.

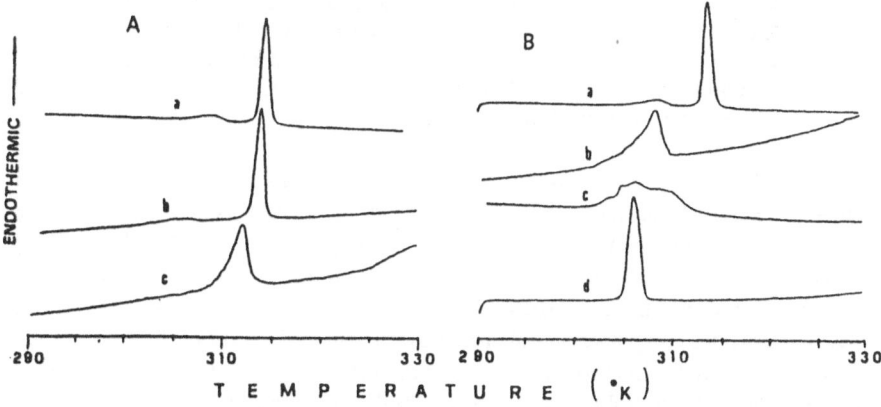

Fig. 2 : **DSC heating curves for DPPC multilamellar liposomes in the absence and in the presence of TBTA (panel A) and TBTC (panel B).**
Panel (A): curve (a): control DPPC; curve (b): DPPC/TBTA molar ratio = 100; curve (c): DPPC/TBTA molar ratio = 10. Panel (B): curve (a): control DPPC; curve (b): DPPC/TBTC molar ratio = 200; curve (c): DPPC/TBTC molar ratio = 100; curve (d) DPPC/TBTC molar ratio = 25.

DPH is a fluorescent probe that localizes in the hydrophobic core of liposomes, whereas its charged derivative TMA-DPH localizes close to the bilayer surface because of its charged amino group.[18,19] These two molecules can thus provide information on two different microenvironments of phospholipid membranes. The effect of different concentrations of TBTA and TBTC on the main phase transition of DPPC liposomes is shown in Fig 3. The higher effect of TBTC as compared with TBTA is evident: the lipid phase transition is considerably broader in the presence of TBTC even at low concentration of pesticide. Fig. 4 shows the steady-state fluorescence polarization for DPH and TMA-DPH in DPPC liposomes.

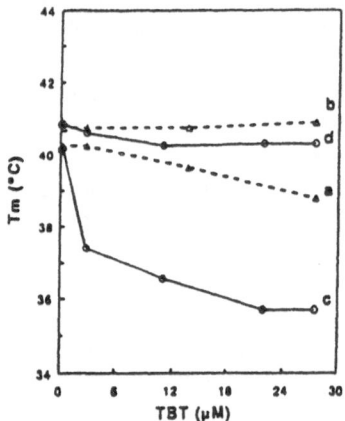

Fig. 3 : **Effect of different concentrations of TBTA and TBTC on DPPC main phase transition (T$_m$) as detected by DPPH steady-state fluorescence polarization.**

The two broken lines (a) and (b) represent the beginning and the end of the lipid phase transition for DPPPC/TBTA liposomes. For this system, the gel and the liquid-crystalline phase exist below line (a) and above line (b). The two continuous lines (c) and (d) represent the beginning and the end of the lipid phase transition for DPPC/TBTC liposomes. For this system, the gel and the liquid-crystalline phase exist below the line (c) and above the line (d). Between each pair of lines (a,b and c,d), gel and liquid-crystalline phases coexist.

The DPH fluorescence data, obtained below (26 °C) and above (46 °C) the main lipid-phase transition, show that TBTC affects to a great extent the gel phase (26 °C), but has no effect on the liquid-crystalline phase (46 °C). The TMA-DPH fluorescence data show that TBTC affects the lipid bilayer to a low extent when present at high concentrations. Above the main phase transition (46 °C), there is no effect of pesticides on lipid structural organization. These data indicate a localization of the pesticides in the hydrophobic region of the bilayer, in agreement with infrared and DSC data. A possible localization of TBTC close to the liposome surface is suggested when the compound is present at high concentration (see Fig. 4, panel B), most likely arising from saturation by the pesticide of the hydrophobic core of liposomes.

TBTC affects the T$_m$ of liposomes in different ways, depending on the nature of phospholipids. Fig. 5 shows that TBTC does not affect the T$_m$ of liposomes made of DMPC, whereas it considerably decreases the T$_m$ of liposomes made of DMPG. Moreover, TBTC inserted in liposomes of DMPC at 12.5 °C (gel phase) decreases DPH fluorescence polarization, whereas at 34 °C (liquid-crystalline phase), it has no effect on the lipid structural organization (Fig. 6). TBTC inserted in DMPG bilayers induces opposite effects on DPH fluorescence polarization (P): at 12.5 °C, P decreases whereas at 34 °C, P increases (Fig. 6). These data indicate that both the length of the acyl chains (compare with Fig. 3) and the nature of the polar head group may modulate the effects of TBTC on phospholipid membranes.

Although it is reported[2] that the anion moiety shows little effects on the biological activity of organotin compounds, the present data suggest that the anion moiety may play an important role in the organotin biocidal activity such as modifying the structural organization of the lipid bilayer and, as a consequence, also modifying the functions signal transduction and ion transport functions of membrane proteins.[20-22]

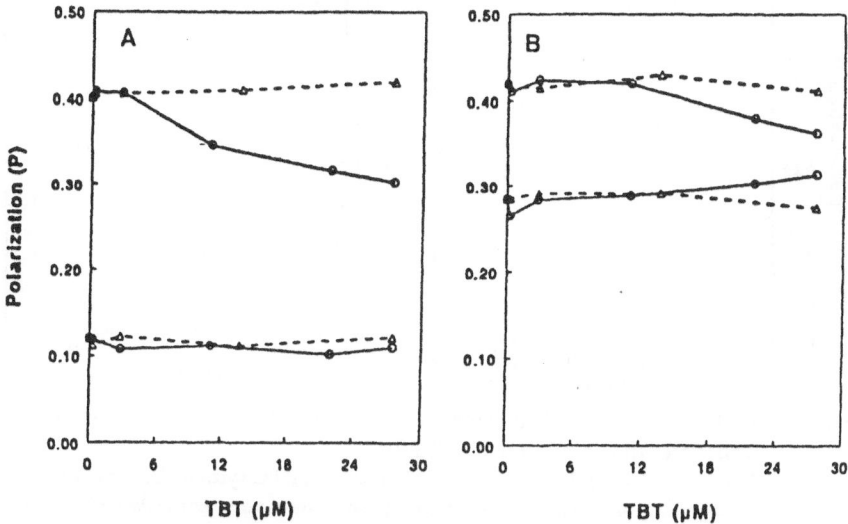

Fig. 4 : Steady-state fluorescence polarization of DPH (panel A) and TMA-DPH (panel B) in DPPC liposomes.
Continuous lines: DPPC/TBTC system; broken line DPPC/TBTA system. Upper lines: data recorded at 26 °C; lower lines: data recorded at 46 °C.

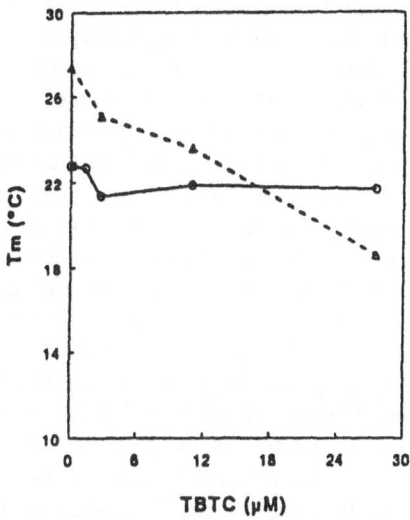

Fig. 5 : Effect of TBTC concentration on the main phase transition (T_m) for DMPC and DMPG liposomes as detected by DPH steady-state fluorescence polarization.
Continuous line: DMPC/TBTC system; broken line: DMPG/TBTC system.

417

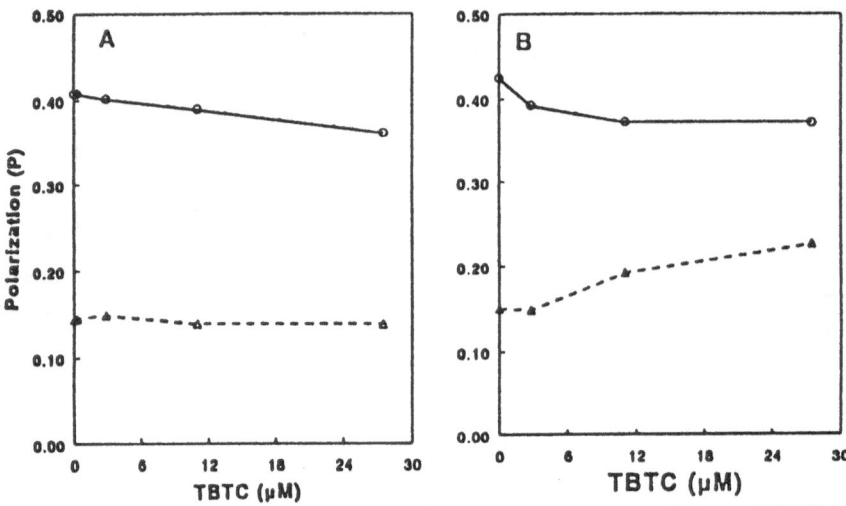

Fig. 6 : DPH steady-state fluorescence polarization in DMPC and DMPG liposomes in the presence of TBTC.
Panel A: DMPC/TBTC system; Panel B: DMPG/TBTC system. Broken lines: data obtained at 34 °C (liquid-crystalline phase); continuous lines: data obtained at 12.5 °C (gel phase).

Molecular interactions of organotin compounds with isolated F_oF_1 ATP synthase complex and mitochondrial membranes

Organotin compounds provide a useful model for exchange diffusion carriers; studies have shown that low molecular weight lipophilic molecules can act as shuttles across a lipid bilayer when the only driving force is the concentration gradient across the membrane.[6,20] The mode of action of organotin compounds in biological membranes can be more complex in the presence of proteins and specific carriers. The interaction of organotin compounds with mitochondrial membranes has been widely investigated to elucidate the mechanism of toxicity of these compounds.[7,21,22] Dialkyltin and trialkyltin compounds inhibit ATP synthesis and hydrolysis. The major site of action of organotin inhibitors is the hydrophobic F_o segment of the F_1F_o ATP synthase complex associated with the inner mitochondrial membrane, but the specific site of action has not been clearly established.[23,24] An analysis of this binding site could be performed by using fluorescent 5-coordinate organotin inhibitors with a specific high affinity site of interaction with the F_o segment of mitochondrial ATP synthase.[25] One of these compounds, dibutyltin-3-hydroxyflavone bromide [$Bu_2SnBr(of)$], has been used as a fluorescent probe of membrane-bound mitochondrial F_1F_o ATP synthase and its potential inhibitory effect on mitochondrial membrane and isolated enzymes has been analyzed. Figure 7A shows the effect of $Bu_2SnBr(of)$ on the inhibition of F_1F_o ATP synthase in the isolated form and in natural mitochondrial membranes. The results clearly demonstrate that $Bu_2SnBr(of)$ binds to the membrane F_1F_o ATP synthase since the isolated and purified enzyme complex is inhibited by the organotin hydroxyflavone compound. In addition, the soluble part (F_1-ATPase) of the enzyme is not sensitive to the inhibition by $Bu_2SnBr(of)$ (data not shown), indicating that the site of action

418

is on the F_o hydrophobic segment of F_1F_o ATP synthase, and is similar to that by trialkyltins and triaryltins.[25] The titration of F_1F_o ATP synthase with $Bu_2SnBr(of)$ leads to increase in fluorescence intensity until the binding site is saturated (Fig. 7B). This fluorescence enhancement is decreased or reversed in the presence of tributyltin chloride (data not shown), which implicates the same site of action and thus competition between this inhibitor and the fluorescent organotin compound. These data indicate that the large fluorescence enhancement that occurs as a consequence of the binding of the fluorescent organotin compound to the F_o sector of ATP synthase complex is of potential interest in quantifying the enzyme associated with membrane preparation or during the purification.

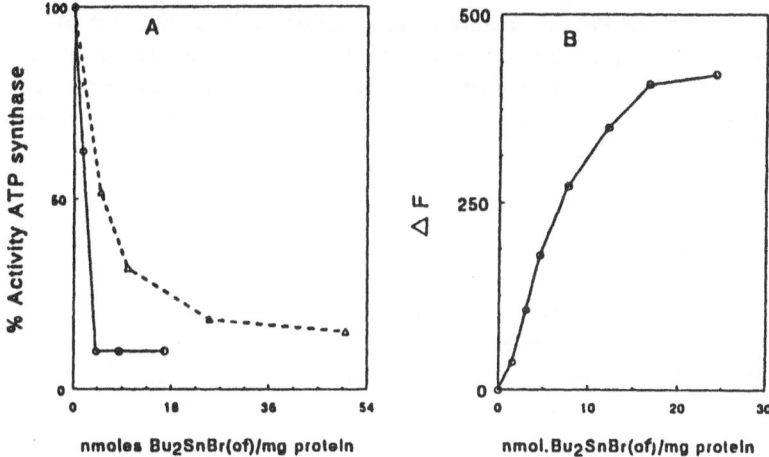

Fig 7 :   **Effect of $Bu_2SnBr(of)$ on the enzymatic activity of F_oF_1 ATP synthase complex (panel A)   and titration of isolated mitochondrial F_oF_1 ATP synthase by $Bu_2SnBr(of)$ (panel B).** Panel (A): continuous line refers to mitochondrial and broken line to ATP synthase complex isolated according to [12]. The isolated enzyme contains residual phospholipids corresponding to 0.07 µmoles Pi/mg protein. Panel (B): Aliquots (1 µl) of ethanolic solutions of $Bu_2SnBr(of)$ were added to the protein suspension (0.33 mg protein in 2.0 ml 10 mM tricine pH 8 buffer). The fluorescence values (ΔF) have been corrected for light scattering and normalized to 1 mg of protein.

On comparing the results from both model and natural membranes we can advance the hypothesis that organotin compounds, after their partition in the membrane, destabilize lipid bilayers and, as a consequence, may labilize lipid-protein interactions. Moreover, they can function as ionophores that facilitate halide/hydroxyl exchange across cell-mitochondrial membranes,[5,6] and would probably interact with specific reactive groups[23,24,26] or subunits [8,9] of membrane proteins. Under such conditions, the optimal enzyme conformation for the catalysis or the conformational changes during catalytic events is altered and/or cannot take place. Membrane molecular organization is the structural basis of its function since membrane catalytic activities are dependent on the physico-chemical properties of the microenvironment where proteins are located. These stringent connections between structure and function are modified by the interaction with the organotin inhibitors as observed by the changes associated with membrane cellular activities.

Fluorescent organotin compounds are promising reagents and useful probes for their utility and versatility in the study of the integrity of the structure-function relationships of membrane proteins and of the elucidation of the molecular mechanism of the toxic action at membrane level. The ability to interact with a specific binding site in the apolar region of proteins can be used as the basis of a simple method to analyze organotin-sensitive sites in other F_1F_o ATP synthases from mammalian and plant mitochondria, or in Na^+/K^+ ATPase and Ca^{2+} ATPase associated with plasma membrane transport. This class of fluorescent probes could also be of potential value in bioenergetic studies of mammalian mitochondrial mutations, which are associated with mitochondrial DNA diseases[27,28] and with molecular defects present mainly in the membrane F_o component of the ATP synthase enzyme complex. The ability of organotin compounds to interact with biological membranes can be of considerable importance for the detection of subtle changes associated with membrane functions and for the prediction of the response of cells to a toxic insult.

Acknowledgements

We thank Prof. D.E. Griffiths for the gift of the fluorescent organotin compounds. This work was supported by grants from the Ministry for University and for Technological and Scientific Research (M.U.R.S.T.) 60% E.B., F.T., G.Z.

References

1. A. Ross, *Ann. N.Y. Acad. Sci.*, **125** (1965) 107.
2. N.J. Snoeij, A.H. Penninks, and W. Seinen, *Environ. Res.*, **44** (1987) 335.
3. R.J. Maguire, *Water Poll. Res. J. Canada*, **26** (1991) 243.
4. P.T.S. Wong, Y.K. Chau, O. Kramar and G.A. Bengert, *Can. J. Fish. Aquat. Sci.*, **39** (1982) 483.
5. I.J. Boyer, *Toxicology*, **55** (1989) 253.
6. M.J. Selwyn, A.P. Dowson, M. Stockdale and N. Gains, *Eur. J. Biochem.*, **14** (1970) 120.
7. W.N. Aldridge and B.W. Street, *Biochem. J.*, **91** (1964) 287.
8. N. Sone and B.J. Hagihara, *Biochem.*, **56** (1964)151.
9. M. Stockdale, A.P. Dawson, and M.J. Selwyn, *Eur. J. Biochem.*, **15** (1970) 342.
10. B.M. Elliot, W.N. Aldridge and J.W. Bridges, *Biochem. J.*, **177** (1979) 461.
11. D.J. Stiggal, Y.M. Galante and Y. Hatefi, *Methods Enzymol.*, **55** (1979) 308.
12. J.A.Berden and M.M. Voorm-Brouwer, *Biochim. Biophys. Acta*, **501** (1978) 424.
13. O.H. Lowry, N.J. Rosenbrough, A.L. Farr and R.J. Randall, *J. Biol. Chem.*, **193** (1951) 680.
14. F. Dabbeni-Sala and P. Palatini, *Biochem. Biophys. Res. Commun.*, **133** (1985) 807.
15. J. Umemura, D.G. Cameron and H.H. Mantsch, *Biochim. Biophys. Acta*, **602** (1980) 44.

16. I.M. Asher and I.W. Levin, *Biochim. Biophys. Acta*, **468** (1977) 63.
17. R. Barbieri and M.T. Musmeci, *J. Inorg. Biochem.*, **32** (1988) 89.
18. L.W. Engel and F.G. Prendergast, *Biochemistry*, **20** (1981) 7338.
19. M. Shinitzky, in: M. Shinitzky, (Ed.), Physiology of Membrane Fluidity, CRC Press, Boca Raton, FL, 1984, pp. 1-52.
20. J.O. Wieth and M.T. Tosteson, *J. Gen. Physiol.*, **73** (1979) 765.
21. M.S. Rose and W.N.Aldridge, *Biochem. J.*, **127** (1972) 51.
22. K. Cain and D.E. Griffiths, *Biochem. J.*, **162** (1977) 575.
23. K. Cain, M.D. Partis and D.E. Griffiths, *Biochem. J.*, **161** (1977) 593.
24. T. Yagy and Y. Hatefi, *Biochemistry*, **23** (1984) 2449.
25. J. Usta and D.E. Griffiths, *Biochem. Biophys. Res. Commun.*, **188** (1992) 365.
26. E.L. Emanuel, M.A. Carver, G.C. Solaini and D.E. Griffiths, *Biochim. Biophys. Acta*, **766** (1984) 209.
27. D.C. Wallace, *Science*, **256** (1992) 628.
28. D.C. Wallace, X. Zheng, M.T. Lott, J.M. Shoffner, J.A. Hodge, R. Kelly, C.M. Epstein and L.C. Hopkins, *Cell*, **55** (1988)601.

Main Group Elements and Their Compounds
V.G. Kumar Das (Ed)
Copyright © 1996 Narosa Publishing House, New Delhi, India

Cellular and Biochemical Aspects of Antitumor Activity of Organotin Compounds

Yasuaki Arakawa

Department of Hygiene & Preventive Medicine, Faculty of Health Sciences, The University of Shizuoka, 52-1 Yada, Shizuoka-shi, Shizuoka 422, Japan

Organotin compounds such as dibutyltin dichloride exhibit *in vivo* antitumor activity towards Ehrlich ascites tumor, IMC-carcinoma, P-388 lymphocytic leukaemia, and Sarcoma 180 systems in descending order of activity in mice. In a two-stage mouse skin carcinogenesis system of initiation and promotion, the optimal dose of the dibutyltin compound inhibits the promotion stage more strongly, especially the first stage of the two-stage promotion system, than the initiation stage. On the other hand, *in vitro* studies reveal that the dibutyltin compound dramatically inhibits the proliferation of malignant cells such as thymic lymphosarcoma cells and HeLa cells, and further inhibits the initiation stage of the two-stage transformation of BALB/c 3T3 cell. Cellular and biochemical studies on the mechanism of these antiproliferative activities reveal that hydrophobic organotin compounds such as dibutyltin and tributyltin selectively accumulate in the Golgi apparatus and endoplasmic reticulum (ER), but not in the nucleus, and destroy the specific stratified structure of Golgi apparatus and the characteristic reticular structure of the ER, and also inhibit Golgi function (e.g. ceramide metabolism) and ER function (e.g. inositol triphosphate (IP3)-induced intracellular Ca^{2+} mobilization). On the other hand, these compounds inhibit total phospholipid synthesis and the breakdown of inositol phospholipids, namely, phosphatidylinositol (PI) turnover, and impair the activation of phospholipid transport between organelles by affecting on the physical properties of membrane (e.g. the change in the membrane order of various phospholipid vesicles, especially such as phosphatidylinositol 4-monophosphate (PIP1) and phosphatidylinositol 4,5-diphosphate (PIP2) vesicles). In the light of these results, it would appear that hydrophobic organotin compounds such as dibutyltin and tributyltin inhibit the intracellular phospholipid transport between organelles by impairing the structure and function of the Golgi apparatus and the ER, and consequently inhibit the intracellular phospholipid metabolism as well as finally the membrane-mediated signal transduction leading to DNA synthesis via phospholipid turnover and Ca^{2+} mobilization.

Organotin compounds such as dibutyltin dichloride (Bu_2SnCl_2) and dioctyltin dichloride (Oct_2SnCl_2) induce a severe thymus atrophy when fed orally to weanling rats. This atrophy is reversible and dose-dependent. Recently, we found that this atrophy was due to the depletion of thymic lymphocytes, which in turn, depended on the degree of inhibition of DNA synthesis and further cell proliferation.[1-10] This prompted the present detailed study on the inhibitory effects of organotin compounds on the proliferation system of various malignant cells using various carcinogenesis systems *in vivo* and *in vitro*. An explanation of the mechanism of organotin inhibition is advanced.

Experimental

Materials

The following organotin compounds were used in this study:[1] monomethyltin trichloride (MeSnCl$_3$), mono-*n*-butyltin trichloride (*n*-BuSnCl$_3$), dimethyltin dichloride (Me$_2$SnCl$_2$), di-*n*-propyltin dichloride (Pr$_2$SnCl$_2$), and di-*n*-butyltin dichloride (Bu$_2$SnCl$_2$) were purchased from K & K Laboratories (Plainview, N.Y.); di-*n*-octyltin dichloride (*n*-Oct$_2$SnCl$_2$), trimethyltin chloride (Me$_3$SnCl), triethyltin chloride (Et$_3$SnCl), tri-*n*-butyltin chloride (Bu$_3$SnCl), and triphenyltin chloride (Ph$_3$SnCl), were obtained from the Aldrich Chemical Co., Inc. (Milwaukee, Wis.). The purity of these compounds was not less than 98%. All radiolabeled compounds were purchased from the Japan Isotope Center. The RPMI medium 1640 (with L-α-Glutamine), fetal bovine serum (heat inactivated) and penicillin-streptomycin solution (5,000 U/ml-5,000 mg/ml) were purchased from Gibco Laboratories (NY, USA). Bactoconcanavalin A (ConA) was purchased from Difco Laboratories (Detroit, Michigan, USA). Standard phospholipids such as L-α -phosphatidylcholine (PC, from frozen egg yolk), L-α-phosphatidylserine (PS, from bovine brain), L-α-phosphatidylinositol (PI, from soybean), L-α-phosphatidylinositol 4-monophosphate (PIP, from bovine brain), L-α-phosphatidylinositol 4,5-diphosphate (PIP2, from bovine brain), 1,2-dioleoylrac-glycerol (DAG), L-α-phosphatidylethanolamine (PE, from egg yolk), L-α-phosphatidic acid (PA, from egg yolk lecithin) and arachidonic acid (AA, from porcine liver) were purchased from Sigma Chemical Company (St.Louis, MO, USA). Dimilume-30 and Soluene-350 were obtained from Packard. 1,6-Diphenyl-1,3,5-hexatriene (DPH) was obtained from the Aldridge Chemical Company (Milwaukee, WI, USA). Other reagents included special-grade organic solvents such as chloroform, methanol, ethanol, and acetone (each provided by Wako Pure Chemicals Co., [Tokyo, Japan]). Culture tubes (12 x 75 mm , clear and with caps, from Falcon Co., [Oxnard, Calif.]), Acrodisc (disposable filter assembly, pore size: 1.2 μm, diameter: 25 mm, from Gelman Science, Inc. [Michigan]) and liquid scintillation vials from Wheaton Scientific (Millville, N.J.) were used as the experimental appliances.

In vivo studies

The antitumor activity of organotin compounds *in vivo* was examined towards sarcoma 180, IMC-carcinoma, P-388 lymphocytic leukemia, and Ehrlich ascites tumor in male mice. The mice were inoculated in the peritoneal cavity with an ascitic tumor at a level of 5 x 10^6 cells. Organotin compounds were suspended in a Tween-80/distilled water/alcohol mixture (1:97:2, v/v). One day after inoculation, the mice were injected intraperitoneally with the suspensions of the compounds. A total of one to five injections were given at daily intervals in one experiment. The dose per injection was kept in the range of 10 to 0.1 mg/kg body weight. The results of the screening were evaluated by computing the T/C ratio, which is the mean survival time of the treated group divided by that of the control group, and the increase in life span of treated animals as compared to the controls, since the increase in life span = (T/C)% - 100. A T/C value of 100 means that the drug has no effect of either increasing or decreasing the tumor, while a T/C value > 115 means significant activity.[1]

Further, the antitumor activity was examined using the two-stage mouse skin carcinogenesis system. Initiation was performed with 100 nanomoles moles of 7,12-dimethyl benzanthracene (DMBA). One week later, twice-weekly applications of 5 μg of 12-0-tetradecanoylphorbol 13-acetate (TPA) or acetone were begun (promotion stage I). Starting on the third week of promotion, the mice received twice-weekly applications of 2.5 μg of mezerein or acetone (promotion stage II) for the rest of the 14-week experiment. Bu_2SnCl_2 (5 μg) was applied 30 min before treatment with the initiator and promoters.[11,12]

In vitro studies

Two-stage in vitro transformation of BALB/c 3T3 cell was measured by the method of IARC/NCI/EPA Working Group.[13,14] Briefly, BALB/c 3T3 A31-1-1 cells, obtained from Dr. T.Kakunaga, were cultured in Eagle's minimal essential medium supplemented with 10% fetal calf serum. For induction of transformed foci, 1×10^4 cells were seeded onto 60-mm dishes, and 20-methylcolanthrene (MCA, 0.1 μg/ml) was added to each culture dish on the following day. After 48 hr of treatment, MCA was removed by washing with fresh medium and by replacement with promoter (TPA, 0.1 μg/ml) containing medium. Dibutyltin (3×10^{-7} M or 3×10^{-8} M) was simultaneously added with either MCA or TPA. The culture medium was changed every 3 or 4 days for 5 weeks. Types of transformed focus were determined under a dissecting microscope after fixation and Giemsa staining. Only densely stained foci with clear criss-crossing of the cells at their periphery were scored as transformed foci.Thymocytes were obtained by mincing rat thymus gland of Wistar-derived rat with scissors in cold RPMI 1640 medium and by gently passing the mince through a stainless steel sieve (220 μm diameter) with the same cold medium.

The viability of the cells (1 to 3×10^6 cells in 1 ml) was determined in hemocytometers with a 0.04% erislosin solution in saline after the incubation with 10^{-4} to 10^{-8} M individual organotin compounds in RPMI 1640 medium containing 10% fetal calf serum, penicillin (100 units/ml), and streptomycin (100 μg/ml) at 37°C in a humidified atmosphere of 5% CO_2 in air for 24 hr. The individual organotin compound was dissolved in ethanol, and 5 ml of it was added to 1 ml of the cell suspension and vesicle solution. The final concentration of ethanol in the medium should be less than 0.5%, the concentration which is without effect on cell survival and polarization measurements of phospholipid vesicles.

DNA and RNA synthesis was determined by measuring the incorporation of [6-^{3}H]thymidine into DNA and of [5-^{3}H]uridine into RNA. Thymocytes (1×10^6 to 3×10^6 cells) were suspended in 1 ml of RPMI 1640 medium containing 10% fetal calf serum, penicillin (100 units/ml), and streptomycin (100 μg/ml) and preincubated in the absence or presence of 10^{-6} to 10^{-7} M organotin compounds (5 μl in EtOH solution) at 37 °C for 1 to 2 hr. For the evaluation of DNA synthesis, the cells were radiolabeled with 1.0 mCi of [6-^{3}H]thymidine (5.0 Ci/m mol, RCC [Amersham, England]) per culture during the final 4 hr of a 24-hr culture. For the evaluation of RNA synthesis, the cells were radiolabeled with 2.0 mCi of [5-^{3}H]uridine (25 to 30 Ci/mmol RCC [Amersham, England]) per culture for variable intervals. When the synthesis of DNA and of RNA in the mitogen-stimulated cells was to be determined, the cells were first stimulated by the addition of concanavalin A (ConA, 5 µg/ml) after the 1- to 2-hr preincubation with organotin compound. This concentration of ConA gave a maximal stimulation of [6-^{3}H]thymidine incorporation into DNA after 24 hr incubation. After the incubation, the cultures were cooled in ice and the cells were harvested by aspiration through an Acrodisc Filter Assembly (pore size: 1.2 µm; diameter: 25 µm) from the culture tubes. The cells collected on the filter discs were washed by aspirating 3 to 4 ml of cold 5% trichloroacetic acid (TCA) aqueous solution for DNA assay and of 10% TCA for RNA assay into the syringe. The TCA-insoluble fractions on the filter discs were transferred to glass scintillation vials (Falcon) by pressure elution with 0.6 ml of Soluene-350 (Packard) prefilled in the syringe, homogenized with 5 ml of Dimilume-30 (Packard), and then counted in a Packard Tri-Carb 3255 liquid scintillation spectrometer.

Influx of ^{45}Ca^{2+} (10 to 40 µCi/mg Ca, [Amersham, England]) was assayed in a modified Hanks balanced solution (phosphate-free). A cell suspension of thymocytes (10^6 cells per ml) was incubated with 10 µCi of ^{45}Ca^{2+} per 1 ml in the absence and presence of 10^{-9} to 10^{-4} M organotin compounds at 37 °C for variable intervals. At specified times, the cultures were harvested by aspiration through an Acrodisc Filter Assembly, as described above. The cell pellets on the filter were dissolved in 0.6 ml of Soluene-350 solubilizer and mixed with 3 ml of Dimilume scintillator, and the radioactivity was determined.

Phospholipid synthesis was measured by the incorporation of carrier-free ^{32}P-phosphoric acid (100 µCi/ml, [Japan Isotope Center]) into the lipid fraction of the cultured cells. Thymocytes (1 x 10^7 to 4 x 10^7 cells in 1 ml) were incubated in RPMI 1640 medium containing 10 % fetal calf serum with 1 to 10 µCi of ^{32}P-phosphoric acid in the absence and presence of 10^{-6} to 10^{-7} M Bu$_2$SnCl$_2$ at 37 °C at variable intervals. The cells were stimulated by the addition of Con A (5 µg/ml) 5 min after preincubation with Bu$_2$SnCl$_2$. After the incubation, the cells were washed twice with iced phosphate buffered saline. The packed cells were then extracted with 3 ml of chloroform-methanol-water (1:2:0.8, v/v) for 30 min with occasional shaking. After centrifuging at 3,000 rpm (1,000 g) for 10 min, the supernatant was carefully aspirated and diluted with 1 ml of chloroform and 1 ml of water. Further centrifugation at 3,000 rpm for 10 min resolved the emulsion into two phases. The lower phase (1 ml) was removed and the radioactivity of total lipid extracts was determined using a Model 3255 Packard liquid scintillation counter. When the incorporation of radioactive label into the individual phosphatides was to be determined, the phospholipid components in the lower phase (1 ml) were first separated by the two dimensional thin-layer chromatography (TCL) on a commercially precoated, silica gel plate (20 cm x 20 cm, Art. 5721, Merk, [Darmstadt]) using solvent systems of chloroform-methanol-28% ammonia (65:25:5, v/v; the first dimension) and chloroform-acetone-methanol-acetic acid-water (50:20:10:10:5, v/v; the second dimension).[16] The separated phosphatides were located by radioautography and extracted. The radioactivity incorporated into the separated phospholipid components from ^{32}P phosphoric acid was counted in the liquid scintillation counter.

Phospholipase activity was measured by the release of $[^{14}C]$ AA from the cellular lipids, mainly phospholipids with a modification of the method described by Hirata.[27,31] Thymocytes (1×10^7 cells/ml) were preincubated in a total volume of 20 ml of RPMI 1640 medium containing 1% fetal calf serum with 20 µCi of $[1-^{14}C]$ AA (55.5 µCi/mmol, Dupont-New England Nuclear [NEN] Products [Billerica, Mass.]) at 37 °C in a humidified atmosphere of 5% CO_2/95% air for 1 hr. The cells were washed twice with fresh media containing 0.5% fatty acid-free albumin in order to remove the excess radioactive AA and resuspended in 20 ml of the same media and divided so as to contain 1 ml of cell suspension (1×10^7 cells) per culture tube. The prelabeled cells of each tube were further incubated in the absence and presence of 10^{-5} to 10^{-8} M organotin compound for variable intervals at 37 °C. The thymocytes stimulation was performed by adding ConA (5 µg/ml) or fMet-Leu-Phe (1×10^{-8} M) after the organotin treatment. The reaction was terminated by adding 1 ml of ice cold 10-mM phosphate-buffered saline, pH 7.4. After centrifugation at 600 x g for 5 min, an aliquot (1 ml) of the supernatant was transferred into a counting vial. To the counting vial was added 8 ml of an aqueous counting scintillant, ACS II (Amersham), and the counts per minute (cpm) of each sample were determined with a Packard Tri-Carb 3255 liquid scintillation spectrometer.

The activities of various phospholipases were measured by the method of Hirata[17] using L-α-dipalmitoyl-[choline-methy-^{14}C-]-phosphatidylcholine (New England Nuclear, 153.0 mCi/mmol)as substrate. The reaction mixtures containing 0 or 4×10^{-7} M individual organotin compound, 0.5 µCi of radioactive substrate, 2 mM L-α -phosphatidylcholine (Liver, Sigma), and 25 mM Tris-glycyl-glycine buffer, pH 8.0, in a total volume of 250 µl, were incubated at 37 °C for 5 min after the addition of individual phospholipases, phospholipase A_2 from porcine pancreas (Sigma,2.5 µg), phospholipase C from B. cereous (Boehringer Mannheim, 5.0 µg). The reaction was terminated by 3 ml of chloroform/methanol/concentrated HCl (2:1:0.01, v/v). After centrifugation at 3000 rpm for 10 min, 2 ml of aliquots were dried under a stream of N_2 gas. The products were separated by means of single-dimensional thin-layer chromatography on Silica Gel G plates with a solvent system of chloroform/methanol/water (65:35:15, v/v). The radioactive spots were localized by radioautography and the radioactivity in each area was determined by liquid scintillation.

Phospholipid vesicles were prepared by the following procedure. Phospholipids (1 mg) dissolved in chloroform-methanol were mixed at the desired composition and dried under a stream of nitrogen, followed by high vacuum pumping for 2 hr. They were then suspended in 5 ml of 20 mM Tris-HCl buffer (pH 7.5, containing 100 mM NaCl) and briefly sonicated at 0 °C for 1 min in a bath type ultrasonicator, followed by vigorous vortexing. Small unilamellar vesicles were prepared by further sonication at 0 °C until the suspension became clear (for about 3 min). Fluorescence polarization was measured at 25 °C by a Perkin-Elmer Model MPF-44B fluorescence spectrophotometer using DPH as a fluorophore. The fluorescence was excited at 365 nm and detected at 425 nm. Lipid concentrations were approximately 21-23 nmol/ml and the ratio of DPH to lipid was 1/200 to 1/100. Fluorescence data were presented by either $I_{/} /I_{\perp}$, $P=(I_{/} /I_{\perp} - 1)/(I_{/} /I_{\perp} + 1)$ as the degree of fluorescence polarization, or $r = (I_{/} /I_{\perp} - 1)/(I_{/} /I_{\perp} + 2)$ as the fluorescence anisotropy, where I_{11} and $I_{\perp}$ are the fluorescence intensities measured at parallel and perpendicular to the direction of polarization of the exciting beam, respectively. Results were expressed as a percentage change in the degree of fluorescence polarization, $100(P-Po)/Po$.

Intracellular organotin compounds were visualized by the simple fluorimetric method using morin (2', 3, 4', 5, 7-pentahydroxyflavone) as a fluorescent indicator.[18] The visualization studies of intracellular organotins were carried out with normal human skin (SF-TY) fibroblasts (Japanese Cancer Research Resource Bank, Tokyo). The Golgi apparatus was stained with fluorescent C_6-NBD-ceramide (C_6-NBD-Cer).[19-22] The morphological studies of the Golgi apparatus were carried out with SF-TY cells.

The Golgi functions were examined by determining the metabolism rate of the fluorescent ceramide(C_6-NBD-Cer) to glycosylceramide(C_6-NBD-GlcCer) and sphingomyelin(C_6-NBD-SM).[21] The function studies of the Golgi apparatus were carried out with the chinese hamster ovary cells(CHO cells).

The reticular structure of the endoplasmic reticulum(ER) was stained with the lipophilic, cationic fluorescent dye $DiOC_6$(3)(3,3'-dihexyl oxacarbocyanine iodide).[23] The morphological studies of the ER were carried out with the African green monkey epithelial cell line, CV-1. The ER functions were examined by determining the IP_3-induced intracellular Ca^{2+} mobilization at ER.[24-26] Glutaraldehyde fixation was carried out as follows. Cells were washed three times with HMEM and incubated with fixative containing 0.5% glutaraldehyde, 5% sucrose (w/v), and 0.1 M piperazine-N,N'-bis(2-ethanesulfonic acid) (PIPES), pH 7.0 for 10 min at 37 °C. Cells were then washed three times with HCMF and incubated with organotin. Perforated cells were prepared according to the method of Simons and Virta.[27]

Results and Discussion

Antitumor activity of organotins

The antitumor activity of dibutyltin dichloride *in vivo* towards various tumor systems, expressed as % T/C, is compared in Table 1. The compound showed the highest activity against the Ehrlich-ascites tumor system and gave T/C value in the range of 98 to 186 with dosage ranging from 0.1 to 3.0 mg/kg, the maximum activity being observed at a single dose of 3 mg/kg. Generally, a single or double injection at high dose (2 to 3 mg/kg), at which the compound does not show any toxicity, proved more effective against any tumor than five injections at low dose levels (0.1 to 0.3 mg/kg). Figure 1 illustrates the anti-tumor effect of Bu_2SnCl_2 on Ehrlich ascites tumor cells. The survival period of the control group was about 22 days. This was increased upon administering the dibutyltin. Table 2 shows the effect of Bu_2SnCl_2 on tumor initiation-promotion systems in mouse skin carcinogenesis. When the optimal dose of dibutyltin was provided at the first phase TPA promotion stage (Stage I promotion) of the two-stage promotion system, the inhibitory effect was most pronounced. This was the case with 5 µg of Bu_2SnCl_2, when the occurrence of papillomas was completely suppressed (Fig. 2).

Table 1 : Antitumour activity of dibutyltin dichloride

Tumour	Dose[a]		Life span[b]	
	mg/kg i.p.	Injection time	Survival time (day)	T/C (%)
Sarcoma 180	0	5	15.1 ± 0.4	100
	0.1	5	17.6 ± 0.3	117
	0.3	5	17.8 ± 0.4	118
	1.0	5	15.8 ± 0.5	105
	2.0	2	15.0 ± 0.3	100
	2.0	4	15.0 ± 0.4	100
	3.0	1	13.2 ± 0.3	87
	3.0	2	13.0 ± 0.3	86
IMC-carcinoma	0	5	16.5 ± 0.3	100
	0.1	5	18.6 ± 0.4	113
	0.3	5	19.1 ± 0.5	116
	1.0	5	19.8 ± 0.5	120
	2.0	2	20.8 ± 0.3	126
	2.0	4	21.3 ± 0.7	129
	3.0	1	21.0 ± 0.6	127
	3.0	2	22.0 ± 0.7	133
Lymphocyte leukemia P-388	0	5	10.4 ± 0.3	100
	0.1	5	10.6 ± 0.2	102
	0.3	5	12.1 ± 0.5	116
	1.0	5	11.8 ± 0.4	114
	2.0	2	12.3 ± 0.5	118
	2.0	4	12.2 ± 0.5	117
	3.0	1	12.1 ± 0.3	116
	3.0	2	12.5 ± 0.3	120
Ehrlich ascites tumour	0	5	21.1 ± 0.4	100
	0.1	5	20.7 ± 0.5	98
	0.3	5	21.8 ± 0.7	104
	1.0	5	22.7 ± 0.5	108
	2.0	2	29.1 ± 0.4	138
	2.0	4	28.1 ± 0.8	133
	3.0	1	39.3 ± 0.7	186
	3.0	2	37.6 ± 0.7	178

[a]A total of 1 to 5 injections were given at daily intervals in one experiment
[b]The increase in survival of treated animals over control is expressed as T/C (%). The values of survival time are the means $\pm$ SE of 10 animals per group.

Table 2 : Effect of Bu$_2$SnCl$_2$ on tumour initiation–promotion systems in mouse skin carcinogenesis†

Initiation		Promotion				Tumour response	
		Stage I		Stage II		Papilloma number/mouse	% of positive control (in Expt. 4)
Acetone	1 wk	TPA	4 times	Mezerein	28 times	0	0
DMBA	1 wk	Acetone	4 times	Mezerein	28 times	0	0
DMBA	1 wk	TPA	4 times	Acetone	28 times	2.4 ± 0.1	25
DMBA	1 wk	TPA	4 times	Mezerein	28 times	9.6 ± 0.3	100
DMBA + Bu$_2$Sn (5 µg)	1 wk	TPA + Bu$_2$Sn (5 µg)	4 times	Mezerein + Bu$_2$Sn (5 µg)	28 times	0	0
DMBA + Bu$_2$Sn (50 µg)	1 wk	TPA + Bu$_2$Sn (50 µg)	4 times	Mezerein + Bu$_2$Sn (5 µg)	28 times	16.8 ± 0.7	175
DMBA + Bu$_2$Sn (0.5 µg)	1 wk	TPA	4 times	Mezerein	28 times	20.2 ± 0.8	210.4
DMBA + Bu$_2$Sn (5 µg)	1 wk	TPA	4 times	Mezerein	28 times	5.6 ± 0.2	58.3
DMBA + Bu$_2$Sn (50 µg)	1 wk	TPA	4 times	Mezerein	28 times	19.2 ± 0.8	200
DMBA	1 wk	TPA + Bu$_2$Sn (0.5 µg)	4 times	Mezerein	28 times	2.9 ± 0.1	30.2
DMBA	1 wk	TPA + Bu$_2$Sn (5 µg)	4 times	Mezerein	28 times	0	0
DMBA	1 wk	TPA + Bu$_2$Sn (50 µg)	4 times	Mezerein	28 times	3.8 ± 0.2	39.6
DMBA	1 wk	TPA	4 times	Mezerein + Bu$_2$Sn (0.5 µg)	28 times	22.1 ± 0.9	230.2
DMBA	1 wk	TPA	4 times	Mezerein + Bu$_2$Sn (5 µg)	28 times	4.5 ± 0.2	46.9
DMBA	1 wk	TPA	4 times	Mezerein + Bu$_2$Sn (50 µg)	28 times	15.4 ± 0.7	160.4
-- Bu$_2$Sn (50 µg)	1 wk	-- Bu$_2$Sn (50 µg)	4 times	-- Bu$_2$Sn (50 µg)	28 times	0	0

†Initiation was with 100 nmoles of DMBA (in Expt 1, only acetone); 1 week later, twice-weekly applications of 5 µg of TPA or acetone were b egun (Stage I). Starting on the third week of promotion, the mice received twice-weekly applications of 2.5 µg of mezerein or acetone (Stage II) for the rest of the 14-week experiment. Bu$_2$SnCl$_2$ was applied 30 min before treatment with the initiator or promoters. At the end of the experimental period, > 95% of the mice were alive. Tumour response is expressed as the means ± SE of papilloma number per mouse and as a percentage of the positive control (in Expt. 4). Each value is the result of tests based on 10 animals per group.

429

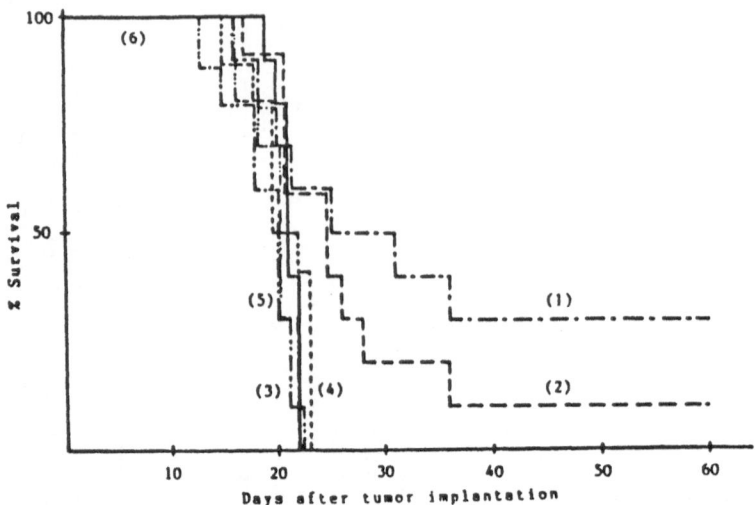

Fig. 1 : Antitumour cell of Bu₂SnCl₂ on Ehrlich ascites tumour cells. Bu₂SnCl₂ was injected i.p. into dd-Y mice daily from days 1 to 5 after Ehrlich tumour cells (5 x 10⁶) were inoculated i.p. into the mice. Survival times are the means of 10 animals per group. (1) 3 mg/kg x 1 or 2 (- •-) , (2) 2 mg/kg x 2 or 4 (– –), (3) 1 mg/kg x 5 (-•••-), (4) 0.3 mg/kg x 5 (——), (5) 0.1 mg/kg x 5 (-••••-), (6) Control (——)

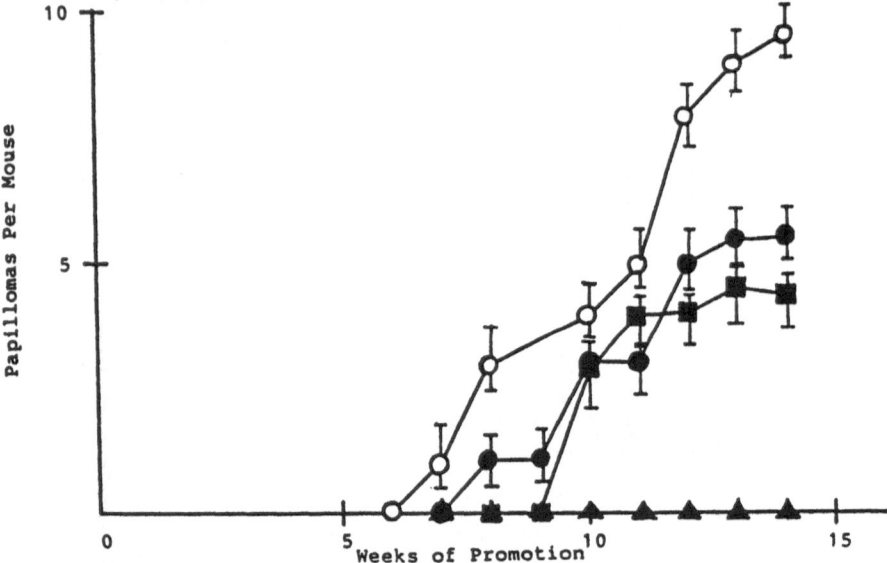

Fig. 2 : Inhibitory effects of Bu₂SnCl₂ on two-stage mouse skin carcinogenesis system of initiation and promotion. The mice were initiated with 100 nmoles of DMBA and promoted with 5 µg of TPA and 2.5 µg of mezerein. Bu₂SnCl₂ (5 µg) was applied 30 min before treatment with the initiator and promoters. Each point is expressed as papilloma number per mouse and vertical bars denote SE of the means for 10 animals per group. (ο) Control, (•) DMBA-Bu₂SnCl₂ (initiation stage), (▲) TPA-Bu₂SnCl₂ (promotion stage I), (■) Mezerein-Bu₂SnCl₂ (promotion stage II).

Figure 3 shows the effect of dibutyltin on two-stage *in vitro* transformation of BALB/c 3T3 cells. The cells were initiated with MCA and promoted with TPA. Dibutyltin chloride was added at approximately the same time as MCA or TPA. Row A represents negative control; row B is with MCA alone, i.e., initiation only; row C is with TPA alone, i.e., promotion only. Under these conditions, the cells do not transform. However, when the cells were initiated with MCA and promoted with TPA, the cells were transformed completely, as in row D. This represents positive control. On the other hand, when dibutyltin at 3×10^{-7} M concentration was present at the MCA initiation stage, the transformation was completely suppressed. This effect is shown in row E when the same amount (3×10^{-7} M) of dibutyltin was present, however, at the TPA promotion stage, the transformation was not suppressed (row F). When low concentrations (3×10^{-8} M) of dibutyltin were present at both the MCA initiation stage and TPA promotion stage, there was no suppression of transformation. From these results, it is seen that Bu_2SnCl_2 is a potent inhibitor of MCA initiation stage in a two-stage system of cell transformation.

In contrast to the above result of *in vitro* two-stage transformation studies, *in vivo* studies of two-stage mouse skin carcinogenesis systems showed consistently that Bu_2SnCl_2 effectively inhibited the TPA promotion stage. The reason for this apparent discrepancy between the *in vivo* and *in vitro* results is presently unknown.

Fig. 3 : Effect of Bu_2SnCl_2 on two-stage *in vitro* transformation of BALB/c 3T3 cells. The cells (1×10^4) were initiated with MCA (0.1 µg/ml) and promoted with TPA (0.1 µg/ml). Bu_2SnCl_2 was added simultaneously with either MCA or TPA. A: whole negative control, B: MCA alone (promotion blank), C: TPA alone (initiation blank), D: MCA-TPA (positive control), E: [MCA + Bu_2SnCl_2 (3×10^{-7} M)]-TPA, F: MCA-[TPA + Bu_2SnCl_2 (3×10^{-7} M)], G: [MCA + Bu_2SnCl_2 (3×10^{-8} M)]-TPA, H: MCA-[TPA + Bu_2SnCl_2 (3×10^{-8} M)].

The effects of Bu_2SnCl_2 on the proliferation of two mature malignant culture cells, thymic lymphosarcoma cells (BW5147) and HeLa cells were also investigated *in vitro* and compared with those of thymocytes in which DNA synthesis is more active than in other normal tissue cells (Fig. 4). The result was that DNA syntheses of thymic lymphosarcoma cells and HeLa cells were more dramatically (and also in dose-related fashion) inhibited by concentrations of Bu_2SnCl_2 greater than 10^{-7} M than was the case with mitogen and nonstimulated thymocytes. At the concentration range from 10^{-7} to 10^{-6} M, Bu_2SnCl_2 affected only minimally the cell viability of thymocytes.

These results, taken together, suggest that Bu_2SnCl_2 inhibits the further growth of malignant cell population by preventing the mitotic division of the cells and thus can be effective as an anticancer agent.

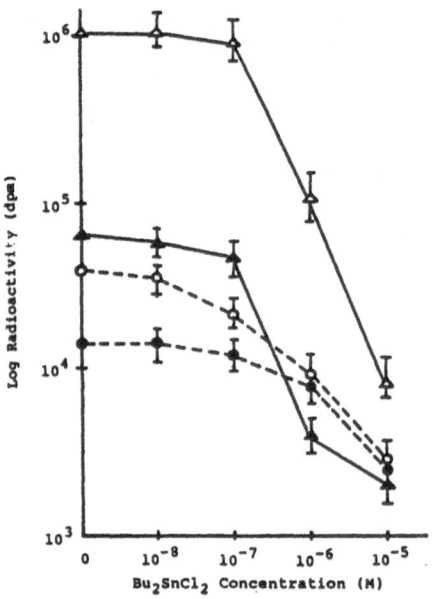

Fig. 4 : Effect of Bu_2SnCl_2 on DNA synthesis of proliferating cells. Cells (each, 1 x 10^6 cells/ml) were cultured with varying amounts of Bu_2SnCl_2 in octuple during 24 hr and [^{3}H]thymidine was present during the last 4 hr of the culture period. Vertical bars denote SE of the mean for 10 determinations. (○) ConA-stimulated thymocytes, (●) non-stimualted thymocytes, (△) Thymic lymphosarcoma cells, (▲) HeLa cells.

Cellular and biochemical aspects of the activity manifestation

Intracellular distribution of organotins

The bioactivity of organotin compounds depends on their solubility in biological fluids and the extent of their incorporation into the cells, i.e., their intracellular distribution. In this respect, the number and nature of organic ligands residing on tin may be anticipated to be a relevant consideration. A convenient means of locating organotins within cells involves the use of the fluorescent labeling technique. Applying this technique has enabled the visualization of hydrophobic organotin compounds such as di- and tributyltin chloride, and triphenyltin chloride, which tend to accumulate selectively in the Golgi apparatus and endoplasmic reticulum (ER), but not in the nucleus (Fig.5).[59-61] This observation was confirmed by using organelle-destroying agents such as monensin or nocodazole. Following treatment with monensin or nocodazole, the hydrophobic organotins were found to be dispersed throughout the cytoplasm, with negligible concentration in the Golgi apparatus and ER. In addition, their accumulation in the Golgi apparatus and ER was inhibited when fixed cells were washed with Triton-X 100 before incubation with organotins (data not shown). These results constitute strong evidence for the selective accumulation of hydrophobic organotins in the Golgi apparatus and ER.

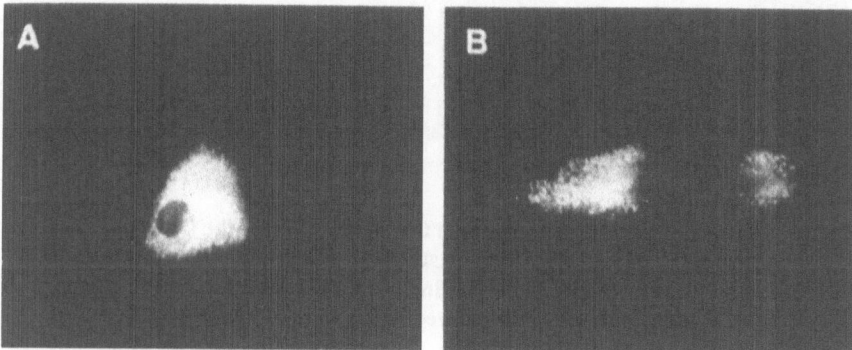

Fig. 5 : Intracellular distribution of various organotin compounds in fixed cell. Human skin fibroblasts were fixed with 0.5% glutaraldehyde for 10 min at room temperature, washed and incubated for 10 min at 37 °C with 50 μM Bu$_2$SnCl$_2$ (A) or 100 μM Me$_2$SnCl$_2$ (B). The cells were then washed, incubated with 100 μg/ml morin for 5 min at 37 °C, washed again and photographed.

Effects on structure and function of Golgi apparatus and endoplasmic reticulum

The specific stratified structure of Golgi apparatus can be observed clearly by using fluorescent ceramide (Fig.6A). However, the structure is completely destroyed by the presence of dibutyltin dichloride (Fig.6B). Furthermore, the dibutyltin suppresses such Golgi functions as lipid metabolism. In particular, the ceramide metabolism to glycosylceramide and sphingomyelin is significantly impaired (Fig.7). These results suggest that the suppression of Golgi functions by organotins may be due to the destruction of the Golgi apparatus structure.[59-62,67]

A similar destruction of the characteristic reticular structure of ER is brought about by di- and tributyltins (Fig.8). The ER network is retracted from the periphery and large cisternae are formed.[61-63] Furthermore, the hydrophobic organotins inhibit ER functions such as inositol 1,4,5-triphosphate(IP$_3$)-induced intracellular Ca^{2+} mobilization by promoting Ca^{2+} release in a similar fashion to IP$_3$ (Fig.9).[60,64] This inhibition appears to be due to the alteration or the destruction of the ER membrane structure. Monobutyltin chloride and tetrabutyltin affect neither the structure nor the function of the ER.[60,62-64]

From the results on the Golgi apparatus and the ER, it appears that the action of organotin compounds on each organelle depends primarily on their intracellular distribution which are provided by their lipotropy, i.e., the number and length of their organic ligands.

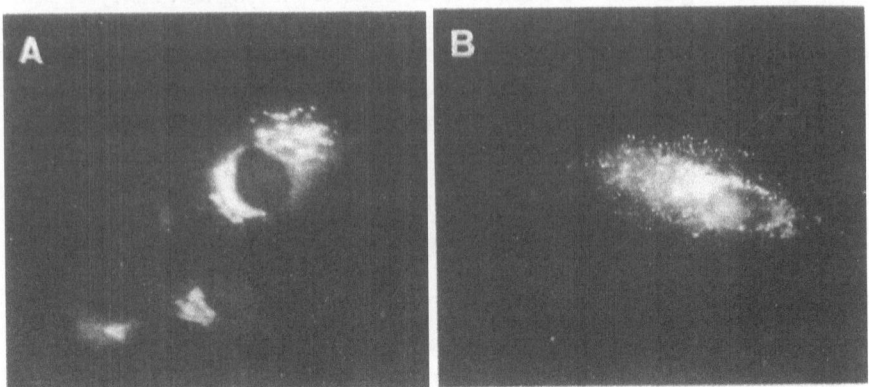

Fig. 6 : **Effects of organotin compound on the morphology of the Golgi apparatus in living cells. Human skin fibroblasts (SF-TY) were incubated in the absence (A) or presence (B) of 1 μM Bu₂SnCl₂ for 2 hr, and then were stained with C₆-NBD-ceramide-BSA and viewed in the fluorescence microscope.**

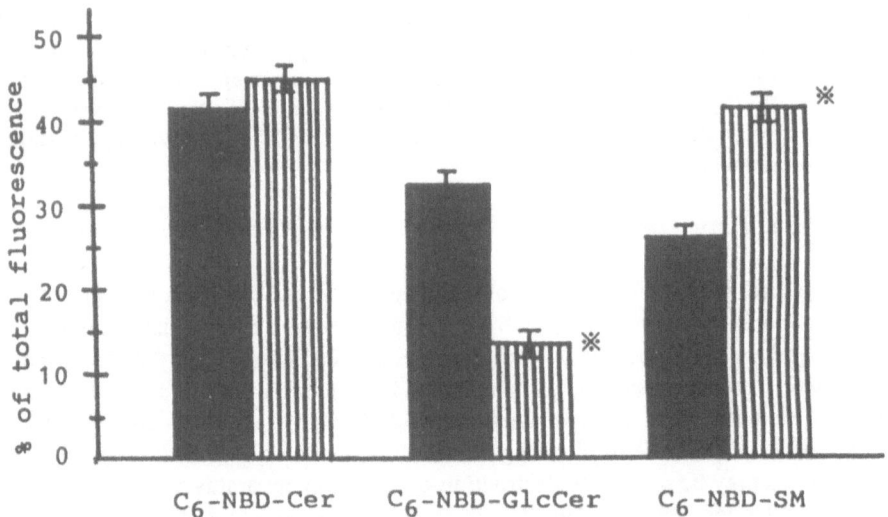

Fig. 7 : **Effects of organotin compound on the metabolism of C₆-NBD-ceramide in Golgi apparatus. CHO cells were incubated without (control, III) and with 10 μM Bu₂SnCl₂ (■) for 3 hr, and then incubated with C₆-NBD-Cer/BSA. Each value represents the mean ± SE of four independent experiments; those marked with asterisks differ significantly (Student's test) from the corresponding control value (* p < 0.001). Cer: cermide, GlcCer: glucosylceramide, SM: sphingomyelin.**

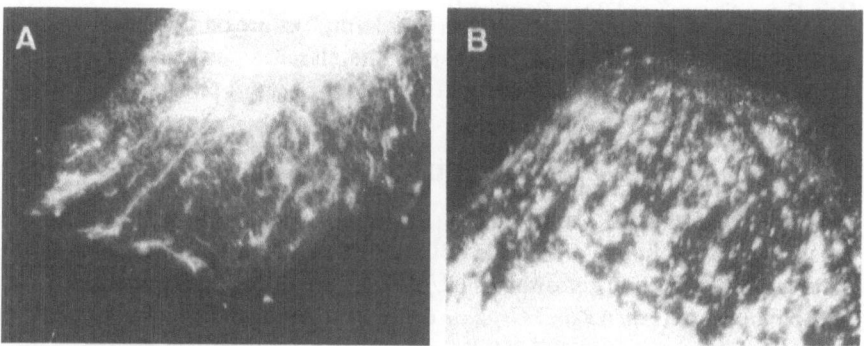

Fig. 8 : Effects of organotin compounds on the morphology of the ER in living cell.
African green monkey kidney epithelial cell line (CV-1) were incubated in the
absence (A) or presence (B) of 5 µM Bu₃SnCl for 10 min at 37 °C. Next, the
cells were fixed with 0.5% glutaraldehyde for 5 min at room temperature,
washed and incubated with 2.5 µg/ml DiOC₆ (3)(3.3'-dihexyloxacarbocyanine
iodide) for 15 sec at room temperature. The cells were then washed and
photographed.

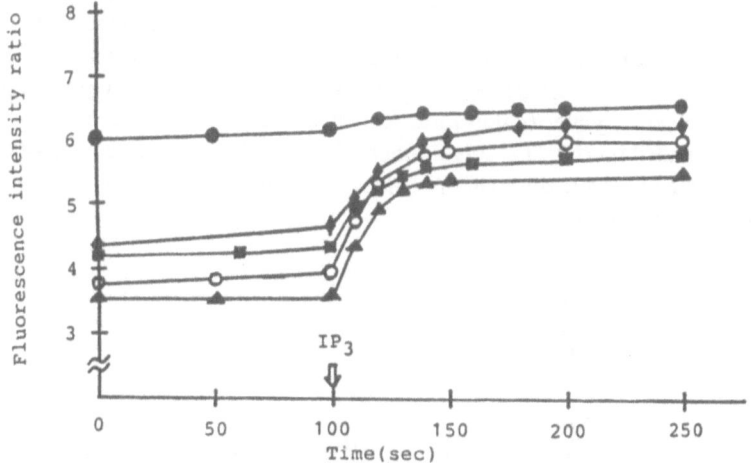

Fig. 9 : Effects of organotin compounds on Ins(1,4,5)P₃-induced intracellular Ca²⁺
mobilization. Saponin-permeabilized RBL-2H3 cells (5 x 10⁶ cells/ml) were
incubated with 5 x 10⁻⁶ M individual organotin compounds for 10 min after
the addition of 1.5 µM Fura-2. The kinetics of Ca²⁺ release were resolved by
determination of fluorescence intensity ratio suing dual excitation mode of
335 nm and 375 nm after the addition of 333 nM IP₃. Each point denotes the
mean for five determinations. Control (○), BuSnCl₃ (▲), Bu₃SnCl (•),
Bu₄Sn (■), (Bu₃Sn)₂O, TBTO (♦).

Effects on physical properties of phospholipid membranes

It is known that the activation of phospholipid transport between organelles is
affected by physical properties of the phospholipid membrane. In particular,
phospholipids of the Golgi apparatus and the ER are considered to be transported
to plasma membrane in vesicle form by vesicle budding and fusion, as shown in

435

Tin-Based Antitumour Drugs

Marcel Gielen

*Free University of Brussels VUB, Faculty of Applied Sciences,
Department of General and Organic Chemistry, Room 8G512, Pleinlaan 2,
B-1050 Brussels, Belgium*

Over the past decade, organotin compounds have emerged as the front runner among organometallics in investigations into their potential as antitumour agents.[1,2]

Several such compounds prepared in our laboratories have indeed shown[3-5] promising *in vitro* antitumour properties against two human tumour cell lines, MCF-7, a breast cancer, and WiDr, a colon carcinoma. Many of these belong to the diorganotin structural class and typically contain esteryl Sn-O linkages. An example is provided by the series, $C_5H_3N(COO)_2SnR''R$, whose member, ethylphenyltin 2,6-pyridine dicarboxylate (chloroform solvated)[6] has been shown to contain heptacoordinated tin (Fig. 1). It is a dimer, like the 1:1 condensation derivatives of carboxylic acids with diorganotin oxides (*vide infra*), with a water molecule covalently linked to tin.

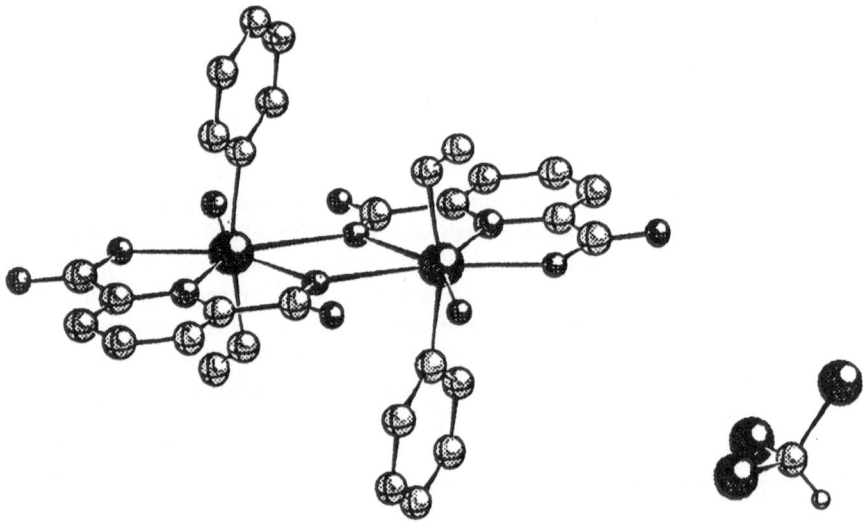

Fig. 1 : **Crystal structure of ethylphenyltin 2,6-pyridine dicarboxylate (chloroform solvated)[6]**

Fig.10. Dibutyltin dichloride exerts an "ordering" effect on the membranes of various phospholipid vesicles (Fig.11);[65,66] the effect is particularly strong for phosphatidylinositol 4-monophosphate (PIP$_1$) and phosphatidylinositol 4,5-diphosphate (PIP$_2$) vesicle membranes than for other phospholipid vesicle membranes. These effects are dose-dependent. This finding provides some clues for the inhibitory mechanism of dibutyltin on PI turnover (discussed below) for it is to be noted that PIP$_2$ is the immediate target in provoking the breakdown of inositol phospholipid. PIP$_2$ is the substrate which is hydrolyzed to diacylglycerol[53-55] and inositol phosphate or inositol polyphosphate[56-58] by phospholipase C (Fig. 12).

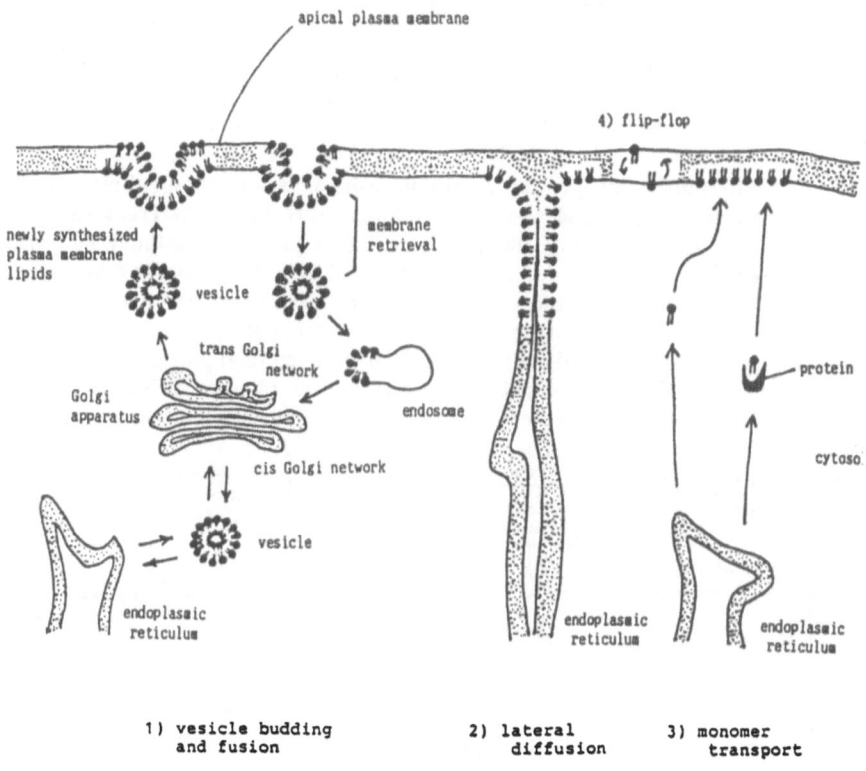

Fig. 10 : The proposed intracellular transport of phospholipids

436

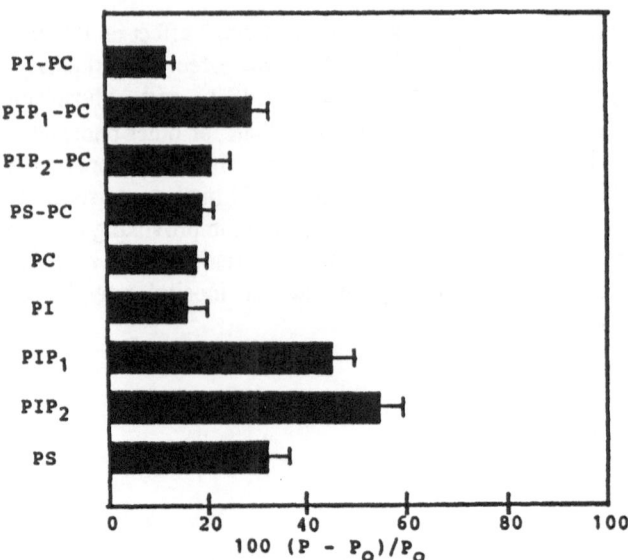

Fig. 11 : Effect of Bu₂SnCl₂ on the membrane order of various phospholipid vesicles. Each phospholipid vesicle containing DPH in 20 μM Tris-HCl buffer (pH 7.5, containing 100 mM NaCl) was incubated without and with Bu₂SnCl₂ (10^{-4} M) for 5-7 min at 25 °C and the membrane order was measured by fluorescence polarization. Each value was expressed as a percentage change in the degree of fluorescence polarization, $100(P - P_o)/P_o$. All the phospholipid vesicles contained 18 μg/ml total phospholipid. In the case of the mixed vesicles, the ratio of each phospholipid to PC is 1/2. Horizontal bars denote SE of the mean for 10 determinations.

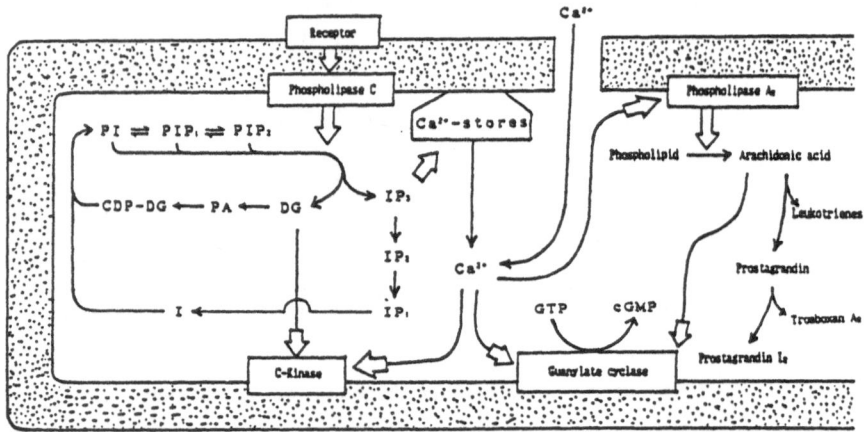

Fig. 12 : Signal transduction via phospholipid turnover and Ca mobilization

Suppressive effects on cell proliferation system

Proliferating cells such as thymic lymphocytes were more sensitive to dialkyltin compounds such as dibutyltin- and dioctyltin dichlorides than to other organotin compounds and also these cells were more susceptible than tissue cells of other organs (Table 3).

437

Table 3 : **Inhibition of thymidine uptake in rat lymphoid cells and tissue cells by dibutyltin dichloride[†]**

Molarity of Bu_2SnCl_2	Uptake of ^3H-labelled thymidine (% of control)				
	Thymocytes	Spleen cells	Liver cells	Kidney cells	Brain cells
1×10^{-4} M	15.8 ± 0.3	16.8 ± 0.4	61.1 ± 1.7	62.1 ± 1.0	95.4 ± 2.8
1×10^{-6} M	26.0 ± 0.5	33.7 ± 0.7	89.5 ± 2.5	92.1 ± 2.7	106.2 ± 3.1

[†]Cells (10^6 cells/ml) of each organ were cultured with or without Bu_2SnCl_2 during 24 hr, and [^3H] thymidine was present during the last 4 hr of the culture period. Each value is expressed as a percentage of the control value obtained in the absence of Bu_2SnCl_2, and is the mean $\pm$ SE of 10 determinations.

As shown in Fig.13, DNA synthesis of the thymic lymphocytes was significantly inhibited by dibutyltin even at the concentration of 10^{-7} M at which cell viability was not yet impaired. Moreover, a parallelism between dose-response curves of dibutyltin for DNA synthesis and cell viability was found. These results indicate that dibutyltin primarily induces an inhibition of cell proliferation and secondarily causes cell death.

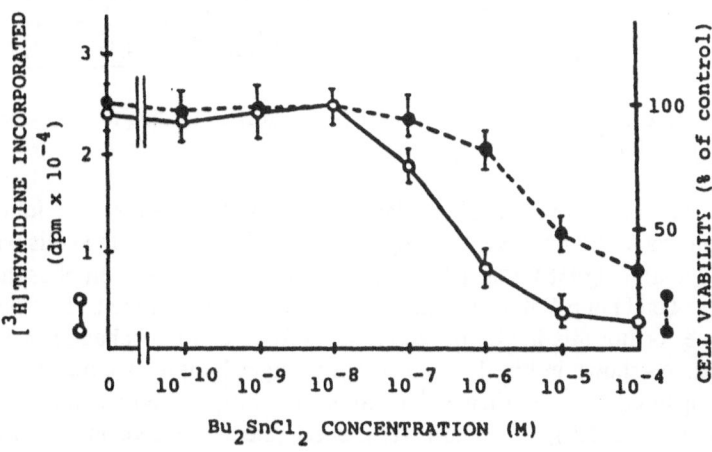

Fig. 13 : Effect of Bu_2SnCl_2 concentrations on DNA synthesis (o) and viability (●) of rat thymocytes. Cells (10^6 cells/ml) were cultured in triplicate during 24 hr and [^3H] thymidine was present during the last 4 hr of the culture period. Vertical bars denote SE of the mean for seven determinations.

For the present, the mechanism of cell proliferation and transformation is not fully understood. However, a number of hypotheses have been proposed for the main pathway leading to DNA synthesis and ultimate mitotic division.[28-52] In any case, a signal tranduction DNA synthesis appears to be initiated by stimulating the phospholipase activation system and provoking PI turnover and arachidonate release (Fig. 14). Therefore, the inihibition mechanism of dibutyltin on cell proliferation was examined by using lymphocyte transformation which has been most fully defined as a model.

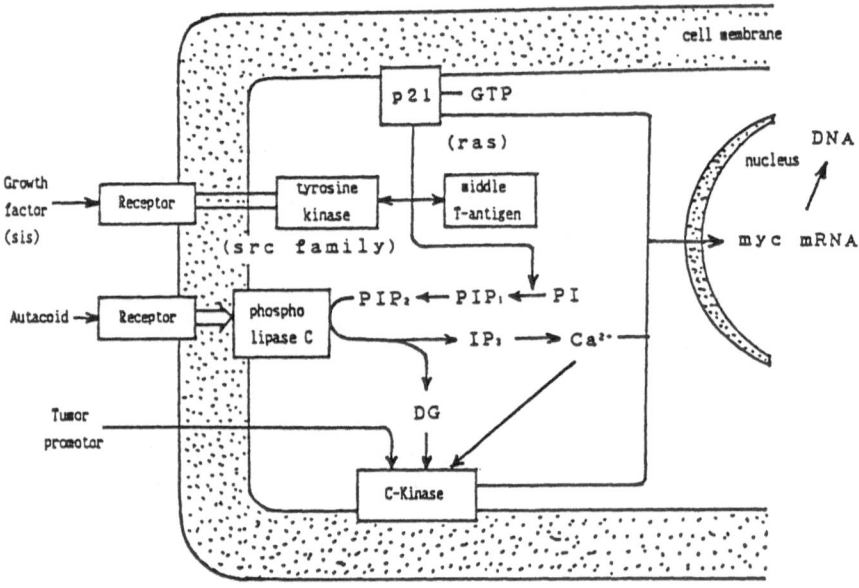

Fig. 14 : Some hypotheses proposed for the main pathway leading to cell growth and transformation

From the results obtained, it is clear that dibutyltin exerts strong inhibitory effects on DNA and RNA synthesis (Fig. 15). Ca^{2+} entry into the cells was not inhibited at all (Fig. 16). Fig. 17 shows the effect of dibutyltin on phospholipid synthesis. Total phopsholipid synthesis was significantly inhibited. The question of whether or not this inhibition is restricted to certain phosphatides was next examined. As shown in Fig. 18, in its earliest stages lymphocyte transformation is accompanied by significant increases in phosphatidylinositol (PI) and phosphatidic acid (PA), i.e., PI turnover is accelerated.[38,40] However, the presence of dibutyltin of more than 10^{-7} M concentration inhibited this remarkable acceleration. Moreover, a parallel was found between the dose-dependent inhibition of acceleration of PI turnover and that of DNA synthesis by dibutyltin. Increases in phosphatidylserine (PS), phosphatidylethanolamine (PE) and phosphatidylcholine (PC) synthesis were not observed during the first 5-30 min of the lymphocyte stimulation, and at least 2-3 hr exposure was required for their significant increases.

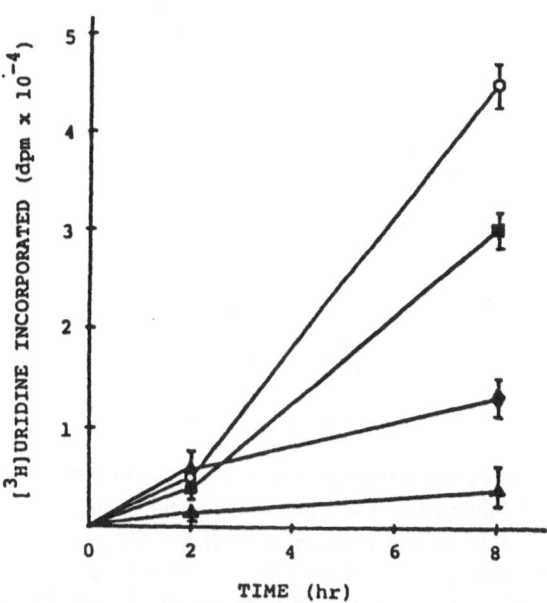

Fig. 15 : Effect of organotin compounds on RNA synthesis. Incorporation of [³H]uridine into RNA of thymocytes (10⁶ cells/ml) after stimulation with ConA (5 µg/ml) was measured in the absence (○) and presence of 10^{-7} M n-Bu$_2$SnCl$_2$ (▲), 10^{-7} M Ph$_3$SnCl (♦), or 10^{-7} M MeSnCl$_3$ (■). Each point is corrected for radioactivity incorporated without ConA at each incubation time. Vertical bars denote the SE of the mean for five determinations.

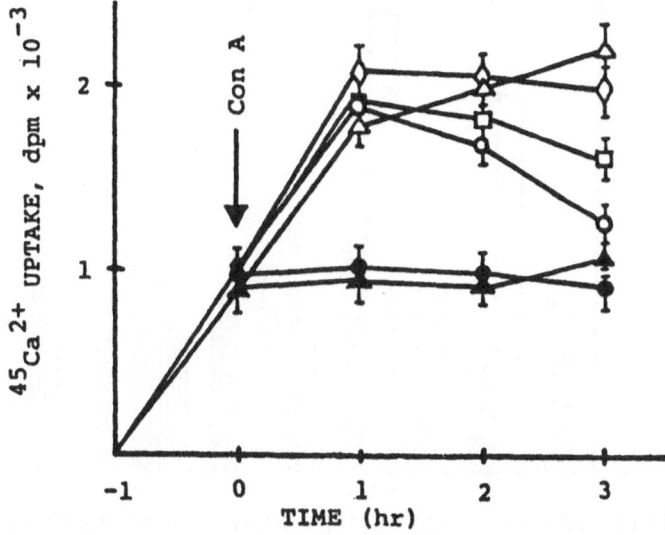

Fig. 16 : Effect of organotin compounds on calcium uptake. Incorporation of ⁴⁵Ca²⁺ into thymocytes (10⁶ cells/ml) after stimulation with ConA (5 µg/ml) was measured in the absence (○) and presence of 10^{-7} M n-Bu$_2$SnCl$_2$ (▲), 10^{-7} M Ph$_3$SnCl (◊), or 10^{-7} M MeSnCl$_3$ (□), and ⁴⁵Ca²⁺ uptake by nonstimulated thymocytes was measured in the absence (•) and presence (▲) of 10^{-7} M n-Bu$_2$SnCl$_2$. Vertical bars denote the SE of the mean for five determinations.

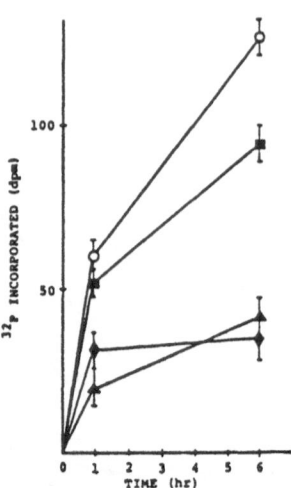

Fig. 17 : Effect of organotin compounds on phospholipid synthesis of rat thymocytes. Phospholipid synthesis was measured by the incorporation of ^{32}P into the lipid fraction of the cultured cells (1.5×10^6 cells per milliliter) after stimulation with Con A (5 μg/ml) in the absence (○) and presence of $10^{-7}M$ n-Bu$_2$SnCl$_2$ (▲), 10^{-7} M MeSnCl$_3$ (■), or $10^{-7}M$ Ph$_3$SnCl (◆). Each point is corrected for radioactivity incorporated without Con A at each incubation time. Vertical bars denote the SE of the mean for five determinations.

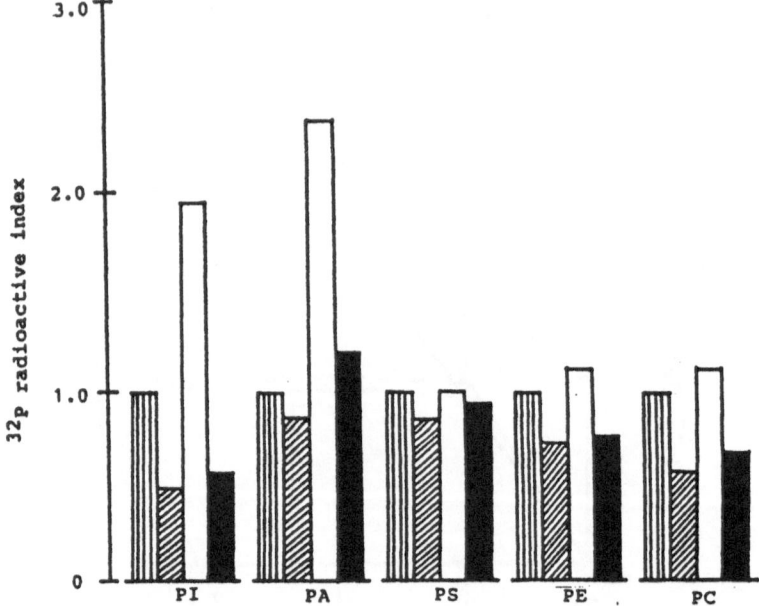

Fig. 18 : Effect of Bu$_2$SnCl$_2$ on phospholipid metabolism. Rat thymocytes (3×10^6 cells/ml) prelabeled with ^{32}P-phosphoric acid (10 μCi/ml) were treated without and with Con A (5 μg/ml) in the absence and presence of Bu$_2$SnCl$_2$ (5×10^{-7} M) for 5 min. The mean radioactivity of each phospholipid component separated from the control culture (▥) was taken as 1.0 and was compared with that of the corresponding phospholipid component from the experimental cultures treated with Bu$_2$SnCl$_2$ (□), Con A (▨) and Con A + Bu$_2$SnCl$_2$ (■)

441

On the other hand, the degradation of phospholipids was also inhibited by the presence of dibutyltin. Figure 19 shows the time course of arachidonate release from phospholipids by phospholipase A_2. The acceleration of arachidonate release was significantly inhibited by dibutyltin of greater than 10^{-7} M concentration. However, no direct inhibitory effects of organotin compounds on phospholipase A_2, C and D were found, although the substrate was hydrolyzed to three main products depending upon the species of phospholipase (Table 4).

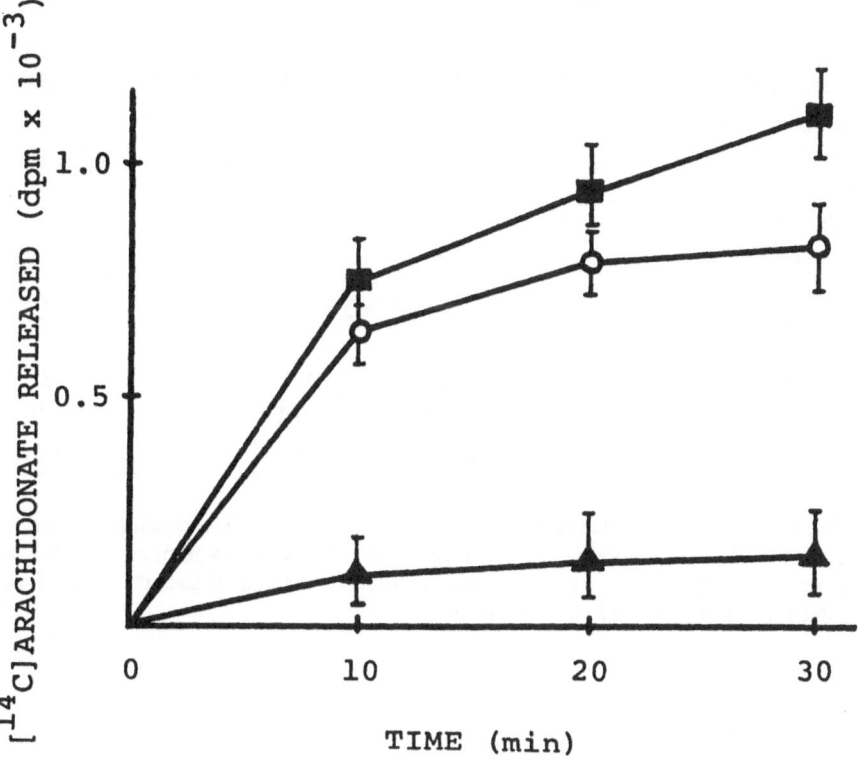

Fig. 19 : Time course of release of arachidonic acid. [1-^{14}C]Arachidonate release was measured after stimulation with 5 µg of Con A in the absence (○) and presence of 10^{-7} M Bu$_2$SnCl$_2$ (▲) or 10^{-7} M MeSnCl$_3$ (■). Each point is corrected for radioactivity released without Con A at each incubation time. Vertical bars denote SE of the mean for five determinations.

Conclusion

Taking the cell biochemical results into considertion with the cell biological results, it is proposed that antiproliferative activity of organotin compounds appears to be due to the inhibition of intracellular phospholipid transport between organelles by impairing the structure and functions of the Golgi apparatus and the ER, and the consequent inhibition of phospholipid metabolism and finally inhibition of the membrane-mediated signal transduction system leading to DNA synthesis via phospholipid turnover and Ca^{2+} mobilization.

Table 4 : *In vivo* effect of organotins on various phospholipase activities

Enzyme source	Spot No[†]	Phosphatidylcholine hydrolyzed (% of total counts)			
		Control	Organotin added $(1 \times 10^{-7}$ M)		
			Bu_2SnCl_2	Bu_3SnCl	Ph_3SnCl
Phospholipase A2	1	25.0 ± 0.7	27.3 ± 0.2	31.3 ± 0.8	30.1 ± 0.7
(porcine pancreas)	2	5.5 ± 0.2	7.3 ± 0.3	8.2 ± 0.9	9.0 ± 1.2
	3	4.8 ± 0.1	5.9 ± 0.5	7.1 ± 1.1	6.5 ± 0.7
Phospholipase C	1	7.5 ± 1.3	8.2 ± 1.0	5.3 ± 0.5	6.8 ± 0.3
(*B. cereus*)	2	0.7 ± 0.2	0.7 ± 0.1	0.6 ± 0.1	1.0 ± 0.2
	3	70.0 ± 4.2	63.0 ± 2.8	75.0 ± 5.9	63.0 ± 3.1
Phospholipase D	1	15.8 ± 0.9	17.4 ± 1.2	17.6 ± 1.2	16.8 ± 0.7
(cabbage)	2	1.0 ± 0.1	1.3 ± 0.1	1.4 ± 0.1	0.9 ± 0.1
	3	1.5 ± 0.3	2.3 ± 0.2	2.1 ± 0.2	1.4 ± 0.1

[†]The spot numbers indicate the hydrolyzed components of substrate, L-α-dipalmitoyl-[choline-methyl-^{14}C-]-phosphatidylcholine. The total recovered radioactivity from one sample was taken as 100% and the relative radioactivity in each spot was determined from this value. Each value is the mean $\pm$ SE of five determinations.

References

1. Y. Arakawa and O. Wada, Suppression of cell proliferation by certain organotin compounds, In: *Tin and Malignant Cell Growth*, ed., J.J. Zuckerman, CRC Press, Boca Raton, Florida, 1988, Chapter 9, p. 83.
2. Y. Arakawa and O. Wada, *Biochem Biophys Res Commun.*, **123** (1984) 543.
3. Y. Arakawa and O. Wada, *Biochem Biophys Res Commun.*, **125** (1984) 59.
4. Y. Arakawa, *J. Pharmacobio-Dyn*, **6** (1983) s-23.
5. Y. Arakawa and O. Wada O, Suppression of cell proliferation by certain organotin compounds. Proceedings of the Second International Symposium on the Effect of Tin upon Malignant Cell Growth, May 19-22, Scranton , PA, USA, 4-5, 1985.
6. Y. Arakawa, T. Abe, T.H. Yu and O. Wada, Anti-tumor activity and anti-inflammatory action of organotin compounds. Proceedings of the Third Symposium on Roles of Metal in Biological Reactions, Biology and Medicine, June 5-6, Nagoya, Japan, 1986.
7. Y. Arakawa, Side effect of organotin compound possessing antitumor activity. Proceedings of the Third International Symposium on the Effect of Tin upon

Malignant Cell Growth, September 5-6, Padua, Italy, 1986, p.8.

8. Y. Arakawa, *Ibid* (1986) p.24.

9. Y. Arakawa, Anti-inflammatory action of organotin compounds. Proceedings of the Fifth International Conference on the Organometalic and Coordination Chemistry of Germanium.Tin and Lead, September 8-12, Padua, Italy, 1986, p.3.

10. Y. Arakawa, T. Abe, T.H. Yu and O. Wada, *J Pharmacobio-Dyn* **10** (1987) s-2.

11. C.E. Weeks, T.J. Slaga, H. Hennings, G.L. Gleason and W.M. Bracken, *J Natl. Cancer Inst.*, **63** (1979) 401.

12. T.J. Slaga, *Environ Health Perspectives*, **50** (1983) 3.

13. IARC/NCI/EPA Working Group, *Cancer Res.*, **45** (1985) 2395.

14. T. Enomoto and H. Yamasaki, *Cancer Res.*, 1985, **45**, 2681.

15. R.W. Walenga, H.J. Showell, M.B. Feinstein and E.L. Becker, *Life Sci.*, **27** (1980) 1047.

16. Y. Masuzawa, T. Osawa, K. Inoue and S. Kojima, *Biochim Biophys Acta*, **326** (1973) 339.

17. F. Hirata, *J Biol Chem.*, **256** (1981) 7730.

18. Y. Arakawa, O. Wada and M. Manabe, *Anal. Chem.*, **55** (1983) 1901.

19. N.C. Lipsky and R.E. Pagano, *Science*, **288** (1985) 745.

20. T. Kobayashi and R.E. Pagano, *Cell*, **55** (1988) 797.

21. T. Kobayashi and R.E. Pagano, *J. Biol. Chem.*, **264** (1989) 5966.

22. T. Kobayashi and Y. Arakawa, *J. Cell. Biol.*, **113** (1991) 235.

23. M. Terasaki, J. Song, J.R. Wong, M.J. Weiss and L.B. Chen, *Cell*, **38** (1984) 101.

24. H. Streh., R.F. Irvine, M.J. Berridge and I. Schulz, *Nature*, **306** (1983) 67.

25. S.K. Joseph, A.P. Thomas, R.J. Williams, R.F. Irvine and J.R. Williamson, *J. Biol. Chem.*, **259** (1984) 3077.

26. M.J. Berridge and R.F. Irvine, *Nature*, **312** (1984) 315.

27. K. Simons and H. Virta, *EMBO J.*, **6** (1987) 2241.

28. M.F. Greaves, S. Bauminger and G. Janossy, *Clin exp. Immunol.*, **10** (1972) 537.

29. J.H. Peters and P. Hausen, *Eur J Biochem.*, **19** (1971) 502.

30. J.H. Peters and P. Hausen, *Eur J Biochem.*, **19** (1971) 509.

31. M.R. Ouastel and J.G. Kaplan, *Exptl Cell Res.*, **63** (1970) 230.

32. R.B. Whitney and R.M. Sutherland, *J Cell Physio.*, **182** (1973) 19.

33. J.F. Whitfield, A.D. Perris and T. Youdale, *J Cell Physiol.*, **73** (1969) 203.

34. V.C. Maino, N.M. Green and M.J. Crumpton, *Nature*, **251** (1974) 324.

35. R.E. Barnett, R.E. Scott, L.T. Furcht and J.H. Kersey, *Nature*, **249** (1974) 465.

36. T. Katagiri, T. Terao and T. Osawa, *J Biochem.*, **79** (1976) 849.

37. F. Hirata, S. Toyoshima, J. Axelrod and M. Waxdal, *Proc Natl Acad Sci USA*, 1980, **77**, 862.

38. D.B. Fischer and G.G. Mueller, *Proc Natl Acad Sci USA*, **60** (1968) 1396.

39. M.M. Billah, E.G. Lapetina and P. Cuatrecasas, *Biochem Biophys Res Commun.*, **90** (1979) 92.

40. D.A. Kennerly, T.J. Sullivan, P. Saylwester and C.W. Parker, *J Exp Med.*, **150** (1979) 1039.
41. F. Hirata, B.A. Corcoran, K. Venkatasubramanian, E. Schiffmann and J. Axelrod, *Proc Natl Acad Sci USA*, **76** (1979) 2640.
42. R.L. Bell, D.A. Kennerly, N. Stanford and P.W. Majerus, *Proc Natl Acad Sci USA*, **76** (1979) 3238.
43. H.L. Cooper and A.D. Rubin, *Blood*, **25** (1965) 1014.
44. T. Ono, H. Terayama, F. Takaku and K. Nakano, *Biochim Biophys Acta*, **161** (1968) 361.
45. B.G.T. Pogo, V.G. Allfrey and A.E. Mirsky, *Proc Natl Acad Sci USA*, **55** (1966) 805.
46. L.J. Kleinsmith, V.G. Allfrey and A.E. Mirsky, *Science*, **154** (1966) 780.
47. E.M. Johnson, J. Karn and V.G. Allfrey, *J Biol Chem.*, **249** (1974) 4990.
48. R. Levy, S. Levy, S.A. Rosenberg and R.T. Simpson, *Biochemistry*, **12** (1973) 224.
49. Y. Takai, A. Kishimoto, U. Kikkawa, T. Mori and Y. Nishizuka, *Biochem Biophys Res Commun.*, **91** (1979) 1218.
50. M.R. Hokin and L.E. Hokin, *J Biol Chem.*, **203** (1953) 967.
51. P. Downes and R.H. Michell, *Cell Calcium*, **3** (1982) 467.
52. M.J. Berridge, R.M.C. Dawson, C.P. Downes, J.P. Heslop and R.F. Irvine, *Biochem J.*, **212** (1983) 473.
53. Y. Kawahara, Y. Takai, R. Minakuchi, K. Sano and Y. Nishizuka, *Biochem. Biophys ResCommun.*, **97** (1980) 309.
54. K. Sano, Y. Takai, J. Yamanishi and Y. Nishizuka, *J Biol Chem.*, **258** (1983) 2010.
55. H. Ieyasu, Y. Takai, K. Kaibuchi, M. Sawamura and Y. Nishizuka, *Biochem Biophys Res Commun.*, **108** (1982) 1701.
56. M.J. Berridge and R.F. Irvine, *Nature*, **312** (1984) 315.
57. K. Suematsu, M. Hirata, T. Hashimoto and H. Kiriyama, *Biochem Biophys Res Commun.*, **120** (1984) 481.
58. P. Volpe, G. Salviati, E.D. Virgilio and T. Pozzan, *Nature*, **316** (1985) 347.
59. Y. Arakawa, Visualization of intracellular distribution of dibutyltin compound, Proceedings of the 6th Int. Conf. on Organometal. & Coord. Chem. of Ge, Sn and Pb, Brussels, 1989.
60. Y. Arakawa, T. Iizuka and C. Matsumoto, *Biomed Res Trace Elements*, **2**(3) (1991) 321.
61. Y. Arakawa and O. Wada, Biological properties of alkyltin compounds. In: *Metal Ions in Biological Systems*, ed., H. Sigel, Volume 29, Marcel Dekker, Inc., New York, 1992, pp. 101-136.
62. T. Iizuka and Y. Arakawa, *Jpn. J. Hyg.*, **46**(1) (1991) 302.
63. T. Iizuka and Y. Arakawa, *Jpn. J. Hyg.*, **47**(1) (1992) 249.
64. T. Iizuka, C. Matsumoto and Y. Arakawa, *Jpn. J. Hyg.*, **46**(1) (1991) 303.
65. Y. Arakawa, *Main Group Metal Chem.*, **12** (1989) 37.
66. Y. Arakawa, *Jpn. J. Hyg.*, **44**(1) (1989) 439.
67. T. Iizuka and Y. Arakawa, *Jpn. J. Hyg.*, **47**(1) (1992) 248.

Within the series, the di-*n*-butyltin compound proved to be the most active in *in vitro* tests, giving inhibition dose ID_{50} values 60 and 106 ng/mL, respectively, against MCF-7 and WiDr.[4,5] By way of comparison, cisplatin gave ID_{50} values of 850 and 624 ng/mL, respectively, against the same tumour cell lines.

We have also synthesized diorganotin derivatives of substituted salicylic acid.[4,5] Only the carboxylic acid function reacts with the diorganotin oxide moiety; the phenolic hydroxyl group remains intact in the organotin compound formed. Two types of such compounds can be prepared depending on the molar ratio of carboxylic acid to the diorganotin oxide used in the preparation.

When a 2:1 ratio is used, the expected distorted octahedral diorganotin dicarboxylate is formed. By way of example, the crystal structure of di-*n*-butyltin bis(5-chlorosalicylate)[7] is given in Figure 2.

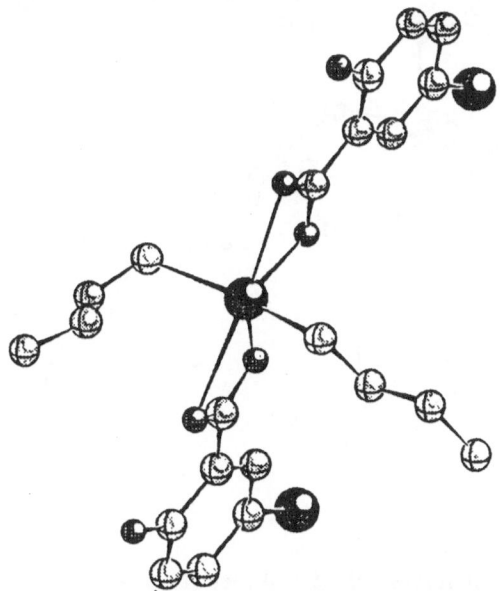

Fig. 2 : Crystal structure of di-*n*-butyltin bis(5-chlorosalicylate)[7]

In the antitumour tests, the di-*n*-butyltin compounds again proved to be the most active ones. For instance, di-*n*-butyltin bis(4-hydroxy-3-methoxybenzoate) is characterized by ID_{50} of 44 and 82 ng/mL against MCF-7 and WiDr, respectively.[4,5]

Antitumour activity of bis[carboxylato(diorganotin)] oxides

When a 1:1 molar ratio of carboxylic acid to diorganotin oxide is used, dimers of bis[carboxylato(diorganotin)] oxides form. The structure of the dimer of di-*n*-butyl(3,4,5-trimethoxybenzoato) tin oxide has been determined both in $CDCl_3$ solution and in the solid state (Fig. 3).[8] Here again, the di-*n*-buyltin compounds proved to be the most active ones, much more active than cisplatin. The 1:1 condensation compound of di-*n*-butyltin oxide and 5-methoxysalicylic acid, {[5-$CH_3OC_6H_3(OH)COOSnBu_2]_2O\}_2$, for instance, scored ID_{50} values of 29 and 122 ng/mL against MCF-7 and WiDr, respectively.[4,5]

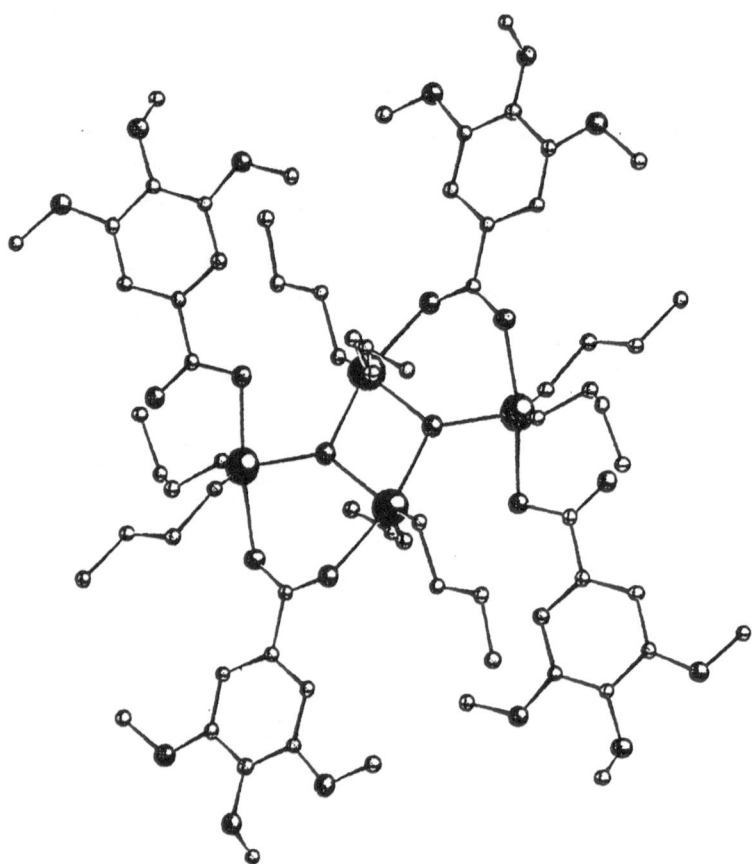

Fig. 3 : Crystal structure of the dimer of di-*n*-butyl(3,4,5-trimethoxy-benzoato)tin oxide[8]

Condensation complex of di-*n*-buyltin oxide with salicylaldoxime

Whereas di-*n*-buyltin oxide reacts with carboxylic acids either in a 1:1 or 1:2 molar ratio, the condensation complex of di-*n*-buyltin oxide with salicylaldoxime gives consistently, after recrystallization, a 3:2 derivative (compound M), independent of the molar ratio used. The molecular structure of compound M in the crystalline state in shown in Figure 4.[9] Inhibition doses ID_{50} of 67 and 215 ng/mL were found for this compound against MCF-7 and WiDr, respectively.

In compound M, one seven-coordinate tin atom and two different five-coordinate tin atoms are present, as shown by X-ray diffraction and by ^{119}Sn NMR spectroscopic measurements; there are also two differently coordinated salicylaldoximate ligands, as shown by ^{1}H and ^{13}C NMR spectroscopy. When M is dissolved in $CDCl_3$ or in C_6D_6, it maintains its solid state structure, as shown by multinuclear 1D NMR and 2D NMR.[9]

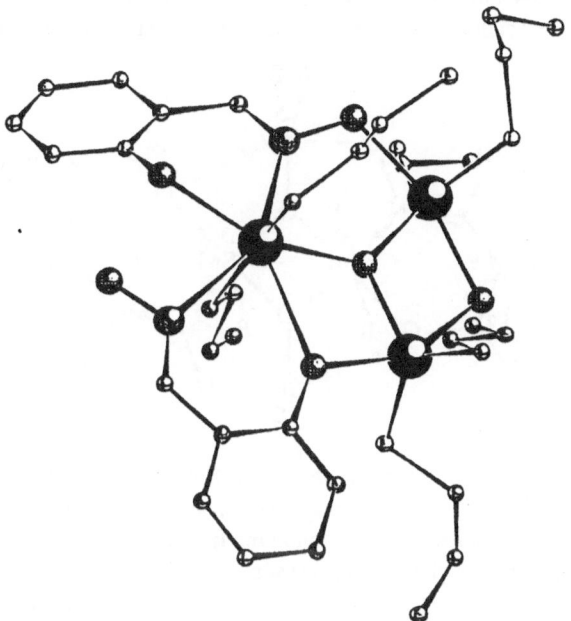

Fig. 4 : Crystal structure of the dimer of the di-*n*-buytltin derivative of salicylaldoxime (compound M)

However, from M, three new compounds are formed as minor species in solution: **m1**, **m2** and **m3**. The following structures have been proposed for these three new products on the basis of NMR and chemical evidence.[10]

Compound **m1** is quite similar to M but contains one more salicylaldoximate moiety coordinated to the two five-coordinate tin atoms.

m1:

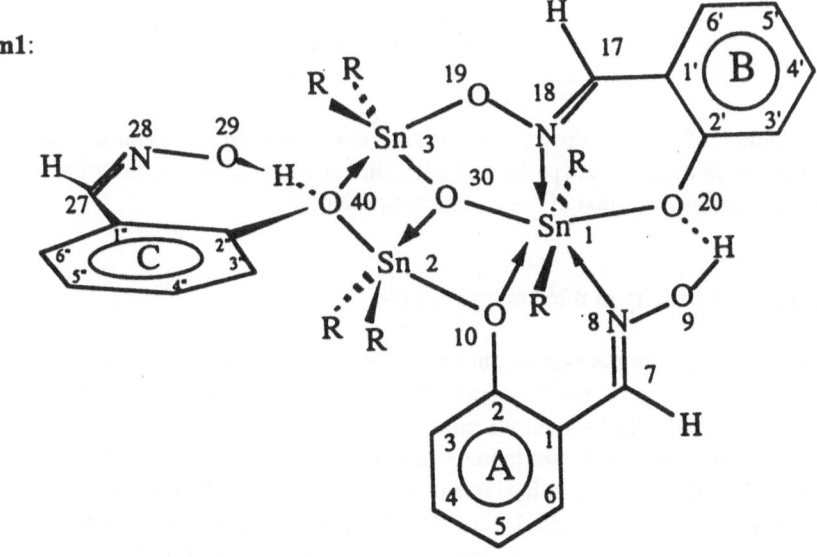

449

Two equivalent six-coordinate tin atoms, two equivalent five-coordinate tin atoms, and two equivalent salicylaldoximate ligands, are present in compound **m2**.

m2:

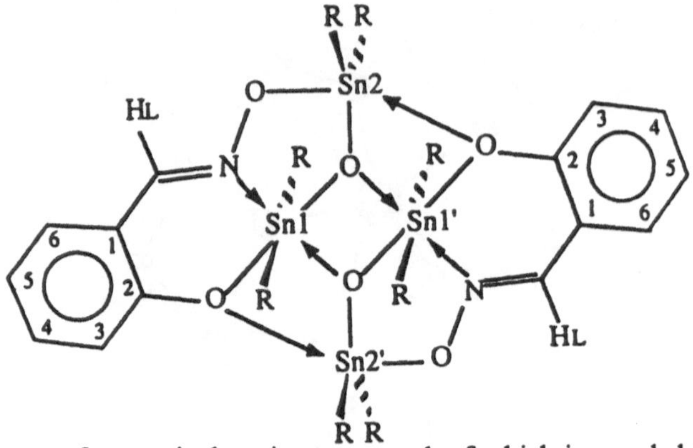

In **m3**, there are four equivalent tin atoms, each of which is coupled to its three neighbours. It has an adamantane-based Sn_4O_6 cluster structure, with two of the oxygen atoms originating from two distannoxane moieties, and four other ones from the four equivalent salicylaldoximate ligands.

m3:

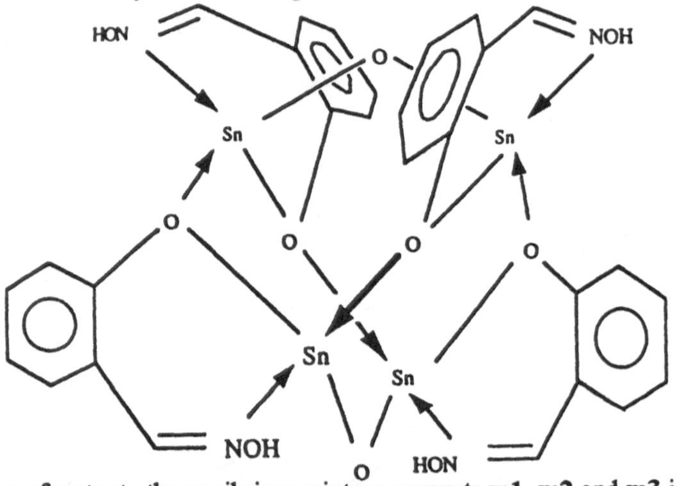

The addition of water to the equibrium mixture converts **m1**, **m2** and **m3** into **M**. Because the *in vitro* tests are performed with diluted water solutions, the antitumour activity determined is that of the parent compound **M**.

Antitumour activity of diorganostannates

In order to increase the water-solubility of these diorganotin derivatives, which might, according to Atassi,[11] increase their antitumour properties, we converted some of them into their tetraethylammonium halide adducts.[12]

However, the *in vitro* antitumour activity of the tetraethylammonium halide adducts of the diorganotin 2,6-pyridine dicarboxylates[12] is no better than that of the parent molecules, even though their solubility in protic and also in less polar solvents is considerably enhanced. The same is true for diorganotin thiosalicylates.[13]

Exceptionally high antitumour activity of triphenyltin carboxylate

Recently, we have discovered that triphenyltin carboxylates are also an active class of antitumour agents, several of which display higher *in vitro* activity (ID_{50} *ca.* 15 ng/mL) than mitomycin C against MCF-7 and WiDr.[3,14]

Exceptionally high antitumour activity of di-*n*-butyltin difluorobenzoates

We also recently synthesized a series of di-*n*-butyltin fluorocarboxylates.[15] The presence of fluorine atoms as substituents the phenyl ring of the benzoate moiety enhances the solubility of these compounds both in water and in non-polar solvents.

The results obtained for di-*n*-butyltin difluorobenzoates are summarized in Table 1. Clearly, the 2:1 compounds **1, 2** and **3** are less active than their 1:1 analogs, **4, 5** and **6**, which exhibit *in vitro* antitumour activities against MCF-7 comparable to those shown by mitomycin C. We have synthesized many more di-*n*-butyltin polyfluorocarboxylates, the activities of whichare the subect of a patent application.

Table 1 : ID_{50} values (ng/mL) of compounds of the type $(F_2C_6H_3COO)_2Sn(n-C_4H_9)_2$ **1, 2** and **3**, and of the type $\{[(F_2C_6H_3COO)_2Sn(n-C_4H_9)_2Sn_2O\}$, **4, 5** and **6**, tested[10] against two human tumour cell lines, MCF-7 and WiDr

		MCF-7	WiDr
1	2,3-F$_2$	223	283
2	2,6-F$_2$	98	326
3	3,5-F$_2$	30	407
4	2,3-F$_2$	9	120
5	2,6-F$_2$	3	174
6	3,5-F$_2$	11	172
Cisplatin[10]		850	624
Etoposide[10]		187	624
Doxorubicin[10]		63	31
Mitomycin C[10]		3	17

Conclusion

To conclude, many di- and triorganotin compounds have been shown to possess higher *in vitro* activity than cisplatin against the two human tumour cell lines,

MCF-7 and WiDr. Of course, one has to await *in vivo* test results on promising members of these before claiming anything about their real interest in the field of cancer chemotherapy. With more intensive development and screening studies on organotins, it is conceivable that they may indeed realise their high potential in this respect.

References

1. V.L. Narayanan, M. Nasr and K.D. Paull, Computer assisted structure - Antileukemic activity correlations of organotin compounds and initial exploration of their potential anti-HIV activity. In: *Tin-based Antitumour Drugs*, ed., M. Gielen, Springer-Verlag, 1990, pp. 201-217.
2. M. Gielen, A. Meriem, M. Boualam, A. Dalmotte, R. Willem and D. de Vos, Tin-based antitumour drugs: past, present, and future. In: *Chemistry and Technology of Silicon and Tin*, eds., V.G. Kumar Das, S.W. Ng and M. Gielen, Oxford Science Publ., 1992, pp. 312-318.
3. M. Boualam, M. Gielen, A. Meriem, D. de Vos and R. Willem, Pharmachemie B.V.): Anti-tumour composition and compounds, *Eur. Pat.* 90202316.7-21/09/90; M. Boualam, M. Gielen, A. El Khloufi, D. de Vos and R. Willem, Pharmachemie B.V., *Eur. Pat.* 91202746.3-, 22.10.91, Novel organotin compounds having antitumour activity and antitumour compositions.
4. M. Gielen, P. Lelieveld, D. de Vos and R. Willem, *In vitro* antitumour activity of organotin compounds. In: *Metal-based Antitumour Drugs*, ed., M. Gielen, vol. 2, Freund Publ. House, Tel Aviv, 1992, pp. 29-54.
5. M. Gielen, P. Lelieveld, D. de Vos and R. Willem, *In vitro* antitumour activity of organotin(IV) derivatives of salicylic acid and related compounds. In: *Metal Complexes in Cancer Chemotherapy*, ed., B. Keppler, VCH, Weinheim, Germany, 1993, pp. 383-390.
6. E.R.T. Tiekinik, M. Acheddad and M. Gielen, unpublished results.
7. E.R.T. Tiekinik, A. El Khloufi and M. Gielen, unpublished results.
8. M. Gielen, J. Meunier-Piret, M. Biesemans, R. Willem and A. El Khloufie, *Appl. Organomet. Chem.*, **6** (1992) 59.
9. F. Kayser, M. Biesemans, M. Boualam, J. Meunier-Piret, A. El Khloufie, A. Bouhdid, K. Jurkschat, M. Gielen and R. Willem, *Organometallics*, in press.
10. M. Boualam, M. Biesemans, J. Meunier-Piret, R. Willem and M. Gielen, *Appl. Organomet. Chem.*, **6** (1992) 197.
11. G. Atassi, *Rev. Si, Ge, Sn and Sn Cpds*, **8** (1985) 219.
12. M. Boualam, R. Willem, M. Biesemans, A. Delmotte, A. El Khloufi and M. Gielen, *Appl. Organomet. Chem.*, in press.
13. R. Willem, M. Biesemans, F. Kayser, M. Boualam and M. Gielen, *Inorg. Chim. Acta*, **197** (1992) 25.
14. M. Gielen, R. Willem, M. Biesemans, M. Boualam, A. El Khloufi and D. de Vos, *Appl. Organomet. Chem.*, **6** (1992) 287.
15. M. Gielen, M. Biesemans, A. El Khloufi, J. Meunier-Piret, F. Kayser and R. Willem, *J. Fluorine Chem.*, **64** (1993) 279.

Main Group Elements and Their Compounds
V.G. Kumar Das (Ed)
Copyright © 1996 Narosa Publishing House, New Delhi, India

Organoantimony(III) Compounds : A New Class of Organometallic Antitumour Agents

Cristian Silvestru[a] and Ionel Haiduc[b]

[a]*Instituto de Química, Universidad Nacional Autónoma de Mexico, Ciudad Universitaria, 04510 Mexico, D.F. and* [b]*Facultatea de Chimie, Unviersitatea "Babes-Bolyai", Ro-3400 Cluj-Napoca, Roumania*

The fight against cancer remains an on-going one, despite the significant strides made in immunotherapy in recent times, and the wide variety of improved as well as new drugs that are constantly being subjected to screening. Among the category of new drugs that are receiving much attention are metal-based compounds.[1,2] Interest in this class of compounds first began in 1969 with the landmark discovery by Rosenberg[3] of the effectiveness of *cisplatin* in the treatment of testicular and ovarian cancers. Previous to this, there had been no systematic studies on the antitumour properties of metal compounds, although some lead, mercury and especially arsenic compounds had been occasionally used, though without spectacular results, in the clinical treatment of some cancers. *Cisplatin, cis-*$[Pt(NH_3)_2Cl_2]$, was first described in 1844 by Peyrone,[4] but it took more than one hundred years to discover its powerful antitumour properties. The consequence of this important discovery was that numerous scientists and laboratories in the world began to direct their efforts not only towards finding other platinum compounds with improved therapeutic properties and lower toxicity, but also to synthesize and test for their antitumour properties many other inorganic metal compounds. Moreover, the investigations were extended to organometallic derivatives. To date compounds of practically all transition and main group metals have been screened for antitumour properties (Figs. 1 and 2), and many of them have been found to exhibit more than marginal activity towards the standard animal tumour systems.

										B		
										Al	Si	
Sc	Ti	V	Cr	Mn	Fe	Co	Ni	Cu	Zn	Ga	Ge	As
Y	Zr	Nb	Mo		Ru	Rh	Pd	Ag	Cd	In	Sn	Sb
La	Hf	Ta	W	Re	Os	Ir	Pt	Au	Hg	Tl	Pb	Bi	Te

Fig. 1 : **Metals and metalloids whose compounds have been screened as anticancer agents**

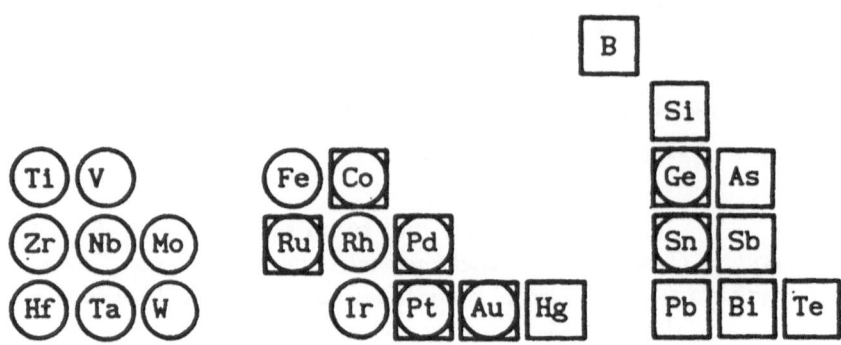

Fig. 2 : **Metals and metalloids whose organometallic compounds have been screened as antitumour agents : circled entries indicate π-complexes, boxed entries indicate metal-carbon σ-bonded complexes**

With main group metals, most of the studies have focussed on inorganic gallium compounds and organometallic germanium and tin derivatives.[1] Thus, carboxyethylgermanium sesquioxide, $(HO_2CCH_2CH_2GeO_{1.5})_n$ (Ge-132), was even used in clinical treatments (however, without spectacular results until now),[5] while, more recently, some diorganotin compounds were reported to exhibit significantly higher *in vitro* activity than *cisplatin*.[6]

By contrast, antimony compounds appear to have been rather negligibly investigated for their anticancer properties, although their spectrum of biological activity and use in the treatment of some tropical diseases have been well documented (Table 1). Among the few that were tested and found to exhibit antitumour properties were antimony(III) acetonates and antimony-containing heteropolytungstates. However, no organoantimony compounds appear to have been screened for their antitumour activity until the 1990's when we initiated work in this direction in our laboratories.[7-10] This paper expands on our reported work, and, in particular, compares the activities of two diphenylantimony(III) derivatives of dithiophosphorus ligands with their diphenyltin(IV) analogues.

Results and Discussion

The diorganoantimony(III) and -tin(IV) compounds selected for antitumour testing were prepared by metathetical reactions according to equations (1) and (2):[11-14]

$$Ph_2SbOCOCH_3 + MS_2PR_2 \rightarrow Ph_2SbS_2PR_2 + MOCOCH_3 \qquad (1)$$
$$M = H \; ; \; R = Ph$$
$$M = NH_4 \; ; \; R = OPr^i$$

$$R'_2SnCl_2 + 2\,MS_2PR_2 \rightarrow R'_2Sn(S_2PR_2)_2 + 2\,MCl \qquad (2)$$

$$M = NH_4\ ;\ R = Ph\ ;\ R' = \textit{n}\text{-Bu, Ph}$$
$$R = OPr^i\ ;\ R' = Ph$$

Table 1 : Inorganic antimony(III) compounds and their biological activities

Compound	Used in the treatment of
$K_2[Sb_2(\textit{d}\text{-}C_4O_6H_2)_2].3H_2O$	Emetic, Leishmaniasis, Granuloma inguinale, Schistosomiasis
$Sb(SCH_2CONH_2)_3$	Lymphogranuloma venerum, Leishmaniasis
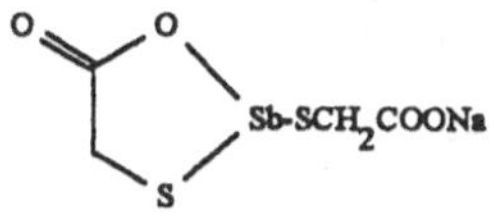	Lymphogranuloma venerum, Leishmaniasis
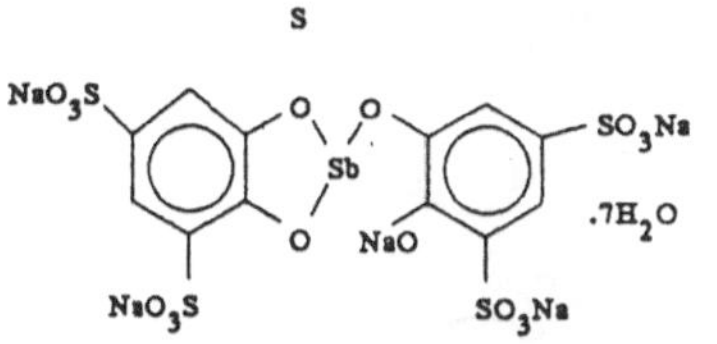	Lymphogranuloma venerum, Schistosomiasis
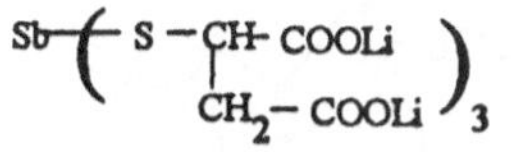	Filariasis
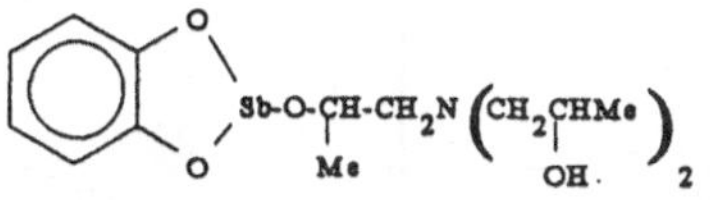	Leishmaniasis, Schistosomiasis

All the compounds are white, crystalline solids which were purified by repeated recrystallizations from organic solvents before being screened for their activity. Their structures were investigated by means of various spectroscopic techniques, including IR, NMR, Mössbauer and Mass spectrometry. Three of the compounds, viz. $Ph_2SbS_2PPh_2$,[11] $Ph_2SbS_2P(OPr^i)_2$,[12] $Ph_2Sn[S_2P(OPr^i)_2]_2$,[14] were also investigated by X-ray diffraction. In both diphenylantimony(III) derivatives, $Ph_2SbS_2PR_2$, the dithiophosphorus ligands exhibit a bimetallic triconnective coordination pattern, leading to dimeric (R = Ph) or chain polymeric (R = OPr^i) associations in the solid state, through secondary sulphur-antimony intermolecular interactions (Figs. 3a and 3b). By contrast, $Ph_2Sn[S_2P(OPr^i)_2]_2$ has a monomeric structure in the solid state, with the dithio ligands acting as symmetric monometallic biconnective (isobidentate) groups; the C-Sn-C angle is 180° (Fig. 3c). The corresponding diorganotin(IV) diphenyldithiophosphinates possess a different monomeric structure in solid state, the infrared and Mossbauer data suggesting asymmetric monometallic biconnective (anisobidentate) dithio ligands and non-linear orientation of the C-Sn bonds (i.e. C-Sn-C angle ≤ 180°). Similar structures were established by single crystal X-ray studies for $Me_2Sn(S_2PR_2)_2$ (R = Me,[15] Et[16]). Despite the structural differences observed in the solid state for the compounds tested for antitumour properties, it is very likely that in solution the secondary sulphur-metal interactions are broken, leading to free monomeric fragments. This assumption is supported by solution NMR studies on these compounds.

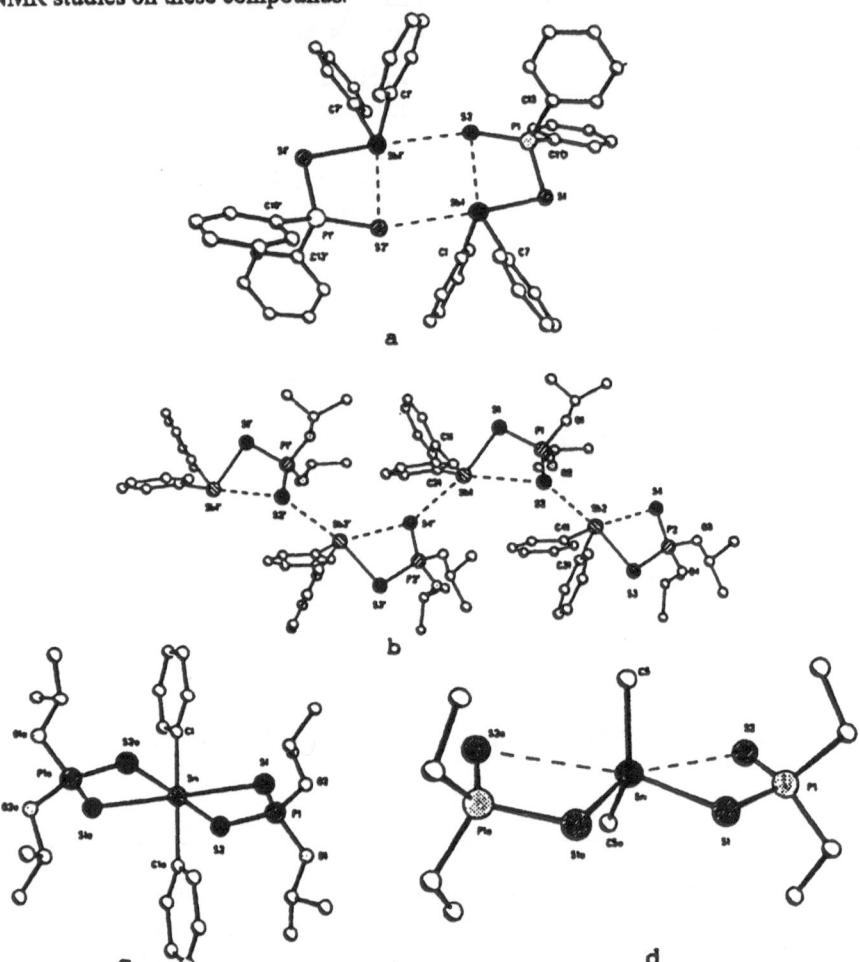

Fig. 3 : Molecular structures (Pluto diagrams) of organoantimony(III) and organotin(IV) derivatives containing dithiophosphorus ligands

The antitumour properties of the diphenylantimony(III) and diphenyltin(IV) derivatives mentioned above were tested *in vitro* against Ehrlich ascites tumour.[8,9] All four compounds were almost equally effective in inhibiting cell proliferation, cell viability and protein synthesis. However, the cell respiration and activity of some enzymes (e.g. Ca-ATPase, LDH) were considerably impaired. The effects were dependent on the doses used as well as the exposure time. In general, the observed effects produced by the test compounds showed no statistically significant differences among them.

The *in vivo* tests using both diphenylantimony(III) and diorganotin(IV) derivatives were carried out on mice bearing Ehrlich ascites tumour and P388 leukemia. A first experiment in which the inhibitory effects of the above organometallic compounds were evaluated against Ehrlich ascites tumour (Table 2) revealed that all four compounds exhibited antitumour properties. At both test doses, the observed inhibition was more than 40%. However, significant differences could be noted among the compounds. Thus regardless of the nature of the dithiophosphorus ligand, the diphenylantimony(III) derivatives were more active than the diphenyltin(IV) analogs, and for any given organometallic centre, the presence of the dithiophosphato ligand leads to a higher degree of tumour growth inhibition than that of the dithiophosphinato ligand.[8]

Table 2 : **Tumour growth inhibitory effects of diphenylantimony(III) and diphenyltin(IV) derivatives towards Ehrlich ascites tumour in Swiss mice[a,b]**

Compound	% Inhibition	
	10 mg/kg[c]	20 mg/kg[c]
$Ph_2SbS_2PPh_2$	60	84
$Ph_2SbS_2P(OPr^i)_2$	74	88
$Ph_2Sn(S_2PPh)_2$	43	46
$Ph_2Sn[S_2P(OPr^i)_2]_2$	53	74

[a]2×10^6 EAT cells were interperitoneally (i.p.) inoculated in male Swiss mice.
[b]I.p. administered on days 1, 3 and 5 after tumour transplantation.
[c]Total dose used (mg/kg body weight).

The *in vivo* studies performed on AKR mice bearing the same tumour system revealed for both the diphenylantimony(III) derivatives a dose-dependent tumour inhibitory response (Table 3). At the lower dose used (5 mg/kg) the effect was minor, while increasing the total dose to 30 mg/kg resulted in a marked reduction in tumour growth. However, significant differences between the dithiophosphato and dithiophosphinato derivatives were obtained only at 20 mg/kg dosage level, with the former again proving to be the more active of the two.

A comparison of the inhibitory effects of the diphenylantimony(III) compounds against Ehrlich ascites tumour, using the two mouse strains showed that at the same total dose and schedule treatment (i.p., on days 1, 3 and 5 after tumour transplantation), the effects were comparable (Table 4).

In another experiment, the increase in the life span of mice inoculated with 4×10^6 EAT cells was evaluated after treatment with organoantimony compounds

457

Table 3 : Tumour growth inhibitory effects of diphenylantimony(III) derivatives against Ehrlich ascites tumour in AKR mice[a]

Total dose[b] (mg/kg)	% Inhibition	
	$Ph_2SbS_2PPh_2$	$Ph_2SbS_2P(OPr^i)_2$
5	20	16
10	54	61
15	59	67
20	74	91
30	93	94

[a]2×10^6 EAT cells were i.p. inoculated in male AKR mice
[b]I.p. administered on days 1, 3 and 5 after tumour transplantation

Table 4 : Comparative inhibitory effects of diphenylantimony(III) derivatives against Ehrlich ascites tumour as a function of mouse strain

Total dose (mg/kg)	% Inhibition	
	$Ph_2SbS_2PPh_2$	$Ph_2SbS_2P(OPr^i)_2$
Swiss mice		
10	60	74
20	84	88
AKR mice		
10	54	61
20	74	91

Table 5 : Median survival time and increase in life span produced by organoantimony(III) compounds in mice bearing Ehrlich ascites tumour[a]

Compound	Dose[b] (mg/kg)	Median survival time (days)	T/C (%)
Control	-[c]	13.2	100
$Ph_2SbS_2PPh_2$	3 x 5	19.4	147
$Ph_2SbS_2P(OPr^i)_2$	3 x 5	24.1	183

[a]4×10^6 Ehrlich ascites tumour cells inoculated i.p. in male Swiss mice
[b]Doses of 5 mg/kg administered on days 1, 3 and 5 after tumour transplantation
[c]Control animals received 3 injections of 0.5 mL solvent mixture on days 1, 3 and 5.

(Table 5). The results of the screening were evaluated by computing the T/C value, which is the mean survival time of the treated group divided by that of the control group. A lifespan (T/C) value greater than 115 indicates significant activity. The antitumour activity of $Ph_2SbS_2P(OPr^i)_2$ was now even more pronounced, the median survival time of the treated animals being increased *ca.* 80%, compared to *ca.* 50% using the dithiophosphinato analogue. Moreover, when the number of long-term surviving animals was evaluated (i.e. treated mice still alive 180 days after the tumour transplant), a cure rate of 30% was obtained with the dithiophosphato derivative, while all the control and dithiophosphinato-treated mice died before days 20 and 40, respectively (Fig. 4).

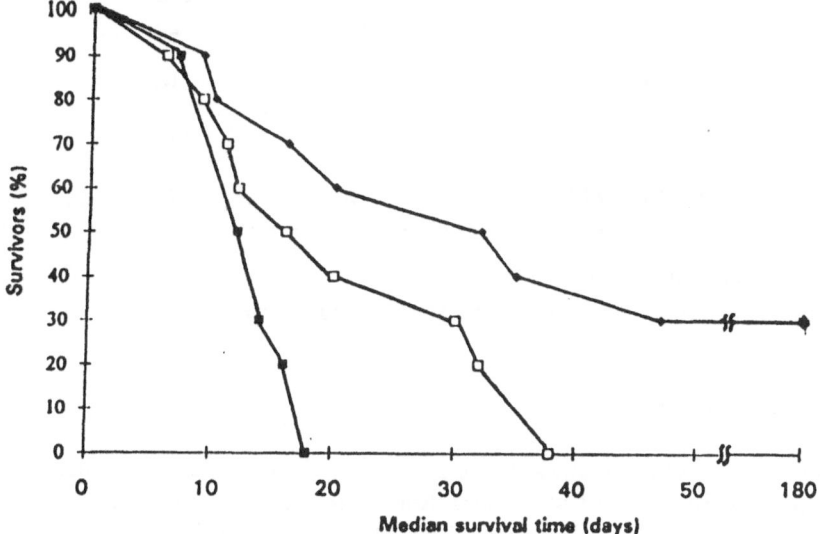

Fig. 4 : Survival rate of organoantimony-treated mice relative to control. Male Swiss mice bearing Ehrlich ascites tumour were used for the test (4×10^6 EAT cells; i.p.; treatment schedule: i.p. administered on days 1, 3 and 5): ■ : control, □ : $Ph_2SbS_2PPh_2$, ♦ : $Ph_2SbS_2P(OPr^i)_2$.

The potential antitumour properties of the two diphenylantimony(III) derivatives and of three related diorganotin(IV) compounds were also evaluated against P388 leukemia in mice.[10] The dibutyltin(IV) derivative was included in the tests because previous studies using compounds containing this organometallic moiety had suggested that the presence of the dibutyltin(IV) group is responsible for an increased antitumour effect. In the present study, however, no significant differences were apparent compared to the diphenyltin(IV) analogue. Although the tested organometallic dithiophosphinates and dithiophosphates were less active than *cisplatin* against this animal tumour system (Table 6), the results nevertheless confirm the higher antitumour effect of $Ph_2SbS_2P(OPr^i)_2$.

Table 6 : Test results of diphenylantimony(III) and diorganotin(IV) compounds against P388 leukemia in BDF$_1$ female mice[a]

Compound	Dose[b] (mg/kg)	T/C (%)
Control	-	100
Cisplatin	4	245
Ph$_2$SbS$_2$P(OPr^i)$_2$	5	136
	10	118
Ph$_2$SbS$_2$PPh$_2$	5	118
	10	123
Ph$_2$Sn[S$_2$P(OPr^i)$_2$]$_2$	5	118
Ph$_2$Sn(S$_2$PPh$_2$)$_2$	5	123
n-Bu$_2$Sn(S$_2$PPh$_2$)$_2$	5	127

[a]10^6 tumour cells were i.p. implanted in mice
[b]I.P. administered on days 1, 2 and 3 after tumour transplantation

Acknowledgements

The authors wish to thank Dr. Carmen Socaciu (Biochemistry Department, Agronomic Institute of Cluj-Napoca, Roumania) and Dr. Adela Bara (Oncology Institute of Cluj-Napoca, Roumania) for the studies on Ehrlich ascites tumour, and Prof. Dr. Bernhard K. Keppler (Anorganisch-Chemisches Institut der Universitat Heidelberg, Germany for the tests on P388 Leukemia.

References

1. I. Haiduc and C. Silvestru, *Organometallics in Cancer Chemotherapy*, CRC Press, Boca Raton, Florida, Vol. I. Main Group Metal Compounds, 1989; Vol. II. Transition Metal Compounds, 1990.
2. M. Gielen (ed.), *Metal-based anti-tumour drugs*, Freund, London, 1988.
3. B. Rosenberg, L. Van Camp, J.E. Trosko and V.H. Mansour, *Nature (London)*, **222** (1969) 385.
4. M. Peyrone, *Justus Liebigs Ann. Chem.*, **51** (1984) 1.
5. R.R. Brutkiewicz and F. Suzuki, *In Vivo*, **1** (1987) 189.
6. a) M. Boualam, R. Willem, J. Geran, A. Sebald, P. Lelieveld, D. De Vos and M. Gielen, *Appl. Organomet. Chem.*, **4** (1990) 335; b) A. Meriem, R. Willem, M. Biesemans, B. Mahieu, D. De Vos, P. Lelieveld and M. Gielen, *Appl. Organomet. Chem.*, **5** (1991) 195; c) M. Boualam. M.Biesemans, J. Meunier-

Piret, R. Willem and M. Gielen, *Appl. Organomet. Chem.*, **6** (1992) 197.

7. C. Silvestru, C. Socaciu, A. Bara and I. Haiduc, *Anticancer Res.*, **10** (1990) 803.

8. A. Bara, C. Socaciu, C. Silvestru and I. Haiduc, *Anticancer Res.*, **11** (1991) 1651.

9. C. Socaciu, A. Bara, C. Silvestru and I. Haiduc, *In Vivo*, **5** (1991) 425.

10. B.K. Keppler. C. Silvestru and I. Haiduc, unpublished results.

11. C. Silvestru, L. Silaghi-Dumitrescu, I. Haiduc, M.J. Begley, M. Nunn and D.B. Sowerby, *J. Chem. Soc., Dalton Trans.*, (1986) 1031.

12. C. Silvestru, M. Curtui, I. Haiduc, M.J. Begley and D.B. Sowerby, *J. Organomet. Chem.*, **426** (1992) 49.

13. C. Silvestru, F. Ilies, I. Haiduc, M. Gielen and J.J. Zuckerman, *J. Organomet. Chem.*, **330** (1987) 315.

14. J.L. Lefferts, K.C. Molloy, J.J. Zuckerman, I. Haiduc, M. Curtui, C. Guta and D. Ruse, *Inorg. Chem.*, **19** (1980) 2861.

15. K.C. Molloy, M.B. Hossain, D. Van der Helm, J.J. Zuckerman and F.P. Mullins, *Inorg. Chem.*, **20** (1981) 2172.

16. C. Silvestru, I. Haiduc, S. Klima, U. Thewalt, M. Gielen and J.J. Zuckerman, *J. Organomet. Chem.*, **327** (1987) 181.

Main Group Elements and Their Compounds
V.G. Kumar Das (Ed)
Copyright © 1996 Narosa Publishing House, New Delhi, India

Synthesis and Biological Properties of Organoboron -germanium, -silicon and -tin derivatives of 2-(1,2-dihydro-2-oxo-3*H*-indol-3-ylidenyl)hydrazinecarbothioamide

R.V. Singh

Department of Chemistry, University of Rajasthan, Jaipur, 302004, India

1*H*-Indol-2,3-dione (isatin) reacts with hydrazinecarbothioamides to give the corresponding hydrazones,[1-3] several of which are known to exhibit antiviral activity.[4,5] The parent ligand (LH) exists in the tautomeric forms depicted below.

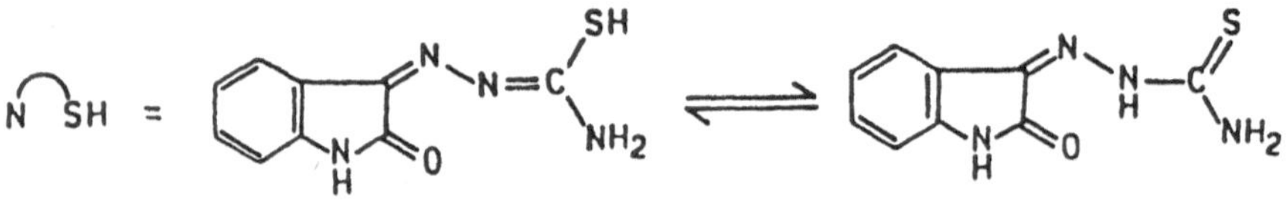

This paper reports the synthesis, structure and biological properties of the organo-boron, -silicon, -germanium and -tin derivatives of this ligand.

Synthesis of ligand and organo-element derivatives

Preparation of 2-(1,2-Dihydro-2-oxo—3H-indol-3-ylidenyl)hydrazinecarbothioamide

An ethanolic solution of 1*H*-indol-2,3-dione was added slowly to an ethanolic solution of hydrazinecarbothioamide. The mixture was then refluxed over a water bath for 3-4 h and allowed to stand overnight. On cooling, 2-(1,2-dihydro-2-oxo-3*H*-indol-3-ylidenyl)hydrazinecarbothioamide separated out as a yellow powder which was collected, dried and recrystallized from ethanol; M.Pt. 246 °C.

462

Synthesis of organoboron complexes

Phenyldihydroxyborane and 2-(1,2-dihydro-2-oxo-3H-indol-3-ylidenyl)hydrazine-carbothioamide in 1/1 or 1/2 molar ratio were condensed in benzene using a Dean and Stark apparatus. The reaction mixture was refluxed for several hours and the progress of the reaction was monitored by the liberation of water. After the completion of the reaction, the solvent was removed using a rotary evaporator to obtain the solid product which was dried in vacuo. It was then washed several times with dry cyclohexane and again dried in vacuo for 3-4 h.

Synthesis of organosilicon complexes

Dimethyldichlorosilane, diphenyldichlorosilane, trimethylchlorosilane and triphenylchlorosilane were reacted with the sodium salt of 2-(1,2-dihydro-2-oxo-3H-indol-3-ylidenyl)hydrazinecarbothioamide in 1:1 or 1:2 molar proportions in methanol as appropriate. The sodium chloride precipitated was filtered off and the solution concentrated. The resulting products were then dried under vacuum for 4 h following washing with a methanol-ether (1:1, v/v) mixture.

Synthesis of organogermanium complexes

Triphenylchlorogermane and a calculated amount of the ligand (1:1 molar ratio) were dissolved in THF. Triethylamine was then added to the mixture. The precipitate of $Et_3N.HCl$ formed was filtered off and the product obtained upon removal of solvent was dried in vacuo. It was purified by repeated washing with cyclohexane.

Synthesis of organotin complexes

The reactions of tributyltin chloride and triphenyltin chloride with the sodium salt of 2-(1,2-dihydro-2-oxo-3H-indol-3-ylidenyl)hydrazinecarbothioamide were carried out in a 1:1 molar ratio of the reactants in methanol. Following removal of precipitated NaCl, work-up of the solution yielded Bu_3SnL and Ph_3SnL.

The elemented analyses and other physical data of the synthesised products are listed in Table 1. Procedural details on the measurements have been reported elsewhere.[6-8]

Antifertility activity

The antifertility activities of the parent ligand LH and its three complexes 7, 8 and 9 were studied using male albino rats. 40 healthy adult male albino rats from an inbred colony were reared in an air-conditioned animal house at 24 ± 2 °C with 14 h light. Water and food was given *ad libitum*. The rats were divided into four groups of ten animals each for the tests.

The ligand at the dosage of 50 mg/kg body weight was administered orally for a period of 60 days in 0.2 mL olive oil. Animals treated merely with olive oil served as control. At the end of experimental period, the animals were screened for fertility

463

Table 1 : Elemental analyses, yields and physical properties of organo-boron, silicon, germanium and tin derivatives of LH[a]

Compound and Empirical Formula	No.	Colour M.P. (°C)	Yield (%)	Analysis [Found (calc.) (%)]					M.W. Found (calc.)
				C	H	N	S	E	
PhB(OH)L ($C_{13}H_{13}N_4O_2SB$)	1	Yellow 219° d	65	57.54 (55.61)	4.12 (4.04)	17.30 (17.29)	9.88 (9.90)	3.31 (3.34)	316 (324)
PhBL$_2$ ($C_{24}H_{19}N_8SB$)	2	Dark yellow 225°	72	55.68 (54.80)	3.84 (3.64)	21.24 (21.30)	12.07 (12.19)	2.01 (2.06)	546 (526)
Me$_2$SiClL ($C_{11}H_{13}N_4OSClSi$)	3	Brown above 300°	78	42.88 (42.21)	4.26 (4.19)	17.79 (17.90)	10.21 (10.24)	8.95 (8.97)	333 (313)
Me$_2$SiL$_2$ ($C_{20}H_{20}N_8O_2S_2Si$)	4	Brown 229°	64	48.79 (48.33)	4.18 (4.06)	22.48 (22.55)	12.83 (12.90)	5.60 (5.65)	506 (476)
Ph$_2$SiClL ($C_{21}H_{17}N_4OSClSi$)	5	Mustard above 300°	75	58.26 (57.72)	3.99 (3.92)	12.76 (12.82)	7.31 (7.34)	6.41 (6.43)	419 (437)
Ph$_2$SiL$_2$ ($C_{20}H_{20}N_8O_2S_2Si$)	6	Dark yellow 233°	82	58.83 (58.02)	4.10 (3.90)	18.11 (18.04)	10.25 (10.33)	4.48 (4.52)	638 (621)
Me$_3$SiL ($C_{12}H_{16}N_4OSSi$)	7	Brown 220° d	78	48.51 (49.36)	5.63 (5.52)	19.09 (19.19)	10.92 (10.98)	9.59 (9.62)	312 (292)
Ph$_3$SiL ($C_{27}H_{22}N_4OSSi$)	8	Yellow 235° d	84	68.26 (67.70)	4.88 (4.63)	11.65 (11.70)	6.58 (6.69)	5.84 (5.86)	459 (479)
Ph$_3$GeL $C_{27}H_{22}N_4OSFe$	9	Yellow 160°	78	62.31 (61.98)	4.56 (4.23)	10.65 (10.70)	6.09 (6.12)	13.82 (13.87)	502 (523)
Bu$_3$SnL ($C_{21}H_{34}N_4OSSn$)	10	Yellow 220° d	64	50.44 (49.53)	6.89 (6.73)	10.94 (11.00)	6.22 (6.29)	23.27 (23.30)	521 (509)
Ph$_3$SnL (CHNOSSn)	11	Orange Yellow 150° d	70	56.36 (56.97)	3.94 (3.89)	9.82 (9.84)	5.71 (5.63)	20.78 (20.84)	588 (569)

a : LH = 2-(1,2-dihydro-2-oxo-3H-indol-3-ylidenyl)hydrazinecarbothioamide and autopsied for the study of histological and biochemical changes. Reproductive organs were excised, blotted free of blood, weighed and were frozen for biochemical estimation. The tests were also fixed in Ousin's fluid for histological examination.

The sperm motility and density of cauda-epididymal spermatozoa were assessed by the method of Prasad *et al.*[9] The total protein[10], sialic acid[11], glycogen[12] and total cholesterol[13] contents were determined by standard methods.

Microbial assay

The *in vitro* antifungal and antibacterial activities of the parent ligand and the organo-element complexes were investigated along with a standard fungicide (Bavistin) and a bactericide (Streptomycin). The organisms selected for antifungal activities were *Macrophomina phaseolina* and *Fusarium oxysporum*; for the antibacterial studies the organisms were *Pseudomonas cepacicola*, *Klebsiella aerogenous*, *Escherichia coli* and *Staphylococcus aureus*. The radial growth method and inhibition zone technique were employed to evaluate the fungicidal and bactericidal activities, respectively.[14]

Results and discussion

Spectroscopic studies

The newly synthesized complexes are coloured solids which dissolve readily in DMSO, DMF, methanol and chloroform. The molar conductance values (7-10 ohm^{-1} cm^{-2} mol^{-1}) in DMF show that they are non-electrolytes. The monomeric nature of these complexes has been confirmed by molecular weight determinations.

In the electronic spectrum of the ligand, 2-(1,2-dihydro-2-oxo-*3H*-indol-3-ylidenyl)hydrazinecarbothioamide, a band arising from the >C=N chromophore at 370 nm shifts to a shorter wavelength for organosilicon, organogermanium and organotin derivatives and to higher wavelengths for the organoboron derivatives. Two bands of medium intensity at 270 nm and 300 nm arising from the $\pi \rightarrow \pi^*$ transition of the ligand remain almost unchanged in the spectra of organoelement derivatives.

In the IR spectra of the ligand a strong band observed at 1660 cm^{-1} arising from ν(C=N) is shifted to lower frequencies in the organosilicon, organogermanium and organotin complexes and to higher frequencies in the organoboron complexes. The appearance of ν(E←N)[6,7,15,16] and ν(E-S) (E = B, Si, Ge and Sn) bands respectively at *ca.* 1535, 570, 650 cm^{-1} and *ca.* 870, 535, 410 and 325 cm^{-1} implicate the azomethine nitrogen and the thiolo sulphur in coordinative interaction at the element centre. The bands at 3460 and 3350 cm^{-1}, arising from the asymmetric and symmetric modes[20] of the amino group are observed at almost the same position in the spectra of organoelement complexes, suggesting the non-involvement of this amino group in bonding.

The ^{1}H-NMR spectra of 2-(1,2-dihydro-2-oxo-*3H*-indol-3- ylidenyl)-hydrazinecarbothioamide and the complexes have been recorded in DMSO-d_6. The chemical shift values are tabulated in Table 2.

465

Table 3 shows ^{13}C NMR chemical shift values of the free ligand and of its complexed form in the two representative compounds, 7 and 8.

Table 2 : ^{1}H NMR data (δ ppm) of LH and its organoelement derivatives

Compound	-NH(1) (b)	-NH (2) (b)	-NH$_2$ (b)	Aromatic (m)	E-CH$_3$/C$_4$H$_9$/ C$_6$H$_5$
LH	11.28	12.56	3.44	8.04	-
PhB(OH)L	-	12.52	3.42	8.20	6.40
PhBL$_2$	-	12.52	3.40	8.20	6.48
Me$_2$Si(Cl)L	-	12.56	3.46	8.14	0.75
Me$_2$SiL$_2$	-	12.60	3.44	8.16	0.94
Ph$_2$Si(Cl)L	-	12.58	3.42	8.12	6.58
Ph$_2$SiL$_2$	-	12.56	3.46	8.16	6.64
Me$_3$SiL	-	12.60	3.46	8.09	0.88
Ph$_3$SiL	-	12.52	3.44	8.09	6.72
Ph$_3$GeL	-	12.54	3.46	8.12	6.56
Bu$_3$SnL	-	12.58	3.48	8.14	0.71 - 2.06
Ph$_3$SnL	-	12.56	3.42	8.11	6.85

Table 3 : ^{13}C NMR data (δ ppm) of isatin-3-thiosemicarbazone and its organosilicon derivatives

Compound	Thiolo carbon	Azomethine carbon	Amido carbon	Aromatic carbons	Si-CH$_3$/C$_6$H$_5$
LH	178.02	161.98	141.61	131.59, 130.78, 121.89, 120.43, 119.19, 110.52	
7	161.82	154.34	140.8	131.26, 129.52, 121.68, 121.41, 119.67, 110.41	16.58
8	165.36	156.84	141. 32	131.45, 130.83, 121.72, 121.39, 119.21, 110.46	132.32, 134.26, 137.18, 139.23

The [11]B nuclear resonance is observed at 4.54 and 7.28 ppm, respectively, for the 1:1 and 1:2 boron complexes. The values suggest a tetracoordinated environment[21] around the boron atom and the presence of a (B←N) dative bond.

On the basis of above spectral observations, the following structures are proposed for the organo-boron, -silicon, -germanium and -tin complexes.

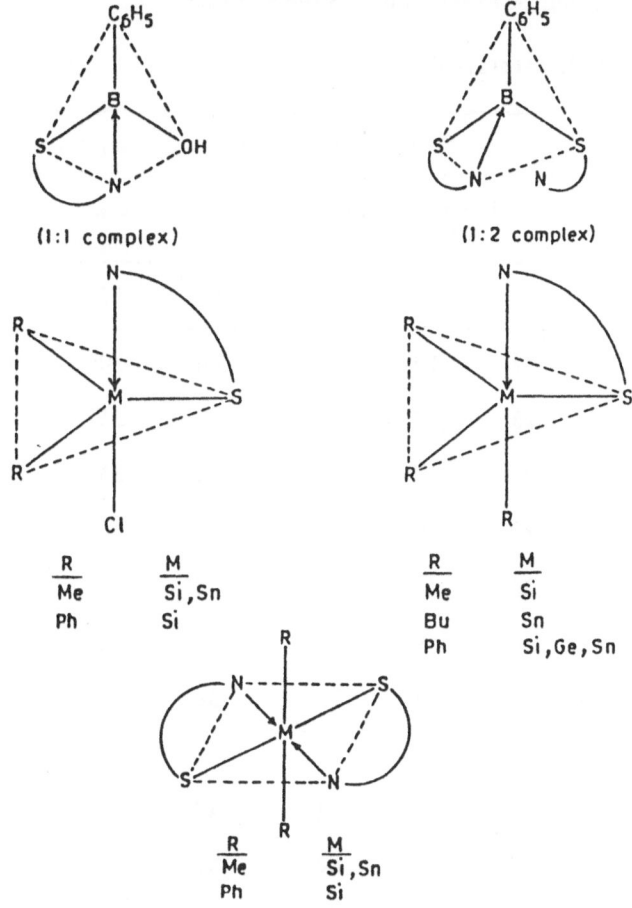

(1:1 complex)

(1:2 complex)

R	M
Me	Si,Sn
Ph	Si

R	M
Me	Si
Bu	Sn
Ph	Si,Ge,Sn

R	M
Me	Si,Sn
Ph	Si

Bioassay

The results of the antifungal and antibacterial tests are summarized in Tables 4 and 5 respectively.

In each case, maximal activity was manifested by the organotin complexes; R_2EL_2 complexes were also slightly more active than R_2ECIL complexes (E = Sn, Si). The order of activity is $R_3Sn>R_3Ge\geq R_3Si>R_2Si>RB$.

A common technique used for correlating structures of molecules with their toxicological activities is that based on the quantitative structure-activity relationship (QSAR) model. The results show that there is a high correlation between the toxicity of the individual series and their total surface area (TSA) values. These results are consistent with previous reported work.[22]

In bactericidal studies, the complexes showed toxicity towards both the Gram (+) and Gram (-) strains, whereas the standard, streptomycin, was toxic only towards the Gram(+) strain. The results show that overall the parent ligand and its

467

complexes were slightly more toxic toward the Gram(+) strain.

The antifertility data reported in Table 6 indicate significant decrease in the body weight and weight of testes, epididymis, ventral prostate and seminal vesicle after treatment with the parent ligand and compounds, 7, 8 and 9. The fertility test was 70 percent negative in ligand-treated animals, whereas it was 90 to 93 percent negative in the animals treated with 7, 8 and 9 (Table 7).

Table 4 : Fungicidal screening data

| Compound | % Inhibition after 96 h (conc. in ppm) | | | | | |
| | *Macrophomina phaseolina* | | | *Fusarium oxysporum* | | |
	50	100	200	50	100	200
Bavistin	82	100	100	86	100	100
LH	15	28	35	18	25	42
PhB(OH)L	22	37	54	24	35	58
PhBL$_2$	28	40	66	41	52	70
Me$_2$Si(Cl)L	26	44	68	32	51	72
Me$_2$SiL$_2$	28	49	72	34	50	74
Me$_3$SiL	25	51	73	38	55	78
Ph$_2$Si(Cl)L	31	55	77	42	59	80
Ph$_2$SiL$_2$	38	59	82	47	63	86
Me$_2$Sn(Cl)L	40	64	86	52	68	88
Ph$_3$SiL	48	65	88	59	71	90
Ph$_3$GeL	54	68	88	60	72	92
Me$_2$SnL$_2$	60	70	94	63	77	98
Bu$_3$SnL	72	84	100	75	92	100
Ph$_3$SnL	80	100	100	88	100	100

Table 5 : Bactericidal screening data

| Compound | Diameter of Inhibition zone (mm) (conc. in ppm) | | | | | | | |
| | Pseudomonas cepacicola | | Klebsiella aerogenous | | Escherichia coli | | Staphylococcus aureus | |
	500	1000	500	1000	500	1000	500	1000
Streptomycin	2	3	-	2	1	2	15	17
LH	5	7	5	6	4	6	6	7
PhB(OH)L	5	7	5	7	5	7	7	8
PhBL$_2$	6	7	6	7	5	6	8	10
Me$_2$Si(Cl)L	6	8	7	9	7	8	7	10
Me$_2$SiL$_2$	7	9	8	10	9	10	8	11
Me$_3$SiL	7	8	8	11	10	11	9	12
Ph$_2$Si(Cl)L	8	10	9	11	10	12	10	14
Ph$_2$SiL$_2$	9	11	9	12	9	12	11	14
Me$_2$Sn(Cl)L	10	12	11	13	10	13	13	15
Ph$_3$SiL	10	12	12	13	11	13	14	15
Ph$_3$GeL	11	13	12	14	10	14	13	16
Me$_2$SnL$_2$	12	14	13	15	12	15	15	17
Bu$_3$SnL	14	16	15	17	14	18	17	19
Ph$_3$SnL	15	19	17	19	17	20	19	22

Table 6 : Changes in body and organ weights in test rats following treatment with LH and compounds 7, 8 and 9

| Treatment | Average body wt. (g) | | Organ weight mg/100 g, body wt. | | | |
	Initial	Final	Testes	Epi-didymis	Seminal vesicle	Ventral prostate
Control	180.0 ± 12.0	200.0 ± 8.0	1015.0 ± 60.5	450.0 ± 30.0	350.0 ± 17.5	260.5 ± 10.5
LH	190.0 ± 10.0	140.0 ± 10.0	805.7 ± 50.0[a]	390.0 ± 35.0[b]	310.0 ± 15.5[a]	210.0 ± 15.0[a]
7	185.0 ± 10.0	145.0 ± 10.5[a]	790.0 ± 40.0[b]	370.0 ± 30.0[b]	300.0 ± 11.3[a]	200.0 ± 17.0[a]
8	190.0 ± 16.0	140.0 ± 13.0	740.0 ± 35.0[b]	368.0 ± 18.0[b]	305.0 ± 17.0[a]	190.3 ± 18.5[b]
9	180.0 ± 15.0[b]	190.0 ± 5.0[b]	750.0 ± 37.0[b]	385.0 ± 25.0[b]	315.0 ± 17.5[b]	210.0 ± 14.0[b]

a : $P \leq 0.01$; b : $P \leq 0.001$. All values are mean ± S.E.M.

Table 7 : **Altered sperm dynamics and fertility test following treatment of test rats with LH and its organoelement derivatives**

Treatment	Sperm Density (million/mL)		Sperm Motility	Fertility Test (%)
	Testes	Cauda epididymis	Cauda epididymis	
Control	4.8 ± 0.55	18.52 ± 1.2	70.0 ± 2.81	90 (+ve)
LH	3.0 ± 0.21^b	9.3 ± 1.5^b	45.0 ± 2.71^b	70(-ve)^b
7	2.9 ± 0.15^b	8.0 ± 0.35^b	28.0 ± 2.50^b	90(-ve)^b
8	2.1 ± 0.13^b	7.9 ± 0.31^b	27.8 ± 3.0^b	90(-ve)^b
9	1.90 ± 0.11^b	6.5 ± 0.42^b	20.0 ± 3.50^b	93(-ve)^b

$^b P \leq 0.001$. All values are mean $\pm$ S.E.M.

The administration of the parent ligand and its complexes 7, 8 and 9 to the rats caused morphological changes to the sperms. This was probably due to damage to the spermatogonial cells. The 90 - 93 percent negative fertility tests obtained with 7, 8 and 9 may be attributed to the lack of forward progression and reduction in density of spermatozoa and altered biochemical millieu of cauda epididymis. The levels of sialic acid and proteins in testes, epididymis, seminal vesicle and ventral prostate were significantly lower in animals treated with the parent ligand as well as with the compounds 7-9 (Table 8).

The reduction in sialic acid levels is probably the result of inhibition of spermatogenesis in the testes.[23] It is also noted that there is a decrease in cholesterol content in the testes following treatment with the compounds. In as much as cholesterol functions as a precursor in androgen biosynthesis in the testes, its decreased concentration may be anticipated to affect the fertility and sperm output.[24,25]

Acknowledgements

The author is thankful to the Department of Science and Technology, Rajasathan State, Jaipur and CSIR, New Delhi for financial assistance and to Drs. S.C. Joshi and C.G. Jain, Department of Zoology, Rajasthan University, Jaipur for recording the antifertility activity of the compounds. The author is also grateful to Mrs. Nighat Fahmi and Miss Chitra Saxena of the Department for assistance in the synthesis and screening of these compounds for their antifungal and antibacterial activities.

Table 8 : Biochemical changes in sialic acid, total protein, glycogen and total cholesterol contents in reproductive organs of test rats

Treatment	Sialic acid (mg/gm)				Total protein (mg/gm)				Total cholesterol (mg/gm) Testes	Glycogen (mg/gm) Testes
	Testes	Epididymis	Seminal vesicle	Ventral prostate	Testes	Epididymis	Seminal vesicle	Ventral prostate		
Control	7.0 ± 0.7	6.5 ± 0.8	6.9 ± 0.9	6.10 ± 0.4	210.5 ± 15.0	190.0 ± 20.0	250.7 ± 15.3	225.0 ± 15.0	8.21 ± 0.52	1.50 ± 0.04
LH	5.0 ± 0.5[a]	4.9 ± 0.7[a]	5.1 ± 0.2[a]	4.8 ± 0.2[a]	160.0 ± 10.0[a]	140.0 ± 10.0[a]	190.0 ± 10.0	160.0 ± 13.0	5.3 ± 0.70[b]	1.00 ± 0.04[b]
7	4.1 ± 0.3[b]	4.2 ± 0.4[b]	4.3 ± 0.5[b]	4.0 ± 0.2[b]	170.0 ± 7.0[a]	145.0 ± 5.3[a]	195.0 ± 7.0	140.0 ± 10.0	4.8 ± 0.50[b]	0.92 ± 0.05[b]
8	4.0 ± 0.4[b]	4.2 ± 0.3[b]	4.1 ± 0.2[b]	3.9 ± 0.4[b]	130.0 ± 15.0[b]	125.0 ± 10.0[b]	170.0 ± 10.0[b]	135.0 ± 10.0[b]	4.8 ± 0.30[b]	0.90 ± 0.03[b]
9	4.0 ± 0.15	3.8 ± 0.10	3.5 ± 0.17	3.6 ± 0.20	120.0 ± 13.2	110.0 ± 9.0	140.0 ± 5.0	108.0 ± 7.3	4.5 ± 0.30	0.80 ± 0.02

a : $P \geq 0.01$; b : $P \geq 0.001$

References

1. D.J. Bauer and P.W. Sadler, *Nature* (Lond), **190** (1961) 1167.
2. J. Bernstein, H.L. Yale, K. Losee, M.Holsing, J. Martins and W.A. Lott, *J. Amer. Chem. Soc.*, **73** (1951) 906.
3. G. Doleschall and K. Lempert, *Acta Chima. Acad. Sci. Hung.*, **64** (1970) 369.
4. D.J. Bauer and P.W. Sdlei, *Brit. J. Pharmacol.*, **15** (1960) 101.
5. B. Lucka-Sobstel and A. Zejc, *Diss. Pharm. Pharmacol.*, **24** (1972) 585.
6. Chitra Saxena and R.V. Singh, *Main Group Met. Chem.*, **95** (1989) 109.
7. K. Singh, R.V. Singh and J.P. Tandon, *Synth. React. Inorg. Met.-Org. Chem.*, **17** (1987) 385.
8. A. Kumari, R.V. Singh and J.P. Tandon, *Main Group Met. Chem.*, **14** (1991) 167.
9. O.H. Lowry, N.J. Rosenbrough, A.L. Farr and R.J. Randall, *J. Biol. Biochem.*, **193** (1951) 265.
10. L. Warren, *J. Biol. Chem.*, **234** (1959) 1971.
11. R. Montgomery, *Bio-phys.*, **67** (1957) 378.
12. J.M. Walker, *Methods in molecular biology, Vol. 1 - Protiens*, 1st edition, Humana Press Inc., Clifton, NJ, U.S.A., 1984.
13. G. Gupta, M. Rajalakshmi, M.R.N. Prasad and N.R. Mudgal, *Andrologia*, **6** (1974) 35.
14. D. Singh and R.V. Singh, *Main Group Met. Chem.*, **13** (1990) 19.
15. R.V. Singh, *Main Group Met. Chem.*, **13** (1990) 55.
16. K. Veno and A.E. Martel, *J. Amer. Chem. Soc.*, **59** (1955) 98.
17. V.P. Singh, R.V. Singh and J.P. Tandon, *J. Inorg. Biochem.*, **39** (1990) 237.
18. K. Singh, R.V. Singh and J.P. Tandon, *Inorg. Chim. Acta*, **151** (1988) 179.
19. A.K. Saxena and J.P. Tandon, *Polyhedron*, **2** (1983) 443.
20. A. Saxena and J.P. Tandon, *Inorg. Chim. Acta*, **84** (1984) 195.
21. H. Noth and B. Wrackmeyer, "NMR Basic Principles and Progress", In: *Nuclear Magnetic Resonance Spectroscopy of Boron Compounds*, eds., P. Diehl, E. Fluck and R. Kosfeld, Springer-Verlag, Berlin, 1978, p. 14.
22. G. Eng, E.J. Tierney, G.J. Olson, F.E. Brinckman and J.M. Bellama, *App. Organomet. Chem.*, **5** (1991) 33.
23. G. Gupta, M. Rajalakshmi, M.R.N. Prasad and N.R. Mudgal, *Andrologia*, **6** (1974) 35.
24. K.B. Eik-Nes and P.F. Hall, *Proc. Soc. Exp. Biol. Med.*, **111** (1962) 280.
25. E. Steinberger, R.K. Technocanin and A. Steinberger, *J. Steroid, Biochem.*, **11** (1979) 185.

Main Group Elements and Their Compounds
V.G. Kumar Das (Ed)
Copyright © 1996 Narosa Publishing House, New Delhi, India

Complexes of Organotin Halides with Bidentate Diphosphoryl Ligands

V.S. Petrosyan[a], N.S. Yashina[a], E.V. Grigor'ev[a], A.A. Prischenko[a], M.V. Livantsov[a], L.A. Aslanov[a], A.V. Yatsenko[a], J. Lorberth[b] and L. Pellerito[c]

[a]Chemistry Department, M.V. Lomonosov University, Moscow, Russia,
[b]Chemistry Department, Philipps-University, Marburg, Germany and
[c]Chemistry Department, Palermo University, Palermo, Italy.

Our recent studies[1] on the coordination compounds of tin tetrahalides and methyltin trihalides with mono- and bidentate phosphines using ^{31}P and ^{119}Sn NMR spectroscopy have shown that the NMR parameters correlate rather well with the electron distribution in the complexes as determined by the nature of ligands and their location in the coordination sphere of the tin atom. The detailed solid state structures of the phosphine complexes could not, however, be obtained because of their amorphous nature.

By extending the range of the bidentate diphosphoryl ligands used, we have now succeeded in isolating several crystalline adducts whose crystal structures have been determined by X-ray diffraction. Solid state Mössbauer spectroscopic and solution NMR studies have also been conducted. These have enabled interesting structural comparisons to be made within this class of complexes, interest in which has recently been aroused on account of their potential as antitumour agents.

^{31}P and ^{119}Sn NMR spectra and structures of organotin trihalide complexes with methylenediphosphonates

Complexation of $RSnHal_3$ (R = Me, Hal = Cl, Br; R = Ph, Hal = Cl) with bidentate tetraalkyl methylenediphosphonates $[(R'O)_2P(O)]_2CHR''$ (R' = Et,Pr^i; R'' = H, NMe$_2$) has been studied by means of ^{31}P and ^{119}Sn NMR spectroscopy[2] in dichloromethane solutions at -90 °C in order to slow down the rapid exchange processes.

$RSnHal_3$ and methylene diphosphonates form 1:1 complexes of the general formulation, $RSnHal_3 \cdot [(R'O)_2P(O)]_2CHR''$. These are shown in Table 1 (compounds 1 - 12).

NMR spectral parameters of complexes 1-12 are listed in Table 2. The $\delta(^{119}Sn)$ values and multiplicities of NMR signals suggest the formation of two isomers of hexacoordinate chelate complexes having equivalent and non-equivalent phosphorus atoms (structures 1a and 1b).

Table 1 : Complexes of RSnHal$_3$·[(R'O)$_2$P(O)]$_2$CHR"

No.	R	R'	R"	Hal	No.	R	R'	R"	Hal
1	Me	Et	H	Cl	7	Me	Et	NMe$_2$	Br
2	Me	Pr^i	H	Cl	8	Me	Pr^i	NMe$_2$	Br
3	Me	Et	NMe$_2$	Cl	9	Ph	Et	H	Cl
4	Me	Pr^i	NMe$_2$	Cl	10	Ph	Pr^i	H	Cl
5	Me	E	H	Br	11	Ph	Et	NMe$_2$	Cl
6	Me	Pr^i	H	Br	12	Ph	Pr^i	NMe$_2$	Cl

Table 2 : ^{31}P and ^{119}Sn NMR parameters for RSnHal$_3$ complexes with methylenediphosphonates (CD$_2$Cl$_2$, -90 °C)^{a,b}

Complex	δ(^{31}P) ppm	δ(^{119}Sn) ppm	^{2}J(^{119}Sn-^{31}P) Hz	^{2}J(^{31}P-^{31}P) Hz
1a	20.0	-463 t	197	-
1b	19.7(A); 18.7(M)	-475 dd	215(AX); 39(MX)	< 2
2a	18.4	-474 t	207	-
2b	17.9(A); 16.7(M)	-483 dd	220(AX); 42(MX)	<2
3a	19.6	-469 t	213	-
3b	18.9(A); 18.1(B)	-480 (X)	222(AX); 62(BX)	82
4a	18.1	-478 t	221	-
4b	17.6(A); 16.4(B)	-485 (X)	230(AX); 63(BX)	82
5	18.8	-705 br	-	-
6a	20.4	-746 t	208	-
6b	20.4(A); 17.7(M)	-697 dd	223(AX); 83(MX)	<2
7a	20.1	-740 t	220	-
7b	19.7(A); 18.0(B)	-697 (X)	226(AX); 104(BX)	80
8a	16.6	-753 t	229	-
8b	16.4(A); 13.9(B)	-705 (X)	236(AX); 111(BX)	83
9a	21.1	-531 t	186	-
9b	21.5(A); 19.4(M)	-529 dd	196(AX); 52(MX)	2
10a	18.7	-535 t	190	-
10b	18.7(A); 16.4(M)	-535 dd	197(AX); 53(MX)	<2
11a	18.9	-533 t	197	-
11b	19.2(A); 17.0(B)	-535 (X)	200(AX); 70(BX)	82
12a	17.3	-534 t	205	-
12b	17.4(A); 15.0(B)	-535 (X)	206(AX); 75(BX)	83

^{a}br - broad signal, d - doublet, t - triplet

^{b}The a-isomers have A$_2$X spin systems; the b-isomers have ABX or AMX spin systems

474

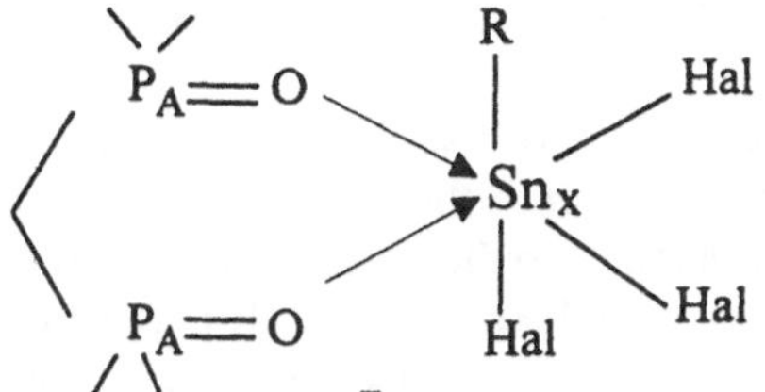

Ia

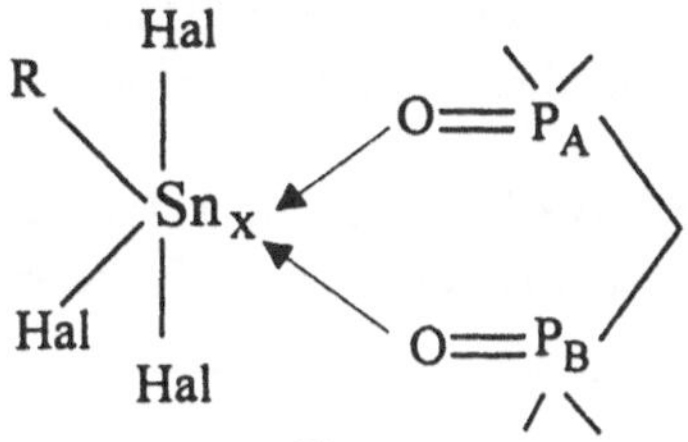

Ib

The a-isomers have A_2X spin systems with triplets in ^{119}Sn spectra and singlets with $^{117/119}Sn$ satellites in ^{31}P spectra. The b-isomers with R'' = NMe_2 have ABX spin systems with AB-quartet and satellites in ^{31}P spectra and X-part in ^{119}Sn NMR spectra. The signals of AMX spin systems appear in the NMR spectra of the b-isomer with R''=H because of the low $^2J(^{31}P-^{31}P)$ values. The NMR parameters were attributed to the non-equivalent phosphorus nuclei so that Pa nuclei in b-isomers have chemical environments and thus NMR parameters similar to those for P_a nuclei in a-isomers.

The $^2J(^{119}Sn-^{31}P)$ values for the phosphoryl groups in *trans*-position to halogens are 120-180 Hz higher than those of the phosphoryl groups in *trans*-positions to organic substituents; both types of coupling constants are of the same sign as verified by simulation spectra. The differences in coupling constants arise from redistribution of electron densities in Sn-O-P fragments influenced by *trans*-ligands.

Adding an excess of the organotin acceptor causes a redistribution reaction to occur, giving R_2SnHal_2 and the respective tin tetrahalide complex:

$$2\,RSnCl_3 + L \rightarrow SnCl_4{\cdot}L + R_2SnCl_2 \leftarrow RSnCl_3{\cdot}L + RSnCl_3$$
(R = Me, Ph; L = bidentate diphosphoryl ligand)

The NMR parameters of some $SnCl_4$ adducts obtained in this way are listed in Table 3.

Table 3 : NMR parameters of complexes $SnCl_4{\cdot}[(EtO)_2P(O)]_2CHR$ (CD_2Cl_2, -90 °C)

R	$\delta(^{31}P)$, ppm	$\delta(^{119}Sn)$, ppm	$^2J(^{119}Sn-^{31}P)$, Hz
H	19.1	-672 t	101
NMe_2	17.1	-673 t	115

^{31}P and ^{119}Sn NMR spectra and structures of RSnHal$_3$·(EtO)$_2$P(O)CH(NMe$_2$)P(O)Pr$_2$ complexes

Based on the ^{31}P and ^{119}Sn NMR spectral analyses, MeSnHal$_3$ (Hal = Cl, Br) and PhSnCl$_3$ react with the asymmetric diphosphoryl ligand, (EtO)$_2$P(O)CH(NMe$_2$)P(O)Pr$_2$, in dichloromethane solution to form two isomeric hexacoordinated chelate complexes. The NMR parameters for these isomers are listed in Table 4. The $^2J(^{119}Sn$-$^{31}P)$ values indicate that one of the PO groups is situated *trans* to the organic ligand in both cases (structures IIa and IIb). The dominant isomer in all cases has the structure IIa. Signals of the third possible isomer IIc have not been found in the NMR spectra.

Table 4 : ^{31}P and ^{119}Sn NMR parameters for RSnHal$_3$ complexes with (EtO)$_2$P(O)CH(NMe$_2$)P(O)Pr$_2$ (CD$_2$Cl$_2$, -90 C)

R	Hal	$\delta(^{31}P)$ ppm	$\delta(^{119}Sn)$ ppm	$^2J(^{119}Sn$-$^{31}P)^a$ Hz	Structure
Me	Cl	24.4(A), 65.8(M)	-480 dd	202(AX), 105(MX)	IIa
		23.3(A), 65.2(M)	-486 dd	75(AX), 201(MX)	IIb
Me	Br	22.3(A), 64.5(M)	-678 br	207(AX), 136(MX)	IIa
		20.1(A), 65(M)	-708 br	122(AX)^b	IIb
Ph	Cl	24.0(A), 64.4(M)	-533 dd	190(AX), 113(MX)	IIa
		21.8(A), 65.5(M)	-538 dd	86(AX), 183(MX)	IIb

^{a} $^2J(^{31}P$-$^{31}P)$ = 40–42 Hz
^{b} J_{MX} could not be determined because of overlapping signals

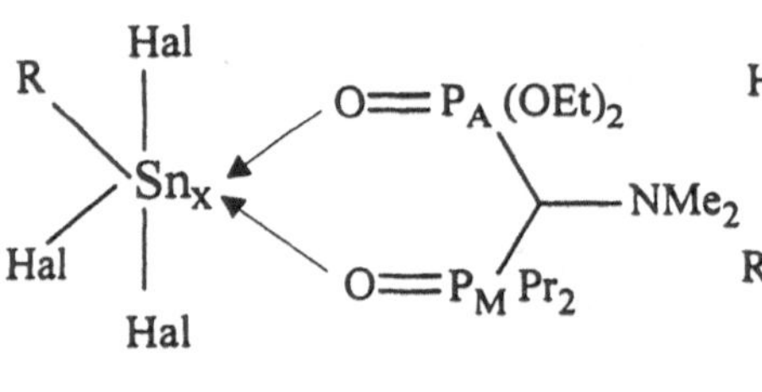

IIa

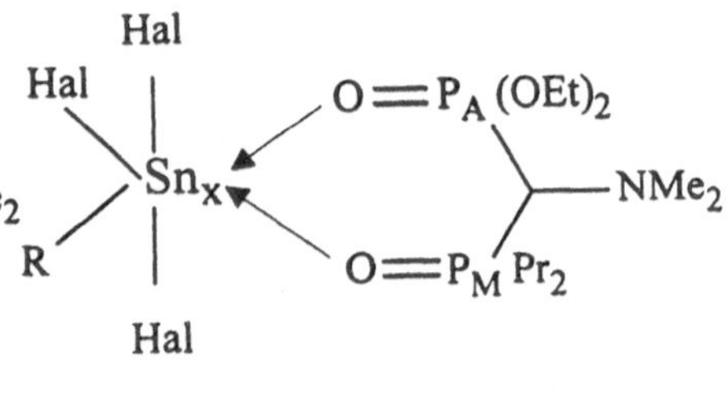

IIb

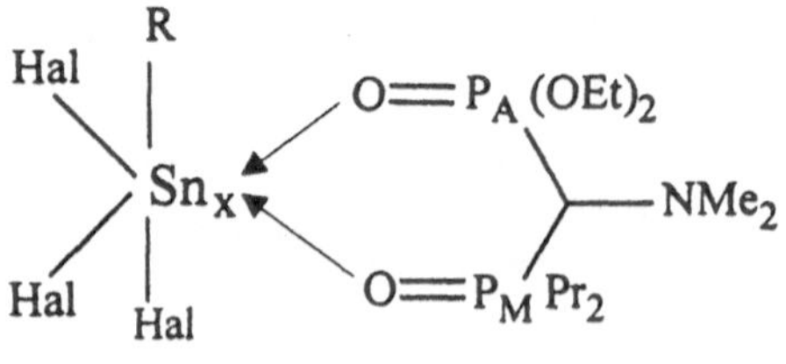

IIc

^{31}P and ^{119}Sn NMR spectra and structures of RSnHal$_3$ complexes with ethylene- and propylene diphosphonates

RSHal$_3$·[(EtO)$_2$P(O)CH$_2$]$_2$ (R = Me, Hal = Cl, Br; R = Ph, Hal = Cl) complexes have been synthesized and studied by means of low temperature ^{31}P and ^{119}Sn NMR spectroscopy. We could analyse only the NMR spectra of PhSnCl$_3$·[EtO)$_2$P(O)CH$_2$]$_2$ and not of the methyltin complexes as the latter were too labile even at -90 °C. There are two types of hexacoordinated organotin species in solution: one with equivalent, and the other with non-equivalent PO-groups. Polymeric structures with the bridging bidentate ligands are considered for these complexes (structures IIIa and IIIb).

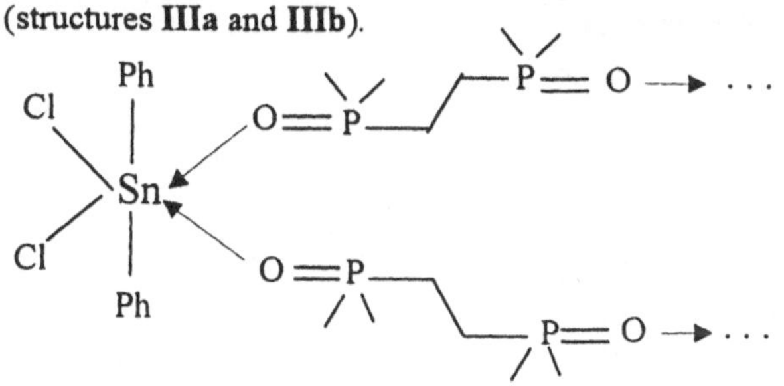

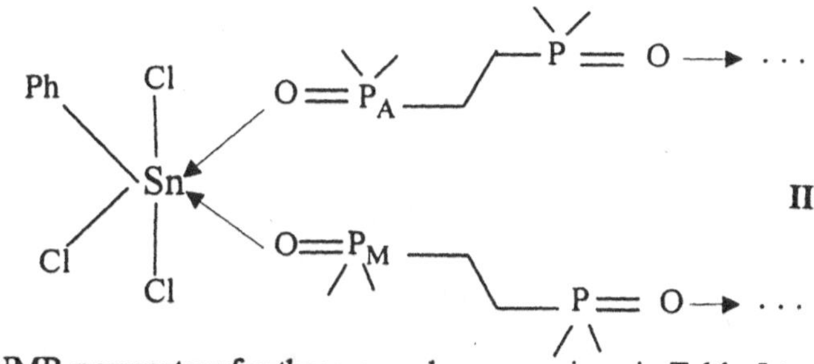

The NMR parameters for these complexes are given in Table 5 together with those for the complex of SnCl$_4$ with ethylenediphosphonates obtained by the redistribution reaction.

Table 5 : ^{31}P and ^{119}Sn NMR parameters for complexes RSnHal$_3$·[(EtO)$_2$P(O)CH$_2$]$_2$, (CD$_2$Cl$_2$, -90 °C)

R	$\delta(^{31}$P), ppm	$\delta(^{119}$Sn), ppm	$^2J(^{119}$Sn-^{31}P)†, Hz
Ph (IIIa)	229.9	-557 t	253
Ph (IIIb)	2879(A); 27.0(M)	-551 dd	227(AX); 76(MX)
Cl	28.1	-689 t	139

† $^3J(^{31}$P-^{31}P) = 0

The *trans*-ligand dependencies of $^2J(^{119}$Sn-^{31}P) values in the ethylenediphosphonate adducts are the same as for the complexes with methylenediphosphonates. The coupling constants in the former adducts are slightly larger than in the latter ones probably due to the larger magnitudes of the Sn-O-P

477

angles in open chain structures (**III**) than in 6-membered chelate rings (**1**).

The complexes, $RSnHal_3 \cdot (EtO)_2P(O)CH_2CH(Me)P(O)(OEt)_2$, (R = Me, Hal = Cl, Br; R = Ph, Hal = Cl), have also been synthesized and studied by means of ^{31}P and ^{119}Sn NMR spectroscopy at low temperatures.

The occurrence of two different phosphonate units in propylenediphosphonate extends the variety of possible modes of ligand coordination to the organotin trihalides. Thus the ^{31}P and ^{119}Sn NMR spectra of the $PhSnCl_3$ adduct contain signals unequivocally attributable to four different kinds of hexacoordinated complexes. The NMR parameters for these complexes are given in Table 6. The three doublet of doublets in the ^{119}Sn NMR spectrum correspond to the three possible chelate structures (**IVa–c**).

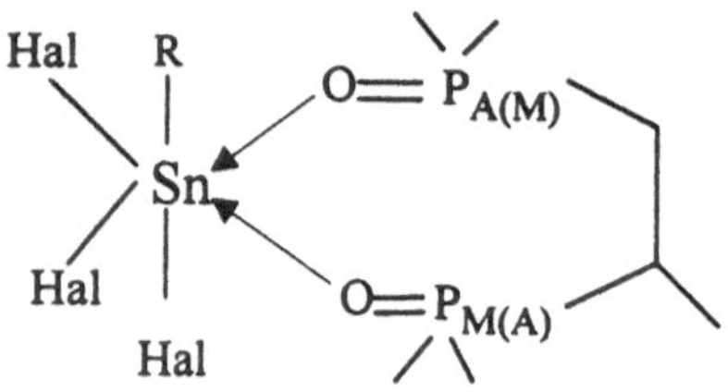

IVa 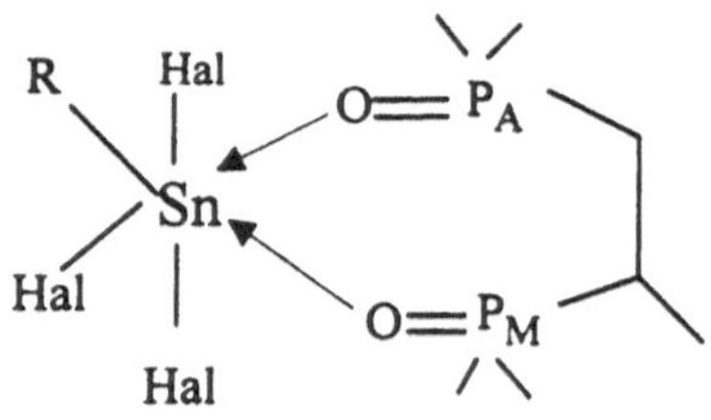   **IVb**

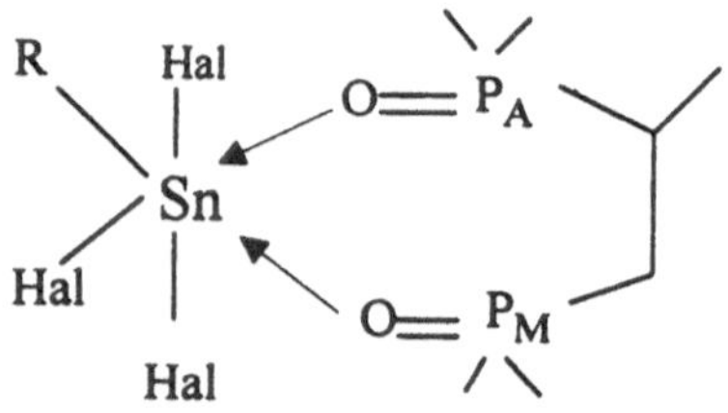

IVc

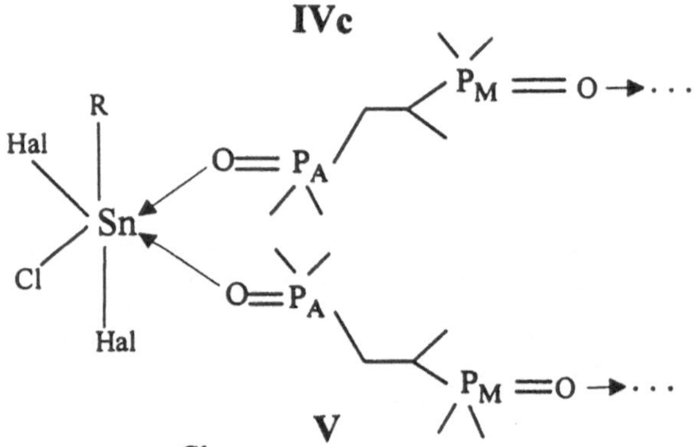

V

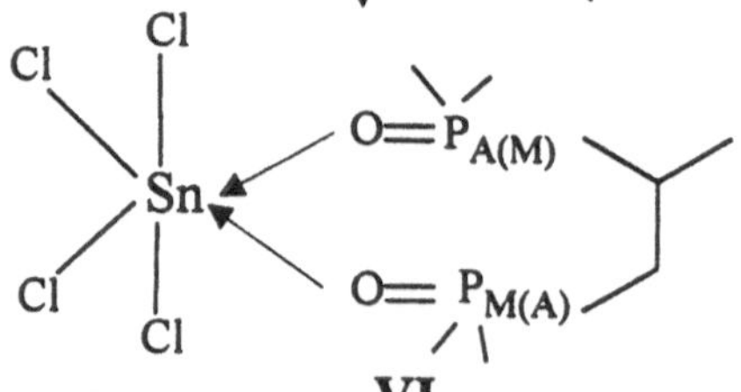

VI

A chelating structure is adduced on the basis of the argument that an alternate polymeric structure for the compound would entail as many as seven types of coordination octahedra, all of which are energetically feasible. The presence of the only one isomer (VI) for the SnCl$_4$ complex with propylenediphosphonate with non-equivalent PO groups also corroborates the chelating bonding mode of this ligand, the preference for which is perhaps explicable in terms of the stabilization of the *gauche*-conformation of the ligand consequent upon chelation to the metal.

The remaining triplet in the ^{119}Sn NMR spectrum probably corresponds to the symmetrical octahedral complex (V) with two monodentate ligands. The NMR parameters for this complex are similar to those for the ethylenediphosphonate adduct with structure 111a.

From the NMR parameters, the MeSnCl$_3$ complex with propylenediphosphonate appears to have the same isomeric composition as the PhSnCl$_3$ complex in solution. On the other hand, NMR parameters corresponding to only two isomers could be recorded for the MeSnBr$_3$ complex, owing presumably to its high lability in solution and broadening of the NMR signals (Table 6).

The foregoing NMR study of the complexes of organotin trihalides with diphosphoryl ligands has revealed a substantial dependence of $^2J(^{119}$Sn-^{31}P) on the nature of the *trans*-ligand: substitution of halogen in *trans*-position to the phosphoryl ligand by alkyl or aryl group leads to a decrease in $^2J(^{119}$Sn-^{31}P) from 180-280 Hz to 30-140 Hz, the sign remaining the same. This coupling constant depends also on the geometry of the complexes. Thus in going from chelates with 6-membered rings to more open polymer structures and 7-membered chelates, the $^2J(^{119}$Sn-^{31}P) values change from 180-236 Hz to 216-280 Hz for organotin trihalides, and from 100-120 Hz to 130-160 Hz for tin tetrachloride. These changes accompany the increased openings of the Sn-O-P angles.

Table 6 : ^{31}P and ^{119}Sn NMR parameters for complexes RSnHal$_3$·(EtO)$_2$P(O)CH$_2$CH(Me)P(O)(OEt)$_2$

R	Hal	Structure	δ(^{31}P), ppm	δ(^{119}Sn), ppm	$^2J(^{119}$Sn-^{31}P), Hz
Ph	Cl	IVa	28.0(A), 27.7(M)	-557dd	235(AX), 266(MX)
		IVb or c	26.3(A), 25.2(M)	-553dd	216(AX), 98(MX)
		IVc or b	27.7(A), 24.3(M)	-552dd	237(AX), 66(MX)
		V	30.1(A), 26.1(M)	-559t	253(AX)
Me	Cl	IVa	27.6(A), 25.6(M)	-496dd	280(AX), 257(MX)
		IVb or c	25.3(A), *(M)	-501dd	248(AX), 84(MX)
		IVc or b	*(A), 24.3(M)	-500dd	262(AX), 52(MX)
		V	29.3(A), 26.8(M)	-497t	270(AX)
Me	Br	IVb or c	24.8(A), 23.4(M)	-719dd	261(AX), 132(MX)
		IVc or b	26.3(A), 22.2(M)	-718dd	270(AX), 102(MX)
Cl	Cl	VI	25.3(A), 26.8(M)	-689dd	128(AX), 160(MX)

*The chemical shifts could not be determined because of overlapping signals.

Crystal structures of complexes of organotin(IV) trihalides with diphosphoryl ligands

The X-ray analysis of four $RSnHal_3$ complexes with methylenediphosphonates[3] has enabled a comparison of the differential effects exerted by the ligands in the primary coordination sphere of the tin atom, more especially the effects of the *trans*-ligand upon the bonding character of the Sn-O-P fragment. This is relevant to the validation of the observed trends in the NMR.

The complexes, $RSnCl_3 \cdot [(Pr^iO)_2CHR'$, (R = Me, Ph, R' = H; R = Ph, R' = NMe_2) crystallize in the form of symmetrical isomers with both phosphoryl groups in *trans*-position to halogen atoms (**1a**), and with similar Sn-O (2.205-2.258 Å), P-O (1.478-1486 Å) bond distances and Sn-O-P (137.2-140.9°) bond angles. The coordination octahedra are slight distorted, O-Sn-O bond angles being of 79-83°, and C-Sn-Cl_{ax} fragment being bent towards phosphoryl groups (C-Sn-Cl_{ax} bond angle does not exceed 168°).

The structure of $MeSnBr_3 \cdot [Pr^iO)_2P(O)]_2CHNMe_2$ is quite different, having one of the P-O fragments in *trans*-position to the methyl group. The corresponding Sn-O bond (2.149 Å) is considerably shorter than that in *trans*-position to the bromine atom (2.216 Å), the P-O bond distances and Sn-O-P bond angles remaining unaffected. This fact can be explained by *trans*-strengthening of the Sn-O bond following the substitution of *trans*-situated halogen by the electron donating alkyl ligand. This leads also to decrease of the $^2J(^{119}Sn-^{31}P)$ value as has been illustrated by NMR data. Thus the strengthening and shortening of Sn-O bonds are followed by lowering of the $^2J(^{119}Sn-^{31}P)$ magnitudes. The increase in acceptor ability in going from organotin trihalides to tin tetrahalides also leads to a decrease in the $^2J(^{119}Sn-^{31}P)$ value.

Structures of diorganotin(IV) dihalide complexes with diphosphoryl ligands

Over 50 complexes of R_2SnHal_2 (R = Me, Et, Bu; Hal = Cl, Br; R = Ph, Hal = Cl) with the following ligands, methylenediphosphonates $\{[(RO)_2P(O)]_2CHR'$ (R = Et, Pr^i; R' = H, NMe_2)$\}$, methylenediphosphinates $\{[RO(R')P(O)]_2CH_2$ (R = Pr^i, R' = Me; R = Et, R' = $(EtO)_2CH)\}$, tetraethylethylene- and propylenediphosphinates, have been synthesised and their structures investigated by means of Mössbauer spectroscopy and/or X-ray analysis.

All 1:1 complexes of R_2SnHal_2 with the diphosphonyl ligand showed quadrupole splittings in their Mössbauer spectra in the range 3.60-4.34 mm s^{-1}, indicating *trans*-R_2SnX_4-octahedral geometry.[4] However, the preferred mode of ligand coordination and the type of complex structure could not be determined from these data.

$(Me_2SnCl_2)_2 \cdot [Pr^iO)_2P(O)]_2CHNMe_2$ and its bromo-analogue showed lower quadrupole splitting values (3.38 and 3.28 mm s^{-1}, respectively) suggesting that the tin atoms are pentacoordinate. Trigonal bipyramidal geometries are envisaged for these complexes with *cis*-equatorial location for the two tin-bound methyl groups as indicated from point charge model calculations.[5] Methylenediphosphonate ligand serves in this case as a bidentate bridging ligand, linking two pentacoordinated organotin fragments in the molecule. X-ray analyses have confirmed the variety of structures that result from the coordinative interaction of R_2SnHal_2 with diphosphoryl ligands. Thus, Ph_2SnCl_2 forms with methylenediphosphonates $[(RO)_2P(O)]_2CHR'$ (R = Et, R' = H; R = Pr^i, R' = H, NMe$_2$) monomeric chelate complexes with distorted coordination octahedra (VII) with *trans*-Ph_2Sn skeletal configuration; the C-Sn-C bond angles are in the range 161-163° and O-Sn-O bond angles do not exceed 80°.[6]

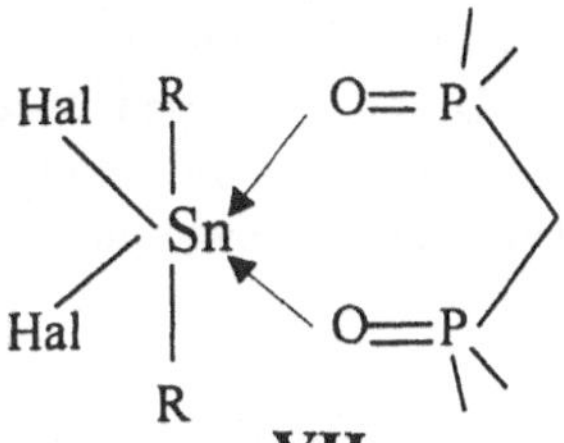

VII

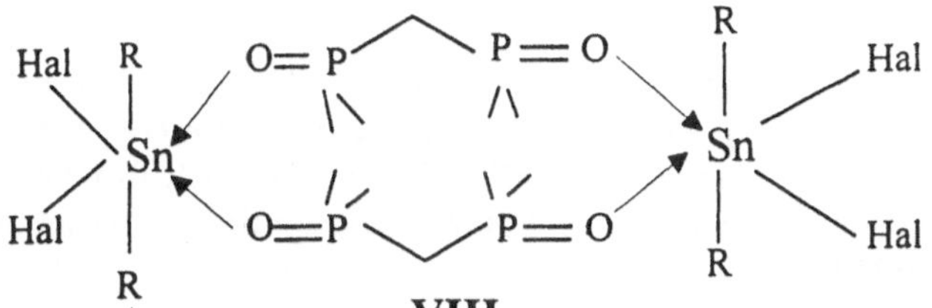

VIII

The complex $Et_2SnCl_2 \cdot [Pr^i(Me)P(O)]_2CH_2$ has also the monomeric chelate structure, but the octahedral environment at tin is highly distorted. The two Sn-O bond distances differ significantly (2.417 and 2.497 Å) while other pairs of bonds Sn-Cl and Sn-C remain equal.

The X-ray analysis of the complex of $MeSnCl_2$ with $[(EtO)_2P(O)]_2CHNMe_2$[7] revealed it to be unexpectedly a dimeric binuclear adduct (**VIII**). Two octahedrally coordinated organotin fragments are *cis*-linked by two bidentate bridging ligands forming the 12-membered ring. The C-Sn-C bond angle of 154.5° indicates considerable distortion of the coordination octahedra. The O-Sn-O bond angles in this more opened ring structure are larger (98.3°) than in chelate complexes, as also the Sn-O-P bond angles of 150° which contrast with values in the range 132-139° for 6-membered chelate rings.

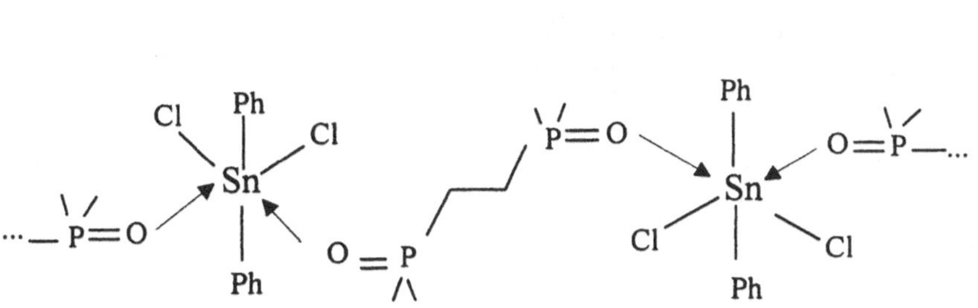

IX

The complex of Ph_2SnCl_2 with tetraethyl ethylenediphosphonate has the

481

polymeric structure (IX), as shown by X-ray analysis. Ethylenediphosphonate acts as bidentate bridging ligand linking the neighbouring tin coordination octahedra in *cis*-fashion. All bond distances and angles in the coordination sphere of the tin atom are within the usual ranges for hexacoordinated complexes of diorganotin dihalides with diphosphoryl ligands, with the exception of the Sn-O-P bond angles (152°) which are greater than those in the chelate adducts with methylenediphosphonates (132-139°). This difference in Sn-O-P bond angles in two structural types of octahedral complexes, the Sn-O and P-O bond distances being equal within experimental error, may cause the observed difference in $^2J(Sn-^{31}P)$ values for $PhSnCl_3$ adducts with methylene- and ethylenediphosphonates.

Biological activity of diphosphoryl complexes

Several mono- and diorganotin dihalide complexes with diphosphoryl ligands have been tested *in vitro* for antitumour and anti-AIDS activities at the National Cancer Institute (Bethesda, USA). Based on preliminary results, $RSnHal_3$ complexes show no activity while the complex $Et_2SnCl_2 \cdot [(EtO)_2P(O)CH_2]_2$ displays antitumour activity against the lung cancer cell NCl-H522 when applied at concentrations in the range 10^{-6}-10^{-5} M.

References

1. V.S. Petrosyan, N.S. Yashina and E.I. Gefel, *Silcion, Germanium, Tin and Lead Compounds*, **9** (1986) 213.
2. E.V. Grigor'ev, N.S. Yashina, M.V. Livantsov, A.A. Prichenko and V.S. Petrosyan, *Koord. Khim.*, **18** (1992) 1150.
3. E.V. Grigor'ev, N.S. Yashina, V.S. Petrosyan, A.V. Yatsenko and L.A. Aslanov, *Metallorg. Khim.*, **6** (1993) 182.
4. G.M. Bancroft and R.H. Platt, *Adv. Inorg. Chem. Radiochem.*, **15** (1972) 59.
5. G.M. Bancroft, V.G. Kumar Das, T.K. Sham and M.G. Clark, *J. Chem. Soc., Dalton Trans.*, (1976) 643.
6. E.V. Grigor'ev, N.S. Yashina, V.S. Petrosyan, A.V. Yatsenko and L.A. Aslanov, *Metallorg. Khim.*, **6** (1993) 175.
7. J. Lorberth, S.-H. Shin, M. Otto, S. Wocadlo, W. Massa and N.S. Yashina, *J. Organomet. Chem.*, **407** (1991) 313.

Main Group Elements and Their Compounds
V.G. Kumar Das (Ed)
Copyright © 1996 Narosa Publishing House, New Delhi, India

Empirical Correlations of Nuclear Spin-spin Coupling Constants with the Structures of Bicyclodiorganotin Compounds

Khoo Lian Ee[a] and George Eng[b]

[a]School of Science, Nanyang Technological University, 469 Bukit Timah Road, 1025 Singapore and [b]DC Agriculture Experimental Station and Department of Chemistry, University of the District of Columbia, Washington DC 20008, USA

Valence bond calculations have been used as a basis to derive a correlation between the vicinal coupling constant, ^3J(H-C-C-H) and the dihedral angle as in the well known Karplus equation.[1] In addition, the ^3J(H-C-C-H) value is also affected by parameters such as electronegativity of the substituents, the position of substitution, hybridization of the intervening and coupled atoms, as well as bond lengths and angles.[2] When a heteroatom is substituted for a hydrogen atom in the coupling path, the theoretical considerations for vicinal coupling constant relationships are complex due to the considerable number of electrons and geometrical parameters involved. Thus many vicinal coupling constant correlations involving heteroatoms such as ^3J(C-C-C-H) are correlated empirically.[3] Even less is known about the vicinal ^{119}Sn-H coupling constant across a C=N imine bond. Allen *et al.*[4] indicated that there was a change in the value of ^3J(SnC=CH-o) of PhSnX$_3$ (X=Cl and Br) in the presence of Lewis bases. Recently Srivastava and co-workers[5] reported that the ^3J(SnC=CH-o) coupling constant can be correlated linearly with the electronegativity of the ligating atom for a series of triphenyltin halide adducts, Ph$_3$SnX:L, according to the equation: ^3J(adduct) = ^3J(Ph$_3$SnX) + A•n•E$_L$, where X = electronegative atom attached to Sn, A = constant, n = number of ligating atoms and E$_L$ the electronegativity of the ligating atom in the ligand (L).

In this study, we report the empirical correlations of ^3J(SnN=CH) with ^2J(SnCH$_3$) and the structure of several bicyclic diorganotin compounds as shown in Fig. 1.

Experimental

Dibutyltin oxide, dimethyltin oxide, α-amino acids and other organic chemicals were purchased commercially and used without further purification. The bicyclic diorganotin compounds 1-15 shown in Fig. 1 were synthesized using published procedures in 60-80% yields. The compounds are listed in Table 1 together with their melting points.

Proton NMR spectra were recorded on a Bruker AC-P300 spectrometer in CDCl$_3$ using TMS as the internal standard. A few drops of DMSO-d$_6$ were added to samples that had low solubility. The ^3J(SnN=CH) values were determined from the distance between the two pairs of tin-coupled satellite peaks which flanked on either side of the intense azomethine proton peak located at *ca.* δ 9 ppm.

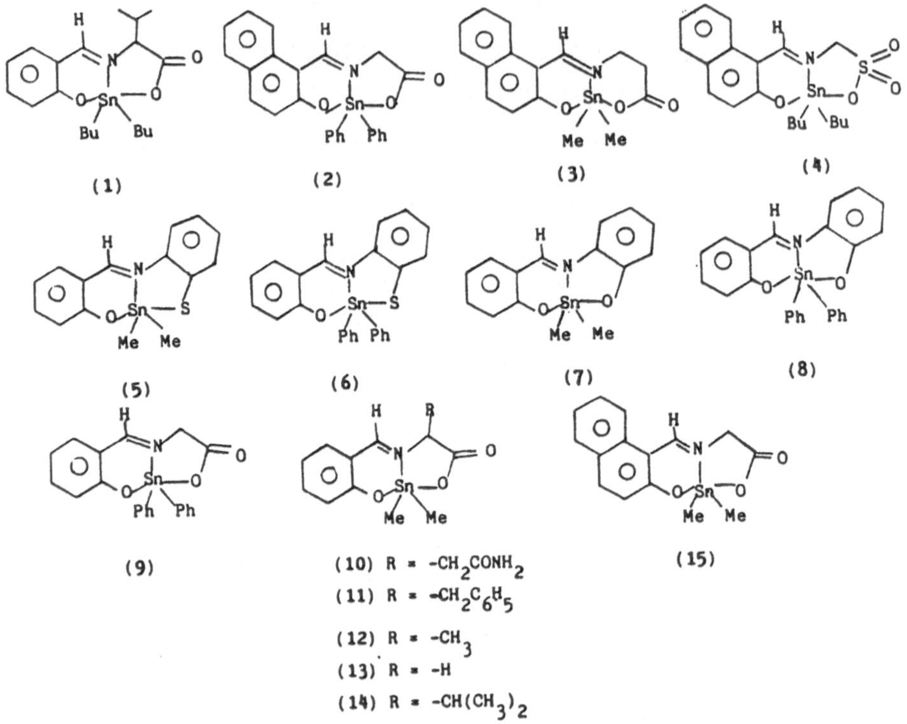

Fig. 1 : Structures of some bicyclo diorganotin compounds

Results and Discussion

The ^{119}Sn-H vicinal coupling constant across an imine bond, ^{3}J(SnN-CH), was first reported by Kawakami and Tanaka[17] for the diorganotin complexes prepared from dimethyltin dichloride and some ONO tridentate Schiff bases derived from substituted *o*-aminophenols and salicylaldehyde. Results from this study support the existence of the Sn←N coordinate bonding in these complexes. The ^{3}J(SnN=CH) and ^{2}J(SnCH$_3$) coupling constant values recorded for the bicyclodimethyltin compounds along with the coupling constants of other bicyclodiorganotin compounds are listed in Table 1. Also tabulated in this table are the Sn←N coordinate bond distances for these compounds determined from X-ray studies. Fig. 2 shows a plot of ^{3}J(SnN=CH) coupling constants vs ^{2}J(SnCH$_3$) coupling constants for the dimethyltin bicyclic compounds.[3,5,7,10-15] Assuming that the coupling mechanism for ^{2}J(SnCH$_3$) is governed by the Fermi contact term and that only rehybridization, and not the effective nuclear charge, is important then an increase in the ^{3}J(SnN=CH) coupling constant should lead to a decrease in the ^{2}J(SnCH$_3$) value for dimethyltin derivatives. Clearly, this is not the case as evident from Fig. 2. More data with a wide range of ^{3}J(SnN=CH) and ^{2}J(SnCH$_3$) values are needed before a conclusion is reached pertaining to the relationship between ^{3}J(SnN=CH) and ^{2}J(SnCH$_3$). However, when the ^{3}J(SnN=CH) coupling constants for compounds

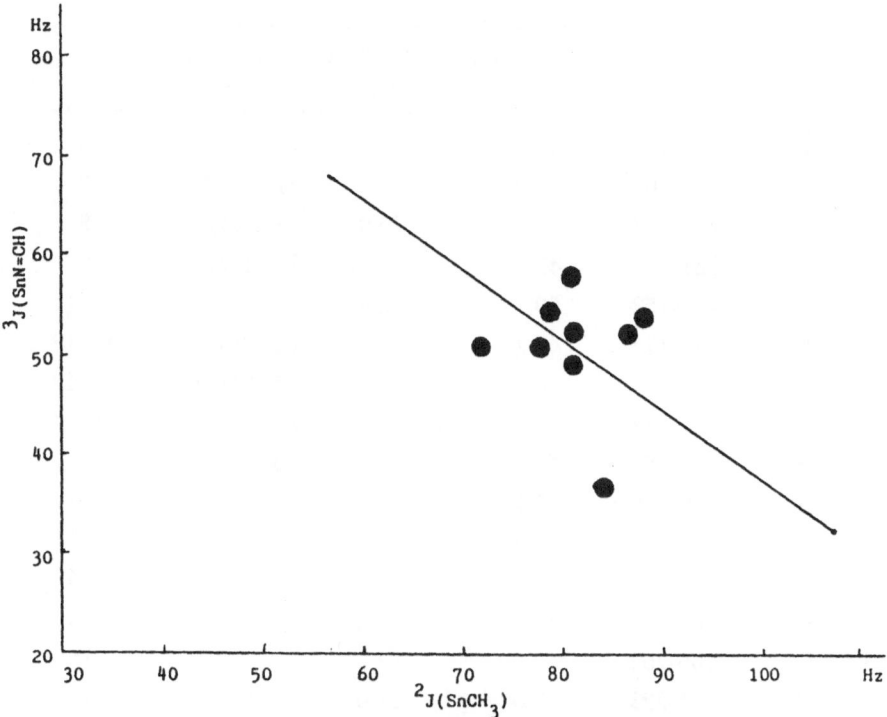

Fig. 2 : Plot of ^3J(SnN=CH) *vs.* ^2J(SnCH$_3$)

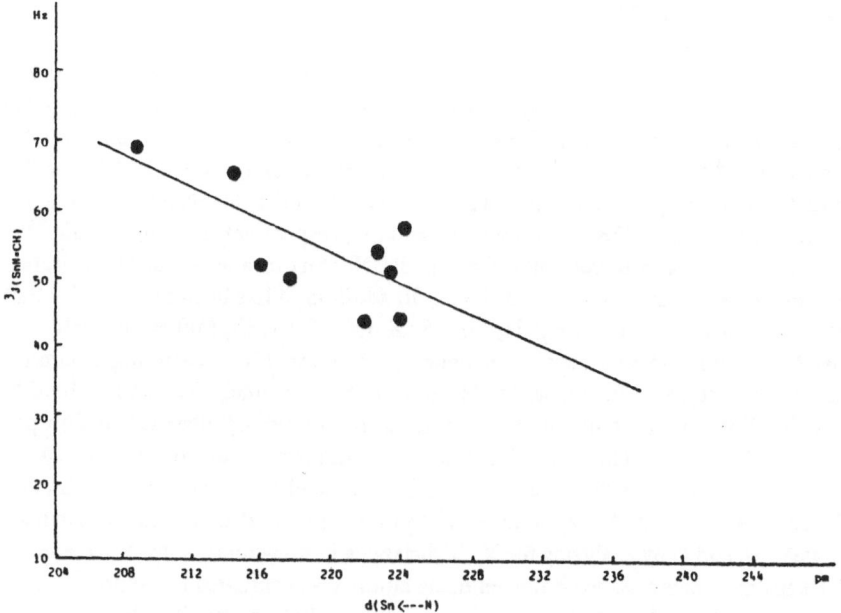

Fig. 3 : Plot of ^3J(SnN=CH) *vs.* d(Sn←N)

Table 1 : Spin-spin coupling constants and Sn-N bond distances of bicyclodiorganotin compounds

Compound	m.p. °C	$^3J(SnN=CH)$ Hz	$^2J(SnCH_2)$ Hz	d(Sn-N) pm	Reference
1	153-155	52.0		215.8(8)	6
2	198-200	64.9		214.2(5)*	7
3	232-234	54.2	79.0	222.4(3)	8
4	161-163	43.7		224.1(7)	9
5	157-158	50.1	72.2	217.6(4)	10, 11
6	175-176	44.2		221.7(3)	10, 12
7	123-124	50.8	78.0	222.9(11)	10, 13
8	157-158	57.5		224.1(13)	10, 14
9	183-184	68.6		208.6(3)	15
10	218-220	53.1	88.6		16
11	160-162	52.2	81.2		16
12	210-212	48.8	81.6		16
13	184-185	36.0	84.0		16
14	200-202	52.0	82.0		16
15	221-223	57.6	81.0		16

*This figure is an average of 4 independent determinations 2.145(5), 2.137(5), 2.150(4) and 2.135(5).

1-9 are plotted against the Sn←N bond distance, a reasonable correlation (r = 0.73) is obtained as shown in Fig. 3. The data showed that an increase in the $^3J(SnN=CH)$ value leads to a decrease in the N→Sn bond distance. This is in agreement with the observation[18] that the trans(9Hz) and cis(34Hz) values of $^3J(SnN=CH)$ for the adduct, n-BuSnCl$_3$: C$_6$H$_5$CH=N-C$_5$H$_4$N-2, depend on the Sn←N coordinate bond distance (232.0 and 228.2 pm for trans and cis, respectively). Other studies[19,20] have reported that a lower value for the $^1J(^{15}N→Sn)$ in a series of stannatranes indicates a weaker Sn←N coordinate bond. In addition, it has been reported[21,22] that the Sn←N distances in N(CH$_2$CH$_2$CH$_2$)$_3$SnX, [X = I, Br, Cl, Me] are linearly correlated with the vibrational frequencies of ʋ(Sn←N). Assuming a similar relationship between the $^3J(SnN=CH)$ and the Sn←N bond distance for the title bicyclic diorganotin compounds, a "minimal non-bonding" distance of 266 pm between Sn and N atoms is obtained. It is interesting to note that Drager[23] considered a value of 308 pm as "minimal non-bonded approach" between Sn and N. Glidewell[24] calculated a value of 302 pm as "minimal non-bonded distance" between Sn and N by analyzing the X...Y distances in non-linear X-M-Y fragments. The van der Waals distance between these atoms was estimated by Bondi[25] as 367 pm, while a typical Sn-N distance[26] is approximately 215 pm. The longest Sn-N distance reported to-date is 296.5 pm for diphenyltin dichloride pyrazine adduct.[27]

Thus the estimated "minimal non-bonding" distance of 266 pm in the present study is within previously reported values for di-, tri- and tetraorganotin compounds containing Sn-N coordinative linkages.[28] However, it is important to note that the correlation between the vicinal ^{119}Sn-H coupling constant across an imine bond and the interatomic distance is empirical. Other similar bicyclic diorganotin compounds must be examined to verify the correlation.

References

1. M. Karplus, *J. Chem. Phys.*, **30** (1959) 11.
2. R.K. Harris, *Nuclear Magnetic Resonance Spectroscopy*, Longmans Scientific and Technical, U.K., 1986.
3. T. Spoormkaer and M.J.A. de Bie, *Recl. Trav. Chim. Pays-Bas*, **99** (1980) 154; L.M. Jackman and S. Sternhell, *Applications of Nuclear Magnetic Resonacne Spectroscopy in Organic Chemistry*, 2nd Ed. Pergamon Press, 1969.
4. C.W. Allen, A.E. Burroughs and R.G. Austey, *Inorg. Nucl. Chem. Lett.*, **9** (1973) 1211.
5. P.C. Srivastava and S.K. Srivastava, *Spectrochim. Acta*, **41A** (1985) 687; P.C. Srivastava and A. Trivedi, *Spectrochim. Acta*, **48A** (1992) 1415.
6. F.E. Smith, R.C. Hynes, T.T. Ang, L.E. Khoo and G. Eng, *Can. J. Chem.*, **70** (1992) 1114; T.T. Ang and L.E. Khoo, In: *Chemistry and Technology of Silicon and Tin*, eds., V.G. Kumar Das, S.W. Ng and M. Gielen, Oxford University Press, 1992, pp. 514-520.
7. F.E. Smith, R.C. Hynes and L.E. Khoo, to be published
8. N.K. Goh, L.E. Khoo and T.C.W. Mak, *Polyhedron*, **12** (1993) 925.
9. L.E. Khoo, Z-Y. Zho and T.C.W. Mak, to be published.
10. R. Cefalu, R. Bosco, F. Bonati, F. Maggio and R. Barbieri, *Z. Anorg. Allg. Chemie*, **370** (1970) 180.
11. G.Y. Yeap, H.K. Fun, S.B. Teo and S.G. Teoh, *Acta Cryst.*, **C48** (1992) 1109.
12. V.H. Preut, H.J. Hautt, F. Huber, R. Cefalu and R. Barbieri, *Z. Anorg. Allg. Chemie*, **407** (1974) 257.
13. V.H. Preut, F. Huber, H.J. Haupt, R. Cefalu and R. Barbieri, *Z. Anorg. Allg. Chemie*, **410** (1974) 88.
14. V.H. Perut, F. Huber, R. Barbieri and N. Bertazzi, *Z. Anorg. Allg. Chemie*, **423** (1976) 75.
15. J-T. Wang, Y-W. Zhang, Y-M. Xu and Z-W. Wang, *Heteroatom Chemistry*, **3** (1992) 599.
16. H.C. Tan, Research Dissertation, Universiti Sains Malaysia, Penang, Malaysia, 1991.
17. K. Kawakami and T. Tanaka, *J. Organometal. Chem.*, **49** (1973) 409.
18. G. Matsubayashi, T. Tanaka, S. Nishigaki and K. Nakatsu, *J.C.S. Dalton Transactions*, (1979) 501.
19. A. Tzschach, H. Weichmann and K. Jurkschat, *J. Organometal. Chem. Library*, **12** (1981) 293.
20. E. Kupce and E. Lukevics, *Russian Chem. Rev.*, **58** (1989) 1011.
21. K. Schenzel, A. Kolbe, A. Sokolovska and P. Reich, *J. Molecular Structure*, **218** (1990) 189.

22. K. Schenzel, A. Kolbe and P. Reich, *Monatsh. für Chemie*, **121** (1990) 615.
23. M. Drager, *Z. Anorg. Allg. Chemie*, **423** (1976) 53.
24. C. Glidewell, *Inorg. Chim. Acta*, **12** (1975) 219.
25. A. Bondi, *J. Phys. Chem.*, **68** (1964) 441.
26. P.G. Harrison, In: *Chemistry of Tin*, ed, Blackie, 1989, 10.
27. D. Cunningham, T. Higgins and P. McArdle, *Chem. Commun.*, (1984) 833.
28. V.G. Kumar Das, K.M. Lo, Chen Wei, S.J. Blunden and T.C.W. Mak, *J. Organomet. Chem.*, **322** (1987) 163; V.G. Kumar Das, K.M. Lo, Chen Wei and T.C.W. Mak, *Organometallics*, **6** (1987) 10.

Main Group Elements and Their Compounds
V.G. Kumar Das (Ed)
Copyright © 1996 Narosa Publishing House, New Delhi, India

Structural Study of Organotin (IV) Cations Using Tin-119*m* Mössbauer Spectroscopy

Leopold May[a], Deborah Whalen[b], Benjamin R. Reed[c]
and George Eng[b]

[a]*Department of Chemistry, The Catholic University of America*
Washington, DC 20064
[b]*The District of Columbia Agricultural Experiment Station and Department of*
Chemistry, The University of the District of Columbia
Washington, DC 20008
[c]*ADA-AF, Gaithesburg, MD 20899, USA*

In an earlier investigation of the interaction between triphenyltin (TPT) compounds and the fungus, *Ceratocystis ulmi*, we found that the Mössbauer spectra of fungal cells incubated with four different TPT compounds were identical.[1] This suggested that the TPT compounds interacted with the fungus in the same manner, possibly through the formation of the TPT$^+$ cation. This prompted a study on the Mössbauer spectrum of TPT$^+$ which could shed light on the mechanism of its interaction with the fungus.

Organotin (IV) cations, known to be present in aqueous solutions, have been studied using spectral techniques such as IR[2], NMR[3,4] and Raman[2]. Because the concentrations of the triaryltin cations in aqueous solutions are too low to be observed directly by Mössbauer spectroscopy, a method was devised to isolate the cations on cation exchange columns. The spectra of the attached cations were observed by examining the column materials directly. Results of the study of the TPT$^+$ cation as well as other aryltin cations are reported here.

Experimental

Preparation of matrix -isolated organotin cations

The Cellex column was prepared by soaking the Cellex beads (Bio-Rad Cellex-CM cation exchange carboxymethylcellulose) in chloroform or methanol for 3 days. Cellex beads were then placed in a 1 cm diameter column and filled to a height of approximately 3 cm. A chloroform or methanol solution saturated with the aryltin chloride parent compound was passed through the column with flow rates adjusted to about 1.0 mL/h. When the absorption band of the phenyl groups at 260 nm was found in the effluent, the column was considered to be saturated with the cation. The entire column material was then transferred to a Mössbauer holder and kept frozen until the Mössbauer spectrum was recorded. Approximately 270 mg of tin was used in each sample.

Mössbauer spectroscopy

The spectra were recorded at 80K on a Mössbauer spectrometer model MS-900 (Ranger Scientific Co., Burelson,TX) in the acceleration mode with a moving source geometry in a liquid nitrogen cryostat (CYRO Industries of America, Inc., Salem, NH). The samples were mounted in teflon holders. The source was 15 mCi $Ca^{119m}SnO_3$, and the velocity was calibrated at room temperature using a combination of $BaSnO_3$ and Sn foil (splitting = 2.52 mm s^{-1}). The resultant spectra were analyzed by a least-square fit to Lorentzian-shaped lines.

Results and Discussion

The Mössbauer parameters (quadrupole splitting (QS) and isomer shift (IS)) of the matrix-isolated aryltin cations and of their parent aryltin chlorides are given in Table 1. The magnitude of the QS gives information concerning the symmetry of the ligands around the tin atom, and the IS is related to the s-electronic density at the tin nucleus. As expected, the QS of the parent aryltin halides are greater than zero.

Table 1 : Mössbauer parameters of matrix-isolated organotin cations and parent compounds[a]

Organotin halide	Organotin cation		Organotin chloride	
	IS	QS	IS	QS
$(C_6H_5)_3SnCl$	0.15(4)	0	1.35(1)	2.52(2)
$(p-ClC_6H_4)_3SnCl$	0.04(4)	0	1.24(1)	2.38(2)
$(m-CH_3C_6H_4)_3SnCl$	0.00(5)	0	1.37(1)	2.72(2)
$(p-CH_3SC_6H_4)_3SnCl$	0.09(6)	0	1.20(1)	2.47(4)
$(p-FC_6H_4)_3SnCl$	0.12(1)	0	1.21(1)	2.26(6)
$(C_6H_5)_2SnCl_2$	0.89(1)	1.72(3)	0.99(1)	1.96(4)

[a] All spectra were taken at 80 K. Parameters are given in mm s^{-1} relative to $BaSnO_3$: IS = isomer shift, QS = quadrupole shift. The numbers in brackets are the errors in the last figure.

Most of the spectra of the Cellex-isolated cations showed only single lines (QS = 0 mm s^{-1}) with near-zero IS values (Table 1). A typical spectrum is shown in Fig. 1. The spectra of all the triaryltin cations contained singlets indicating that the tin atoms are in symmetrical fields. This is consistent with a geometric structure for the triaryltin cation in which the three aryl groups are coplanar with the tin atom. The IS values of these cations are near zero and are more negative than the IS of the parent triaryltin halide (1.2 - 1.4 mm s^{-1}) or tetraaryltins. This is expected because the s-electron density at the tin nucleus decreases when it becomes more positively charged in the cation. The cations can be divided into three groups based upon their IS values : 0.00 - 0.04 mm s^{-1} ($m-CH_3$, $p-Cl$), 0.09 mm s^{-1} ($p-CH_3S$), and 0.12 - 0.15 mm s^{-1} ($p-F$, H), whereas the triaryltin chlorides can be divided into two groups: 1.36 mm s⁻¹ (*m*-CH₃, H) and 1.23 mm s⁻¹ for the remainder. These observations suggest that the electronic and resonance effects of the substituents in the phenyl rings are complicated and different in the cations than in the parent neutral compounds. The results are also in sharp contrast to the Mössbauer data reported on triorganotin cations isolated from solution as tetraphenylborate salts in the presence of donor ligands, which specify unequivocally pentacoordinate tin sites in the compounds[5,6] as result of inner-sphere ligand coordination.

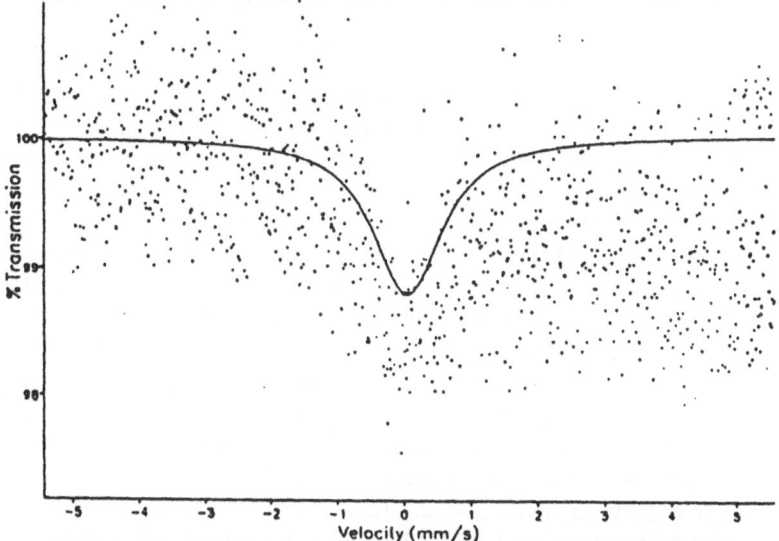

Fig. 1 : Mössbauer spectrum of the Cellex-isolated triphenyltin cation

The structure of diphenyltin cation, presumed to be similarly "naked", is not linear because its Mössbauer spectrum shows that there is a doublet with a QS of 1.72 mm s⁻¹. If the cation were linear, the phenyl group would be arranged symmetrically around the tin nucleus, and there would be no QS. In contrast, the dimethyltin cation in aqueous medium has been shown to be linear.[2] This conceivably arises on account of solvation and a *trans*-Me₂Sn(H₂O)₄²⁺ octahedral geometry for this ion. A bent Ph₂Sn skeletal arrangement for the Ph₂Sn²⁺ cation in the present study could be occasioned by some degree of coordinative interaction of the cation with the Cellex matrix.

References

1. L. May, G. Eng, S.P. Coddington and L.L. Stockton, *Hyperfine Interact.*, **42** (1988) 909.
2. M.M. McGrady and R.S. Tobias, *Inorg. Chem.*, **3** (1964) 1157.
3. B. Wrackmeyer, G. Kehr, A. Sebald and J. Kümmerlenas, *Chem. Ber.*, **125** (1992) 1597.
4. J.B. Lambert and B. Kuhlman, *J. Chem. Soc., Chem. Commun.*, (1992) 931.
5. G.M. Bancroft, V.G. Kumar Das, T.K. Sham and M.G. Clark, *J. Chem. Soc., Dalton Trans.*, (1976) 643.
6. W.K. Ng, Chen Wei, V.G. Kumar Das and R.-J. Butcher, *J. Organomet. Chem.*, **361** (1989) 53.

Main Group Elements and Their Compounds
V.G. Kumar Das (Ed)
Copyright © 1996 Narosa Publishing House, New Delhi, India

The Impact of Organotins on the Environment

I.P. Beletskaya and S.L. Davydova

Moscow State University, Russian Academy of Sciences, Russia

Organotins are important chemicals in modern technology. Their exposure to man and the environment comes through their diverse applications encompassing such well established uses as thermal stabilizers for polyvinylchloride plastics, as catalysts for the production of polyurethane foams and room temperature-vulcanizing silicone elastomers, and as industrial biocides (wood preservatives and marine antifouling paints) and crop protectants. Although not well recognized, the burning of petroleum and coal provides a major source for the entry of tin in chemically bound form into the environment. The tin content in the biosphere has been measured by several analytical methods: in air, food or biosamples by atomic emission spectroscopy or neutron activation analysis; in aqueous solution by voltamperometry; in soil by fluorometry. The general decomposition schemes in the environment for alkyltin and aryltin compounds of low and high molecular weights, as well as the distribution levels of tin in the atmosphere, hydrosphere and lithosphere have been well reported. The biomethylation of organotins, which appears scant in comparison with the biomethylation of mercury and other heavy metals, is, however, a major pertinent consideration in evaluating the environmental impact of organotins.

Organotin compounds are also widely used in organic synthesis. There are a number of reasons for this. In the first place a wide variety of organic groups can be covalently bonded to the tin atom, which in the ubiquitous +4 oxidation state, yields the 4 principal classes of tetra-, tri-, di- and mono-organotin(IV) compounds. Secondly, organotin compounds are known to display some specificity in the cleavage of the Sn-C bonds by a number of elctrophilic reagents and, indeed, have served as model substrates for mechanistic studies on S_E reactions.

Lastly, and this is especially relevant to the field of organic synthesis, organotins manifest moderate reactivity when compared to organolithium and organomagnesium reagents. This allows for a greater tolerance of functional groups by organotins in their reactions, as well as the use of tin compounds containing functional substituents in the tin-bound organic groups.

The increased use of organotins in organic synthesis followed closely on the heels of the discovery[2] of carbon-carbon bond forming cross-coupling reactions between other types of organometallic compounds and organic halides, particularly, aryl- and vinyl-halides catalyzed by transition-metal complexes.

The cross-coupling reaction between an organotin compound and an organic halide is best catalyzed by palladium complexes, and goes by the name of Stille-reaction[3].

It was shown that the coupling could be realised under very mild conditions, if palladium salts containing no phosphinic ligands are used as catalysts[4], viz.

$$PdCl_2 (MeCN)_2 , (\eta^3-C_3H_7PdCl)_2 \text{ and } Pd(OAc)_2$$

The catalysts are sometimes loosely referred to as "ligandless" palladium (denoted by 'Pd'). Exemplifying the use of these catalysts are the cross-coupling reactions illustrated below with organic halides[5] (equation 1) and acid halides[6] (equation 2),

$$RX + R^I SnR_3^{II} \xrightarrow{['Pd']} R - R^I \qquad (1)$$

R = aryl, vinyl, het, allyl
R' = alkyl, allyl, alkynyl, alkenyl, aryl, het
R" = alkyl

$$RCOCl + R^I SnR_3^{II} \xrightarrow{['Pd']} RCOR^I \qquad (2)$$

R = alkyl, aryl, vinyl

and the carbonylation of aromatic and heteroaromatic halides wherein the organotin serves as a carbanionic or other nucleophilic donor[7,8] (equations 3 and 4)

$$RX + CO + R^I SnR_3^{II} \xrightarrow{['Pd']} RCOR^I \qquad (3)$$

$$RX + CO + R_3^I SnNu \xrightarrow{['Pd']} RCONu \qquad (4)$$

R = aryl, benzyl, het, alkenyl, alkynyl
Nu = OMe, NR_2^{II}, SR^{II}

The organotin compounds themselves can be obtained through reaction of aryl- or vinyl-halides with hexaalkylditins, catalyzed by "ligandless" palladium[9] (equation 5).

$$RX + R_6^I Sn_2 \xrightarrow{['Pd']} RSnR_3^I \qquad (5)$$

It was shown recently that many reactions catalyzed by palladium complexes (for example, cross-coupling of arylboranic acids or sodium tetraphenylborates with aryl(vinyl)halides, arylation of olefins by the Check-reaction, carbonylation of aromatic halides) are accelerated if they are conducted in mixture of organic solvent and water or, (in the case of water soluble substrates) in the pure aqueous medium[10]. Clearly, the use of water as solvent and the absence of phosphine ligands have both an economic and ecological appeal for these reactions.

The strategy has been applied with success in the high-yielding cross-coupling reactions of tetraorganotins with *in-situ* generated diazonium salts[10] (equation 6) and iodonium salts[11] (equation 7).

$$ArN_2X + RSnMe_3 \xrightarrow[H_2O/MeCN]{Pd(OAc)_2} Ar\text{-}R + Me_3SnX \qquad (6)$$

$Ar = p\text{-}NO_2C_6H_4,\ o\text{-}NO_2C_6H_4,\ p\text{-}IC_6H_4,\ p\text{-}BuC_6H_4,\ p\text{-}MeOC_6H_4,$
$p\text{-}MeC_6H_4;\ X = HSO_4^-,\ Cl^-;\ R = Me,\ p\text{-}CH_3C_6H_4,\ p\text{-}ClC_6H_4$

$$Ar_2IX + RSnMe_3 \xrightarrow{cat} Ar\text{-}R + ArI + Me_3SnX \qquad (7)$$

$Ar = p\text{-}NO_2C_6H_4,\ C_6H_5\ ;\ X = BF_4^-,\ CF_3COO^-,\ HSO_4^-;\ R = Me,$
$m\text{-}CH_3C_6H_4,\ Ph,\ PhCH = CH,\ p\text{-}ClC_6H_4$

The principal routes for the introduction of industrial organotin chemicals into the environment are well known: as agricultural insecticides or as antifouling paints, as smoke emitted from the incineration of organotin waste, through the production of organotin-treated glass, paper, textiles and polymer products (polyvinylchlorides, polyurethanes, silicones, polyesters, polyolefines), and others[12,14].

Another source of pollution by tin chemicals, probably a major one, comes from the burning of petroleum and coal. For petroleum, the burning and consequent chemical transformations take place on a huge scale in many countries[15].

The inherent magnitude of the problem is arguably reflected in the following estimates of oil deposits (percentage) available in the world.

Table 1 : Percentage deposits of crude oil (1990)

USA	2.9	Latin America	14.2
Near East	63.0	USSR	6.4
Africa	6.3	China	2.6
Western Europe	2.0	Asia	2.4

The metal content in crude oil has the following general profile: V, Na $= 10^{-1}$; Fe, Ca, Al, Ni $= 10^{-2}$; K, Mg, Si, Cr, Mo, Hg, Co, Zn $= 10^{-3}$; Ba, Sr, Mn, Cu, Ga, Se, As, Ag $= 10^{-4}$; Sb, In, U $= 10^{-5}$; Pb, Sc, La, Eu $= 10^{-6}$; Be, Ti, Sn, Au $= 10^{-7}$; others $< 10^{-7}$ mass %.[6]

The emission of tin into the environment by the burning of coal and dung in 1970 was 1.6×10^3 tonnes (or 11.0 g/km^2 of earth surface). In 1980 it was 3.3×10^3 tonnes (or 23.1 g/km^2 of earth) and the projected value in the year 2000 is 5.2×10^3 tonnes (or 35 g/km^2 of earth). The silicon content in petroleum ashes reaches some several percent. Also the content of lead in ashes of different oils is relatively high, ranging from 10^{-7} to 10^{-4} %. However, the tin content in petroleum ash is relatively constant at 10^{-3} %.[15,16]

Whereas the heavy oil fractions from Russian crude oils tended to be rich in lead, this was not the case for tin.[15] Furthermore, the tin containing organic products in these fractions were those of low volatility, with attached esteryl, dihydroxy, SH- or NH-groups.[17,18]

When introduced into the environment, these organic tin compounds can influence the biosphere as well as living organisms. Their effect on humans, in particular, is via several of the following processes:[19,20]

- interaction with mitochondrial membranes to cause swelling;
- suppression of mitochondrial oxidative phosphorylation;
- inhibition of fundamental energy transfer processes involved with ATP and α-ketoamino acid oxidation;
- blocking of sulfuryl-groups in cells.

To differentiate the processes and to quantify the impact on the environment, one has to have a reliable method for determining the tin content in any sphere that is being polluted. Modern methods for tin determination[9] in the biosphere include AES and NAA, mainly for air and biological samples, fluoro- and spectrometry for soil samples, voltamperometry for water samples, AAS and AES for food and drink samples, and ICP for many types of samples. The determination is often limited by, for example, the high thermal stability of tin oxide in AAS. The presence of Al, Cu, alkali and alkaline - earth metals also present difficulties in AAS and AES measurements. Hot flames, namely air - hydrogen or nitrogen oxide flames, are often used for this reason. ICP, NAA and X-ray fluorescent analyses are also recommended.[21]

After being introduced into the environment, the tin-containing compounds circulate in the atmosphere and hydrosphere. According to the Word Health Organization[22], food (without preservatives) contains roughly 1-4 mg of Sn, water about 30 mg of Sn and air about 1 mg of Sn per daily dose. Tin is easily excreted; it accumulates with age in the lungs and bones.[23]

Tin is an essential element for animals (2 mg/kg for rats) and even for man.[19] Reilly[24] has suggested that the absorption of tin from food by man is minimal, being only 1-2% of the amount consumed. On the other hand, Sn(II) is absorbed four times more than Sn(IV); the half-life of tin in living organisms for both oxidation states of metal (II and IV) is about 100 days.

Most industrially used organotins show high thermal stability. The decomposition of tetra-, tri-, di-butyltin compounds takes place at 180 - 200 °C, and mono-butyltin compounds at 120 °C. At 280 °C they all break down completely after 5 hours.[25]

Organotin compounds are readily decomposed by ultraviolet radiation, and this appears to be one of the most significant modes of decomposition in the environment.

Fungi and bacteria are also able to break the tin-carbon bond, for example, the tributyl- and triphenyl-tin derivatives can be broken down by the bacterium, *Pseudomonas aeruginosa* and by the fungus, *Coniophora puteana*.[26]

Sheldon[27] has proposed a generalized scheme for the decomposition of di- and tri-alkyl (aryl) tin compounds in the environment.

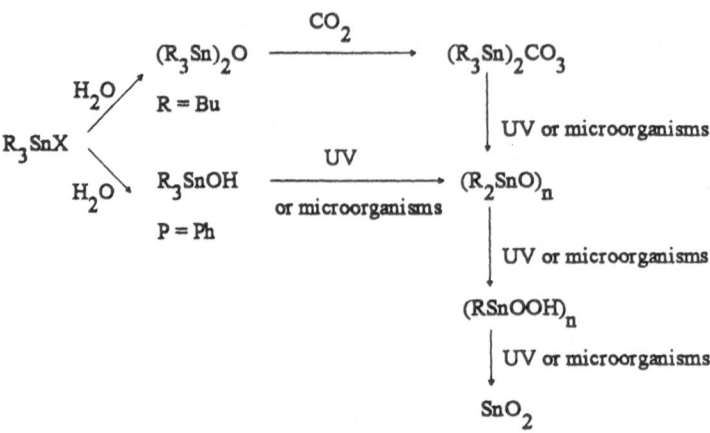

Scheme 1 : Generalized scheme for the decomposition of organotin compounds in the environment[27]

The majority of symmetrical R_4Sn compounds is stable to water and alkali solutions, but unsymmetrical organotins easily hydrolyse to SnO_2 under the same conditions. The photochemical sensitivity of the tin-carbon bond is different for symmetrical and for unsymmetrical organotins. Studies have indicated that the above degradation scheme involving sequential Sn-C cleavage also occurs in higher plants and animals[28]. Of considerable interest to environmental scientists is whether the reverse process - the alkylation (methylation) of tin occurs in the presence of natural agents, and whether the process is as facile as for some heavy metals such as mercury.

It has been reported that water-plants such as *Entermorpha* sp. or *Sargassum* sp. transform inorganic tin to mono- or di-methylated tin. The biomethylation of tin can lead to an accumulation of the metal by *Pseudomonas* up to 18% of the cell mass[26]. The Minamata Bay (Japan) was poisoned by methyl mercury in the 1950's: the *Pseudomonada* bacteria converted the large accumulated levels of mercury in the water to the extremely poisonous mono- and di-methyl mercury.[29]

Looking through literature data one can easily see that at least two groups of elements have the possibility of being methylated bacteriologically:

	^{50}Sn			
^{80}Hg	^{82}Pb		^{33}As	^{34}Se
atomic mass > 118			atomic mass < 78	

The left hand group (atomic mass above 118) consists of Hg, Pb, Sn and the right hand group (atomic mass under 78) consists of As and Se. These elements can be transformed to mono-, di-, tri-methylated derivatives, which are acutely poisonous to living organisms.

Biomethylation has only be documented for the elements of the left hand group: methylmercury, methyltin and methyllead have been detected in water and sediments.[30] The main analytical problem here has concerned the very low levels at which these chemicals are found: methylmercury exists at not more than one part per million level ($\mu g.g^{-1}$) and methyltin at about one part per trillion ($pg.g^{-1}$). Despite such low concentrations the accumulation of methylated compounds in the environment (e.g. via food chains)leads to much higher levels in man.[30] Accurate analytical methods for their determination have been described recently by Craig and co-workers.[31]

References

1. O.A. Reutov and I.P. Beletskaya, *Reaction Mechanism of Organometallic Compounds*, Worth-Holland Publ. Company, Amsterdam, 1968.
2. J.P. Collman, L.S. Hegedus, J.R. Norton and R.G. Finke, *Principles and Applications of Organotransition Metal Chemistry*, University Science Books, Nill Valley, California, 1987.
3. D. Milstein and J.K. Stille, *J. Amer. Chem. Soc.*, **101** (1979) 4992.
4. I.P. Beletskaya, *Organic Synthesis. An Interdisciplinary Challenge* Blackwell Scientific Publ.,1984, 43; *J. Organometal. Chem.*, **241** (1983) 13; *Fundamental Research in Homogeneous Catalysis*, Gordon and Breach Science Publ., 1986, 281; *Izvestija Ac. Sci., USSR, serie chem.* (1990), 2211(in Russian).
5. N.A. Bumagin, I.G. Bumagina and I.P. Beletskaya, *Doklady Ac. Sci. USSR*, **818** (1984) 274 (in Russian).
6. N.A. Bumagin, I.G. Bumagina and I.P. Beletskaya, *J. Org. Chem.*, **18** (1982) 1131 (in Russian).
7. N.A. Bumagin, I.G. Bumagina and I.P. Beletskaya, *Doklady Ac. Sci., USSR*, **261** (1981) 1141 (in Russian).
8. N.A. Bumagin, Yu.V. Gulevich and I.P. Beletskaya, *J. Organomet. Chem.*, **285** (1985) 415; *Izvestija Ac. Sci., USSR, seria chem.*, (1984) 953.
9. N.A. Bumagin, I.G. Bumagina, I.P. Beletskaya, *Doklady Ac. Sci. of USSR*, **274** (1984) 1103 (in Russian); *J. Izvestija Ac. Sci., USSR, serie chem.*, (1984) 1137 (in Russian).
10. I.P. Beletskaya, Palladium Catalyzed Organic Reactions in Aqueous Media, in *New Aspects of Organic Chemistry*, eds, Z. Ioshida and I. Ohshiro, Kodansha - VCH, 1992.
11. N.A. Bumagin, N.A. Suchomlinova and I.P. Beletskaya, *Izvestija Ac. Sci.*, *serie chem.*, **N11** (1990) 2665 (in Russian).
12. A.G. Davies and P.J. Smith, in *Comprehensive Organometallic Chemistry*, eds, G. Wilkinson, F.G.A. Stone and E.A. Abel, Pergamon Press, U.K., 1982, Chap. 2, p.520.
13. I.R. Beletskaya, in *Chemistry and Technology of Si and Sn*, eds, V.G. Kumar Das and M. Gielen, Oxford Sci. Publication, 1992, p.136.
14. C.J. Evans and S. Karpel, *Organotin Compounds in Modern Technology*, Elsevier, 1985, p.279.
15. Proceedings of IRUJAS-93 (Russia-Japan Joint Symposium on Petrochemistry) Y.F. Soboleva *et al.*, Sakhalin, 1993, p.42.

16. V.F. Kamjanov, V.S. Akensov and V.F. Titov, Heteroatomic Compounds of Oil, Novosibirsk, *Science*, 1983, p.175 (in Russian).

17. N.K. Nadirov (ed.), Metals in Oils, Alma-Ata, *Science*, 1984, p.163 (in Russian).

18. T. Yen, The Role of Metals in Petroleum, *Ann. Arbor Sci.*, *Publ.* 1975, p.30.

19. P.J. Craig, Environmental Aspects of Organometallic Chemistry in *Comprehensive Organometallic Chemistry*, 1985, p.979.

20. Ju. A. Ershov and T.V. Pleteneva, *Mechanisms of Toxicological Action of Inorganic Compounds*, Moscow, Medicina, 1989, p.196 (in Russian).

21. S.V. Davydova and N.V. Rastova, *J. Analyt. Chem.*, **47** (1992) 1033 (in Russian).

22. Report, *Tin and Organotin Compounds*, of World Health Organisation, Geneva, 1991, p.115 (in Russian).

23. W. Eichler, Gift in Unserer Nahrung, *Kilda Verlag*, **39** (1982) 27.

24. C. Reilly, *Metal Contamination of Food*, Moscow, Agroprom., 1985, p.116 (in Russian).

25. S.M. Rasaev *et al.*, *Macromolecular Compounds*, **8** (1966) 552, 771,1890; **9** (1967) 150; **10** (1968) 1004 (in Russian).

26. V. Dragutan, in Abstracts of 2nd International Symposium on Applied Bioinorganic Chemistry, Guangzhou (PR China), 1992, p.155.

27. A. Sheldon, *J. Paint Technol.*, **47** (1975) 54.

28. K.D. Freitag and R. Bock, *Pestic. Sci.*, **5** (1974) 731.

29. Swedish Expert Group, Methyl Mercury in Fish, *Toxicological-Epidemiological Appraisal of Risks, Suppl.*, 4, 1971.

30. H. Siegel (ed,) , *Metal Ions in Biological Systems*, Concepts of Metal Ion Toxicity, Moscow, Editorial House, Mir, **20** (1993) p.336 (in Russian).

31. P.J. Craig and F. Glockling, Biological Alkylation of Heavy Elements, Royal Soc. of Chemistry, Cambridge, U.K., 1988.

Main Group Elements and Their Compounds
V.G. Kumar Das (Ed)
Copyright © 1996 Narosa Publishing House, New Delhi, India

A New Manufacturing Process for Phosphatic Fertilizer

Chen Tianlang and Xiao Shexiu

Department of Chemistry, Sichuan University, Chengdu 610064,
Sichuan, P.R. China

Fertilizers form the cornerstone of agriculture. Their global consumption has been growing at the rate of 6 to 10% per annum since 1945. Of the fertilizers, the phosphatic fertilizer has special significance for almost all plants. Up to now, there are only two processes to produce the phosphatic fertilizer, viz. a "wet process" and a "hot process". Products of the former include single superphosphate, nitrophosphate, MAP (monoammonium phosphate) and DAP (diammonium phosphate), while those of the latter are fused calcium magnesium phosphate, steel sediment phosphate, etc. The above mentioned processes, however, are high in energy consumption and material wastage, besides being also tediously long. In order to overcome these problems, we have proposed a new and simple manufacturing process for producing phosphatic fertilizers which we have termed the "dry process".

Materials and Methods

Materials

Inorganic mineral acids (sulphuric, nitric, etc.) urea, powdered rock phosphorus.

Method

The dry process for producing phosphatic fertilizer is characterized by the following steps:
First, a compound called solid phosphorus-releasing agent (SPRA) is produced in accordance with the following scheme:

Inorganic mineral acid + Urea → Solid phosphorus-releasing agent (SPRA)
(or mixtures of inorganic acids)

The solid phosphorus-releasing agents are white powders possessing excellent water solubility. They are a homogenous mixture of salts or complexes in which the molar ratio of acid and urea is usually 1:1 or 2:1.
Secondly, according to the desired N/P ratio, the SPRA is mixed in sufficient

quantities at room temperature with the float, i.e., powdered rock phosphorus.

Finally, the whole is granulated to the desired size, ready for direct application as fertilizer (labeled DPF, "dry process" phosphate fertilizer).

It is to be noted that the phosphorus-releasing agent undergoes no chemical reaction with the float in the mixing process as verified by X-ray diffraction analysis. A reaction only occurs when water is added. Thus when the fertilizer is applied in the field, contact with the water in the soil promotes the phosphorus-releasing reaction between the acid in SPRA and the float. This in effect means that the field becomes the fertilizer factory!

Results and Discussion

Using the "dry process", about 30 tons of compound fertilizer containing 30% NP were manufactured in the laboratory, and agricultural tests performed with it over a 3-year period. The tests were carried out in the wheat fields of Sichuan. The results are summarized in Table 1.

Table 1 : Results of agricultural tests on wheat using phosphatic fertilizer manufactured by the "dry process"

Treatment[b,c]	Yields (kg/Mu)[a]						
	Mei-shan county	Sui-nin county		Wei-yan county	Shuang-liu county		average
		(1)	(2)		(1)	(2)	
CK	154.9	195.9	192.1	177.5	263.3	193.3	196.1
Urea	266.7	291.7	297.1	231.6	361.7	353.7	301.2
Urea + SPF	274.7	329.2	307.1	272.5	360.0	386.7	321.7
Urea + FCM	262.7	333.4	315.4	227.5	360.0	370.0	331.5
Urea + PRP	260.0	308.4	298.8	239.5	336.7	350.0	298.9
DPF	277.3	300.0	332.1	305.8	363.3	393.3	328.6
L.S.D.							
1%	23.2	54.2	60.0	6.5	68.7	30.3	23.4
5%	33.2	77.1	85.8	9.3	97.7	93.2	31.6

[a]Mu is a measure of unit land area in P.R. China. 1 Mu = 0.1647 acre.
[b]Abbrev.: CK - no treatment. SPF - single superphosphate fertilizer. FCM - fused calcium magnesium phosphate. PRP - phosphatic rock powder. DPF - "dry process" phosphatic fertilizer (acid in SPRA used here is nitric).
[c]Fertilizer used for each treatment amounts to about N 20 kg, P_2O_5 10 kg per Mu, except for treatment cases involving CK and Urea only.

It is apparent from Table 1 that the beneficial effects of the fertilizer produced by the "dry process", in equal amounts of NP, are analogous to those of single superphosphate, and obviously superior to those obtained using fused calcium magnesium phosphate or phosphatic rock powder.

It should not noted that the fertilizer produced by the "dry process" is not only a phosphatic fertilizer, but also a compound fertilizer containing NP (N coming from urea). The total content of available phosphorus (as P_2O_5) is about 7 to 10%. The NP ratio can be adjusted to the needs of plants. If it is desired to produce ternary NPK or multi-component fertilizers, the potassium salts or some minor elements required for the plants can be easily incorporated in the "dry process".

Another important merit of the "dry process" is that it is possible to use phosphorus rocks with middle and low grade P_2O_5 contents for its preparation.

Although we are currently engaged in more tests using the fertilizer in other crops, it is apparent that the "dry process" which combines simplicity with cost-advantages will have special appeal to developing countries. Already, two pilot plants of capacity 3000 tonnes per annum have been built in Chengdu and Chongqing in Sichuan.

References

1. Chen Tianlang, *Patent of China*, No. 87103006 (1987).
2. Chen Tianlang and Xiao Shenxiu, *Phosphatic and Compound Fertilizers*, **3** (1989) 11.
3. Chen Tianlang and Ziao Shenxiu, *Chemical Industry of Sichuan*, **2** (1990) 23.

Subject Index